TURBULENT TRANSPORT IN FUSION PLASMAS

TURBULENT TRANSPORT IN FUSION PLASMAS

First ITER International Summer School

Aix en Provence, France 16 – 20 July 2007

EDITOR
Sadruddin Benkadda
CNRS-Université de Provence
Provence, France

SPONSORING ORGANIZATIONS

ITER Orgnisation
CNRS (Centre National de la Recherché Scientifique)
Université de Provence
CEA (Commissariat à l'Energie Atomique)
MEXT (Japanese Ministry of Education, Culture, Sports, Science and Technology)
French Ministry of Foreign Affairs
Region PACA (Region of Provence-Alpes and Cote d'Azur)
Conseil General des Bouches du Rhône
Communauté du Pays d'Aix
Institute of Political Sciences at Aix en Provence

AMERICAN INSTITUTE OF PHYSICS

Melville, New York, 2008
AIP CONFERENCE PROCEEDINGS VOLUME ■ 1013

Editor

Sadruddin Benkadda
France-Japan Magnetic Fusion Laboratory
LIA 336 CNRS-Université de Provence
Centre Universitaire St Jérôme. Case 321
13397 Marseille Cedex 20. France

E-mail: benkadda@up.univ-mrs.fr

L.C. Catalog Card No. 2008926443

ISBN 978-0-7354-0534-9
ISSN 0094-243X

Printed in the United States of America

CONTENTS

Preface

In November 2006, representatives of the world's 7 major fusion research communities signed an agreement to embark on the construction and operation of ITER, the next major, and critical, step in the programme to demonstrate the feasibility of harnessing fusion power for peaceful purposes. As ITER construction gets underway at Cadarache, it is important to emphasize that ITER will not simply be a large construction project, but also a major scientific and technological research programme. ITER is both a significant energy Research & Developement project, and an important research facility and resource for the international fusion community. As ITER Director-General, I consider it essential that ITER develops into a centre of excellence in fusion research in order to fulfil its role in the development of fusion energy.

Harnessing the potential of fusion energy is one of mankind's great scientific challenges, but it is a long term challenge, and so it essential that we equip the scientific researchers of the future to carry fusion research forward and to lead the scientific exploitation of ITER. It is therefore a great pleasure for me to be associated with the series of ITER Physics Summer Schools launched jointly by CNRS and the University of Provence. These Summer Schools aim to offer a complete picture of experimental and theoretical aspects of tokamak physics to prepare young researchers to tackle the current and future challenges in magnetic fusion - in essence equipping young researchers with the tools to meet the physics challenges which will need to be addressed in exploiting ITER's potential.

At the first School, which took place during 16-20 July, 2007 at Aix-en-Provence, leading researchers from the international fusion community gathered in Aix to present a series of lectures to advanced graduate students and recent PhD graduates on "Turbulence in Fusion Plasmas", dealing with the physics of turbulent transport in magnetically confined plasmas and the implications of turbulence for plasma confinement. These lectures and accompanying informal discussions provided a comprehensive introduction to the theoretical and phenomenological framework of turbulent transport in fusion devices. They dealt both with experimental aspects of turbulence measurements in present fusion devices and with the frontier physics involved in theoretical models of plasma turbulence. The lectures which are collected in these proceedings will be a valuable resource for the fusion community in the future.

The Chairman of the Organizing Committee for the Summer School, Professor Sadruddin Benkadda, of the University of Provence, and the Chairman of the Scientific Committee, Professor Osamu Motojima of the National Institute for Fusion Sciences in Toki, Japan, are to be congratulated for their vision in drawing up the programme for this First Summer School and for assembling some of the world's leading experts in this area of fusion research as lecturers. This provided an auspicious beginning for an initiative which should make an important contribution to training fusion researchers on the global scale for the challenges that lie ahead.

Kaname Ikeda
ITER Director-General

Editor's Foreword

The first edition of the ITER International Summer School was held from 16th till 20th of July 2007 at the Institut d'Etudes Politiques (IEP) of Aix in Provence. The Local Organizing Committee was chaired by Sadruddin Benkadda and composed by members of University of Provence as well as CEA Cadarache. The scientific programme and participant selection were the responsibility of the International Scientifique Committee which was co-chaired by Osamu Motojima (NIFS, Japan) and Sadruddin Benkadda (CNRS-University of Provence). 80 participants resulting from about twenty countries attended the meeting. They divided up as follows: 53 from the U.E (Austria, Germany, Belgium, Great Britain, France, Greece, Hungary, Ireland, Italy, Sweden, Czech Republic) and 27 resulting except U.E (Japan, United States, India, Russia, Brazil, Ukraine, Georgia, Morocco). 22 international specialists gave lectures in this school. The speakers were from Germany (Max Planck Institut Garching, University of Dusseldorf), Denmark (Risö National Laboratory), Spain (CIEMAT), United States (University of Texas at Austin, University of California at San Diego, Oakdrige National Laboratory, University of California at Los Angeles, Princeton University), France (Commissariat à l'Energie Atomique, Ecole Polytechnique, University of Nancy), United Kingdom (U.K.A.E.A), India (Institute for Plasma Research, Ahmedabad) and Japan (National Institute for Fusion Science, Japanese Atomic Energy Authority, Osaka University).

This first school organized under the aegis of the Centre National de la Recherche Scientifique (C.N.R.S) and Université de Provence knew a frank success. It corresponds to an important need of training within the framework of the project ITER. The choice of launching this school corresponds to the needs of training young researchers coming from various horizons and the multidisciplinary skills of which are appreciated within the framework of the complex physics of ITER which besides the physics of hot plasmas appeals to diverse skills in material science, physics of turbulence and nonlinear dynamics as well as in mathematics. This school aims at preparing young researchers to tackle the current and anticipated challenges at magnetic fusion devices, and spreading the global knowledge required for a timely and competent exploitation of the ITER physics potential.

The emphasis of the first school has been on the physics of turbulent transport and its implications on confinement. A series of lectures and informal discussions included an introduction to the theoretical and phenomenological framework of turbulent transport in fusion devices, and current theoretical models of frontier physics, as well as an overview of the main diagnostics used in experiments. Explicit examples of physics analysis drawn from current tokamak experiments helped inform these exchanges.

The organizing committee is pleased to thank the ITER organization for its sponsorship, the organizations : CNRS, Université de Provence, C.E.A, Conseil Général des Bouches du Rhône, Région Provence Alpes Côte d'Azur, Communauté du Pays d'Aix, le Ministère des Affaires Etrangères, M.E.X.T (Japanese Ministry of Education, Culture, Sports, Science and Technology) who granted financial support to organize this meeting, and the Institut d'Etudes Politiques d'Aix en Provence for hosting the summer school and for its logistic support.

I would like to thank Mr Shimpei Futatani for his valuable help in the editorial process. Finally, I gratefully acknowledge the authors of the enclosed contributions, who willingly took their part in the creative work of writing a timely book.

Happy reading!

Sadruddin Benkadda

Drift Wave Turbulence

W. Horton*, J.-H. Kim*, E. Asp†, T. Hoang**, T.-H. Watanabe‡ and H. Sugama‡

*Institute for Fusion Studies, the University of Texas at Austin, USA
†Ecole Polytechnique Fédérale de Lausanne, Centre de Recherches en Physique des Plasmas Association Euratom-Confédération Suisse, CH-1015 Lausanne, Switzerland
**Assoc. Euratom-CEA, CEA/DSM/DRFC Cadarache, 13108 St Paul-Lez-Durance, France
‡National Institute for Fusion Science/Graduate University for Advanced Studies, Japan

Abstract. Drift waves occur universally in magnetized plasmas producing the dominant mechanism for transport of particles, energy and momentum across magnetic field lines. A wealth of information obtained from laboratory experiments for plasma confinement is reviewed for drift waves driven unstable by density gradients, temperature gradients and trapped particle effects. The modern understanding of origin of the scaling laws for Bohm and gyro-Bohm transport fluxes is discussed. The role of sheared flows and magnetic shear in reducing the transport fluxes is discussed and illustrated with large scale computer simulations. Plasmas turbulence models are derived with reduced magnetized fluid descriptions. The types of theoretical descriptions reviewed include weak turbulence theory and anisotropic Kolmogorov-like spectral indices, and the mixing length. A number of standard turbulent diffusivity formulas are given for the various space-time scales of the drift-wave turbulent mixing.

Keywords: Drift-waves, ITG, TEM, turbulence, transport, vortices, tokamaks
PACS: 52.35.Kt, 52.25.Fi, 52.25.Gj, 52.55.Fa, 52.35.Py

1. DRIFT WAVE MECHANISM

One of the most intellectually interesting problems in plasma physics is the problem of turbulence and the associated transport of the plasma properties including density, temperature and momentum. Plasma turbulence can be rather different from that in neutral fluids due to the numerous waves in the plasma. Plasma turbulence is closer to that of geostrophic flows on rotating planets where the Rossby wave is closely analogous to the plasma drift wave. At finite amplitude the wave dispersion often balances the wave nonlinearity to form solitary structures and coherent vortex structures. This aspect of plasma turbulence has been the topic of workshops and conferences both at ITP Santa Barbara [1] and the Marseille Workshops published as Benkadda et al. [2]. A monograph largely devoted to these problems is Horton and Ichikawa [3]. Drift waves are the most widely studied form of plasma turbulence in magnetized, confined plasmas so this first Chapter of the ITER School Proceedings is devoted to the study of these drift waves.

The search for controlled thermonuclear fusion has been dominated by the use of strong magnetic fields to confine laboratory plasmas. Through the 1970s to the 1990s the dominant thought was that the low-amplitude drift-wave turbulence would not seriously impede reaching the critical Lawson condition in MHD stable machines and sufficient auxiliary (external) plasma heating, $P_{\rm aux}$. Exceeding the Lawson condition, $n_e\tau_E \sim 1.5\times 10^{20}\,{\rm m}^{-3}{\rm s}$ at 25 KeV, is required for D-T ignition. In the Lawson condition on $n_e\tau_E$, $\tau_E = W/(P_{\rm in} - dW/dt)$ is the energy confinement time defined by stored plasma energy (W), the power input ($P_{\rm in}$) and the time variation of the stored energy (dW/dt).

The plasma community now realizes that merely increasing the auxiliary heating is not sufficient to reach ignition. As external heating is added to the plasma, the turbulence level increases dramatically and more coherent structures appear so that the confinement time is seriously degraded compared to that of lower levels of auxiliary heating.

A less stringent condition derived from the fusion power P_f amplification factor, $Q_f = P_f/P_{\rm loss}$, is maximized at a lower temperature between $10-20$ KeV where the reaction rate $\langle\sigma v\rangle_{\rm DT} \propto T_i^2$ so that $Q_f \sim n_e\tau_E T \gtrsim 10^{21}\,{\rm m}^{-3}{\rm KeV\,s}$. Here, to reduce the amplification Q_f to the fusion triple product $n_e\tau_E T$, we use $P_f = (1/4)E_f n_e^2\langle\sigma v\rangle_{\rm DT}$, where $E_f = 17.6$ MeV is the fusion energy release by a D-T reaction and $P_{\rm loss} = nT_E/\tau_E$.

Reference values for the drift wave turbulence parameters in the ITER device are given in Table 1. Coherent structures are now understood to produce intermittent plasma turbulence. There is now a broader level of recognition in the fusion community that an understanding of the chaos and coherent structures created by plasma drift-wave instabilities is required to reach practical fusion energy producing regimes. Let us now examine in detail how the basic

CP1013, *Turbulent Transport in Fusion Plasmas, First ITER International Summer School*
edited by S. Benkadda

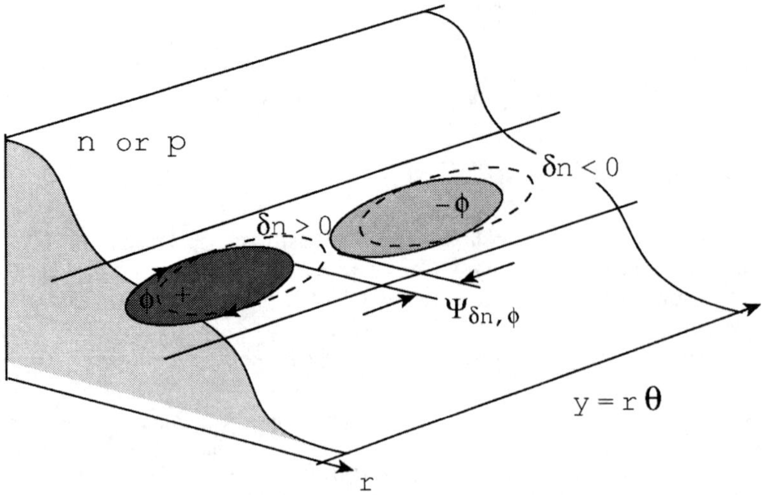

FIGURE 1. The Drift Wave Mechanism

drift-wave mechanism works.

In Fig. 1, we show the profiles of electron density, n_e, and pressure, p_e, along with a drift wave localized over a radial region $\Delta r = \ell_c$, between the two cells in the correlation length. The electrostatic potential that gives rise to the electric field, $\boldsymbol{E} = -\boldsymbol{\nabla}\phi$, is shown by the positive (red) and negative (blue) ovals. Ahead of the potential there are contours of enhanced density ($\delta n_e > 0$) and lower density ($\delta n_e < 0$) phase shifted by $\psi_{\boldsymbol{k}} = \psi(\delta n, \phi)$. The magnetic field is taken to be in the vertical z direction. Thus, the positive potential cell gives a clockwise rotation and the negative cell a counter-clockwise rotation of $\boldsymbol{E} \times \boldsymbol{B}$ convected plasma. Clearly, due the phase shift there is a greater transport of plasma down rather than up the density gradient. Thus, there is a net outward flux of density or pressure given by the y-integrated or time averaged flux $\langle n_e v_r \rangle = \langle n_e E_y / B \rangle$. To be explicit, the model in Fig. 1 and calculation is characterized by the electric potential wave $\phi_k \sin(k_y y - \omega_k t)$ and by the density wave $\delta n_e \sin(k_y y - \omega_k t - \psi_k)$. Calculating the local flux, $\delta n_e v_r$, gives Γ and averaging over y or time gives $\Gamma = \sum_{\boldsymbol{k}} [k_y \phi_k \delta n_k / B] \sin \psi_k$. We will see that the electron response to the potential determines the phase shift ψ_k and amplitude ratio $\delta n_e / \phi_k$. The special case where the phase shift is zero gives $\delta n_e / n_e = e\phi_k / T_e$ and is called the adiabatic response. This adiabatic response is used in many of the classical drift-wave models, e.g. when the temperature gradient drives the drift waves unstable rather than the density gradient ($n \to T$ in Fig. 1). In the turbulent plasma this convective flux occurs at each $\boldsymbol{k}$ with the total being a sum over all scales of k.

We shall see that the typical size of the turbulent cells $\lambda_\perp = 2\pi / k_\perp \sim 6\rho_s$ and the frequency is approximately $\omega_k = k_y v_{de}$, where $v_{de} = -(T_e / eBn_e)(dn_e/dr) = T_e / eBL_n = c_s \rho_s / L_n$. The scale length, $\rho_s = c_s / \omega_{ci}$, is called the ion inertial scale length and determines the scale of the waves and vortex structures. In the quasi-geostrophic flow in planetary atmospheres the equivalent quantity is the Rossby radius [3, p.225]. For reference, a present-day tokamak, e.g. Tore Supra [4] has $R/a = 2.4\,\mathrm{m}/0.72\,\mathrm{m}$, $B = 2\,\mathrm{T}$, $n_e = 3 \times 10^{19}\,\mathrm{m}^{-3}$ and $T_e = 1\,\mathrm{KeV}$. For these parameters $\rho_s / L_n = 0.1\,\mathrm{cm}/80\,\mathrm{cm}$, the sound speed $c_s = 3 \times 10^7\,\mathrm{cm/s}$ and thus $v_{de} = 3 \times 10^4\,\mathrm{cm/s}$ and the drift wave frequency for $k_y = 20\,\mathrm{cm}^{-1}$ is like $6 \times 10^5\,\mathrm{rad/s}$ or $100\,\mathrm{kHz}$. Reference values for ITER included in Table 1. These drift-wave fluctuations have been studied in detail in a wide variety of laboratory machines.

Drift waves are sufficiently slow so that the electrons adjust to the potential fluctuations such that

$$E_\parallel + \frac{1}{en} \boldsymbol{\nabla} p_e = 0. \tag{1}$$

Thus, the waves have the opposite polarization of the MHD waves, which have $E_\parallel = 0$. The parallel electric field, measured in detail in the experiments, is a key part of the drift waves and the related trapped particle mode instabilities. The fluctuating electron current, $\delta j_\parallel$, driven by the parallel electric field takes the energy $\langle \delta j_\parallel E_\parallel \rangle < 0$ out of the nonuniform electron distributions and produces the wave growth. The effect is as though the plasma had a negative resistivity/conductivity $\sigma_\parallel(\omega, \boldsymbol{k})$ for the drift wave.

Detailed measurements and comparisons with theory suggest that, in fact, one of the most serious limiting factor for the success of ITER and similar machines is the control of the drift-wave turbulent transport. The seriousness of the

TABLE 1. ITER [9] and drift wave parameters

Installed auxiliary heating	73 MW
Fusion power	500 MW
Plasma major radius(R)	6.2 m
Plasma minor radius(a)	2.0 m
Plasma volume	830 m^3
Plasma current(I_p)	15 MA
Toroidal field at 6.2 m radius	5.3 T
Density gradient length (L_n)	1 m
Electron temperature (T_e)	10 KeV
Electron Diamagnetic Drift Velocity (v_{de})	1.9 km/s
Ion inertial radius (ρ_s)	0.3 cm
gyro-Bohm diffusivity ($D_{gB} = \rho_s v_{de}$)	5.7 m^2/s
Drift wave frequency ($\omega = k_y v_{de}$)	190 KHz at $k_y \rho_s = 0.3$

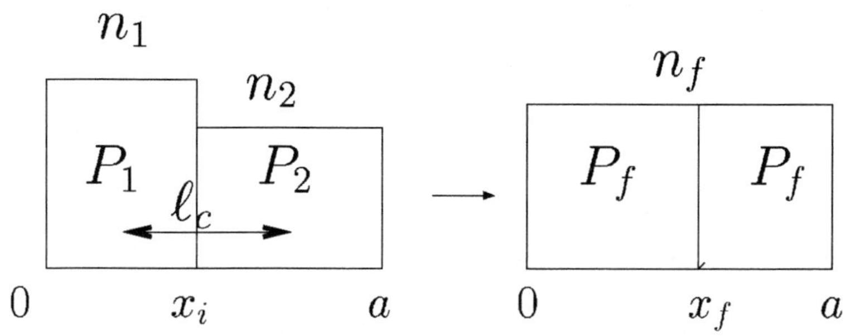

FIGURE 2. Expansion diagram under density gradient $n_1 > n_2$.

turbulence became more evident as machines with increased injected power P_{aux} have reached core power densities of $1\,\text{MW/m}^3$. The plasma temperature rises only as a fractional power law, like $P_{\text{aux}}^{1/2}$, owing to the associated increased turbulence level. The energy confinement time, τ_E, thus decreases even though the temperature increases. The triple product may remain essentially constant.

As a consequence of this stronger heating power, P_{aux}, the turbulence is amplified, which shortens the plasma energy confinement time, $\tau_E = W/P_{\text{aux}}$, in agreement with the ITER database [5, 6, 7]. The drift-wave turbulence has explained the empirical database laws for both helical systems [8] and Tokamaks [6, 7], showing that $\tau_E = \tau_E^0 (P_0/P_{\text{aux}})^{2/3}$. Thus, doubling the P_{aux} shortens the confinement time, $\tau_E \to \tau_E^0/1.59 = 0.64\tau_E^0$. For example, the reference design for ITER (see Table 1) has $P_{\text{ICRF}} = P_{\text{ECE}} = 20\,\text{MW}$ and $P_{\text{NBI}} = 16\,\text{MW}$ injected into the plasma volume of $V = 2\pi^2 Rab \simeq 800\,\text{m}^3$. The average power density is then only $0.07\,\text{MW/m}^3$. To reach the power density of $1\,\text{MW/m}^3$ at the core as in current large tokamaks, the ICRH and ECE power need to be focused in the central core volume of $40\,\text{m}^3$, corresponding to the core radius $r_c = 0.6\,\text{m}$.

2. ENERGY BOUNDS FOR TURBULENCE

2.1. Density Gradients

The electron drift wave and trapped electron instability (trapped particle mode [10])are driven by the square of the plasma density gradient in plasmas with no background temperature gradient. The wave with a temperature gradient is analyzed in Sec.2.2. In the early literature the electron drift wave with $\gamma_k = (\pi/2)^{1/2}\, \omega_{*e}^2 k_\perp^2 \rho_s^2 / |k_\parallel| v_e \propto (dn/dx)^2$ was called the universal instability owing to the drive from $(\nabla n)^2$, which is always present in confined plasmas.

Fig. 2 shows the model for the energy release for the density gradient. The left higher density n_1 (and pressure $p_1 = n_1 T$) box from $x = [0, x_i]$ is allowed to expand to $[0, x_f]$ and compress the gas in the right box until the force

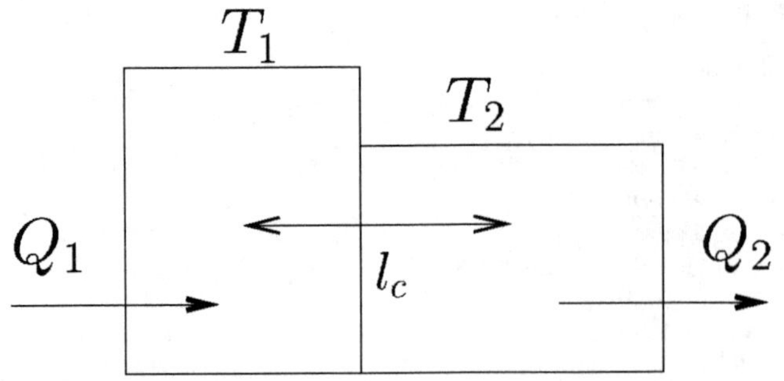

FIGURE 3. Carnot cycle diagram for temperature-gradient driven drift waves

balance is reached with

$$P_f = P_1\left(\frac{x_i}{x_f}\right)^\Gamma = P_2\left(\frac{a-x_i}{a-x_f}\right)^\Gamma . \tag{2}$$

Solving for the new position x_f and calculating the work,

$$W = \frac{1}{a}\int_{x_i}^{x_f}(P_1 - P_2)dx,$$

per unit volume done by the gas during the expansion gives the free energy. For $\Delta x/a \ll 1$, the turbulent energy per unit volume is

$$W_{\rm turb} \le W_{\rm max} = \frac{(\Delta P)^2 x_i(a-x_i)}{\Gamma P a^2} \le \frac{(\Delta P)^2}{4\Gamma P}. \tag{3}$$

If the drift wave energy $W_{\rm turb}$ is almost entirely in the $\boldsymbol{E}\times\boldsymbol{B}$ kinetic energy of the ions, then we have the bound for the average turbulent convection velocity,

$$\langle v_E^2\rangle^{1/2} \le \left(\frac{\ell_c}{L_n}\right) c_s . \tag{4}$$

When the correlation length ℓ_c is limited by the ion inertial scale length $\ell_c = \rho_s$, the bound for the mean square turbulent plasma velocity is

$$\langle v_E^2\rangle \le \left(\frac{\rho_s}{L_n}\right)^2 c_s^2 \equiv v_{de}^2 . \tag{5}$$

The limit of $\langle v_E^2\rangle \le v_{de}^2$ follows from another argument called the mixing length limit. This argument states that when the gradient of the fluctuating density, $\nabla\delta n$, reaches the value $dn/dx = -n/L_n$ of the ambient density gradient, the gradient source, on average, has been eliminated and growth stops. We then have $\delta n/n = 1/k_x L_n$. With $\ell_c = 1/k_x = \rho_s$, The same amplitude limit applies and the scale of the turbulent diffusion $D_{gB} = \langle v_E^2\rangle^{1/2}\ell_c = (\rho_s/L_n)(T_e/eB)$ is defined as the gyro-Bohm diffusivity. To derive the famous Bohm diffusivity, $D_B = T_e/eB$, one instead assumes ℓ_c to be the geometric mean of L_n and ρ_s [11].

2.2. Temperature Gradients

For auxiliarily heated plasmas that develop a temperature gradient ∇T (either ion or electron) from input power $P_{\rm aux}$, the relevant thermodynamic argument is the Carnot engine.

As shown in Fig. 3, there is an input of energy Q_1 in the left box. The vortices and streamers cycle the plasma over the correlation length ℓ_c, connecting T_1 and T_2. Now, for each cycle, the temperature gradient drives the drift wave to release the energy,

$$W_{\rm turb} \le \Delta W_{\rm carnot} = \Delta S \Delta T, \tag{6}$$

where ΔS is the change in the entropy of the gas. From kinetic theory, the value of the gas constant, Γ, depends on the type of drift wave. $\Gamma = 5/3$ where the wave-particle resonance is $\omega_k = \omega_{\nabla B} + \omega_{\rm curv}$ and $\Gamma = 3$ where $\omega_k = k_\parallel v_\parallel$ and

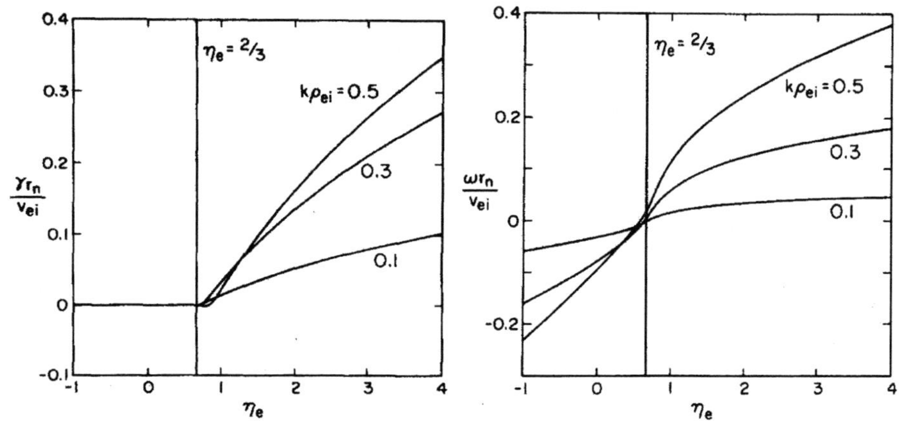

FIGURE 4. The Vlasov (GKE) equation growth rate for the ratio of the temperature gradient length to the density gradient length[12].

$k_\perp \rho_i \ll 1$. For toroidal ITG and ETG, a simple calculation for modes with $\Gamma = 5/3$ yields

$$\Delta S = n\left(\frac{3}{2}\frac{\Delta T}{T} - \frac{\Delta n}{n}\right), \tag{7}$$

which in Eq. (6) together with $\Delta T = -\ell_c dT/dr$ give

$$\Delta W_{\text{carnot}} = \frac{3}{2} nT\ell_c^2 \frac{\partial \ln T}{\partial r}\left[\frac{\partial \ln T}{\partial r} - \frac{2}{3}\frac{\partial \ln n}{\partial r}\right]. \tag{8}$$

Fig. 4 shows the roots of the Vlasov-Poisson dispersion relation, $D_{\mathbf{k}}(\omega,\mu) = 0$, with increasing $\eta \equiv \nabla \ln T / \nabla \ln n$. Thus, for the kinetic dispersion relation,

$$D_{\mathbf{k}}(\omega,\mu) = \frac{1}{T_e} + \frac{1}{T_i}\left[1 - \left\langle \frac{\omega - \omega_{*i}(\eta,\varepsilon)}{\omega - \omega_{Di}\varepsilon + i0^+}\right\rangle\right], \tag{9}$$

where $\varepsilon = E/T_i$ and μ denotes system parameters such as η_i and T_e/T_i. The average $< \cdot >$ is over the local Maxwell phase-space density $f(\varepsilon)$

We find from the Nyquist analysis that the critical temperature gradient is given by

$$\eta_{\text{crit}} = \frac{\partial \ln T}{\partial \ln n} = \frac{2}{3}, \tag{10}$$

as shown in Fig. 4. The agreement between the thermodynamical formula and the stability condition of the kinetic dispersion relations can be understood when we see that the wave frequency, $\omega_{\mathbf{k}} \simeq 0$ at the critical point. Here the relevant gradient of the phase-space density function $f(r)$ is proportional to $\partial/\partial r[n/T^{d/2}]$ where

$$\frac{\partial f}{\partial r} = \frac{\partial}{\partial r}\left(\frac{n(r)}{T^{d/2}(r)}\right) \quad \text{for} \quad E \ll T, \tag{11}$$

with $\Gamma = (d+2)/d$ for a plasma with d degrees of freedom. For one-dimensional waves like Langmuir waves, $d = 1$ and $\Gamma = 3$, and for toroidal drift waves $d = 3$ and $\Gamma = 5/3$.

The key, universal result given by Eq. (8) for W_{carnot} is that there is only a window of η values, $0 < \eta \leq \Gamma - 1$ which are generally stable. Drift waves are unstable for both inverted gradients $\nabla n \cdot \nabla T < 0$, and strong temperature gradients $\nabla T \gg \nabla n$.

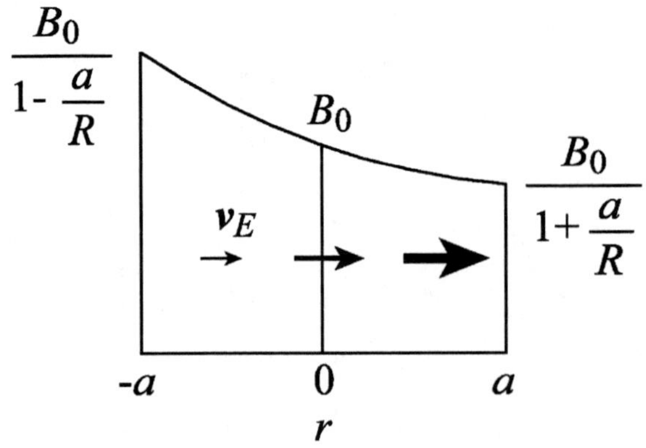

FIGURE 5. Compression/Rarefraction in Convection

3. DRIFT WAVE EIGEN-MODES IN TOROIDAL GEOMETRY

The first effect of toroidal geometry that changes the stability of the drift waves is the unfavorable magnetic field line curvature vector, $\boldsymbol{\kappa} = (\hat{\mathbf{b}}\cdot\nabla)\hat{\mathbf{b}} \simeq (\hat{\mathbf{r}}\cos\theta + \hat{\theta}\sin\theta)/R$, which covers the outer half ($\cos\theta > 0$) of the toroidal surfaces. This magnetic curvature creates small scale interchange turbulence with waves that are only partially stabilized by the magnetic shear, $s = rq'/q$, where $q(r) = rB_\phi/RB_\theta$ gives the helical pitch of the field lines. The E_r- shear can also stabilize the modes as discussed in Sec. 4.2. In addition, and closely related, the weakening of the magnetic field with increasing radius (see Fig. 5) leads to an outward acceleration given by the force, $F = -\mu\nabla B$ acting on the plasma.

In the toroidal drift wave equations the local gradients of the density and pressure are normalized with respect to the gradient length, R, of the magnetic field. Thus, the key parameters are

$$\varepsilon_n = \frac{L_n}{R} \quad \text{and} \quad \varepsilon_p = \frac{L_p}{R} \tag{12}$$

throughout the toroidal equations.

The faster radial flows shown in Fig. 5 gives the plasma compression,

$$\nabla\cdot\boldsymbol{v}_E = \frac{2v_x\cos\theta}{R}, \tag{13}$$

compared to the plasma convection,

$$\boldsymbol{v}_E\cdot\nabla n = -\frac{nv_x}{L_n} = -\frac{n}{L_nB}\frac{\partial\Phi}{\partial y}, \tag{14}$$

gives rise to the two terms proportional to $(1-2\varepsilon_n\cos\theta)\omega_{*e}$ in the drift-wave eigen-mode equation.

The density and potential fluctuations set up along the magnetic field line give rise to ion acoustic waves with $\omega^2 = k_\parallel^2 c_s^2$, where

$$k_\parallel^2 \to -\frac{1}{q^2R^2}\left(\frac{\partial}{\partial\theta} + q(r)\frac{\partial}{\partial\phi}\right)^2 \tag{15}$$

is a differential operator.

Similarly, the shear-Alfvén wave $\omega^2 = k_\parallel^2 v_A^2$ propagates with the $k_\parallel^2$ operator. The periodicity of the torus means that a physical field is the same at $(\theta + 2\pi m, \phi + 2\pi n)$ for all integer values of m, n. These rational surfaces appear when the pitch of the magnetic field is the same as that of the mode $\exp(im\theta - in\phi)$, which occurs when the RHS of Eq. (15) vanishes. The compressional term in Eq. (13) and convection term in Eq. (14) combine with the ion acoustic waves to make the drift wave eigen-mode equation for waves of the form,

$$\Psi(x,t) = \sum_m \Phi_{m,n}(r)e^{im\theta - in\phi - i\omega t} + c.c. = \Phi(r,\theta)e^{-in\phi - i\omega t} + c.c. \tag{16}$$

The toroidal mode solution, $\Phi(r,\theta)$ of this Mathieu-type equation is for each given r determined by

$$\frac{c_s^2}{\omega^2 q^2 R^2}\left(\frac{\partial}{\partial\theta} - inq(r)\right)^2 \Phi + \left[1 - i\delta_{\boldsymbol{k}} - \rho_s^2 \nabla_\perp^2 - \frac{\omega_{*e}}{\omega}(1 - 2\varepsilon_n \cos\theta)\right]\Phi = 0. \tag{17}$$

Here we used the simple phase-shifted, non-adiabatic electron response, $\delta n_e = n_e(e\phi/T_e)(1 - i\delta_{\boldsymbol{k}})$. Introducing the phase variation $\Phi \rightarrow \Phi(\theta)e^{inq(r)\theta}$, required to keep $k_\parallel^2\Phi$ small, we find that $\nabla_\perp^2\Phi = (\partial_r^2 + r^{-2}\partial_\theta^2)\Phi e^{inq\theta} = -(k_r^2 + k_\theta^2) = -k_\theta^2(1 + s^2\theta^2)\Phi$, where $k_r = sk_\theta\theta$ and $k_\theta = nq/r$.

The drift-eigen-mode equation for the one-dimensional ansatz $\Phi(\theta)$ is called the "ballooning mode" equation, which from Eqs.(15 - 17) is

$$\frac{c_s^2}{\omega^2 q^2 R^2}\frac{\partial^2\Phi}{\partial\theta^2} + \left[1 - i\delta_{\boldsymbol{k}} + k_\theta^2\rho_s^2(1 + s^2\theta^2) - \frac{\omega_{*e}}{\omega}(1 - 2\varepsilon_n\cos\theta)\right]\Phi = 0. \tag{18}$$

The solution of Eq. (18) with a shooting code requires as a boundary condition such that when $\theta^2 \rightarrow \infty$ the solution $\Phi \rightarrow \exp(i\frac{\sigma}{2}\theta^2)$, where $\sigma = \omega qR|sk_\theta|\rho_s/c_s$ with $\mathrm{Im}\,\sigma \propto \mathrm{Im}\,\omega > 0$. When the mode is strongly localized to the small θ^2 region, we may use $\cos\theta \simeq 1 - \theta^2/2$ to reduce Eq. (18) to the analytic harmonic oscillator problem with eigenvalues,

$$\omega_{k_\theta,n} = \frac{\omega_{*e}(1 - 2\varepsilon_n) - ic_s|sk_\theta|\rho_s/(qR)(2n+1)}{1 + k_\theta^2\rho_s^2 - i\delta_{\boldsymbol{k}}} \tag{19}$$

and eigen-functions, $\Phi_n(\theta) = \exp(i\frac{\sigma}{2}\theta^2)H_n((-i\sigma)^{1/2}\theta)$. Thus, the magnetic shear, $|s|$, introduces a damping of the waves due to the coupling to the ion acoustic waves that the propagate wave energy to region of Landau damping. Marginal stability conditions follow from Eq. (19) once the details of the dissipative $\delta_{\boldsymbol{k}}(\mu)$ function are given.

4. LABORATORY DRIFT WAVE EXPERIMENTS

4.1. Identification of Drift Waves

Let us now examine how these drift waves arise in simple geometries and what has been measured with probes in university-scale drift wave experiments.

The first detailed study of the drift wave growth, saturation and anomalous transport was in a series of Q-machine experiments [13]. Hendel et al. [13] and Chen [14] describe the identification of the mode frequencies and wavelengths in detail with variation of the system parameters in the Q-machine. The linear growth rate, $\gamma_{\boldsymbol{k}} = \mathrm{Im}\,\omega$, is produced by the phase shift, $\psi_{\delta n,\phi}$, from the electron collisions, ν_e, resulting in a density response,

$$\frac{\delta n_e}{n_e} = \frac{\omega - \omega_{*e} + ik_\parallel^2 v_e^2/\nu_e}{\omega + ik_\parallel^2 v_e^2/\nu_e}\frac{e\phi_{\boldsymbol{k}}}{T_e}, \tag{20}$$

from the potential wave $\phi_{\boldsymbol{k}}$. The growth rate, γ_k, was controlled in the experiments by varying the ion viscosity μ, where $\mu/\rho_m = (3/10)\nu_i\rho_i^2$, ν_i is the ion collision frequency, ρ_i the ion gyroradius and ρ_m the mass density. Since $\mu/\rho_m \propto 1/B^2$, the experimentalists varied the B-field through the critical value, $\boldsymbol{B}_{\mathrm{crit}}$, to measure the growth rate directly as well as the wave frequency. The nonlinear saturation occurs when the nonlinearity in the ion dynamics balances the growth rate from the electrons. This type of nonlinear saturation is generic to drift waves since the rapid electron motion leaves the electron phase as in the linear response theory. In section 6 we show these nonlinear features in more detail.

The first clear measurements of the ITG drift mode in a tokamak was provided by Brower et al. [15] (see Fig. 6) using laser scattering in the TEXT tokamak. This TEXT data used a Heavy Ion Beam Probe (HIBP) to measure both the DC radial electric field, E_r, and the local fluctuating potential, $\delta\phi(\boldsymbol{x},t)$. The plasma density fluctuation, δn_e, was measured with a 6-channel CO_2 laser-scattering diagnostic that yields $\boldsymbol{k} = \boldsymbol{k}_{\mathrm{out}} - \boldsymbol{k}_{\mathrm{in}}$ from the Bragg scattering condition of the outgoing $\boldsymbol{E}_{\boldsymbol{k}_{\mathrm{out}}}$ and incoming $\boldsymbol{E}_{\boldsymbol{k}_{\mathrm{in}}}$ electromagnetic waves from the plasma wave $\delta n_e(\boldsymbol{k})$. The wave number, k_θ, is shown on the axis in Fig. 7. The frequency in the laboratory frame is shown on the $\boldsymbol{y}$-axis, which includes the Doppler shift $\boldsymbol{k}\cdot\boldsymbol{v}_E = -(E_r/B)k_\theta$, where $\boldsymbol{v}_E$ is derived from HIBP data that measures E_r.

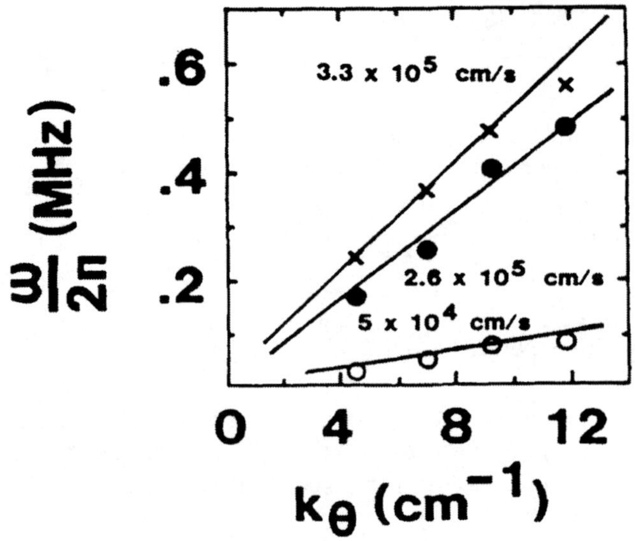

FIGURE 6. Microturbulence statistical dispersion for electron components, $n_e = 2 \times 10^{13}\,\text{cm}^{-3}$(crossed), $2 \times 10^{13}\,\text{cm}^{-3}$(filled circles) and ion component $n_i = 8 \times 10^{13}\,\text{cm}^{-3}$(open circles)[15]

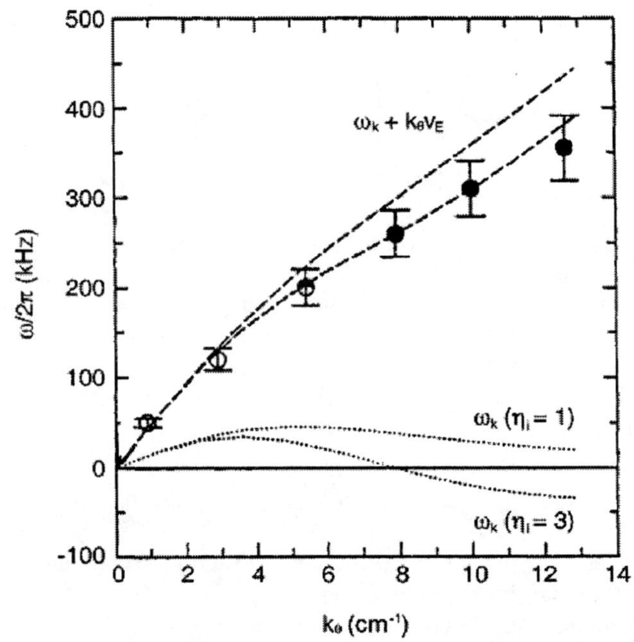

FIGURE 7. The frequency of the electron density fluctuation in the laboratory versus the azimuthal wave number k_θ for scattering.

For the TFTR machine Fig. 8 gives the estimate of the density fluctuation level $\delta n/n$ as a function of increasing neutral beam (NB) heating power P_{NB}. At the left, the data for $P_{\text{NB}} = 0$ is for the ohmic discharge with very low $\delta n/n \simeq 0.001$. As the beam heating power increases, the fractional fluctuations rise to 0.003 at 5 MW, 0.006 at 9 W and 0.015 at 19 MW. This increase of $\delta n/n$ is consistent with power balance $< nT > /\tau_E = (P_{\text{NB}} + P_{\text{Oh}})/\text{Vol}$ with

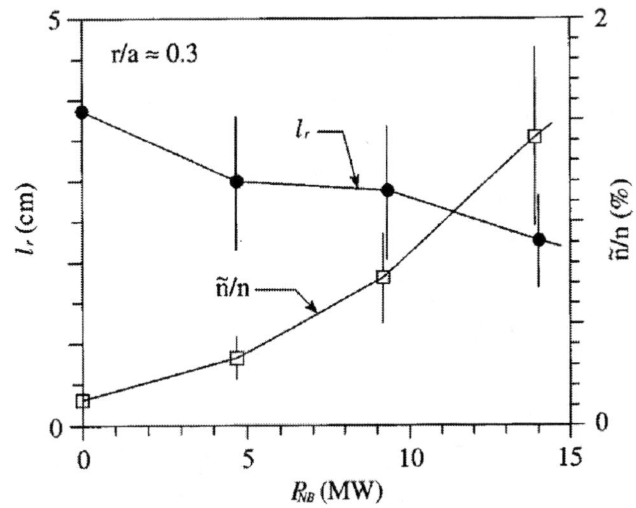

FIGURE 8. Power scan in TFTR microwave scattering for the density radial scale length (circles) and density fluctuation level (squares) as a function of NB power at $r/a \simeq 0.3$[16].

$1/\tau_E \propto \chi_{gB}/a^2 \propto T^{3/2}$. Here Vol is the mean plasma volume determined by the mean value theorem,

$$\int \frac{3}{2} nT \mathrm{d}^3 r = \frac{3}{2} < nT > \cdot Vol. \tag{21}$$

In tokamaks with higher temperatures and thus stronger radial temperature gradients, the drift wave changes character and is destabilized by the ion temperature gradient, $\boldsymbol{\nabla} T_i$. In this regime, called the ion temperature gradient (ITG) instability, the electron dissipation $i\delta_{\boldsymbol{k}}$ can be dropped and the electron response taken as adiabatic and is therefore defined by $\delta n_e / n_e = e\phi_{\boldsymbol{k}}/T_e$.

Simpler basic experiments validating the ITG model were carried out in the Columbia Linear Machine (CLM) as given in Sen et al. [17]. These plasmas, produced by RF heating, have $n_e \sim 2 \times 10^{14} \mathrm{m}^{-3}$ and $T_i \sim 10 \mathrm{eV}$. There is a temperature anisotropy $T_{i\parallel} \neq T_{i\perp}$ that complicates the linear dispersion but is useful for both NBI or RF heated toroidal experiments and the central cell of the tandem mirror. In the GAMMA-10 tandem mirror resonant ion cyclotron heating in the central cell results in $T_{i\perp} \sim 6.5 \mathrm{KeV}$ and $T_{i\parallel} \sim 2.5 \mathrm{KeV}$[18]. The details of the two temperature components of the ITG theory with the simple cylindrical central-cell plasma data can be found in Hojo [19] and Sen et al. [17].

In section 6, we will discuss the free energy of the ITG modes and their simulation in the gyro-kinetic Vlasov system. The literature concerning the numerical modeling of ITG modes is extensive due to its role as the standard model for explaining the anomalous ion thermal diffusivity, χ_i.

4.2. The effect of magnetic and E_r shear on drift waves

The amplitude of the ITG turbulence is controlled by the magnetic shear, $s = rq'/q$, and the shear in the azimuthal (or poloidal) flow velocity, $v_E = -E_r/B$. For example, Hamaguchi and Horton [20, 21, 22] show, with an FLR fluid model, the variation of the amplitudes and the resulting turbulent ion heat flux, $q_i = \langle n_i T_i v_r \rangle$, with magnetic and E_r shears. Kishimoto et al. [23] showed similar results with a PIC code. In the toroidal code, the shear, v_E', from E_r stifles the χ_i transport, whereas suppression of χ_i in the local slab model is given by

$$\chi_i(v_E') = \frac{\chi_i}{1 + (L_s v_E'/c_s)^2}, \tag{22}$$

as shown in Terry [24] and Newman et al. [25]. Here $L_s = qR/s(r)$ is the shear length defined by the distance spanned by the magnetic field when the $\boldsymbol{B}$-vector rotates by approximately $\pi/4$. The physical picture of the velocity-shear effect is given in Biglari et al. [26] where the large eddies or vortices of size $\Delta r = \rho_s(L_s/L_T)$ in the absence of shear flows are reduced in the shearing time, $\tau_{\rm sh} = 1/v_E'$, to small eddies. The details of the shear flow stabilization are complicated,

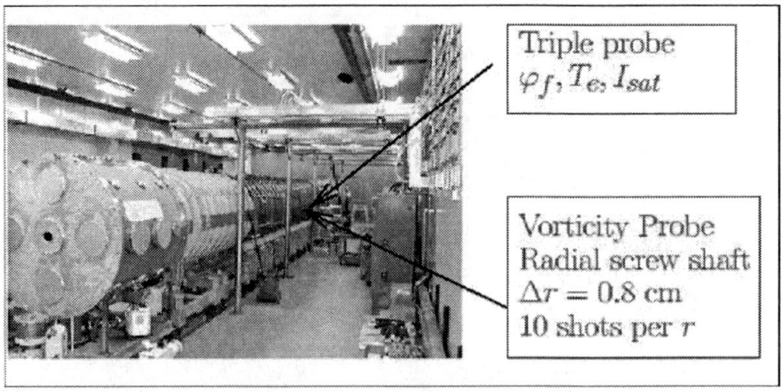

FIGURE 9. LAPD machine ($L = 17$ m, $a = 0.5$ m) with the location of the triple probe and the vorticity probe driven on a a shift taking 10 shots per $\Delta r = 0.8$cm through the region of drift wave and Kevin Helmholtz turbulence.

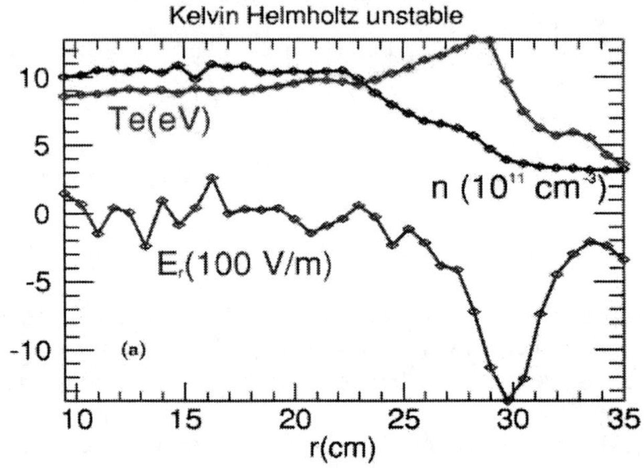

FIGURE 10. Mean profiles obtained as an average in 25 experimental shots during the stationary turbulent state during the wall-bias pulse [30]

see P.H. Diamond's chapter of the proceedings. A kinetic theory calculation is given by Dong et al. [27] and the role of shear flow in magnetized plasma is reviewed by Itoh and Itoh [28], Terry [24], Diamond et al. [29].

Experiments for validation of drift-wave theory were performed on LAPD (Fig. 9). These regimes were created by varying the radial electric field, E_r, from weak to moderate and then to strong. In the moderate and strong E_r regime, the electron-density transport barrier was measured as shown in Fig. 10. The main characteristic of the drift wave was identified clearly in the weak E_r-shear regime and the mixed Kelvin-Helmholtz drift-wave regime.

In the strong E_r-shear regime there is a deep well of negative E_r – as occurs in the edge of H-mode tokamak plasmas. The vorticity probe measures $\omega_z = \hat{\mathbf{b}} \cdot \nabla \times \mathbf{v}_E$ (see Fig. 11).

As the probe sweeps across the shear layer, the vorticity changes from counter-clockwise to clockwise rotating vortices, just as expected from simple considerations and computer simulations. The region between these two oppositely spinning vortices forms the partial transport barrier as suggested by Biglari et al. [26] and Terry [24].

The strong shear flow in a plasma creates temporal and spatial intermittency as shown by the probability distribution function with large skewness and excess probability levels over the gaussian level (red parabola) in Fig. 11. In the central cell of the GAMMA-10 tandem mirror, where the electron temperature $T_e \sim 800$eV, there is sufficient soft-X-ray emission to image the growth and decay of the vortices using two micro-channel plates and tomographic image reconstruction [31, 18]. In Fig. 12, the X-ray tomography shows such an event over the rotational period of $50\,\mu$s. The growth and decay of vortex structures are clearly seen. Moreover, they appear consistent with simulations by Kim et al. [32], which show vortex intermittency in sheared flows.

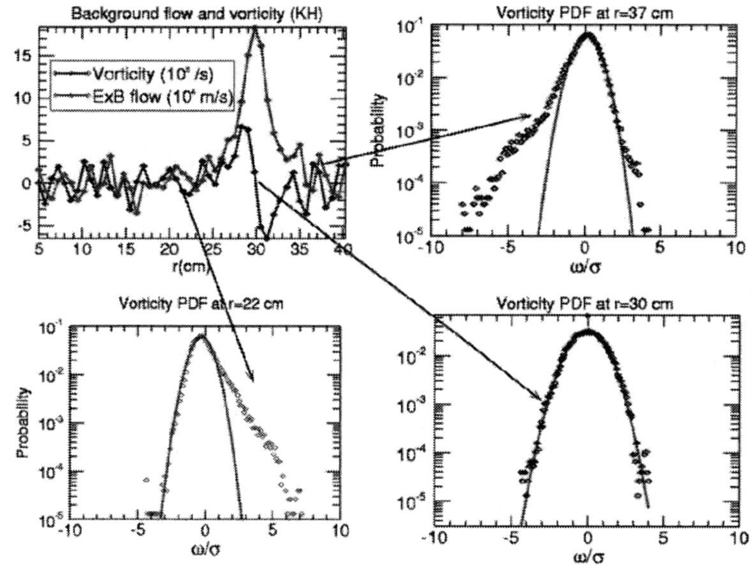

FIGURE 11. Vorticity probabiliy distribution functions for representative radius in strong the E_r-shear regime[30].

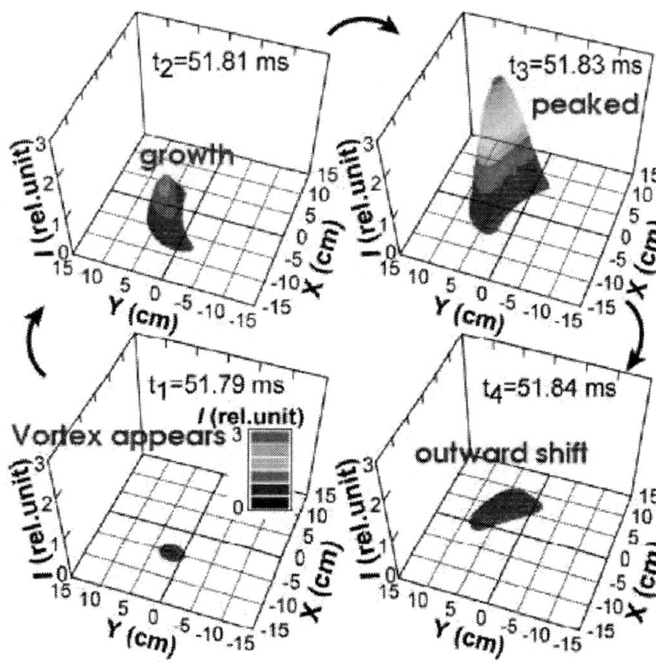

FIGURE 12. X-ray tomography of a detailed vortex structure of an intermittent object showing its typical lifetime ranging around a hundred microseconds, approximately with the rotational motion of an $\boldsymbol{E} \times \boldsymbol{B}$ drift[31].

4.3. Electromagnetic scattering from turbulent density fluctuations

Non-perturbative measurements of the plasma fluctuation are carried out with Bragg scattering, $\boldsymbol{k} = \boldsymbol{k}_{\text{out}} - \boldsymbol{k}_{\text{in}}$, where an in coming micro-wave or lower wave, $\boldsymbol{E}_{\text{in}} \exp(i\boldsymbol{k}_{\text{in}} \cdot \boldsymbol{x} - i\omega_{\text{in}} t)$, sets the electron into quiver motion with $\delta v_e = -e\boldsymbol{E}_{\text{in}}/im_e\omega_{\text{in}}$ that radiates $\boldsymbol{E}_{\text{out}} \exp(i\boldsymbol{k}_{\text{out}} \cdot \boldsymbol{x} - i\omega_{\text{out}} t)$, with current $\delta \boldsymbol{j} = -e\delta n_{\boldsymbol{k}} \delta \boldsymbol{v}_{\boldsymbol{e}}(\boldsymbol{k}_{\text{in}})$. The scattered wave, $\boldsymbol{E}_{\text{out}} \exp(i\boldsymbol{k}_{\text{out}} \cdot \boldsymbol{x} - i\omega_0 t)$ has $\boldsymbol{k}_{\text{out}} = \boldsymbol{k} + \boldsymbol{k}_{\text{in}}$ and $\omega_{\text{out}} = \omega_{\boldsymbol{k}} + \omega_{\text{in}}$. Since $\omega_{\boldsymbol{k}} \ll k_{\text{in}} c$, the frequency shift is often too small to be useful. The scattering cross-section, $d\sigma/d\Omega = (e^2/m_e c^2)^2 \left\langle |\delta n_e(\boldsymbol{k})|^2 \right\rangle \Omega_{scat}/n_e$, where Ω_{scat} is the scattering volume defined by the insertion of the two electromagnetic waves giving a non-perturbative measurement of $\langle |\delta n_e(\boldsymbol{k})|^2 \rangle$. Both X- and O-mode polarizations are useful [33].

In 1976 electromagnetic scattering experiments, by Mazzucato [16] on the ATC machine and by Slusher and Surko [34] on Alcator, showed clearly the presence of substantial drift wave turbulence. The weak turbulence mode coupling theory was used in Horton [35] to give, within error bars, the level of $|\delta n(\boldsymbol{k}_\perp)|$ at the six $k_\perp$ ranging from $1/2.3$ to $1/0.4\,\mathrm{cm}^{-1}$ used in ATC. This type of scattering data is shown in Fig. 8 for TFTR. Use of scattering from O- to X-mode polarization gives a measure of $\left\langle |\delta B_r(\boldsymbol{k})|^2 \right\rangle$ in Zou et al. [36] and in Colas et al. [37] in the core of Tore Supra.

The cross-polarization experiments by Colas et al. [37] show $\delta B_x^2/B^2$ increasing up to levels of $\delta B_x/B_0 \simeq 5\times 10^{-5}$ with higher heating power and $\boldsymbol{\nabla} T_e$. The interpretation of the cross-polarizing experiments is difficult since part of the O→X-mode conversion can also be scattered power that can arise from internal reflection in the machine. These experiments and their interpretation are discussed further in the chapter by P. Hennequin.

5. WEAK TURBULENCE THEORY FOR DRIFT WAVES

For turbulent states with a broad range of wavenumbers, $\Delta\boldsymbol{k}$, and frequencies, $\Delta\omega_{\boldsymbol{k}}$, in the fluctuation spectrum the Weak Turbulence Theory is developed by assuming that the deviation of the fluctuations from Gaussian statistics is weak. Under this condition the chain of coupled correlation functions is broken off at fourth order with what is called the "quasi-normal closure approximation". This means that the ensemble average of the product of four $\phi_{\boldsymbol{k}}$ fields is factorized as

$$\left\langle \phi_{\boldsymbol{k}_1}\phi_{\boldsymbol{k}_2}\phi_{\boldsymbol{k}_3}\phi_{\boldsymbol{k}_4=-\boldsymbol{k}_1-\boldsymbol{k}_2-\boldsymbol{k}_3}\right\rangle = \left\langle \phi_{\boldsymbol{k}_1}\phi_{\boldsymbol{k}_2}\right\rangle\left\langle \phi_{\boldsymbol{k}_3}\phi_{\boldsymbol{k}_4}\right\rangle + \left\langle \phi_{\boldsymbol{k}_1}\phi_{\boldsymbol{k}_3}\right\rangle\left\langle \phi_{\boldsymbol{k}_2}\phi_{\boldsymbol{k}_4}\right\rangle + \left\langle \phi_{\boldsymbol{k}_1}\phi_{\boldsymbol{k}_4}\right\rangle\left\langle \phi_{\boldsymbol{k}_2}\phi_{\boldsymbol{k}_3}\right\rangle, \tag{23}$$

which closes the equations for the dynamics of $I_{\boldsymbol{k}}(t) = \langle \phi_{\boldsymbol{k}}^* \phi_{\boldsymbol{k}} \rangle$.

For drift waves the energy spectrum,

$$W_{\boldsymbol{k}} = (1 + k_\perp^2 \rho_s^2) I_{\boldsymbol{k}}(t), \tag{24}$$

and the resulting weak-turbulence wave-kinetic equation is

$$\begin{aligned} \frac{dW_{\boldsymbol{k}}}{dt} &= 2\gamma_{\boldsymbol{k}} W_{\boldsymbol{k}} + \sum_{\boldsymbol{k}_1,\boldsymbol{k}_2=\boldsymbol{k}-\boldsymbol{k}_1} \frac{(\boldsymbol{k}\times\boldsymbol{k}_1\cdot\hat{\boldsymbol{z}})^2}{B^2}(k_2^2-k_1^2) R_{\boldsymbol{k},\boldsymbol{k}_1,\boldsymbol{k}_2}\cdot \\ &\quad \left[(k_2^2-k_1^2)W_{\boldsymbol{k}_1}W_{\boldsymbol{k}_2} - (k^2-k_1^2)W_{\boldsymbol{k}}W_{\boldsymbol{k}_1} + (k^2-k_2^2)W_{\boldsymbol{k}}W_{\boldsymbol{k}_2}\right], \end{aligned} \tag{25}$$

with the resonance function given by

$$R_{\boldsymbol{k},\boldsymbol{k}_1,\boldsymbol{k}_2} = -\mathrm{Im}\left[\frac{1}{\omega_{\boldsymbol{k}} - \omega_{\boldsymbol{k}_1} - \omega_{\boldsymbol{k}_2} + i(\nu_{\boldsymbol{k}} + \nu_{\boldsymbol{k}_1} + \nu_{\boldsymbol{k}_2})}\right] \simeq \pi\delta(\omega_{\boldsymbol{k}} - \omega_{\boldsymbol{k}_1} - \omega_{\boldsymbol{k}_2}), \tag{26}$$

for $\nu_{\boldsymbol{k}} \ll \max(\omega_{\boldsymbol{k}})$. In the classical model with adiabatic electrons, the wave-kinetic equation obeys the conservation laws of energy, $d/dt \sum_{\boldsymbol{k}} W_{\boldsymbol{k}} = 0$, and enstropy, $d/dt \sum_{\boldsymbol{k}} (1 + k_\perp^2) W_{\boldsymbol{k}} = 0$.

These two invariants lead to the cascade of enstropy, $\Omega = \sum_{\boldsymbol{k}} k_\perp^2 W_{\boldsymbol{k}}(t)$, to high $\boldsymbol{k}_\perp$ and the inverse cascade of energy W to low $\boldsymbol{k}_\perp$. The equilibrium solution of the wave-kinetic equation with no growth/damping, i.e. $\gamma_{\boldsymbol{k}} \equiv 0$, is the equipartition of the energy $W_{\boldsymbol{k}} = W_{\mathrm{total}}/N_{\mathrm{modes}}$, where N_{modes} is the total number of modes.

In reality, there are limited regions of $\boldsymbol{k}$-space where $\gamma_{\boldsymbol{k}} > 0$ and hence where modes can grow from the gradient mechanism in Sec. 1. In all other regions modes are damped. While some approximate analytic solutions to Eqs. (24-26) are known in terms of anisotropic Kolmogorov-type spectra $1/|k_x|^{m_x}|k_y|^{m_y}$, the general solution to the problem can only be found numerically.

Fig. 13 shows by the shaded loop in $\boldsymbol{k}$-space the region of resonant three-wave interaction defined by the subspace $\omega(\boldsymbol{k}) - \omega(\boldsymbol{k}_1) - \omega(\boldsymbol{k}_2) = 0$. There are two special parts of this resonant manifold : one where $\boldsymbol{k}_1 = (2k_x, 0)$, which is a zonal flow type of state with $u_y(x) = k_x\phi_{\boldsymbol{k}_1}\sin(2k_x x)$ and $u_x = 0$, and the the second is the quasi-adiabatic part of the resonant manifold where two neighboring waves $(\boldsymbol{k}_1, \boldsymbol{k}_2)$ resonate, i.e. $(\boldsymbol{k}, \omega_{\boldsymbol{k}}) - (\boldsymbol{k}_2, \omega_{\boldsymbol{k}_2}) = (\boldsymbol{k}_1, \omega_{\boldsymbol{k}_1})$ with small values of $\boldsymbol{k}_1$ and $\omega_{\boldsymbol{k}_1}$. This second class gives a scale separation where the symbols $\boldsymbol{q} = \boldsymbol{k}_1 \ll |\boldsymbol{k}|$ and $\Omega = \omega_{\boldsymbol{k}_1} \ll |\omega_{\boldsymbol{k}}|$

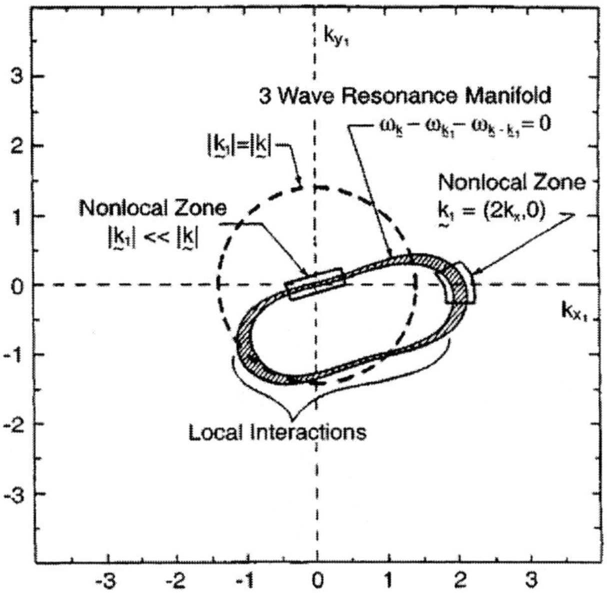

FIGURE 13. Wave-resonance manifold

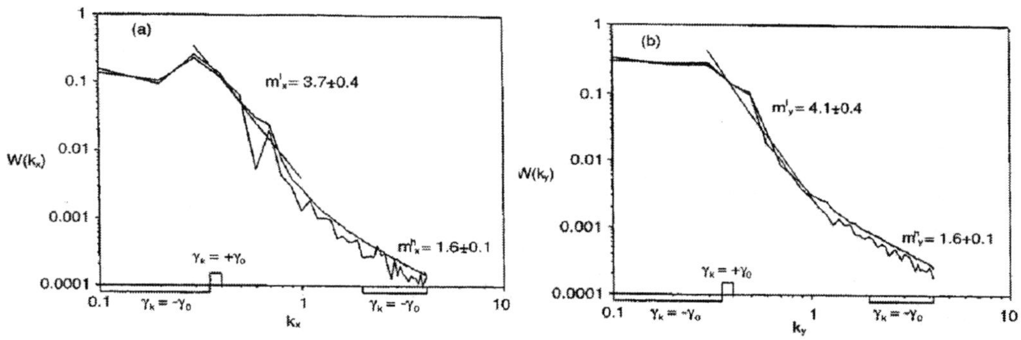

FIGURE 14. Drift-wave spectrum for model with isotropic ring of unstable modes, $\gamma_0 > 0$, for $0.35 < k < 0.4$ and high k damping for $k > 2$.

are often introduced for the large scale part of the spectrum. These cross-scale $\boldsymbol{k} \leftrightarrow \boldsymbol{q}$ interactions give a long-range energy transfer in $\boldsymbol{k}$-space. These two regions of special interest are marked by rectangles in Fig. 13.

Fig. 14 shows an example of the anisotropic Kolmogorov spectrum applied, where there is only an unstable ring for $0.35 < (k_x^2 + k_y^2)^{1/2} \leq 0.4$ with small $\gamma_0 > 0$ and damping of $-\gamma_0$ at both high $\boldsymbol{k}$ and low $\boldsymbol{k}$. The steady state spectra of $W(k_x) = \sum_{k_y} W(k_x, k_y) \simeq 1/|k_x|^{m_x}$ and $W(k_y) = \sum_{k_x} W(k_x, k_y) = 1/|k_y|^{m_y}$ are shown in Fig. 14. We see that most of the turbulent energy has accumulated in $\boldsymbol{k}$'s below the growth region ($k_\perp \sim 0.35$), while the region above the growth region has a power-law decay with $m_x = 3.7 \pm 0.4$ and $m_y = 4.1 \pm 0.4$. These results are typical but not universal for drift wave turbulence.

The adiabatic electron model leads to that the nonlinear $\boldsymbol{E} \times \boldsymbol{B}$ convection of the density by $\boldsymbol{v}_E \cdot \nabla n = 0$ vanishes. In general however, there is an appreciable non-adiabatic part of the density response to ϕ as in Eq. (20). The non-adiabatic part, for example, $\delta n = (1 + \delta_0 \partial/\partial y)\phi \to (1 + i\delta_0 k_y)\phi_{\boldsymbol{k}}$ leads to a nonlinear $\boldsymbol{E} \times \boldsymbol{B}$ drift $\boldsymbol{v}_E \cdot \nabla n \neq 0$ and turbulent transport for $\delta_0 \neq 0$. Waltz has shown that the additional nonlinear term from $\boldsymbol{v}_E \cdot \nabla n$ leads to a lower amplitude as shown in Fig. 15. The values of $\phi_{\text{rms}} = \left\langle \phi^2 \right\rangle^{1/2}$ and the diffusivity D are shown as a function of the non-adiabatic parameter δ_0.

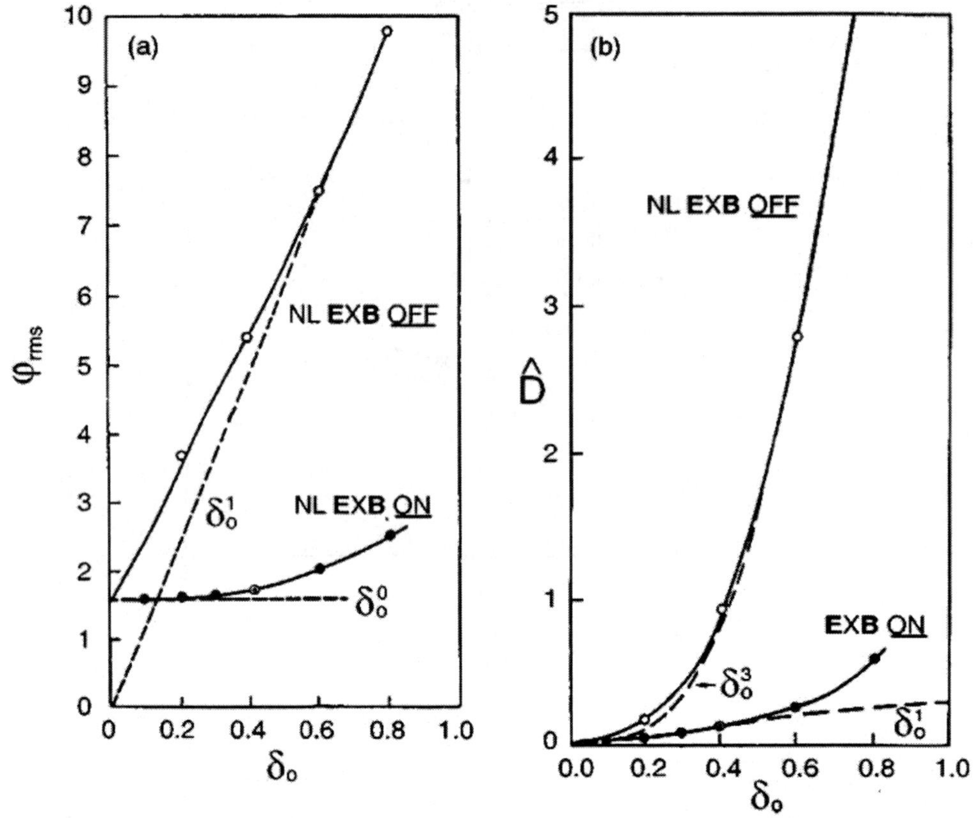

FIGURE 15. Waltz simulation comparing the saturation levels of the Hasegawa-Mima model with NL (nonlinear $v_E \cdot \nabla \delta n$) off and the Terry-Horton model with $v_E \cdot \nabla \delta n$ turned on.

6. ION TEMPERATURE GRADIENT MODE

There are numerous nonlinear formulations of the ion temperature gradient mode in the literature. The models range from one-field models (e.g. Hasegawa-Mima) with one nonlinearity $v_E \cdot \nabla$, to two field-models; both of them include the fluctuating electrostatic potential ϕ. The one-field model, however, uses the the adiabatic electron response $\delta n_e/n_e = e\phi/T_e$ for the density fluctuation whereas the two-field model uses either the fluctuating ion pressure, δp_i, or the fluctuating ion temperature, δT_i, dynamical equation. These gyro-fluid models assume a closure that can be either an equation of state with adiabatic gas constant, Γ, for the ion fluid or closure formulas for the parallel ion-thermal-heat flux, $\delta q_{i\parallel}$ [38, 39, 40]

References describing different models are Horton et al. [41], Hamaguchi and Horton [20], Dimits et al. [42] for the slab ITG mode and Horton et al. [43], Waltz et al. [44], Beer and Hammett [45] for the toroidal ITG mode. The toroidal ITG mode is typically stronger than the slab ITG mode and has a mathematical structure similar to a small scale ($\lambda_\perp \gtrsim \rho_i$) Rayleigh-Taylor instability. The Rayleigh-Taylor MHD instability,

$$\omega^2 = -\gamma_{\rm MHD}^2 = \frac{k_y^2}{k_\perp^2}\left(\frac{1}{\rho_s}\frac{dp}{dx}\frac{1}{R_c}\right) \gg k_\parallel^2 v_A^2, \tag{27}$$

is changed in drift-wave theory to a slower growing mode with $\gamma(k_y = 0) = 0$ and a well-defined maximum growth rate, $\gamma_{\max}$, at a finite $k_y\rho_s \simeq [(1-2\varepsilon_n)/(1+\eta_i)]^{1/2}$. Here $\varepsilon_n = L_n/R$ and $\eta_i = {\rm dln}T_i/{\rm dln}n_i$. The electrostatic shielding by the adiabatic electron fluid that has a dielectric constant, $\varepsilon = 1 + 1/k_\parallel^2\lambda_{De}^2 \gg 1$, since $k_\parallel \lambda_{De} = k_\parallel v_e/\omega_{pe} \ll 1$. The energy source of the instability is explained in Sec. 2.2. The typical $\gamma_{\max} = 0.1c_s/(L_{Ti}R)^{1/2}$ for a toroidal system with a magnetic radius of curvature, $R_c = R$, on the outside of the torus where the mode balloons. The ballooning-mode equation and a simple nonlinear model for the toroidal ITG are given in Horton et al. [43]. Here we give the

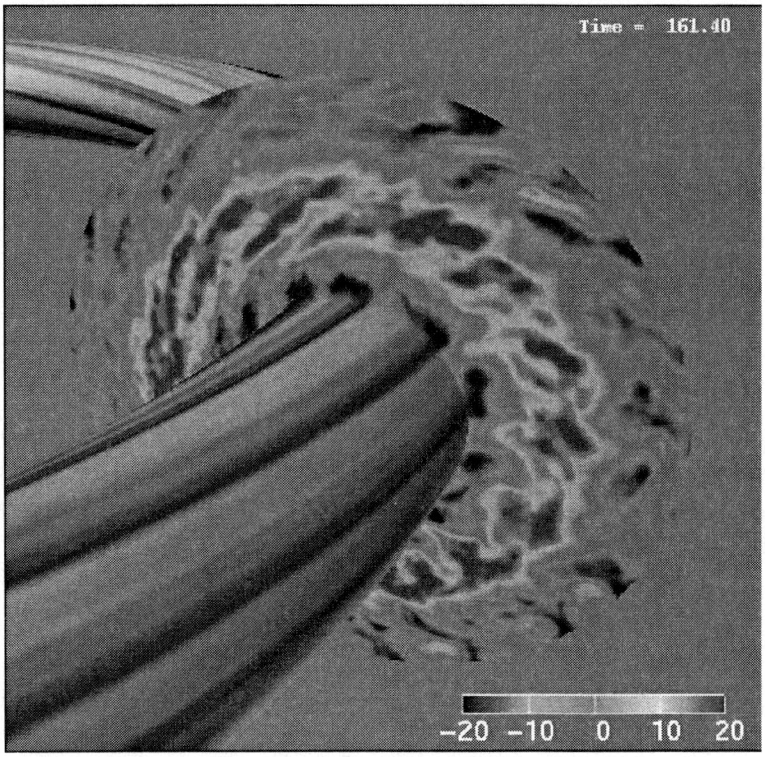

FIGURE 16. ITG simulation from Wakanabe GKV

Balescu version of the Hamaguchi-Horton slab ITG model as an example of a new model that needs numerical solving. Following Balescu [46, p.33] we find that the ITG mode has four partial differential equations,

$$\frac{\partial \delta n_i}{\partial t} = -v_{de}\frac{\partial \delta\phi}{\partial y} - \nabla_{\parallel}\delta v_{\parallel} - \frac{c_s^2}{\Omega_i}\left[\delta\phi, \delta n_i\right] \tag{28}$$

$$\frac{\partial}{\partial t}\left(\delta n_i - \rho_s^2\nabla_{\perp}^2\delta\phi\right) = -v_{de}\frac{\partial}{\partial y}(1-\tau(1+\eta_i)\rho_s^2\nabla_{\perp}^2)\delta\phi - \nabla_{\parallel}\delta v_{\parallel}$$
$$-\frac{c_s^2}{\Omega^i}\left[\delta\phi, (\delta n_i - \rho_s^2\nabla_{\perp}^2\delta\phi)\right] \tag{29}$$

$$\frac{\partial \delta v_{\parallel}}{\partial t} = -\frac{1}{m_i n_0}\nabla_{\parallel}\delta p_i - c_s^2\nabla_{\parallel}\delta\phi - \frac{c_s^2}{\Omega_i}\left[\delta\phi, \delta v_{\parallel}\right] \tag{30}$$

$$\frac{\partial \delta p_i}{\partial t} = -v_{de}\tau(1+\eta_i)\frac{\partial \delta\phi}{\partial y} - \Gamma p_{i0}\nabla_{\parallel}\delta v_{\parallel} - \frac{c_s^2}{\Omega_i}\left[\delta\phi, \delta p_i\right] \tag{31}$$

where $\tau = T_i/T_e$, $v_{de} = (\rho_s/L_n)c_s$ and Γ is the ion adiabatic gas constant. Each equation has a convective nonlinearity.

Currently there are a number of global and flux-tube codes that simulate the full gyro-kinetic equations for the ITG turbulence Lin et al. [47], Dimits et al. [48], Jenko and Dorland [49]. Here we show the distribution of the electrostatic potential fluctuations in all three dimensions in Fig. 16. In this figure where saturation has occurred, the tendency for the fluctuations to establish a zonal-flow pattern is evident.

6.1. Closure models

When the velocity moments of the original gyrokinetic equation are taken to derive the gyrofluid equations, we need to construct closure models which describe high-order moment variables arising from kinetic effects such as parallel streaming, finite gyroradii, and toroidal magnetic drift. Closure models for these kinetic effects were proposed in a series of works by Hammett & Perkins [38], Dorland & Hammett [50], and Beer *et. al.* [51]. In these models, high-order moment variables are written in dissipative forms given by linear combination of low-order moment variables so as to reproduce well the linear kinetic dispersion relation. In the Hammett-Perkins model, the parallel heat flux $q_{\|\mathbf{k}}$ with the wave number vector $\mathbf{k}$ is given in terms of the temperature fluctuation $T_{\mathbf{k}}$ as

$$q_{\|\mathbf{k}} = -2\sqrt{\frac{2}{\pi}}\, n_0 v_t\, i \frac{k_\|}{|k_\||} T_{\mathbf{k}}, \tag{32}$$

where $k_\|$ is the wave number in the direction parallel to the magnetic field. The Hammett-Perkins model is successfully applied to calculation of frequencies and growth rates of linear ITG modes while it does not discribe time-reversal symmetric behaviors of the three-mode nonlinear kinetic ITG system [39] which consists of two unstable slab ITG modes and a stable mode representing a quasilinear flattening of the background temperature profile. Mattor & Parker [39] presented a nonlinear closure model and successfully reproduced the solution of the three-mode ITG problem although it cannot be easily extended to turbulent systems with a large number of degrees of freedom.

Realizing that the time-reversal symmetric solution of the three-mode ITG problem can be represented by the superposition of an unstable normal-mode distribution function and its complex conjugate, Sugama, Watanabe & Horton [40] proposed the nondissipative closure model (NCM) which relates the parallel heat flux $q_{\|\mathbf{k}}$ in the linearly unstable wave number region to the temperature fluctuation $T_{\mathbf{k}}$ and the parallel flow $u_{\mathbf{k}}$ in terms of real-valued coefficients,

$$q_{\|\mathbf{k}} = C_{T\mathbf{k}} n_0 v_t T_{\mathbf{k}} + C_{u\mathbf{k}} n_0 T_i u_{\mathbf{k}} \qquad \text{(for linearly unstable modes)}, \tag{33}$$

where the real-valued coefficients $C_{T\mathbf{k}}$ and $C_{u\mathbf{k}}$ are determined by requiring that the kinetic dispersion relation should be derived from the closure relation. The NCM gives the correct relation between $q_{\|\mathbf{k}}$, $T_{\mathbf{k}}$, and $u_{\mathbf{k}}$ both for the unstable normal-mode solution and its complex-conjugate solution as well as for any linear combination of these solutions.

Since the time reversal symmetry of the original collisionless kinetic equation is retained in Eq. (33), the NCM can exactly reproduce the solution of the three-mode ITG problem.

Furthermore, the NCM was used in two-dimensional fluid simulation of the strong turbulence driven the slab ITG modes and its validity for quantitative prediction of the anomalous ion thermal transport in the slab ITG turbulence was confirmed by comparison to collisionless kinetic simulation results as shown in Fig. 17 [52].

We should note that the NCM can be applicable only to linearly unstable modes and that the Hammett-Perkins-type dissipative closure model was still used for linearly stable modes even for the NCM fluid simulation.

In the slab ITG turbulence shown in Fig. 17, the zonal-flow component was not included whereas only quiet steady states with turbulence transport suppressed by large zonal-flow generation were obtained by both of our kinetic and fluid simulations when including the zonal flow component [53].

In toroidal configurations, zonal-flow generation is weaker than in slab cases although zonal flows are still an important factor to regulate turbulent transport and a wrong description of zonal flows was once suspected by Rosenbluth & Hinton [54] as a cause of difference between gyrokinetic and gyrofluid simulation results of toroidal ITG turbulence.

Improved closure models, which give residual zonal flows predicted by Rosenbluth & Hinton, were proposed by Beer & Hammett [55] and recently by Sugama, Watanabe, & Horton [56].

7. CONCLUSIONS

The basic physical mechanisms of drift-wave instabilities and the associated plasma transport are described in detail. Drift waves develop into a mixture of wave turbulence and coherent structures with details of the mixture depending on the driving forces ∇n_e, ∇T_e and ∇T_i and the amount of magnetic shear and background velocity shear in the system. The mixture of zonal flows, waves and vortices has certain similarities to the Rossby waves in planetary atmospheres and the equatorial-to-polar temperature gradient driven turbulent heat. The principal difference is in the replacement of the gravitational force on planets with a electric force from charge separation in the curved magnetic fields.

The free energy sources created by the radial gradients of the plasma density and the temperature are useful to characterize the strength of the drift wave turbulence. We have suggested that real time diagnostic signals from microwave interferometers and Thompson scattering data can be used to compute the thermodynamic formulas for the upper bounds of the turbulence given in Sec.2 for the density and temperature fluctuations. The derivations of the

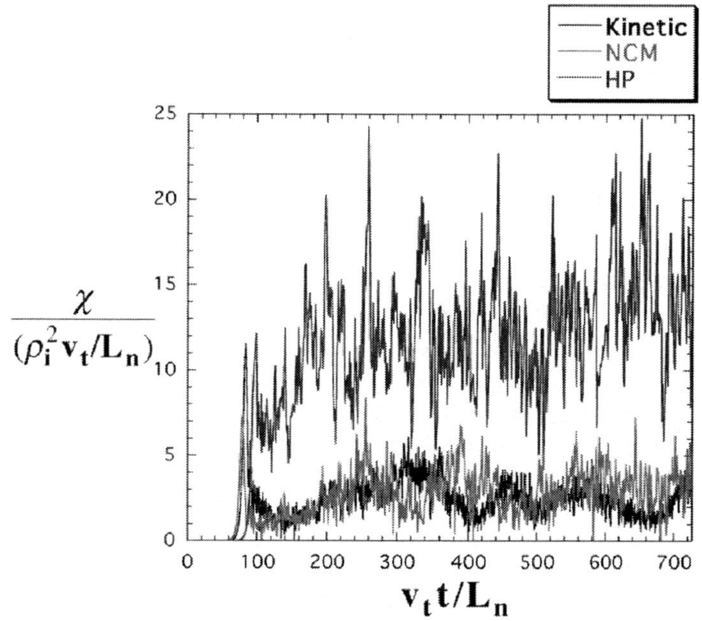

FIGURE 17. Normalized turbulent ion heat diffusivity $\chi/(\rho_i^2 v_t/L_n)$ as a function of normalized time $v_t t/L_n$ obtained by the two-dimensional slab ITG turbulence simulation [52]. Black, red, and blue lines correspond to results from the kinetic simulation, the fluid simulation using the NCM, and that using the Hammett-Perkins (HP) model, respectively.

free energy formulas treat the plasma as an ideal gas with a compressibility gas constant, $\Gamma = (d+2)/d$, where d is the number of degrees of freedom associated with the type of instability under investigation. The Carnot-cycle description of the auxiliary heated plasma is particularly relevant and was used in Horton et al. [57] for the electron temperature profiles in Tore Supra and is being used for real-time data in NSTX.

The structure of the drift wave eigen-modes in the double-periodic toroidal geometry is described after a brief review of the early experiments on drift waves in steady state cylindrical Q-machines. The recent experiments in the Columbia Linear Machine [17] devoted to the ion temperature gradient (ITG) mode and the 6-channel CO_2 laser scattering experiments in TEXT [15] clearly show the role of the hybrid ITG mode and trapped electron mode TEM.

The ITG mode has become the standard model for describing ion transport in basically all toroidal experiments with strong ion temperature gradients. The basic toroidal mechanism and nonlinear models are given in Horton et al. [43]. Nowadays large-scale Tera-flop computers commonly compute the ITG turbulence from direct numerical solutions of the gyrokinetic-Vlasov-Poisson system and as an example the state of the turbulence is shown in section 6 from the NIFS GKV simulation. The standard article on the comparison of direct numerical simulation for the ITG turbulence is Dimits et al. [48]. Review of the state of the numerical simulations of the ETG turbulence is Nevins et al. [58]. The electron turbulent transport is more demanding to compute but shows very large enhancements over the quasi-linear fluxes due to the formation of extended radial structures with correlation lengths, $\ell_c \gg \rho_e$, and the structures are insensitive to the radial electric-field shear but do depend sensitively on the magnetic-shear profile.

The empirical energy confinement scaling laws for both helical system [8] and tokamaks [6, 7] follow the drift-wave formulas and simulations to within the accuracy of the experimental data. Some of the more detailed comparisons of theoretical models [59, 60, 61, 62] with tokamak databases are referenced but a detailed discussion is beyond the scope of this chapter.

REFERENCES

1. W. Horton, and L. E. Reichl, editors, *Statistical Physics and Chaos in Fusion Plasmas*, John Wiley & Sons, New York, 1984.
2. S. Benkadda, Y. Elskens, and F. Doveil, editors, *Transport, Chaos and Plasma Physics:Institut Mediterraneen De Technologie, Marseille, France 10-21 July, 1995*, Advanced Series in Nonlinear Dynamics, World Scientific Pub Co Inc, 1995.
3. W. Horton, and Y. Ichikawa, *Chaos and Structures in Nonlinear Plasmas*, World Scientific, 1996.

4. G. T. Hoang, W. Horton, C. Bourdelle, B. Hu, X.Garbet, and M. Ottaviani, *Phys. Plasmas* **10**, 405–412 (2003).
5. I. H. mode Database Working Group, *Nucl. Fusion* **34**, 131–167 (1994).
6. S. Kaye, M. Greenwald, U. Stroth, O. Kardaun, A. Kus, D. Schissel, J. DeBoo, G. Bracco, K. Thomsen, J. Cordey, Y. Miura, T. Matsuda, H. Tamai, T. Takizuda, T. Hirayama, H. Kikuchi, O. Naito, A. Chudnovskij, J. Ongena, and G. Hoang, *Nucl. Fusion* **32**, 1303–1328 (1997).
7. I. P. E. Group, *Nucl. Fusion* **39**, 2175 (1999).
8. H. Yamada, J. H. Harris, A. Dinkalge, E. Ascasibar, F. Sano, S. Okamura, J. Talmadge, U. Stroth, A. Kus, S. Murakami, M. Yokoyama, C. D. Beidler, V. Tribaldos, K. Y. Wantanabe, and Y. Suzuki, *Nucl. Fusion* **45**, 1684–1693 (2004).
9. E. of 'Progress of ITER Physics Basis', *Nucl. Fusion* **47**, S1–S17 (2007).
10. B. Kadomtsev, and O. Pogutse, "Turbulence in Toroidal systems," in *Reviews of Plasma Physics, Volume 5*, edited by A. Leontovich, 1970, pp. 249–400.
11. D. Bohm, E. H. S. Burhop, and H. S. W. Massey, "The use of probes for plasma exploration in strong magnetic fields," in *The characteristics of electrical discharges in magnetic field*, edited by A. Guthrie, and R. K. Wakerling, New York, McGraw-Hill, 1949.
12. W. Horton, B. G. Hong, and W. M. Tang, *Phys. Fluids* **31**, 2971 (1988).
13. H. W. Hendel, T. K. Chu, and P. A. Politzer, *Phys. Fluids* **11**, 2424–2439 (1968).
14. F. F. Chen, *Introduction to plasma physics and controlled fusion*, Plenum Press, New York, 1974.
15. D. L. Brower, W. A. Peebles, S. K. Kim, N. C. Luhmann, W. M. Tang, and P. E. Phillips, *Phys. Rev. Lett.* **59**, 48–51 (1987).
16. E. Mazzucato, *Phys. Rev. Lett.* **36**, 792–794 (1976).
17. A. K. Sen, J. Chen, and M. Mauel, *Phys. Rev. Lett.* **66**, 429–432 (1991).
18. T. Cho, J. Kohagura, T. Numakura, M. Hirata, H. Higaki, H. Hojo, M. Ichimura, O. Watanabe, M. Yoshikawa, A. Kojima, Y. Miyake, Y. Miyata, K. Shimizu, Y. Tomii, M. Yoshida, K. Sakamoto, T. Imai, V. P. Pastukhov, S. Miyoshi, and G. . Group, *Phys. Rev. Lett.* **97**, 055001 (2006).
19. H. Hojo, *Jpn. J. Appl. Phys.* **34**, 6501 (1995).
20. S. Hamaguchi, and W. Horton, *Phys. Fluids B* **2**, 1833–1851 (1990).
21. S. Hamaguchi, and W. Horton, *Phys. Fluids B* **4**, 319–328 (1991).
22. S. Hamaguchi, and W. Horton, *Plasma Phys. Contr. Fusion* **35**, 203–233 (1992).
23. Y. Kishimoto, T. Tajima, W. Horton, M. J. LeBrun, and J. Y. Kim, *Phys. Plasmas* **3**, 1289–1307 (1996).
24. P. Terry, *Rev. Mod. Phys.* **72**, 109–165 (2000).
25. D. Newman, B. Carreras, D. Lopez-Bruna, P. Diamond, and V. Lebedev, *Phys. Plasmas* **05**, 938 (1998).
26. H. Biglari, P. H. Diamond, and P. W. Terry, *Phys. Rev. Lett.* **60**, 200–203 (1988).
27. J. Q. Dong, W. Horton, and J. Y. Kim, *Phys. Plasmas* **4**, 1867–1876 (1992).
28. K. Itoh, and S.-I. Itoh, *Plasma Phys. Contr. Fusion* **38**, 1–49 (1996).
29. P. H. Diamond, I. S-I, K. Itoh, and T. S. Hahm, *Plasma Phys. Contr. Fusion* **47**, R35–R161 (2005).
30. J. Perez, W. Horton, K. Gentle, W. Rowan, K. Lee, and R. Dahlburg, *Phys. Plasmas* **13**, 2101–2110 (2006).
31. T. Cho, J. Kohagura, M. Hirata, T. Numakura, H. Higaki, H. Hojo, M. Ichimura, K. Ishii, K. M. Islam, A. Itakura, I. Katanuma, Y. Nakashima, T. Saito, Y. Tatematsu, M. Yoshikawa, Y. Takemura, A. Kojima, T. Kobayashi, Y. Yamaguchi, Y. Miyata, N. Yokoyama, Y. Tomii, Y. Miyake, S. Kiminami, K. Shimizu, Y. Kubota, H. Saimaru, Y. Higashizono, A. Mase, Y. Yasaka, K. Ogura, K. Sakamoto, M. Yoshida, V. P. Pastukhov, T. Imai, S. Miyoshi, and G. . Group, *Nucl. Fusion* **45**, 1650–1657 (2005).
32. J.-H. Kim, J. C. Perez, W. Horton, G. Chagelishvili, R. G. Chanishvili, J. G. Lominadze, and J. C. Bowman, *Phys. Plasmas* **13**, 2304 (2006).
33. T. H. Stix, *Waves in Plasmas*, Springer, 1992.
34. R. E. Slusher, and C. M. Surko, *Phys. Rev. Lett.* **40**, 400–403 (1976).
35. W. Horton, *Phys. Rev. Lett.* **37**, 1269–1272 (1976).
36. X. L. Zou, L. Colas, M. Paume, J. M. Chareau, L. Laurent, P. Devynck, and D. Gresillon, *Phys. Rev. Lett.* **75**, 1090–1093 (1995).
37. L. Colas, X. L. Zou, M. Paume, J. M. Chareau, L. Guiziou, G. T. Hoang, Y. Michelot, and D. Gresillon, *Nucl. Fusion* **38**, 903–918 (1998).
38. G. W. Hammett, and F. W. Perkins, *Phys. Rev. Lett.* **64**, 3019–3022 (1990).
39. N. Mattor, and S. E. Parker, *Phys. Rev. Lett.* **79**, 3419–3422 (1997).
40. H. Sugama, T. H. Watanabe, and W. Horton, *Phys. Fluids* **8**, 26172628 (2001).
41. W. Horton, R. Estes, and D. Biskamp, *Plasma Physics* **22**, 663–678 (1980).
42. A. M. Dimits, J. F. Drake, P. N. Guzdar, and A. B. Hassam, *Phys. Plasmas* **3**, 620–626 (1991).
43. W. Horton, D. I. Choi, and W. M. Tang, *Phys. Fluids* **24**, 1077–1085 (1981).
44. R. E. Waltz, G. D. Kerbel, and J. Milovich, *Phys. Plasmas* **1**, 2229–2244 (1994).
45. M. A. Beer, and G. W. Hammett, *Phys. Plasmas* **3**, 4046–4064 (1996).
46. R. Balescu, *Aspects of anomalous transport in plasmas*, Series in Plasma Physics, Institute of Physics Publishing, 2005.
47. Z. Lin, T. S. Hahm, W. W. Lee, W. M. Tang, and R. B. White, *Science* **281**, 1835–1837 (1998).
48. A. M. Dimits, G.Bateman, M. A. Beer, B. I. Cohen, W. Dorland, G. W. Hammett, C. Kim, J. E. Kinsey, M. Kotschenreuther, A. H. Kritz, L. L. Lao, J. Mandrekas, W. M. Nevins, S. E. Parker, A. J. Redd, D. E. Shumaker, R. Sydora, and J. Weiland, *Phys. Plasmas* **7**, 969–983 (2000).
49. F. Jenko, and W. Dorland, *PPCF* **43**, A141–A150 (2001).

50. W. Dorland, and G. W. Hammett, *Phys. Fluids B* **5**, 812–835 (1993).
51. M. A. Beer, G. W. Hammett, G. Rewoldt, E. J. Synakowski, and M. C. Zarnstorff, *Phys. Plasmas* **4**, 1792–1799 (1997).
52. H. Sugama, T.-H. Watanabe, and W. Horton, *Phys. Plasmas* **10**, 726–736 (2003).
53. T. H. Watanabe, and H. Sugama, *Phys. Plasmas* **9**, 3659–3662 (2002).
54. M. N. Rosenbluth, and F. L. Hinton, *Phys. Rev. Lett.* **80**, 724 – 727 (1998).
55. M. A. Beer, and G. W. Hammett, "," in *Proceedings of the Joint Varenna-Lausanne International Workshop on Theory of Fusion Plasmas, Varenna, 1998*, edited by J. W. Connor, E. Sindoni, and J. Vaclavik, Bologna, Societa Italiana de Fisca, 1999.
56. H. Sugama, T.-H. Watanabe, and W. Horton, *Phys. Plasmas* **14**, 022502 (2007).
57. W. Horton, H. V. Wong, P. J. Morrison, A. Wurm, J. H. Kim, J. C. Perez, J. Pratt, G. T. Hoang, B. P. LeBlanc, and R. Ball, *Nuclear Fusion* **45** (2005).
58. W. M. Nevins, J. Candy, S. Cowley, T. Dannert, A. Dimits, W. Dorland, C. Estrada-Milla, G. W. Hammett, F. Jenko, M. J. Pueschel, and D. E. Shumaker, *Phys. Plasmas* **13**, 122306 (2006).
59. J. E. Kinsey, G. Bateman, A. H. Kritz, and A. Redd, *Phys. Plasmas* **3**, 561–570 (1996).
60. J. E. Kinsey, and G. Bateman, *Phys. Plasmas* **3**, 3344–3357 (1996).
61. A. J. Redd, A. H. Kritz, G. Bateman, and W. Horton, *Phys. Plasmas* **5**, 1369–1379 (1998).
62. G. Bateman, A. H. Kritz, J. E. Kinsey, and A. J. Redd, *Phys. Plasmas* **5**, 2355–2362 (1998).

Introduction to Turbulence in Magnetised Plasmas

Bruce D. Scott

Max-Planck-Institut für Plasmaphysik, Euratom Association, D-85748 Garching, Germany

Abstract. The ideas of turbulence of small fluctuations on a background as a statistical phenomenon are outlined. Basic properties such as three-wave interactions and spatial scale cascades are derived from the basic equations. Passive scalar dynamics is treated. The special case of dissipative coupling between the fluid and an otherwise passive scalar, of central relevance to magnetised plasmas, is used as an example.

Keywords: turbulence
PACS: 52.65.Tt, 52.35.Ra, 52.30.-q, 52.25.Fi

PART I — STATISTICAL NONLINEARITY AND CASCADE DYNAMICS

In general situations the dynamics of a magnetised plasma is not comprised solely of a single wave or instability. This is due to the general nonlinearity of the system when two or more eigenfunctions are excited; the various waves are coupled together and interact with each other as well as with the background. Turbulence due to a fluid flow like the basic plasma ExB velocity has as its dominant coupling mechanism a quadratic nonlinearity: advection of each dependent variable by the velocity, itself a dependent variable. Quadratic nonlinearities exhibit a phenomenon known as three wave coupling: in a Fourier decomposition in terms of wavenumbers, each wave $\mathbf{k}$ interacts with the other ones $\mathbf{k}'$ through beat waves $\mathbf{k}''$ satisfying the constraint $\mathbf{k}+\mathbf{k}'+\mathbf{k}''=0$. One can consider this as a process of modes $\mathbf{k}'$ and $\mathbf{k}''$ beating together to drive mode $\mathbf{k}$ (in the shorthand in which the word "mode" is used to signify an individual eigenmode, wave, or Fourier component). We are thinking of a relatively homogeneous system with many degrees of freedom; even if the background has gradients, the turbulence can be thought of as homogeneous if the typical scale of motion is smaller than any gradient scale length of the background.

We can distinguish between quasilinear and nonlinear effects. Quasilinear dynamics refers to a single wave interacting with the background; in other words, mode zero (zero wavenumber, nonzero frequency) and mode $-\mathbf{k}$ driving mode $\mathbf{k}$, as in a linear system, but now together with modes $\mathbf{k}$ and $-\mathbf{k}$ driving mode zero. The waves excite changes in the background which affect the subsequent evolution of the waves. Here, $\mathbf{k}$ and $-\mathbf{k}$ are actually the same mode, since for a real scalar field $\rho(\mathbf{x})$ we have $\rho_{-\mathbf{k}}=\rho_{\mathbf{k}}^*$ in the (complex) Fourier components, where the superscript asterisk denotes the complex conjugate. The quasilinear system therefore involves the same set of modes the linear system does; the only exception is that mode zero is allowed to evolve. When many modes are excited, the dynamics could still be quasilinear if this particular wave/background interaction were still dominant. Such dynamics tends to produce relaxation oscillations in conservative systems: the action of the wave on the background is stabilising but overshoots the point at which the growth rate goes to zero, and then the free energy in the wave all goes back to the background, reestablishing the original unstable state, and so on. The interaction typically has a coherent character, with long term memory of initial conditions.

Turbulence *per se* is very different from this. The same three wave coupling mechanism works as well between all possible triplets in the full set of waves. Each three wave interaction works independently of all the others, but many such interactions affect each individual wave. The dynamics involves many degrees of freedom (one for each dependent variable or eigenmode at each $\mathbf{k}$) all exchanging energy, and while each three wave interaction tends towards coherence, the fact that there are many of them going on simultaneously results in a weak degree of overall coherence between waves at differing $\mathbf{k}$. When these three wave interactions are as strong as the quasilinear interactions, they result in an incoherent type of dynamics, which must be treated statistically. Moreover, it has a short time memory: dynamics at the same spatial point but separated in time become uncorrelated beyond a time range called the correlation time. The correlation time is somewhat longer than the time required for individual eddies to turn over, but it is shorter than the time

CP1013, ***Turbulent Transport in Fusion Plasmas, First ITER International Summer School***
edited by S. Benkadda

required for the collective disturbance free energy to respond to changes in the background parameters. The dynamics therefore quickly loses memory of any initial state, and somewhat more slowly reaches a state called saturation: the individual waves fluctuate vigorously in amplitude and phase, but their free energy (or amplitude) level remains statistically stationary over time scales of several correlation times. This type of dynamics is what we call turbulence. It is a nonequilibrium, dissipative system with many degrees of freedom, with short correlation times and lengths. It is characterised by its statistical properties, not so much by the evolutionary fate of individual structures. For those familiar with geophysical fluid dynamics, the distinction between individual eddies, or structures, within the turbulence and the statistical properties of the fluctuating dynamics is analogous to the distinction between weather and climate.

The basic picture of turbulence one starts with is an "eddy mitosis" model. Large eddies are driven by some "stirring" process and a balance is formed as they break up into roughly two smaller ones each, on average, then these smaller ones into two smaller ones, each, and then so forth down to scales at which dissipation becomes stronger than the nonlinear dynamics. This energy transfer is called a "cascade" because it occurs in a chain of interactions, each of which is between eddies at not very disparate scales. In terms of the turbulence "spectrum," the distribution of energy in terms of wavenumbers, the cascade process is "local" in the sense that the ratio of wavenumbers involved in most of the transfer process is rarely larger than two.

The separation between the main energy containing scales and the small scale dissipation range is characterised by the "Reynolds number," given by the ratio of the product of typical scale length and velocity, and the viscous diffusion coefficient. The turbulence is called "high Reynolds number turbulence" if the cascade process and not the details of the dissipation process controls the balance between drive and dissipation. In terms of an expansion, the small parameter is the inverse Reynolds number, and the turbulence is thought of as random to lowest order, with the energy transfer and any transport processes entering at next order.

Turbulence in fluids is generally three dimensional, but under certain circumstances the vorticity of the motion is ordered along a particular axis. This is either the rotation axis in a rapidly spinning system, like a thin accretion disk of material orbiting a central object like a star, or in a thin-atmosphere system with some rotation like geophysical turbulence, or it is the magnetic field in a plasma whose kinetic energy of motion is not sufficient to strongly disturb the magnetic field. Two dimensional turbulence has the special property that its vorticity is strictly perpendicular to its velocity. As a result, a quantity called the "enstrophy" is conserved in addition to the energy. The energy density is one half the square of the velocity, and the enstrophy density is one half the square of the vorticity. With both quantities arising from the single velocity variable conserved, the one with the highest wavenumber dependence, namely enstrophy, controls the statistics and effects the direct cascade to small scales, and the other one, namely energy, then cascades to larger scales. This phenomenon is called a "dual cascade" and the transfer of energy to large scales is called an "inverse cascade."

Turbulence in an idealised sense is incompressible, with all the motion of the shearing type. It is important to note this in light of the common concept of one dimensional nonlinear chaos. Turbulence is not only more statistical than the results of deterministic chaos models, it also has this shearing character. Neutral fluid turbulence is kept incompressible by the pressure, so long as the velocity of the flow eddies is slower than the sound speed. Disturbances in the density can in certain circumstances (e.g., thermal convection) remain, but are merely "passively" advected, with any coupling back to the flows rather weak in comparison to the time scale represented by the action of the eddies upon each other. A quantity whose dynamics is simple incompressible advection by the turbulence but with no back reaction upon the turbulence is called a "passive scalar." In a magnetised plasma, incompressibility is maintained by the magnetic field if the velocity is slower than the Alfvén velocity, in the usual "low beta" situation in which plasma pressure is much smaller than the effective magnetic field pressure which is part of the magnetohydrodynamic (MHD) force. The pressure evolves freely, and like the density it is under the MHD approximations a passive scalar.

The above concepts comprise Part I of this lecture. The following Sections will add substance to this Introduction, starting with the eddy mitosis model, continuing with the statistical character, the fluid model, three wave coupling and the cascades, the contrast between two and three dimensional models, and finally MHD turbulence and passive scalar dynamics. Part II will turn to the special case of gradient driven turbulence in a magnetised plasma, whose dynamics parallel to the magnetic field invokes dissipative coupling mechanisms between the fluid vorticity and the transported quantities, such that the latter no longer behave as passive scalars. This type of turbulence is very important for two reasons: (1) it is beyond the realm of the MHD model, and (2) it is found in many instances to be responsible for the dominant transport processes in a confined plasma.

EDDY MITOSIS AND THE CASCADE MODEL

The basic picture of turbulence in fluids is the one introduced by Kolmogorov in 1941 and independently by von Weizäcker and Heisenberg in different language in subsequent years. At a certain range of spatial scales, energy is supposed to be introduced into the fluid motions. This is called "stirring" and beyond the characterisation of a drive rate it is independent of the details. The turbulence is supposed to be homogeneous and isotropic, so that spatial scale and spectral wavenumber magnitude, related through $k \leftrightarrow 2\pi/\Delta$, can be used interchangeably. A collection of eddies at the driven spectral range $k \sim k_0$ finds shearing motion tearing each of the eddies apart at about the same rate at which they are driven, so that there is a statistical equilibrium. This "nonlinear decorrelation" process conserves energy, so the energy taken out of the driven range is passed to other scales. Since the dominant instability of a sheared flow has a wavelength slightly shorter than the scale of the flow, most of this energy is passed to the next smallest scale in a logarithmic hierarchy, with a scale ratio of about two. With all of the energy transfer taking place between neighbours in the hierarchy, the picture we have in mind is a "local cascade." At each scale below the driven range, there is a statistical balance mostly between energy passed in from the next larger scale, and nonlinear decorrelation passing the energy further to the next smaller scale. The shearing rate at each scale is proportional to the vorticity of the flow at that scale, so that the coherent lifetime of a typical eddy is comparable to the "eddy turnover time" the time the eddy takes to transport the flow energy and momentum across itself.

Following these ideas, at each level n with wavenumber magnitude k_n a balance is struck between incoming energy due to nonlinear decorrelation at level $n-1$ and stirring at level n, and outgoing energy due to nonlinear decorrelation at level n. This is expressed as

$$(kv)_{n-1}\,E_{n-1} + \gamma_n E_n = (kv)_n\,E_n \tag{1}$$

where E_n and $(kv)_n$ are the energy and vorticity within level n, and γ_n is the rate of stirring at level n. We can take $E_n = v_n^2/2$.

At the low-k end of the spectrum the turbulence is generated by stirring and hence $\gamma > 0$ there. At the high-k end lies the dissipation range, where $\gamma < 0$. Since the separation of scales is supposed to be arbitrarily large, there is a wide spectral range in between for which γ is negligible compared to (kv). This is called the "inertial range" of the turbulence and is the range in which the turbulence is most pure, that is, the conservative nonlinearity is the only part of the dynamics which is active. Within the inertial range, we then have

$$(kv)_{n-1}\,E_{n-1} = (kv)_n\,E_n = T_0 \tag{2}$$

where T_0 is a constant giving the energy "throughput" of the cascade. With $E_n = v_n^2/2$ we find the velocity and energy density spectrum,

$$v_n^3 = 2T_0 k_n^{-1} \qquad\qquad E_n = (2T_0)^{2/3}\,k_n^{-2/3} \tag{3}$$

To find E_k, the energy per unit range of k, we have to take into account the fact that each level n has a range of k which is proportional to k (i.e., we have a logarithmic spacing). In integral form we have

$$\sum_n E_n = \int d\log k\, E_n = \int dk\, E_k \tag{4}$$

which identifies E_k as E_n/k, or $E_n(k_0/k)$ to preserve the dimensions. This yields

$$E_k = E_0\,k^{-5/3} \tag{5}$$

with $E_0 = (2T_0)^{2/3}k_0$. This 5/3 power law is known as the Kolmogorov spectrum. Within the inertial range, this spectrum is "universal" and the dynamics is "self similar," indicating that the scales of motion in time and space can be arbitrarily renormalised.

If the dissipation is caused by viscosity or some other diffusive process, we have in the dissipation range $\gamma(k) = -\chi k^2$, where χ is called the diffusivity. The Reynolds number is defined by comparing the typical velocity and space scales in the *driven range* with the diffusivity,

$$\mathrm{Re} = LV/\chi \tag{6}$$

The "high Reynolds number" regime is the one for which the inertial range involves several levels of the cascade hierarchy, and an important result of that is that the energy balances of both the energy containing range and the inertial range are independent of the strength or form of the dissipation process. For diffusion, the amount of energy consumed by the dissipation should be independent of the diffusion coefficient.

Clearly, the existence of an inertial range, with several levels in the hierarchy for which $\gamma \ll (kv)$, is connected with a large value for Re. It does however take a rather large Reynolds number before the high Reynolds number regime is entered. To see this, we may take a simple situation for which Eq. (1) can be applied: assume that $\gamma_0 > 0$ only for the first level in the hierarchy, the driven range k_0. For all other scales, let $\gamma_n = -\chi k_n^2$. The amplitude of the driven range is given by the "mixing condition,"

$$\gamma_0 = (kv)_0 \qquad \text{hence} \qquad v_0 = \frac{\gamma_0}{k_0} \tag{7}$$

It is a property in the inertial range that the shearing rate is progressively faster at smaller scale, due to the conservation of energy: with Eq. (1) satisfied for $\gamma = 0$ we have

$$\frac{(kv)_{n+1}}{(kv)_n} = \left(\frac{k_{n+1}}{k_n}\right)^{2/3} \tag{8}$$

and hence with $k_{n+1}/k_n = 2$ the ratio is $2^{2/3}$ or about 1.6. Even with a process like convection for which γ is roughly constant below the scale of inhomogeneity, one does therefore reach an inertial range if χ is small enough, that is, if Re is large enough. But the factor of 1.6 is small enough that Re must be very large in order to have a real inertial range. Typically, a flow is not even turbulent for Re lower than a few thousand, and for flow in three dimensions the high Reynolds number regime is not reached until Re is at least 10^6. In astrophysical situations Re can be in the range 10^{12} to 10^{15}. For atmospheric turbulence one finds values some orders of magnitude smaller than that but still large; for air at STP the kinematic viscosity is about $1.5 \times 10^{-5}\,\mathrm{m^2/sec}$, so for $L > 15\,\mathrm{m}$ and $V > 1\,\mathrm{m/sec}$ we have $\mathrm{Re} > 10^6$. The highest resolution three dimensional fluid turbulence computations to date are barely into the high Reynolds number regime, although the mere demonstration of a local cascade is possible with a discretised domain of only 64^3 grid nodes. In two dimensions, as discussed below, the cascade dynamics is different and turbulence may set in for Re values rather lower than this; typically, at least two decades between driven and dissipation range in the spectrum are required ($k_N/k_0 \sim 10^2$). For magnetised plasma turbulence the Reynolds number is not as large as this: the dissipative coupling process discussed in Part II inhibits the cascade process sufficiently that only one decade of scale separation is needed to reach the high Reynolds number regime.

THE STATISTICAL NATURE OF TURBULENCE

One is motivated to ask "nonlinear dynamics, OK, but what makes it turbulence?" The answer is that the motion involved exhibits short space/time correlation and is "ultimately nonperiodic" in the sense that no matter how long the time domain in an experiment or computation is allowed to persist, the longest time periods become dominant in a frequency spectrum, the standard deviation of a time series saturates instead of falling with increasing time intervals, and the statistical description of the turbulent flow amplitudes and decorrelation rates in terms of a mean and a standard deviation is independent of how the dynamical system was started. In terms of a computation, the particular initial conditions chosen should have no effect on the statistical result, even though certain initial scenarios can save time by getting the system to statistical equilibrium in a comparatively shorter time than others. Fully developed turbulence is characterised by a "saturated state" in which any relevant statistical measure is stationary in this sense. This is never certain; one should accept that results are only known within error bars characterised by that standard deviation, and in this case one should always quote the standard deviation of the distribution, not that of the mean (the latter falls with the square root of the number of samples if the former is stationary). In this sense turbulence "converges" only statistically, not monotonically as a coherent dissipative process would do.

Current research is occupied with the question of rare but strong events: those well down the tails of the "probability distribution function" (PDF). The PDF is essentially a histogram. The amplitude of a certain variable is sampled and considered as a series. The mean and standard deviation of the series are measured, and the PDF is computed by dividing the abscissa into intervals which are small compared to the standard deviation. The fraction of the events falling into each of these intervals is plotted as the ordinate, normalised in terms of the number of intervals per standard deviation. A purely random process has a Gaussian PDF for any of its measures, but many features of the turbulence are "intermittent" in the sense that the first one or two standard deviations of the PDF appear Gaussian but the tails are different, usually above the Gaussian curve of the same mean and standard deviation. Nevertheless, the basic statistical character of turbulence can be understood in terms of this near-Gaussian PDF. The PDF curves for turbulence are usually compared to the corresponding Gaussian which has the same mean and standard deviation.

Statistical memory in turbulence can be characterised in terms of autocorrelation functions in the principal fluctuating variables. Both time and space correlations are described, with the $1/e$ half-width of these curves defined as the correlation time or correlation length. The time autocorrelation function of a variable ϕ is defined as

$$F(\tau) = \lim_{T\to\infty} \int_{-T/2}^{T/2} \frac{dt}{T}\,\phi(t)\phi(t-\tau) \qquad (9)$$

also averaged over space. For a purely wavelike process (e.g., a linear oscillation), $F(\tau)$ is sinusoidal. For a purely random process, it is Gaussian. If the spatial domain is periodic on a length L, it is simple to compute the space autocorrelation function in terms of Fourier components,

$$F(\Delta) = c_0 \sum_{l\neq 0} |\phi_l|^2 \cos(2\pi l\Delta/L) \qquad (10)$$

also averaged over time, with the normalising constant c_0 chosen such that $F(0) = 1$. The curve for $F(\Delta)$ is itself periodic and therefore is only significant for $-L/2 < \Delta < L/2$. In turbulence within a bounded domain (e.g., convection, or tokamak edge turbulence), the correlation length in the relevant direction (e.g., down the gradient) can be comparable to the domain size, but it should be much smaller than the domain size in the "horizontal" direction (e.g., East-West for geophysical fluid turbulence, or perpendicular to the magnetic field lines but within the flux surfaces for gradient driven turbulence within a closed, toroidal MHD equilibrium). Most importantly, the correlation time must be short compared to the time intervals required for the statistical measures to reach saturation. In relevantly scaled units, tokamak turbulence typically has a dynamical scale of order 1, a correlation time of order 10, a saturation time at least of order several hundred, and an transport equilibration time at least of order several thousand. This level of temporal scale separation is what makes plausible the model of thermal transport by turbulent mixing on a slowly varying background.

QUADRATIC NONLINEARITY AND THREE WAVE COUPLING FOR SMALL DISTURBANCES

When treating turbulence as such one usually either assumes an homogeneous background, or one with a gradient which drives the turbulence while nevertheless leading to disturbances which are small compared to background quantities. The flow velocity, as in certain circumstances the magnetic field, belongs to the disturbances. The pressure and density belong to the background, and since the disturbances are small these two quantities are treated as constant coefficients except where acted upon by the gradient operator. The scale of motion is also expected to be small compared to the scale of the background, and this simplification is used as a model even when in the case of a thin boundary region (e.g., convection in a thin atmosphere, or tokamak edge turbulence) the scales of inhomogeneity and of the turbulence are comparable but both small compared to the overall size of the equilibrium.

In a situation like this the equations governing the dynamics are linearised except for the terms representing advection or, in the case of turbulence on a magnetised plasma background, the disturbed parallel gradient. These terms have the special character that they are quadratic in the dependent variables. The simplest example of such an effect is $\mathbf{v}\cdot\nabla\mathbf{v}$, the self-advection of the velocity, which is the basis of what gives turbulence its character. The thermodynamic state variables are also advected through quadratic nonlinearities. The simplest example of an equation they satisfy is the continuity equation for the mass density ρ,

$$\frac{\partial \rho}{\partial t} + \nabla\cdot\rho\mathbf{v} = 0 \tag{11}$$

expressing mass conservation. This is rearranged by separating the advection and the divergence terms, and linearising the latter, formally writing

$$\left(\frac{\partial}{\partial t} + \mathbf{v}\cdot\nabla\right)\log\rho + \nabla\cdot\mathbf{v} = 0 \tag{12}$$

which we can do with no loss of generality, but then also considering small disturbances so that $\delta\log\rho \rightarrow \delta\rho/\rho$, writing

$$\frac{\partial\widetilde{\rho}}{\partial t} + \mathbf{v}\cdot\nabla\widetilde{\rho} + \rho\nabla\cdot\mathbf{v} = 0 \tag{13}$$

and treating the factor of ρ multiplying the divergence as a constant parameter while using the tilde symbol to mark the dependent variable. The advection term $\mathbf{v}\cdot\nabla\widetilde{\rho}$ is recognised as a quadratic nonlinearity. The nonlinear part of the divergence term is smaller than this because we are treating situations under the "small scale mixing" condition:

$$\widetilde{\rho} \ll \rho \qquad \text{although} \qquad \nabla\widetilde{\rho} \sim \nabla\rho \tag{14}$$

In an incompressible model, of course, the advection term is the only effect controlling time evolution in this equation.

The three wave coupling condition follows from the properties of Fourier transforms. We treat the background as homogeneous in understanding the properties of the turbulence. The dependent variables are Fourier decomposed according to

$$\widetilde{\rho}(\mathbf{x}) = \sum_{\mathbf{k}} \rho_{\mathbf{k}} e^{i\mathbf{k}\cdot\mathbf{x}} \qquad\qquad \rho_{\mathbf{k}} = \oint \frac{d^n x}{L^n} e^{-i\mathbf{k}\cdot\mathbf{x}}\widetilde{\rho}(\mathbf{x}) \tag{15}$$

where n is the number of dimensions (usually 2 or 3) and the spatial domain is taken to be periodic on length L in each dimension (this is easily generalised). The incompressible continuity equation (with $\nabla\cdot\mathbf{v}=0$) then becomes

$$\frac{\partial\rho_{\mathbf{k}}}{\partial t} = -\oint\frac{d^n x}{L^n} e^{-i\mathbf{k}\cdot\mathbf{x}} \sum_{-\mathbf{k}'}\sum_{-\mathbf{k}''} \mathbf{v}_{-\mathbf{k}'}\cdot(-i\mathbf{k}'')\rho_{-\mathbf{k}''} e^{-i\mathbf{k}'\cdot\mathbf{x}} e^{-i\mathbf{k}''\cdot\mathbf{x}} \tag{16}$$

where we have used the decomposition in terms of the sum for both dependent variables in the nonlinearity and the integral operation to isolate the single component under the time derivative. Note that the choice of signs in $\mathbf{k}$ is arbitrary due to the condition that $\rho(\mathbf{x})$ is real valued, $\rho_{-\mathbf{k}} = \rho_{\mathbf{k}}^{*}$, since we sum over all the components. The factors involving $\mathbf{x}$ and the integral can be isolated from the rest of the right hand side,

$$\oint\frac{d^n x}{L^n} e^{-i(\mathbf{k}+\mathbf{k}'+\mathbf{k}'')\cdot\mathbf{x}} = \delta_{(\mathbf{k}+\mathbf{k}'+\mathbf{k}''),0} \tag{17}$$

yielding the Kronecker delta. The two components are said to beat against one another (the parlance for waves) to drive component $\mathbf{k}$. The condition under which there is a finite interaction is then

$$\mathbf{k}+\mathbf{k}'+\mathbf{k}'' = 0 \tag{18}$$

which is called the three wave condition: the three wavenumbers must vectorially add to zero. Interactions of this type are called "three wave interactions" and they form the basis for the interactions in turbulence. The three wave condition is therefore a relatively trivial consequence of the quadratic nonlinearity and the orthogonality of Fourier transforms. Even if the spatial domain is not periodic, the basic character of turbulence has this qualitative form.

Three wave models are constructed using a single triplet $\{\mathbf{k}_1, \mathbf{k}_2, \mathbf{k}_3\}$, satisfying the three wave condition, for explanatory purposes. The model consists of an evolution equation for each Fourier component which is a member of the triplet. Turbulence, however, derives its character from the fact that there are very many such triplets and the vast majority of them are mutually incoherent, with the three members having no special relation to each other. To lowest order of approximation (the small parameter is usually Re^{-1} or some equivalent) in the mathematical theories of turbulence the various triplets are supposed to be mutually random, with the PDF of cross correlation having Gaussian form with zero mean. All the energy transfer dynamics comes from the corrections, which are computed according to one or another model. The details are very complicated and are left to the references listed at the end. We will return to the three wave interactions and what they say about the transfer dynamics generally, after introducing the incompressible turbulence models.

INCOMPRESSIBLE HYDRODYNAMIC TURBULENCE — ENERGY AND ENSTROPHY

We start with simple hydrodynamic turbulence, following the ideal equation of motion in MHD without the magnetic field,

$$\frac{\partial \mathbf{v}}{\partial t} + \mathbf{v}\cdot\nabla\mathbf{v} = -\frac{1}{\rho}\nabla p \tag{19}$$

We recognise the role of the pressure in maintaining incompressibility by taking the divergence and assuming that $\nabla\cdot\mathbf{v}=0$,

$$\nabla\cdot\frac{1}{\rho}\nabla p = -\nabla\cdot(\mathbf{v}\cdot\nabla\mathbf{v}) \tag{20}$$

This condition controls the evolution of the pressure, and so even in an inhomogeneous model the only other equation controlling evolution would be an incompressible advection equation for ρ. In an homogeneous model we take ρ to be constant in Eq. (19) since $\rho\nabla^2 p \gg \nabla\rho\cdot\nabla p$ due to the small fluctuation approximation (cf. Eq. 14). Taking the curl, we then have

$$\frac{\partial}{\partial t}(\nabla\times\mathbf{v}) + \mathbf{v}\cdot\nabla(\nabla\times\mathbf{v}) = (\nabla\times\mathbf{v})\cdot\nabla\mathbf{v} \tag{21}$$

This equation says that the fluid vorticity $(\nabla\times\mathbf{v})$ evolves not only due to advection but also due to a finite component of the velocity gradient along the vorticity. The squared magnitude of the vorticity is the density of a quantity called "enstrophy" in hydrodynamics. Defining it as $W=(1/2)|\nabla\times\mathbf{v}|^2$, we may form its equation by contracting Eq. (21) with the vorticity,

$$\frac{\partial W}{\partial t} + \nabla\cdot(W\mathbf{v}) = (\nabla\times\mathbf{v})(\nabla\times\mathbf{v}):\nabla\mathbf{v} \tag{22}$$

using the condition $\nabla\cdot\mathbf{v}=0$. This equation says that the enstrophy is conserved except for a class of motions which involves stretching the velocity component along the vorticity further in the direction of the vorticity. Since the motion is incompressible, the components of the velocity making up most of the vorticity must have a convergence to make up the divergence involved in the stretching.

To find out what this type of motion signifies and what its consequence is, we reconsider the results arising from the turbulent cascade model (Sec. II). Recall that a cascade towards small scale which conserves energy must exhibit a shearing rate which increases towards small scale. Given the energy conservation condition, it is a simple matter to conclude that the enstrophy must grow during this cascade, simply from the extra factors of k in W compared to the energy density $U=v^2/2$. The type of motion which we concluded above allows W to grow may be displayed in a simple sketch: Most of the velocity is in eddies each of which rotates in a local two dimensional plane, to which the vorticity is perpendicular. The third dimension has to have a finite velocity component, along the vorticity vector, and it must stretch in this direction. The eddy motion in the local plane must also contract to maintain incompressibility. This basically involves a vortex tube structure which is stretched in the time it maintains its coherence (not more than a few eddy turnover times), before it is sheared apart by the motion both in the local plane and along the local vorticity vector. We therefore conclude simply on the basis of (1) the incompressibility constraint, (2) energy conservation, (3) the cascade to smaller scales, and (4) the growth of the enstrophy, that the turbulent cascade process behind the eddy mitosis model must involve this "vortex tube stretching" in three dimensions.

The foregoing is significant because in several important applications (e.g., motion at large horizontal scale in a thin atmosphere, "geostrophic" motion in a rapidly rotating background, or motion perpendicular to the magnetic field in a magnetised plasma) the turbulence is quite strictly two dimensional (2D). We will see in Part II how dissipative coupling does this for gradient driven turbulence in a magnetised plasma. In a 2D model, we have the additional constraint that the vorticity is strictly perpendicular to the plane in which the velocity and the gradient operator both lie. In this case we have both $(\nabla\times\mathbf{v})\cdot\mathbf{v}=0$ and $(\nabla\times\mathbf{v})\cdot\nabla=0$, so that the right hand side of Eq. (22) is zero. Therefore, the enstrophy as well as the energy is conserved. With W conserved we cannot have the cascade of energy to small scale. To find out how this situation is resolved we may examine three wave coupling in 2D turbulence.

In a 2D incompressible flow, the velocity is given in terms of a stream function, which we label ψ, and the vorticity is a scalar, Ω, times the unit normal to the plane, which we label $\hat{\mathbf{s}}$, so that

$$\mathbf{v}=\hat{\mathbf{s}}\times\nabla\psi \qquad \nabla\times\mathbf{v}=\Omega\hat{\mathbf{s}} \qquad \Omega=\nabla_\perp^2\psi \tag{23}$$

with the perp symbol used to denote the fact that only the two dimensions in the plane are involved. Eq. (21) becomes

$$\frac{\partial\Omega}{\partial t}+\mathbf{v}\cdot\nabla\Omega=0 \tag{24}$$

which is called the 2D Euler equation. Multiplying by $-\psi$ or by Ω, we find

$$\frac{\partial U}{\partial t}+\nabla\cdot\left(U\mathbf{v}-\left[\psi\left\{\frac{\partial}{\partial t}+\mathbf{v}\cdot\nabla\right\}\nabla_\perp\psi\right]\right)=0 \qquad \frac{\partial W}{\partial t}+\nabla\cdot(W\mathbf{v})=0 \tag{25}$$

that is, simultaneous energy and enstrophy conservation. [The second term in the first equation is a quantity related to compression and becomes important only if the density is strongly variable, at which point the assumption of incompressibility itself breaks down.]

Three wave interactions set up by Fourier decomposition of Eq. (24) will also have these conservation properties, therefore representing a conservative mutual transfer of energy and enstrophy from wave to wave. For Eq. (24), for example, one can easily verify that

$$\frac{\partial}{\partial t}\left(\frac{\Omega_{\mathbf{k}}^*\Omega_{\mathbf{k}}+\Omega_{\mathbf{k}'}^*\Omega_{\mathbf{k}'}+\Omega_{\mathbf{k}''}^*\Omega_{\mathbf{k}''}}{2}\right)=0 \tag{26}$$

for any particular triplet $\{\mathbf{k},\mathbf{k}',\mathbf{k}''\}$. The equations satisfied by the members of the triplet are

$$\frac{\partial\Omega_{\mathbf{k}}}{\partial t}=C_{\mathbf{kk}'}\left(\Omega_{-\mathbf{k}''}\psi_{-\mathbf{k}'}-\Omega_{\mathbf{k}'}\psi_{-\mathbf{k}''}\right) \tag{27}$$

$$\frac{\partial\Omega_{\mathbf{k}'}}{\partial t}=C_{\mathbf{kk}'}\left(\Omega_{-\mathbf{k}}\psi_{-\mathbf{k}''}-\Omega_{\mathbf{k}''}\psi_{-\mathbf{k}}\right) \tag{28}$$

$$\frac{\partial\Omega_{\mathbf{k}''}}{\partial t}=C_{\mathbf{kk}'}\left(\Omega_{-\mathbf{k}'}\psi_{-\mathbf{k}}-\Omega_{\mathbf{k}}\psi_{-\mathbf{k}'}\right) \tag{29}$$

which we can form by simple permutation among the members of the triplet. The coupling constant is given by

$$C_{\mathbf{kk}'}=\frac{1}{2}\hat{\mathbf{s}}\cdot(\mathbf{k}\times\mathbf{k}') \tag{30}$$

where we use the relative symmetry in the interactions,

$$\hat{\mathbf{s}}\cdot(\mathbf{k}\times\mathbf{k}')=\hat{\mathbf{s}}\cdot(\mathbf{k}'\times\mathbf{k}'')=\hat{\mathbf{s}}\cdot(\mathbf{k}''\times\mathbf{k}) \tag{31}$$

since $\mathbf{k}''=-\mathbf{k}-\mathbf{k}'$.

Using these forms we can build the energy equations for each of the three components (and their complex conjugates) in the triplet, noting that for each one we have the reality condition, e.g., $\psi_{\mathbf{k}}=\psi_{-\mathbf{k}}^*$. For each energy component given by

$$U_{\mathbf{k}}=|\mathbf{k}\psi_{\mathbf{k}}|^2=-\psi_{\mathbf{k}}\Omega_{-\mathbf{k}} \tag{32}$$

we obtain

$$\frac{\partial U_{\mathbf{k}}}{\partial t} = 2C_{\mathbf{kk}'}\,\mathrm{Re}\,[\psi_{\mathbf{k}}\Omega_{\mathbf{k}'}\psi_{\mathbf{k}''} - \psi_{\mathbf{k}}\psi_{\mathbf{k}'}\Omega_{\mathbf{k}''}] \tag{33}$$

$$\frac{\partial U_{\mathbf{k}'}}{\partial t} = 2C_{\mathbf{kk}'}\,\mathrm{Re}\,[\psi_{\mathbf{k}'}\Omega_{\mathbf{k}''}\psi_{\mathbf{k}} - \psi_{\mathbf{k}'}\psi_{\mathbf{k}''}\Omega_{\mathbf{k}}] \tag{34}$$

$$\frac{\partial U_{\mathbf{k}''}}{\partial t} = 2C_{\mathbf{kk}'}\,\mathrm{Re}\,[\psi_{\mathbf{k}''}\Omega_{\mathbf{k}}\psi_{\mathbf{k}'} - \psi_{\mathbf{k}''}\psi_{\mathbf{k}}\Omega_{\mathbf{k}'}] \tag{35}$$

Pairs of equal terms appearing with opposite sign give the energy transfer effects. Considering the equations as a permutation among the three modes in the triplet, the second term in each equation gives the energy transfer from the equation below it, and the first term in each equation gives the energy transfer from the equation above it. For example, the energy transfer from mode $\mathbf{k}'$ to mode $\mathbf{k}$ is given by

$$T_U(\mathbf{k} \leftarrow \mathbf{k}') = 2C_{\mathbf{kk}'}\,\mathrm{Re}\,[-\psi_{\mathbf{k}}\psi_{\mathbf{k}'}\Omega_{\mathbf{k}''}] = 2C_{\mathbf{kk}'}\,\mathrm{Re}\left[(k'')^2\psi_{\mathbf{k}}\psi_{\mathbf{k}'}\psi_{\mathbf{k}''}\right] \tag{36}$$

Similarly, for each enstrophy component given by

$$W_{\mathbf{k}} = |\Omega_{\mathbf{k}}|^2 = k^4\,|\psi_{\mathbf{k}}|^2 \tag{37}$$

we build the enstrophy transfer equations,

$$\frac{\partial W_{\mathbf{k}}}{\partial t} = 2C_{\mathbf{kk}'}\,\mathrm{Re}\,[\Omega_{\mathbf{k}}\psi_{\mathbf{k}'}\Omega_{\mathbf{k}''} - \Omega_{\mathbf{k}}\Omega_{\mathbf{k}'}\psi_{\mathbf{k}''}] \tag{38}$$

$$\frac{\partial W_{\mathbf{k}'}}{\partial t} = 2C_{\mathbf{kk}'}\,\mathrm{Re}\,[\Omega_{\mathbf{k}'}\psi_{\mathbf{k}''}\Omega_{\mathbf{k}} - \Omega_{\mathbf{k}'}\Omega_{\mathbf{k}''}\psi_{\mathbf{k}}] \tag{39}$$

$$\frac{\partial W_{\mathbf{k}''}}{\partial t} = 2C_{\mathbf{kk}'}\,\mathrm{Re}\,[\Omega_{\mathbf{k}''}\psi_{\mathbf{k}}\Omega_{\mathbf{k}'} - \Omega_{\mathbf{k}''}\Omega_{\mathbf{k}}\psi_{\mathbf{k}'}] \tag{40}$$

Hence, the transfer of enstrophy from mode $\mathbf{k}'$ to mode $\mathbf{k}$ is given by

$$T_W(\mathbf{k} \leftarrow \mathbf{k}') = 2C_{\mathbf{kk}'}\,\mathrm{Re}\,[-\Omega_{\mathbf{k}}\Omega_{\mathbf{k}'}\psi_{\mathbf{k}''}] = 2C_{\mathbf{kk}'}\,\mathrm{Re}\left[-k^2(k')^2\psi_{\mathbf{k}}\psi_{\mathbf{k}'}\psi_{\mathbf{k}''}\right] \tag{41}$$

For the three wave system to have an average spectral transfer, the triple correlation $\langle\psi_{\mathbf{k}}\psi_{\mathbf{k}'}\psi_{\mathbf{k}''}\rangle$ will have a definite sign. If only a single triplet is present, we expect the energy or enstrophy to simply redistribute to reach a statistical equilibrium with a triple correlation of zero. But in turbulence, with many triplets present all with differing relationships between the wavenumbers, we expect an asymmetry to develop since there are more available states towards high k than towards low for any physically realisable instantaneous configuration (i.e., with finite total energy content at $k \to \infty$). The triple correlation and hence the average transfer dynamics will be constrained by the simultaneous conservation of both energy and enstrophy. We can consider the effects of interactions which spread energy and enstrophy towards both ends in the spectrum, out of some intermediate spectral range. Examining Eqs. (36) and (41), noting that the squared factors are positive definite and the coefficient $C_{\mathbf{kk}'}$ is the same in both, we find that for any definite sign of the triple correlation, the signs of the transfer are opposite. If $k < k' < k''$ then the direction of energy transfer from k'' to k' and from k' to k is the same, and similarly for the oppositely directed enstrophy transfer. For both energy and enstrophy, then, the intermediate range k' sees a throughput.

The overall process is called a "dual cascade" since the two quantities arising from the same dependent variable ψ are transferred in opposite directions in the spectrum. The direction of the cascade dynamics will be determined by which one of these transfer effects controls the overall statistics. It is expected to be a local cascade since the coupling coefficients are largest when the magnitudes $\{k, k', k''\}$ are all comparable. The enstrophy and enstrophy transfer contain extra wavenumber factors compared to the energy and energy transfer, so the enstrophy can be expected to mix faster. The statistical behaviour of the whole should therefore be controlled by the enstrophy. The enstrophy should be redistributed towards higher k due to the fact that there are more available states there given any physically realisable initial state (i.e., with finite total energy content at $k \to \infty$). Besides, we already found that for the energy cascade to be "direct" (towards smaller scale) the enstrophy must grow. So, with enstrophy also conserved, we expect the energy cascade to be "inverse" (towards larger scale). We therefore expect the dual cascade to send enstrophy preferentially to higher k and energy preferentially to lower k.

The main points of this Section are that (1) a direct energy cascade requires the vortex tube stretching process in three dimensions since it requires a growing enstrophy, (2) two dimensional incompressible hydrodynamics conserves enstrophy as well as energy, (3) the directions of spectral transfer for energy and enstrophy are opposite given a definite sign of the triple correlation, and therefore (4) we expect to find a dual cascade in two dimensional incompressible hydrodynamic turbulence. We can learn all this simply from the properties of the equations the dynamical model satisfies. We will show that it actually occurs in the computations in Part II.

MHD TURBULENCE

Magnetohydrodynamic (MHD) turbulence is similar to its neutral fluid counterpart, with two differences: incompressibility is maintained by the magnetic field strength (magnetic rather than plasma pressure), and the additional magnetic nonlinearities cause special effects through which flow eddies and Alfvén waves interact with each other to allow long range spectral transfer.

To examine these effects we start with the MHD force equation under the same assumptions as in the previous Section,

$$\frac{\partial \mathbf{v}}{\partial t}+\mathbf{v}\cdot\nabla\mathbf{v}=-\frac{1}{\rho}\nabla\left(p+\frac{B^2}{8\pi}\right)+\frac{\mathbf{B}\cdot\nabla\mathbf{B}}{4\pi\rho} \tag{42}$$

where we have used Ampere's law $\nabla\times\mathbf{B}=(4\pi/c)\mathbf{J}$ to replace the $c^{-1}\mathbf{J}\times\mathbf{B}$ force adding to $-\nabla p$ in the MHD model. In most cases one is interested in the "low beta regime" characterised by

$$\beta=\frac{8\pi p}{B^2}\ll 1 \tag{43}$$

Here, the plasma pressure p is insignificant to the magnetic pressure $B^2/8\pi$ in the maintenance of incompressibility, which we can see by taking the divergence of Eq. (42) assuming $\nabla\cdot\mathbf{v}=0$,

$$\nabla\cdot\frac{1}{8\pi\rho}\nabla B^2=-\nabla\cdot\left(\frac{1}{\rho}\nabla p+\mathbf{v}\cdot\nabla\mathbf{v}\right) \tag{44}$$

It is therefore possible for p as well as ρ to enter the dynamics of background disturbances while B^2 maintains the dynamical incompressibility as long as $\beta\ll 1$, and the dynamics is slow compared to the compressional Alfvén frequency at the same scale, $\omega\ll k_\perp v_A$, and the velocity is sub-Alfvénic, $v\ll v_A$ (the last two conditions are different if the motion is driven by the magnetic field). The perp symbol is now reckoned against the background magnetic field. The Alfvén velocity v_A is defined by

$$v_A^2=\frac{B^2}{4\pi\rho} \tag{45}$$

Most models satisfy these conditions, with the principal exception being the kinematic dynamo process (thermal convection with $\beta\gg 1$, with the magnetic field passively advected).

The equations of incompressible MHD turbulence also include the kinematics of the magnetic field, which assuming $\nabla\cdot\mathbf{v}=0$ satisfies

$$\frac{\partial \mathbf{B}}{\partial t}+\mathbf{v}\cdot\nabla\mathbf{B}=\mathbf{B}\cdot\nabla\mathbf{v} \tag{46}$$

Shear Alfvén dynamics basically involves a magnetic field whose direction is fluctuating but whose field strength is constant except for the small changes required to maintain the incompressibility. The dependent variable then becomes the unit vector. We can also scale the velocity against the background Alfvén velocity v_{A0}, which is v_A defined in terms of the background field strength B_0. The dependent variables are then

$$\mathbf{u}=\frac{\mathbf{v}}{v_{A0}}\qquad\qquad \mathbf{b}=\frac{\mathbf{B}}{B_0} \tag{47}$$

in whose terms Eqs. (42,46) become

$$\frac{1}{v_{A0}}\frac{\partial \mathbf{u}}{\partial t}+\mathbf{u}\cdot\nabla\mathbf{u}=-\nabla I+\mathbf{b}\cdot\nabla\mathbf{b} \tag{48}$$

$$\frac{1}{v_{A0}}\frac{\partial \mathbf{b}}{\partial t}+\mathbf{u}\cdot\nabla\mathbf{b}=\mathbf{b}\cdot\nabla\mathbf{u} \tag{49}$$

where I is now simply a potential which maintains $\nabla\cdot\mathbf{u}=0$. On the other hand, $\nabla\cdot\mathbf{b}=0$ is maintained by the antisymmetry of $\mathbf{ub}-\mathbf{bu}$ as a tensor. We may define the "Elsässer variables"

$$\mathbf{u}_{\pm}=\mathbf{u}\pm\mathbf{b} \tag{50}$$

which satisfy

$$\frac{1}{v_{A0}}\frac{\partial \mathbf{u}_{\pm}}{\partial t}+\mathbf{u}_{\mp}\cdot\nabla\mathbf{u}_{\pm}=-\nabla I \tag{51}$$

Computations of MHD turbulence often use this model, which allows for very simple numerical schemes such as are used for the hydrodynamic model of Eq. (19). The force potential gradient ∇I need not be computed specifically, though it is often used to subtract off the part of the nonlinearity which has a finite divergence (in computational fluid dynamics this is called the "Laplacian pressure method"). We note that there is only one force potential since

$$\nabla\cdot(\mathbf{u}_{-}\cdot\nabla\mathbf{u}_{+})=\nabla\cdot(\mathbf{u}_{+}\cdot\nabla\mathbf{u}_{-}) \tag{52}$$

although some numerical schemes apply two of them independently.

The main difference between MHD and fluid turbulence is that there are two equations each describing advection of one of the Elsässer variables by the other. The three wave coupling condition is no longer among different wavenumber components of the same variable but is now also among the variables. Several consequences regarding the relationship between the velocity and the magnetic field dynamics follow. One consequence is known as "helicity conservation," by which the magnetic helicity density $\mathbf{A}\cdot\mathbf{B}$, where $\mathbf{B}=\nabla\times\mathbf{A}$, can be shown to satisfy a transport equation. This exercise, using the divergence free character of all three vector fields $\mathbf{v}$, $\mathbf{B}$, and $\mathbf{A}$, is left to the reader. Helicity conservation leads to a particularly strong inverse cascade for the magnetic energy, due to the fewer factors of the wavenumbers in its three wave coupling terms. Second, the form of the Elsässer variable nonlinearities results in the squared amplitude of each of them being separately conserved:

$$\frac{1}{v_{A0}}\frac{\partial}{\partial t}\frac{u_{\pm}^2}{2}+\nabla\cdot\left(\frac{u_{\pm}^2}{2}\mathbf{u}_{\mp}+I\mathbf{u}_{\pm}\right)=0 \tag{53}$$

due to the linearity of the ∇ operator and the incompressibility of $\mathbf{u}_{\pm}$. With both of these Elsässer amplitudes conserved, not only the energy but also the "cross helicity" given by $\mathbf{u}\cdot\mathbf{b}$ is conserved. There is therefore some memory of the initial conditions: an initial state with low cross helicity produces a final state with low cross helicity, for example. Third, the spectral transfer of both the energy and the Elsässer amplitudes to high-k dissipation results in an aligning effect: the nonlinear transfer rate of $\mathbf{u}_{+}$ is given by the vigour of $\mathbf{u}_{-}$, and vice versa. So if there is an initial inequality in the Elsässer amplitudes, the one which is smaller is transferred to dissipation faster, leaving all the energy in the one which was larger. As a result, $\mathbf{b}$ becomes either aligned or anti-aligned to $\mathbf{u}$. This tendency has been actually observed in the solar wind. Finally, we observe that if one of the Elsässer amplitudes is zero the "mode structure" is that of a shear Alfvén wave, with $\omega=kv_{A0}$, and the nonlinearity vanishes. Therefore, pure shear Alfvén waves have no nonlinear interaction. This can be described as a cancellation between the "Reynolds stress," from $\mathbf{u}\cdot\nabla\mathbf{u}$, and the "Maxwell stress," from $\mathbf{b}\cdot\nabla\mathbf{b}$. We will not go further into the details of these points, except to note that their basis is a possibility for a small scale vortex and a small scale magnetic disturbance to drive a rather larger scale magnetic disturbance, or for two similarly small scale magnetic disturbances to drive a rather larger scale flow, and for the flows to be particular sensitive to the direction of the local magnetic field. This affords the possibility of a long range spectral interaction, over about one decade corresponding to $k_1\sim k_2\sim 10k_3$. Interactions like this are thought to be responsible for the generation of large scale magnetic field energy (the dynamo process) or large scale flows (magnetic reconnection) starting with small scale turbulence.

PART II — GRADIENT DRIVEN TURBULENCE IN MAGNETISED PLASMAS

Gradient driven turbulence in a magnetically confined plasma is basically ExB fluid turbulence driven by the background pressure gradient. A low beta MHD equilibrium is assumed. The motion is two dimensional, and is kept incompressible by magnetic pressure, which holds since there is not enough energy in the background pressure to drive motions sufficiently vigorous to compress the magnetic field. So far, we have encountered this combination through 2D MHD turbulence. Here, we will go one step further and assume also that there is insufficient energy to bend the magnetic field lines. The motion is fluidlike and two dimensional, with the fluid resistively slipping past the field lines, with dissipative processes acting along the magnetic field to suppress turbulent activity in that direction. This parallel dynamics acts instead to couple the various quantities together as part of the linear forcing effects to which the turbulence is subject. The dynamics is treated as "electrostatic," which means that although there are finite currents and a finite ExB flow, disturbances in the magnetic field are small enough to be neglected.

We do have to treat the third dimension. The fluid part of the dynamics occurs in what we can call the "drift plane" which is locally perpendicular to the magnetic field. The drift velocities and fluxes occur in this drift plane. The dynamics in the third dimension, along the magnetic field, is shaped principally by the electrons and the parallel current. Although the turbulence is largely two dimensional, its interaction with the parallel dynamics make this a three dimensional phenomenon. The stream function for the ExB velocity is the electrostatic potential. A parallel gradient in that same potential may occur as a result of the turbulence itself or may be set by the magnetic geometry. In ideal MHD the parallel component of the electric field vanishes, to there must be a finite resistivity in order to maintain nonzero disturbances in the potential. The character of this turbulence is therefore always dissipative. In MHD, the balance in the electron parallel dynamics is between the electric field and resistive dissipation acting upon a finite parallel current. Seen another way, finite parallel currents and ExB turbulence, hence finite parallel electric fields, imply each other through the finite resistivity.

But in the usual situation for gradient driven turbulence the electron pressure gradient is not small compared to the electric field, and the MHD approximations are broken. In this situation the principal balance in the parallel dynamics is between these two forces, with parallel currents (through resistivity or electron inertia) equalising any differences. There is concurrently an energetic coupling between the electron pressure and the ExB flow eddies, mediated by the parallel current. This parallel dynamics is called the "adiabatic response" and in general it has the property of "dissipative coupling" since (1) the pressure and electrostatic potential are thereby coupled, and (2) with the finite resistivity the coupling acts to damp each quantity toward the other. The corresponding linear eigenmodes are called "drift waves," and the fully developed nonlinear state is called "drift wave turbulence." Drift wave turbulence differs from its fluid counterpart principally by this tendency towards coupling between the flow eddies themselves and the quantities they transport. Gradient driven turbulence in magnetised plasmas may also involve other effects, but this drift wave dynamics is never absent.

The concept of the high Reynolds number regime is still relevant to the case of dissipative coupling. Even though the coupling process is active at all scales, there is enough energy and enstrophy being transferred to smaller scales that diffusive processes still have a role in maintaining the statistical equilibrium. Indeed, the mutual coupling allows the direct cascade of what would otherwise be a passive scalar to indirectly affect the entire system. In this type of turbulence the small scale dissipation is collisional diffusion, with a coefficient of $\rho_e^2\nu_e$, combining the electron gyroradius and collision frequency. With the typical scales of motion we will find below, the Reynolds number has values of order 10^4. The high Reynolds number regime is reached when the energy containing scales are no longer sensitive to the value of the collisional diffusion coefficient. Because the nonlinear transfer processes are moderately inhibited by the dissipative coupling, the high Reynolds number regime is reached for rather small Reynolds number values, about 10^3, in comparison to the pure hydrodynamic 2D fluid case.

This second part of the lecture treats gradient driven turbulence in the simplest model in which it can be studied. Many complicating effects are left out so that the basic character is in the forefront. The adiabatic response could be electromagnetic, the temperatures could be separately involved, finite ion gyroradius effects could enter, but the basic drift wave dynamics always underlies. The coupling maintains the "mode structure" of the turbulence in the presence of this varying dynamics in each nonlinearity. The basic properties of turbulence and passive scalar cascade dynamics remain active, so that for the overall system both energy cascade tendencies are simultaneously present: direct through the electron pressure, and inverse through the ExB turbulence. The statistical equilibrium in the presence of sources and sinks is called "saturated turbulence" and is regarded as independent of its initial conditions. In this case, saturation occurs when both linear and nonlinear transfer mechanisms are all in statistical balance.

PASSIVE SCALAR DYNAMICS

In incompressible neutral fluid turbulence the density follows a pure advection equation,

$$\frac{\partial \widetilde{\rho}}{\partial t} + \mathbf{v} \cdot \nabla \widetilde{\rho} = 0 \tag{54}$$

which we derive from Eq. (13) by taking $\nabla \cdot \mathbf{v} = 0$. The pressure is not free, but is determined by the incompressibility condition in Eq. (20). In low beta MHD turbulence the incompressibility is maintained by the magnetic field, as in Eq. (44), so the pressure also follows a pure advection equation. With the rest of the dynamics satisfying the MHD turbulence equations (Eq. 51), we have a closed system in which although the density and pressure disturbances are advected by $\mathbf{v}$ there is no back reaction of the disturbances upon the turbulence. This property is called "passive advection," and the dependent variable involved is called a "passive scalar."

A passive scalar also obeys the three wave coupling condition, even though two of the wavenumber components apply to the passive scalar and one to the flow. In two dimensions, with the flow itself obeying a dual cascade, the passive scalar conserves only its amplitude. The "energylike quadratic quantity" conserved by Eq. (54) is simply $T = \widetilde{\rho}^2/2$, with the factor of 2 chosen to follow the definitions of energy and enstrophy for the flow. Of course, any power of $\widetilde{\rho}$ is also conserved, but there are no extra factors of k for any such quantity. Specifically, the counterpart to flow energy, $|\nabla \widetilde{\rho}|^2$, is *not* conserved by Eq. (54). In this sense the quantity which is conserved is more like enstrophy than energy. We can call it "free energy" or "entropy." The three wave equations for a specific triplet $\{\mathbf{k}, \mathbf{k}', \mathbf{k}''\}$ for Eq. (54) are given by

$$\frac{\partial \widetilde{\rho}_{\mathbf{k}}}{\partial t} = C_{\mathbf{k}\mathbf{k}'} \left(\widetilde{\rho}_{-\mathbf{k}''} \psi_{-\mathbf{k}'} - \widetilde{\rho}_{\mathbf{k}'} \psi_{-\mathbf{k}''} \right) \tag{55}$$

$$\frac{\partial \widetilde{\rho}_{\mathbf{k}'}}{\partial t} = C_{\mathbf{k}\mathbf{k}'} \left(\widetilde{\rho}_{-\mathbf{k}} \psi_{-\mathbf{k}''} - \widetilde{\rho}_{\mathbf{k}''} \psi_{-\mathbf{k}} \right) \tag{56}$$

$$\frac{\partial \widetilde{\rho}_{\mathbf{k}''}}{\partial t} = C_{\mathbf{k}\mathbf{k}'} \left(\widetilde{\rho}_{-\mathbf{k}'} \psi_{-\mathbf{k}} - \widetilde{\rho}_{\mathbf{k}} \psi_{-\mathbf{k}'} \right) \tag{57}$$

The important thing to note is that they have the same form as the equations satisfied by the vorticity (Eqs. 33–35). The three wave transfer equations will therefore be the same as Eqs. (38–40), becoming

$$\frac{\partial T_{\mathbf{k}}}{\partial t} = 2C_{\mathbf{k}\mathbf{k}'} \operatorname{Re}\left[\widetilde{\rho}_{\mathbf{k}} \psi_{\mathbf{k}'} \widetilde{\rho}_{\mathbf{k}''} - \widetilde{\rho}_{\mathbf{k}} \widetilde{\rho}_{\mathbf{k}'} \psi_{\mathbf{k}''} \right] \tag{58}$$

$$\frac{\partial T_{\mathbf{k}'}}{\partial t} = 2C_{\mathbf{k}\mathbf{k}'} \operatorname{Re}\left[\widetilde{\rho}_{\mathbf{k}'} \psi_{\mathbf{k}''} \widetilde{\rho}_{\mathbf{k}} - \widetilde{\rho}_{\mathbf{k}'} \widetilde{\rho}_{\mathbf{k}''} \psi_{\mathbf{k}} \right] \tag{59}$$

$$\frac{\partial T_{\mathbf{k}''}}{\partial t} = 2C_{\mathbf{k}\mathbf{k}'} \operatorname{Re}\left[\widetilde{\rho}_{\mathbf{k}''} \psi_{\mathbf{k}} \widetilde{\rho}_{\mathbf{k}'} - \widetilde{\rho}_{\mathbf{k}''} \widetilde{\rho}_{\mathbf{k}} \psi_{\mathbf{k}'} \right] \tag{60}$$

Hence, the transfer of enstrophy from mode $\mathbf{k}'$ to mode $\mathbf{k}$ is given by

$$T_T(\mathbf{k} \leftarrow \mathbf{k}') = 2C_{\mathbf{k}\mathbf{k}'} \operatorname{Re}\left[-\widetilde{\rho}_{\mathbf{k}} \widetilde{\rho}_{\mathbf{k}'} \psi_{\mathbf{k}''} \right] \tag{61}$$

With only the single conserved quantity, and with the passive scalar triple correlation $\langle \widetilde{\rho}_{\mathbf{k}} \widetilde{\rho}_{\mathbf{k}'} \psi_{\mathbf{k}''} \rangle$ independent of the one for the flow, $\langle \psi_{\mathbf{k}} \psi_{\mathbf{k}'} \psi_{\mathbf{k}''} \rangle$, we expect its cascade dynamics to favour the direction towards small scales simply through statistical mixing.

The interesting property of turbulence with passive scalar mixing is all this variation in the properties of nonlinear spectral transfer. We expect free energy and enstrophy to go towards small scale but flow energy towards large scale. If we start the flow stream function ψ and the passive scalar $\widetilde{\rho}$ with identical scale, we should find them rather different at later times. If dissipative coupling is present, however, they should evolve together. We will show this in the Section covering computations, below.

DISSIPATIVE COUPLING AND THE ADIABATIC RESPONSE

In gradient driven turbulence in a magnetised plasma, we have the complication that the MHD approximations, which allow the neglect of pressure gradient effects in the MHD kinematics, are usually broken. We can no longer take the ideal MHD approximation that $E_\parallel$, the component of the electric field parallel to $\mathbf{B}$, to be zero. The parallel electron pressure gradient, $\nabla_\parallel p_e$ is in general not negligible in comparison. The parallel gradient is defined as

$$\nabla_\parallel = \frac{1}{B}\mathbf{B}\cdot\nabla \tag{62}$$

Neglecting special structure of the magnetic equilibrium, $\mathbf{B}$ is taken as homogeneous, its field strength B is constant, and its unit vector $\mathbf{b}$ is a constant coordinate direction. With this, the parallel divergence of the parallel current has the same form as a parallel gradient, simply $\nabla_\parallel J_\parallel$. We will find that the existence of a finite $\nabla_\parallel p_e$ in the electron force balance works together with a finite $\nabla_\parallel J_\parallel$ in the pressure equation. The parallel dynamics is therefore not incompressible.

With massless electrons with finite isotropic resistivity the electron force balance is given by

$$\mathbf{E}+\frac{\mathbf{v}}{c}\times\mathbf{B} = \eta\mathbf{J} - \frac{1}{n_e e}\nabla p_e \tag{63}$$

where we note that the velocity here is specifically that of the electrons. Substitution of this electric field into the Maxwell equation for induction ($\partial\mathbf{B}/\partial t$) we obtain a rather more complicated equation than Eq. (46). In a constant density model the ∇p_e term vanishes under the curl operation, but with $\mathbf{v}$ different for electrons and ions we have an extra $\mathbf{J}\times\mathbf{B}$ term, often called the "Hall term" in the literature, which gives rise to additional kinematic effects. Under low frequency conditions, the current and pressure are especially closely related, since the MHD equilibrium condition,

$$\frac{\mathbf{J}\times\mathbf{B}}{c} = \nabla(p_e+p_i) \tag{64}$$

will hold "quasistatically" perpendicular to the field lines: The physical system including pressure and current evolves on slow time scales with the perpendicular balance evolving through "successive equilibria" since the balance is maintained by dynamics on much faster time scales. This holds if the characteristic frequencies of the turbulence satisfy $\omega \ll k_\perp v_A$, which in this context defines the low frequency approximation.

With low frequency motion in quasistatic perpendicular force balance we enter a different regime than MHD. It is related to incompressible MHD, since the concepts of quasistatic perpendicular force balance and incompressibility maintained by the magnetic field basically mean the same thing for perpendicular dynamics. The parallel dynamics is what differs. Here, we enter a situation that in all detail is three dimensional: turbulence very similar to the 2D fluid model of Eq. (24) with passive scalar advection of densities and pressures as in Eq. (54), in planes locally perpendicular to the magnetic field, and dynamics satisfying the parallel component of Eq. (63),

$$E_\parallel + \frac{1}{n_e e}\nabla_\parallel p_e = \eta J_\parallel \tag{65}$$

parallel to the magnetic field. A model like this is called "drift wave" more or less for historical reasons (waves in this physical system "drift" in the direction of $\nabla p_e\times\mathbf{B}$). If the plasma beta is low enough ($\omega \ll k_\parallel v_A$) we may assume that $E_\parallel$ is electrostatic,

$$E_\parallel = -\nabla_\parallel\phi \tag{66}$$

where ϕ is the electrostatic potential. Substituting this into Eq. (65), we find the relationship between the parallel current and the parallel forces,

$$J_\parallel = \frac{1}{\eta}\left(\frac{1}{n_e e}\nabla_\parallel p_e - \nabla_\parallel\phi\right) \tag{67}$$

This is called the "adiabatic response," since dissipation of $J_\parallel$ towards zero tends to enforce quasistatic parallel force balance in the electrons, with both the pressure gradient and the electric field finite. Strict maintenance of this force balance is called "adiabatic electrons" for historical reasons (note it is not the same thing as an adiabatic equation of state for a fluid). For small disturbances with finite $k_\parallel$, the difference to MHD is that $\widetilde{\phi}$ is dissipated towards $\widetilde{p}_e$ by the resistive parallel dynamics, rather than towards zero.

In general the dissipative coupling is greater the larger the gradients in the parallel direction. Ultimately, this is the reason the turbulence is largely two dimensional: the dissipative dynamics involves only the parallel direction, and disturbances and flows in the 2D drift plane are free to evolve in a way similar to high Reynolds number turbulence. In magnetically confined plasmas, the shear of the magnetic field guarantees that the parallel gradients of finite sized disturbances are nonzero (this is a consequence of the combination of magnetic shear and toroidal topology of the closed magnetic flux surfaces which describe the equilibrium). As a result of these properties we also have what is called "flute mode ordering" in the dynamics even with the finite parallel gradient: in terms of wavenumbers,

$$k_\parallel \ll k_\perp \tag{68}$$

wherein it is important to note that we still have $k_\parallel \neq 0$. The magnetic field both orders the motion to occur in the drift plane and provides the dissipative mechanisms to couple all the parts of the dynamical system together.

We will examine the consequences of this parallel dynamics, which acts to couple ϕ directly back to p_e in the turbulence, i.e., the electron pressure is no longer a passive scalar. The simplest model in which to study this is a 2D one, since we can model the parallel dynamics through a set of coupling terms. The perpendicular dynamics is still incompressible, but the parallel dynamics for both the vorticity and pressure is influenced by the parallel compression of the current. Here, the vorticity is the ExB vorticity, the pressure is the electron pressure, and we simplify to an isothermal model wherein $p_e = n_e T_e$ with T_e constant. The role of $J_\parallel$, the parallel current, in the pressure dynamics arises from the electron velocity, which since we neglect sound wave effects is given by

$$v_\parallel = -\frac{J_\parallel}{n_e e} \tag{69}$$

so that

$$\nabla\cdot\mathbf{v} = \nabla\cdot(\mathbf{b}v_\parallel) = \nabla_\parallel v_\parallel = -\frac{1}{n_e e}\nabla_\parallel J_\parallel \tag{70}$$

where due to the small fluctuation rules the term is linearised. The "small fluctuation rules" arise from the ordering described in Eq. (14). The derivation is regarded as an expansion in the small parameter δ, which we will quantify once we find the characteristic scale of motion. At this point we order the relative amplitude and scale of motion with δ, assuming the ordering

$$\frac{e\widetilde{\phi}}{T_e} \sim \frac{\widetilde{p}_e}{p_e} \sim \frac{k_\parallel}{k_\perp} \sim \frac{\Delta}{L_\perp} \sim \delta \ll 1 \tag{71}$$

where Δ is the scale of motion, $L_\perp$ is the scale of the background gradient, and the tilde denotes the disturbances on the equilibrium. The basic rules are that the equations are linearised except for the terms involving $\mathbf{v}_E\cdot\nabla$, the advection by the ExB velocity, since the perpendicular *gradients* of the background and the disturbances are regarded as comparable. We may drop the tilde symbols momentarily, if we recall that ϕ and $J_\parallel$ belong strictly to the disturbances, and that generally second and higher order derivatives are to act solely upon the disturbances (the first derivative carries the scale of $L_\perp$, while higher ones carry the scale Δ).

The role of $J_\parallel$ in the vorticity dynamics arises through the $\mathbf{J}\times\mathbf{B}$ force on the plasma. We start with Eq. (42), and assume that for the ions $\mathbf{v}$ is $\mathbf{v}_E$, and further following the incompressibility rules in the perpendicular plane, that $\mathbf{v}$ is given by a stream function. This stream function is none other than the electrostatic potential, with an extra factor of c/B:

$$\mathbf{v} = \frac{c}{B^2}\mathbf{B}\times\nabla\phi \qquad \frac{\rho c}{B}\nabla\times\mathbf{v} = \Omega\mathbf{b} \qquad \Omega = \frac{\rho c^2}{B^2}\nabla_\perp^2\phi \tag{72}$$

which is the same as in 2D hydrodynamics except for the units. Taking the curl of Eq. (42), treating ρ as constant, and contracting with $\mathbf{b}$, we have

$$\frac{\partial\Omega}{\partial t} + \mathbf{v}_E\cdot\nabla\Omega = \nabla_\parallel J_\parallel \tag{73}$$

This equation is also the one for the *total* divergence free nature of $\mathbf{J}$, so the inertial effects on the left hand side can be regarded as the perpendicular current divergence (from the "polarisation current" arising from inertial delays in the ExB drift motion due to the finite gyrofrequency). Although $J_\parallel \mathbf{b}$ itself does not contribute to the cross product with $\mathbf{B}$, its divergence still enters the vorticity in this manner, because $\nabla_\perp \cdot \mathbf{J}_\perp + \nabla_\parallel J_\parallel = 0$, and $\nabla_\perp \cdot \mathbf{J}_\perp$ is what gives rise to the inertia term on the left hand side of Eq. (73).

Now, following the same rules for the electron pressure, that is, adding the parallel compressibility to a 2D advection equation (Eq. 54) for p_e, we have

$$\frac{\partial p_e}{\partial t} + \mathbf{v}_E \cdot \nabla p_e = \frac{T_e}{e} \nabla_\parallel J_\parallel \tag{74}$$

Finally, we follow the simplest model for the parallel dynamics. Each of Ω and p_e is forced upon by $\nabla_\parallel J_\parallel$. Taking the parallel divergence of $J_\parallel$ in Eq. (67), we find

$$\nabla_\parallel J_\parallel = \frac{1}{\eta}\left(\frac{1}{n_e e} \nabla_\parallel^2 p_e - \nabla_\parallel^2 \phi\right) \tag{75}$$

as with the small fluctuation rules the only non-negligible terms are the ones with the highest-order derivatives (cf. Eq. 14). The model which we substitute for this to get a 2D dynamical system is to replace $-\nabla_\parallel^2$ with a positive constant, $k_\parallel^2$, giving the characteristic parallel wavenumber associated with the parallel dynamics. This sets the parallel current divergence to a simple difference between p_e and ϕ, suitably normalised, with a single parameter resulting from all the multiplying factors,

$$\frac{1}{n_e e} \nabla_\parallel \widetilde{J}_\parallel = D_\parallel \left(\frac{e\widetilde{\phi}}{T_e} - \frac{\widetilde{p}_e}{p_e}\right) \tag{76}$$

where we now explicitly treat the dependent variables as small disturbances on an equilibrium. The coupling parameter is given by

$$D_\parallel = \frac{T_e k_\parallel^2}{n_e e^2 \eta} \tag{77}$$

which has the units of a frequency, that is, the inverse of a characteristic time constant. Coupling through a constant gives rise to real eigenvalues corresponding to exponential growth or damping. In a physically realistic model this is usually a damping mechanism although in combination with other processes it can lead to instabilities. This process is called "dissipative coupling."

We rewrite Eqs. (73,74) explicitly in terms of small disturbances and then substitute Eq. (76) for the parallel current divergence. We remember to keep the background gradient given by

$$\nabla p_e = -\frac{p_e}{L_\perp} \nabla x \tag{78}$$

where in whichever geometry is assumed, x becomes a coordinate whose direction is down the gradient. In deriving these equations we strictly follow the small disturbance ordering in Eq. (14), and also the flute mode ordering in Eq. (68). In addition, a characteristic perpendicular scale arises when we compare Eq. (76) and the form of Ω. Scaling ϕ as $e\widetilde{\phi}/T_e$ and allowing the action of the coupling to force $e\widetilde{\phi}/T_e$ and $\widetilde{p}_e/p_e$ towards each other, we find that the natural scale of $\nabla_\perp^2$ is given by a quantity called the "drift scale," ρ_s, defined as

$$\rho_s^2 = \frac{c^2 T_e M_i}{e^2 B^2} \tag{79}$$

where we have used $n_e = n_i$ and $\rho = n_i M_i$. This has the form of a gyroradius, but it arises from electron mobility and ion inertia, so it is simply the natural scale of the vorticity. Its origin is the combination of the sound speed c_s and the gyrofrequency Ω_i, given by

$$c_s^2 = \frac{T_e}{M_i} \qquad \Omega_i = \frac{eB}{M_i c} \qquad \text{hence} \quad \rho_s = \frac{c_s}{\Omega_i} \tag{80}$$

The turbulence will occupy a range of scales typically of order and larger than ρ_s. We characterise these by Δ or a perpendicular wavenumber $k_\perp$, but remember that $k_\perp$ like the dynamical frequency ω takes a range of values that can span one or two orders of magnitude. The scale of the disturbances will only remain small if $\rho_s \ll L_\perp$. We can now evaluate the small ordering parameter δ, as we are now ordering the scale of motion $\Delta_\perp$ with ρ_s, so that

$$\delta \equiv \frac{\rho_s}{L_\perp} \ll 1 \tag{81}$$

It is a result of the dynamics occupying frequencies up to $c_s/L_\perp$, that the gyrofrequency can be ordered as large also only if $\rho_s \ll L_\perp$. In the literature, Eqs. (71,81), the specific application to these problems of Eq. (14), is called variously "drift ordering" or "gyrokinetic ordering." It is a "maximal ordering" since all the small parameters potentially entering, including that in Eq. (68), are taken together.

The equations (Eqs. 73, 74) now become

$$\frac{\partial\widetilde{\Omega}}{\partial t} + \mathbf{v}_E \cdot \nabla\widetilde{\Omega} = D_\parallel \left(\frac{e\widetilde{\phi}}{T_e} - \frac{\widetilde{p}_e}{p_e} \right) \tag{82}$$

$$\frac{\partial}{\partial t}\frac{\widetilde{p}_e}{p_e} + \mathbf{v}_E \cdot \nabla \frac{\widetilde{p}_e}{p_e} = -\mathbf{v}_E \cdot \nabla \log p_e + D_\parallel \left(\frac{e\widetilde{\phi}}{T_e} - \frac{\widetilde{p}_e}{p_e} \right) \tag{83}$$

where the vorticity is given by

$$\widetilde{\Omega} = \rho_s^2 \nabla_\perp^2 \frac{e\widetilde{\phi}}{T_e} \tag{84}$$

We identify the dissipative coupling mechanism parameterised by $D_\parallel$, explicitly in the form of an inverse time constant, and the background gradient forcing given by $\nabla \log p_e$. These are the linear forcing effects. The rest of the system is just 2D incompressible ExB flow turbulence plus passive advection of the pressure disturbances. Of course, with $D_\parallel$ nonzero, the pressure is no longer passive.

COMPUTATIONS IN THE DISSIPATIVE COUPLING MODEL FOR DRIFT WAVE TURBULENCE

We now develop some physical insight into these phenomena by means of computations. Due to the statistical nature of the turbulence, a deterministic analytical model is not useful. Moreover, due to the fact that it is the nonrandom part of the turbulence which gives rise to all the energetic phenomena — spectral transfer, free energy drive and damping rates, average transport — a purely statistical calculation which takes the turbulence as random to lowest order in the inverse Reynolds number, and then finds corrections due to the existence of the forcing effects, will have the problem that as soon as the physical process ceases to be self similar and there is more than one nonlinear transfer phenomenon active, that it will be too complicate to solve without additional assumptions that predetermine the results. There is no substitute for direct numerical simulation in this situation: dissipative coupling and the existence of several cascade tendency possibilities, not to mention the fact that we still have to show that the spectral transfer process is in fact a local cascade, make it essential that we work from first principles, even in this simplest of gradient driven turbulence models.

We take the dissipative coupling model of the previous Section, Eqs. (82,83), normalising the equations as suggested by the appearance of the scaled units there. It is possible to assume $\rho_s/L_\perp$ to be small and renormalise it into the dependent variables, so that these are scaled as

$$\phi \leftarrow \frac{L_\perp}{\rho_s}\frac{e\widetilde{\phi}}{T_e} \qquad\qquad p \leftarrow \frac{L_\perp}{\rho_s}\frac{\widetilde{p}_e}{p_e} \tag{85}$$

The space coordinates are x and y, with x in the direction down the gradient, and the orientation given by $\nabla x \times \nabla y \cdot \mathbf{b} = 1$. Space and time scales are normalised in terms of

$$t \leftarrow \frac{c_s}{L_\perp} t \qquad\qquad \{x,y\} \leftarrow \rho_s^{-1}\{x,y\} \tag{86}$$

This results in the equations being cast in terms of the single dissipative coupling parameter given by

$$D \leftarrow \frac{L_\perp}{c_s} D_\parallel \tag{87}$$

In these terms, Eqs. (82–83) become

$$\frac{\partial \Omega}{\partial t} + \mathbf{v}_E \cdot \nabla \Omega = D(\phi - p) \tag{88}$$

$$\frac{\partial p}{\partial t} + \mathbf{v}_E \cdot \nabla p = \omega_p \mathbf{v}_E \cdot \nabla x + D(\phi - p) \tag{89}$$

and the vorticity and ExB advective derivative are

$$\Omega = \nabla_\perp^2 \phi \qquad\qquad \mathbf{v}_E \cdot \nabla = \frac{\partial \phi}{\partial x}\frac{\partial}{\partial y} - \frac{\partial \phi}{\partial y}\frac{\partial}{\partial x} \tag{90}$$

The first term on the right side of Eq. (88) is the gradient drive. The coefficient ω_p allows setting the gradient drive to zero for simple tests. This is the prototypical drift wave turbulence model, and it is a useful model for learning the physical character of gradient driven turbulence, although as can be seen from the list of assumptions used to derive it (cf. Eqs. 63–84), it is quite simplified. It is also called the Hasegawa-Wakatani model, after its original authors. The normalisation scheme is called "gyro-Bohm" in the literature. In these units the unit of diffusion is $D_{GB} = \delta \times D_B$, where $D_B = cT_e/eB$ is the "Bohm diffusion coefficient" up to a numerical factor.

There is one more thing to say about turbulence computations, in basically any situation one ever faces: subgrid dissipation still has to be treated. In treating high Reynolds number turbulence, we do not try to find the resulting turbulence and transport in the presence of a given level of dissipation, and not even in the absence of dissipation, but in the presence of an arbitrarily small level of dissipation. One or more of the nonlinear transfer processes will always send enough free energy to arbitrarily small scales that a computation will not be able to represent a statistically saturated state unless this free energy is absorbed. In the literature there are many ways to do this. One may take an actual physical diffusion (for passive scalar variables) or viscosity (for the vorticity or velocity variables) or resistivity (for the magnetic field in MHD turbulence) and try to achieve enough spatial resolution to make the results independent of the value of the dissipation coefficient. Then the value may be set artificially large, allowing computations to carry moderate resolution and hence a larger set of cases to disassemble the various physical effects. Another method is to use a different operator, for example "hyperviscosity" or ∇^4 instead of ∇^2, which for some situations may reach the high Reynolds number regime with coarser resolution. A final method is to use a sophisticated numerical scheme which incorporates dissipation of structures with large second derivative but switches it off for second derivatives below a certain limit; these schemes are called "high order upwind" in the literature. Which method one chooses may be optimised with regard to computational expense, or can even be a matter of personal taste. Either way, one should apply enough resolution that the high Reynolds number regime is reached: the amount of free energy absorbed at large wavenumber should be independent of the form or size of the subgrid dissipation operator used.

We note that the need for subgrid dissipation is additional to the presence or absence of dissipative coupling. In the model of Eqs. (88,89) we still require subgrid diffusion of both Ω and p regardless of the value of D. Recall that the presence of a cascade tendency in a particular direction does not rule out the presence of spectral transfer in the other direction. Indeed, even in 2D Euler turbulence (Eq. 24), the enstrophy which tends toward high k contains finite energy and the energy which tends toward low k contains finite enstrophy. The presence of dissipative coupling can indeed diminish the cascade dynamics somewhat, as we will see, but it never completely assumes the overall role of energy sink processes in balancing the gradient source drive.

In the computations presented below, an advanced upwind scheme has been used, the one presented in Section I of Colella (1990). Details are given therein. Both Ω and the combination $(p - \omega_p x)$ are advected by $\mathbf{v}_E$ as passive scalars, with

$$v_E^x = -\frac{\partial \phi}{\partial y} \qquad\qquad v_E^y = \frac{\partial \phi}{\partial x} \tag{91}$$

This part of the scheme is "explicit" since this part of the right hand side is evaluated with information already to hand, namely, the values of Ω and p on all nodes at the start of a given time step. Subgrid dissipation is effected by this scheme.

The terms involving D are evaluated "implicitly" due to the potentially large damping rates for ϕ towards p, particularly at large scale, since the inverse time constant of the dissipative coupling is $Dk_\perp^{-2}$. That a part of a scheme is implicit means that information at the future time step is required to evaluate the corresponding subset of the terms. Advection of Ω and p by $\mathbf{v}_E$ leads to auxiliaries defined by

$$S_w = \Omega - \tau[\mathbf{v}_E \cdot \nabla\Omega] \qquad S_p = p - \tau[\mathbf{v}_E \cdot \nabla(p - \omega_p x)] \tag{92}$$

where τ is the value of the timestep. Then, the dissipative coupling terms are solved together with $\nabla_\perp^2$, according to

$$\left(\frac{\tau D}{1+\tau D} - \nabla_\perp^2\right)\phi = \frac{\tau D\, S_p}{1+\tau D} - S_w \tag{93}$$

and

$$p = \frac{\tau D\phi + S_p}{1+\tau D} \tag{94}$$

and then the vorticity is recovered via

$$\Omega = S_w - \frac{\tau D}{1+\tau D}(S_p - \phi) \tag{95}$$

avoiding operation by $\nabla_\perp^2$ directly. In this case τ and D are constants, so that Eq. (93) is solved by means of Fourier transforms and then Eqs. (94,95) are applied directly.

To avoid the issue of coupling to large scale "zonal flows," which is beyond the scope of this lecture, the Fourier components with $k_y = 0$ were set to zero at the end of each time step. The grid was equidistant in both x and y, on a doubly periodic domain of dimensions $L \times L$, with $L = 20\pi$. Three values of the resolution, with grid node counts of 64^2, and 128^2, and 256^2 were used. The timestep was 0.05. The turbulence was initialised as such, i.e., at finite amplitude, with $p = \phi = p_0$, with

$$p_0 = a_0 \sum_{k_x,k_y} \left[1 + (k_\perp^2/0.32)^4\right]^{-1/2} e^{i\Theta}\, e^{ik_x x}\, e^{ik_y y} \tag{96}$$

where k_x and k_y are the wavenumbers involved in a discrete Fourier transform over the domain, Θ is a random variable on $[-\pi, \pi]$, and the amplitude a_0 was chosen to set the rms amplitude to 3.0.

No Coupling — The Hydrodynamic Limit

The difference between a simple scalar quantity which is passively advected by the flow and the stream function for an incompressible 2D flow is well illustrated by the simplest version of the model, which is $D = 0$ and $\omega_p = 0$. This limit is called "hydrodynamic" since the turbulence is then just the 2D Euler model of Eq. (24), for Ω and ϕ, and p just follows the basic passive scalar model of Eq. (54). Here, the variables ϕ and p are arranged equal, with the same spectrum, so we may watch them evolve apart due to their differing spectral transfer tendencies as indicated by the three wave analyses above.

The diagnostics used are time traces of selected quantities averaged over each node, logarithmic spectra of selected quantities in k_y averaged over x and sometimes also over time, and spatial morphology of the dependent variables displayed as level contours on the spatial domain. The half squared amplitude of ϕ is defined as A_p and the ExB energy density as U_E, according to

$$A_p = \frac{\phi^2}{2} \qquad U_E = \frac{|\nabla\phi|^2}{2} \tag{97}$$

and A_n and A_w are defined similarly for p and Ω, noting that the thermal free energy density is $U_n = A_n$ and the enstrophy density is $W = A_w$. The transport is defined as

$$Q_e = p v_E^x \tag{98}$$

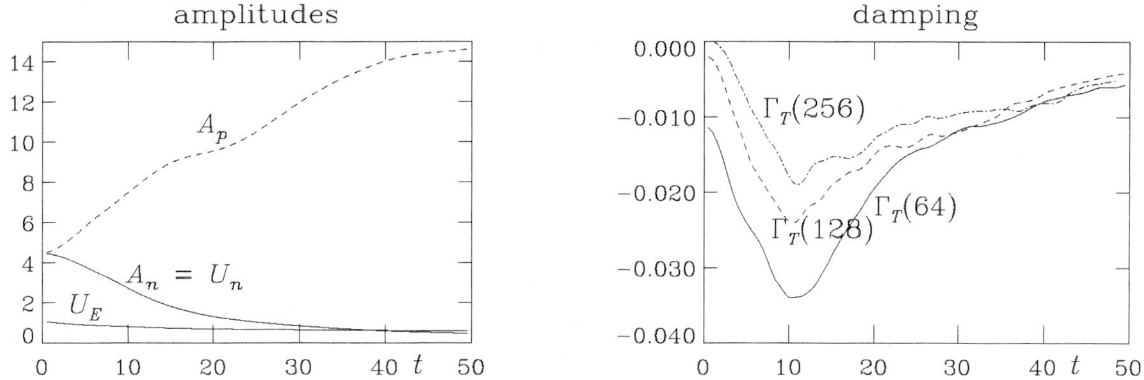

FIGURE 1. Time evolution of the hydrodynamic model. (left) Initial decay of half squared amplitudes of p and ϕ, denoted A_n and A_p, respectively. The free energy density components are also given, U_E for the ExB eddies (ϕ) and U_n for the density (also pressure in this model), noting that U_n is also A_n. (right) Energetic losses (mostly in p due to the direct cascade) for three values of the resolution.

with v_E^x defined in Eq. (91). These are given as both time traces and k_y-spectra. The spectra are defined as the contribution by each k_y to the total quantity, defined for the transport as

$$Q_e(k_y) = p^*(k_y) v_E^x(k_y) \tag{99}$$

averaged over x, with both k_y and $-k_y$ modes contributing, either as a snapshot at a particular time or averaged over a range of time. The decay rate of the total energy is also given as a time trace, defined as

$$\Gamma_T = \frac{1}{2U}\frac{\partial U}{\partial t} \tag{100}$$

where U is $U_E + U_n$ integrated over the spatial domain. Note that in the description of the computations U and W are given as totals, not as densities.

The amplitudes and decay rate are shown in Fig. 1. They show monotonic decay of all quantities, with $\Gamma_T < 0$ at all times. All of the decay in this case is due to the subgrid dissipation at high k. In this case, with the dissipation being switched on for relative second derivatives beyond a certain limit defined by the scheme, the dissipation coefficient is a function of the spatial resolution. A resolution test is therefore the same thing as a Reynolds number test. In the hydrodynamic model, the direct cascade in U_n is the most powerful nonlinear interaction, and we see that in the faster decay of U_n compared to U_E. We see also that the average half squared amplitude of ϕ, shown by A_p, actually rises, as the inverse cascade in U_E moves the ExB energy to larger scales where the factor of $k_\perp^2$ is smaller. The decay rates are shown for each of the three values of the resolution. The initial transient phase, during which the nonlinear transfer dynamics destroys the alignment between p and ϕ, shows a virulent cascade in which free energy is thrown to the highest $k_\perp$ and hence to the numerical dissipation. At later times, however, the decay rates are comparable, since most of the energy resides at lower $k_\perp$, and there are several factors of two over which the local cascade operates.

A high Reynolds number means that there is enough spectral separation between the largest scales and the dissipation range for the nonlinear interactions to operate unconstrained, whether the dissipation is physical or numerical. The high Reynolds number regime is characterised by an dissipation rate which is independent of the resolution, that is, the Reynolds number. Fig. 2 shows the evolution of the ϕ and p spectra, and the vorticity spectrum, out of the initial state. Most of the redistribution occurs over the short time interval of $0 < t < 10$. The spectra of A_n and A_w become much flatter, but that of A_p steepens. This follows from the cascade dynamics for each of these three quantities. Once most of the ExB energy is at larger scales, the cascade dynamics in the middle range slows down, and then the decaying turbulence enters the high Reynolds number regime.

The evolution of the morphology of the disturbances is shown in Fig. 3, in which the top row of frames gives the initial state. The physical nature of the nonlinear interactions is vortex merger for the ExB eddies and shearing apart of the structures for the quantities directly advected by the eddies. As this includes the

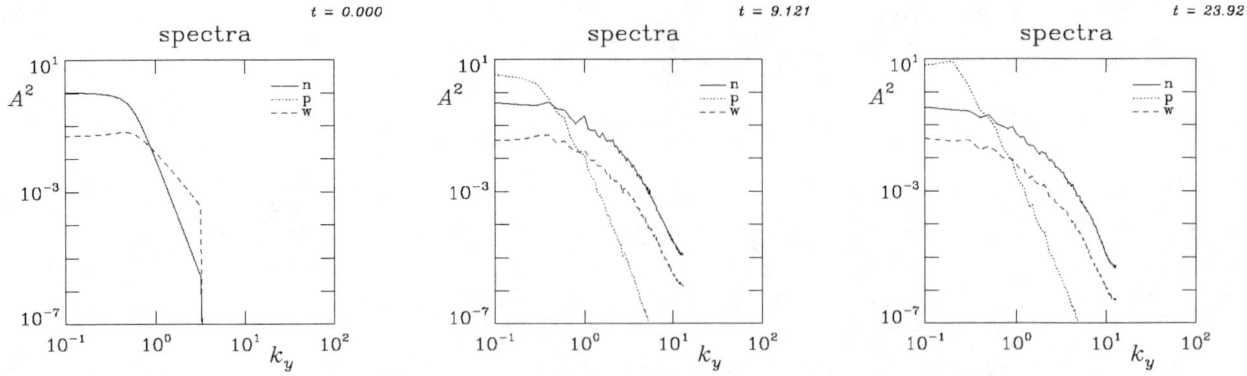

FIGURE 2. Amplitude spectra in the hydrodynamic model, for p, ϕ, and the vorticity, labelled by 'n', 'p', and 'w', respectively. The times of the snapshots are $t = 0$ (left), $t = 9.8$ (center), and $t = 24$ (right). The spectra evolve rapidly apart due to the differing cascade dynamics for p and ϕ. Note $p = \phi$ at $t = 0$, while at late times p and Ω acquire the same spectrum.

vorticity, the ϕ disturbances are involved in both the flows and the advected quantities. The Euler equation's dynamics is essentially this merger/shearing process, until at late times (if the Reynolds number is high enough) one is left with a shear-free flow field. The fate of a passively advected quantity, usually called a passive scalar, in this case p, is for the disturbances to be sheared apart into sheets, much like the vorticity in the Euler equation, whose narrowness is limited only by the Reynolds number. These interactions represent the physical manifestations of the cascade dynamics in the hydrodynamic model. Hence p evolves away from ϕ and towards Ω, as ϕ and Ω evolve apart, as seen.

The Effects of Dissipative Coupling

We now examine what a continuous level of driving and dissipation has on the "pure" turbulence we have just seen in the purely decaying hydrodynamic limit. With this continuous forcing, there are always energy sources and sinks available to the turbulence. The free energy will therefore redistribute itself among the available degrees of freedom until these forcing effects and the various transfer effects are all in statistical balance. This balance is called saturation, and in a dissipatively forced system such as gradient driven turbulence one is mostly interested in the maintenance of this saturated state. The approach to saturation is not as important as the properties the turbulence has once it is statistically stationary. This is why we have not been more concerned with the Reynolds number effects encountered in the hydrodynamic limit. In saturation with dissipative coupling, they are much less pronounced.

With $D \neq 0$ and $\omega_p = 1$ we can examine the persistence of the basic nonlinear ExB flow and passive scalar dynamics in a situation with dissipative forcing. We highlight an intermediate case with $D = 0.1$ in order to show how the various pieces of the dynamics interact, and then examine the energy transfer dynamics in all limits. This case is referred to as nominal. It was taken to $t = 400$ for the three values of the resolution.

The turbulence evolves initially as it would do in the hydrodynamic limit, but then the dissipative coupling prevents p and ϕ from separating completely. One interesting result is that the cases with various resolution do not differ to the extent they do in the purely hydrodynamic model. Most of the activity is in a driven range of the spectrum, and the action of the dissipation involving D prevents the "long term" or "pure" forms of freely decaying turbulence from establishing themselves. The turbulence approaches saturation after about $t = 100$, and averaged quantities are measured over the interval $200 < t < 400$.

Time traces of the free energy components of the turbulence are shown in Fig. 4, and in the right frame the resulting transport is given for the three values of the resolution. The values of Q_e vary within about one standard deviation. It is also noteworthy that the numerical dissipation rate, Γ_E, was 0.028, 0.028, and 0.026, for 64^2, 128^2, and 256^2 grid nodes, respectively. In this case Γ_E is not Γ_T, since ω_p and D are nonzero. Saturation results in $\Gamma_T = 0$ within the uncertainty (the mean is usually less than 0.1 standard deviation).

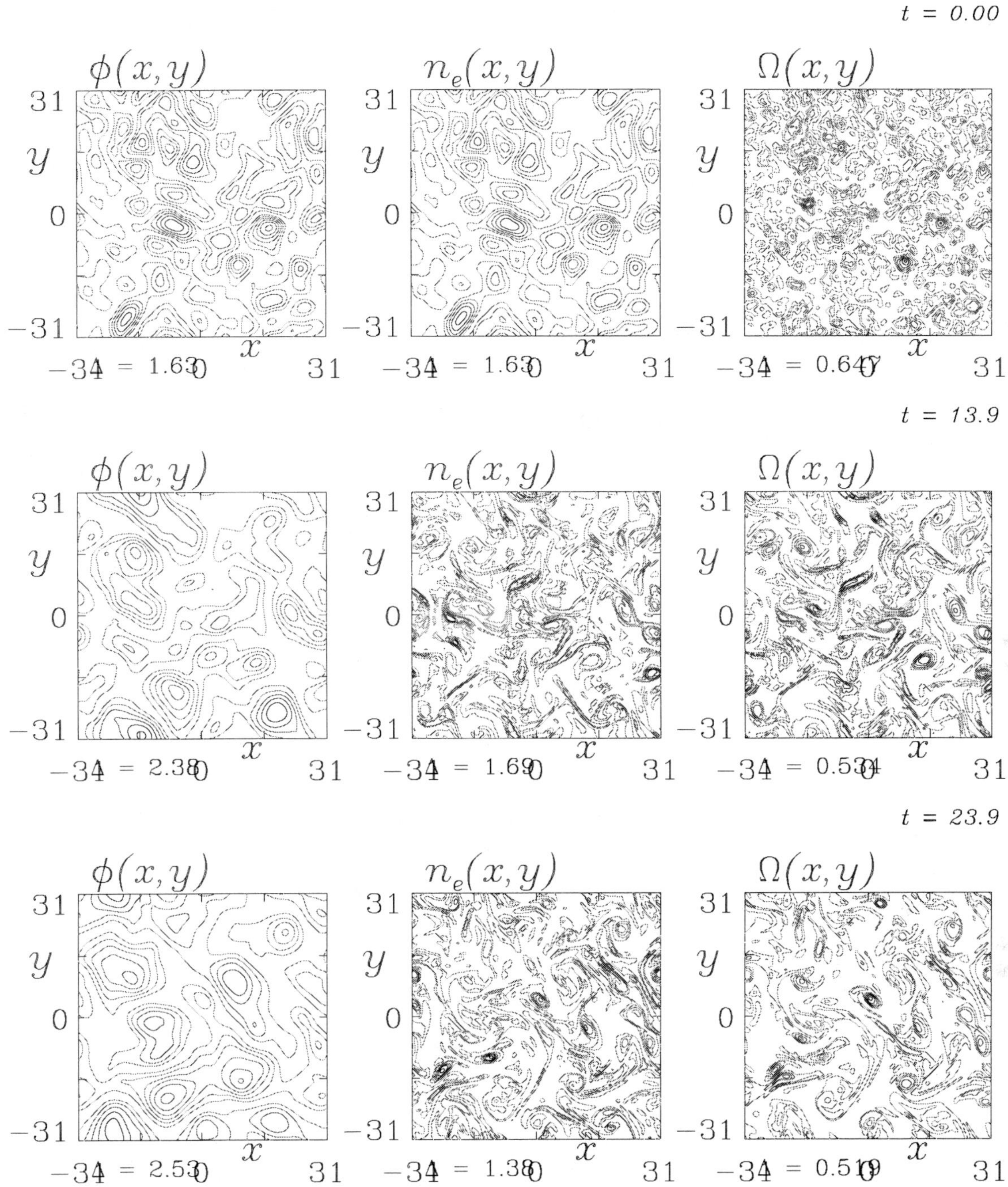

FIGURE 3. Evolution of the disturbances for the hydrodynamic model. The nonlinear interactions represent ExB vortex merger and shearing apart of advected quantities, including the ExB flow vorticity itself.

The energy theorem for these equations is

$$\frac{\partial U}{\partial t} = \oint \frac{d^2x}{L^2}\left(p\frac{\partial p}{\partial t} - \phi\frac{\partial \Omega}{\partial t}\right) = \oint \frac{d^2x}{L^2}\left(\omega_p p v_E^x - D h^2\right) - \text{numerical dissipation} \qquad (101)$$

where $h = p - \phi$ and Γ_E is this numerical dissipation normalised as a damping rate (found by measuring Γ_T and the integral over the terms in parentheses divided by $2U$). This shows that the high Reynolds regime

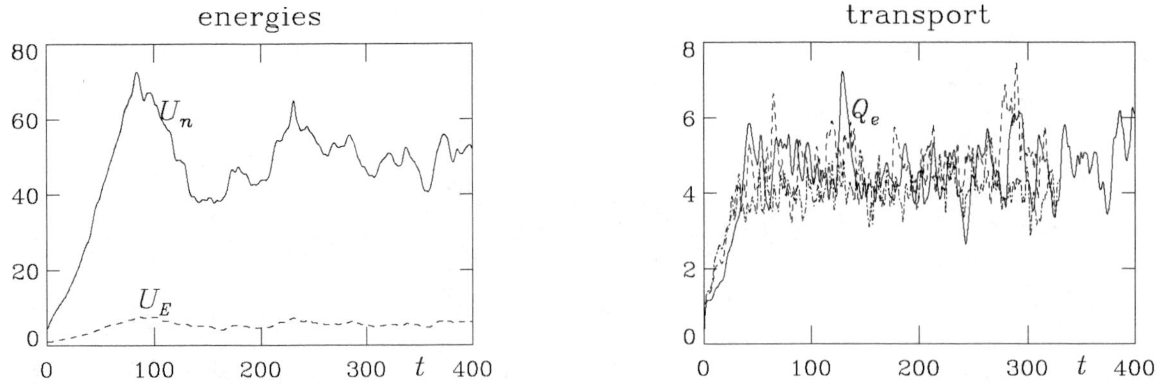

FIGURE 4. Time evolution of the dissipative coupling model, for the nominal case of $D = 0.1$. (left) Half squared amplitudes of p and ϕ, denoted A_n and A_p, respectively, and the free energy density components, U_E for the ExB eddies (ϕ) and U_n for the density, noting that U_n is also A_n. (right) The transport caused by the turbulence, for three values of the resolution. For 64^2, 128^2, and 256^2 grid nodes, the values are 4.69 ± 0.80 and 4.89 ± 0.74 and 4.14 ± 0.51, respectively.

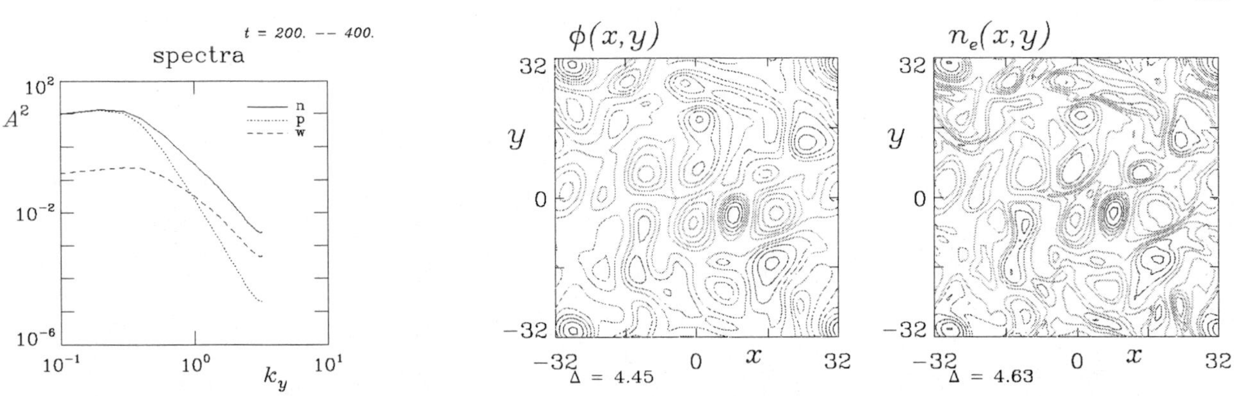

FIGURE 5. (left) Averaged amplitude spectra in the dissipative coupling model, with $D = 0.1$, for p, ϕ, and the vorticity, labelled by 'n', 'p', and 'w', respectively. (center, right) Contours of ϕ and $n_e = p$ at $t = 400$, showing the close coupling at larger scales but differences on smaller scales, corresponding to the spectra. The nonlinear interactions affecting p are stronger relative to the coupling at higher $k_\perp$.

has been reached, and so the resolution consideration is not grievous. A grid of 64^2 nodes is sufficient to study phenomenology as we are doing here. The main result from the time traces is that most of the free energy resides in p, and so the main dissipation process is actually the direct cascade of the associated free energy component, U_n, to arbitrarily short wavelengths.

The amplitude spectra and the disturbances themselves are shown in Fig. 5. Here we can easily see the varying effectiveness of the dissipative coupling at larger and smaller scales. The spectra look more like the "adiabatic" state of $\phi = p$ at low k_y but then separate at smaller scales for which the nonlinear interactions are more able to compete with the coupling. The high degree of correlation between p and ϕ is clear from the spatial distributions, but a the level of details there is more to see in p. The tendency to form sheets is still evident at the smaller scales, and the dissipative coupling reduces the basic tendency for all the ExB energy to go to larger scales.

The energy transfer dynamics is shown in Fig. 6. In each case the transfer is defined as positive if the free energy leaves a spectral range k' and enters the spectral range k. The transfer for each three wave triplet

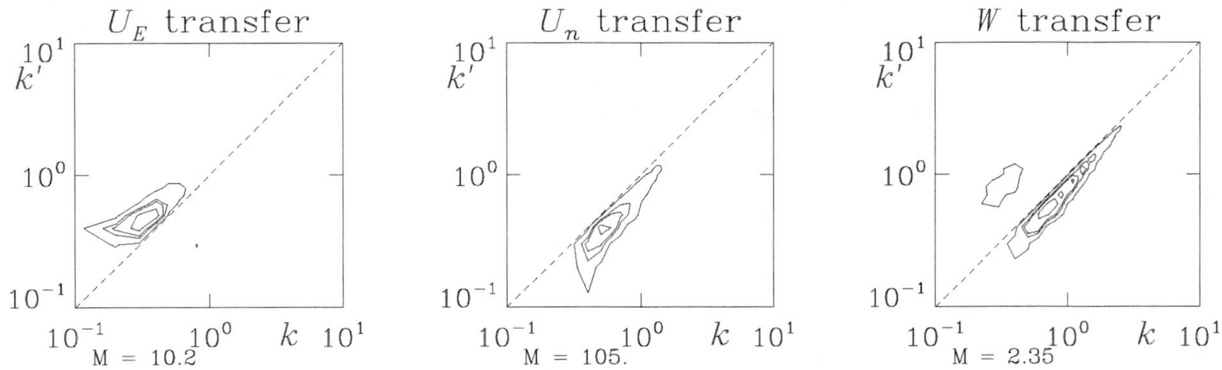

FIGURE 6. Energy and enstrophy transfer in the dissipative coupling model, with $D = 0.1$. Left to right, the transfer from spectral range k' to k is shown for the ExB energy U_E, the thermal free energy U_n, and the enstrophy. The maximum value for each frame is given by M. Contours are drawn only where the transfer is positive, noting the symmetry about $k' = k$, so that activity above the dashed line shows an inverse transfer while activity below the dashed line shows a direct transfer. The cascade is local, with most of the activity near $k' = k$.

is given by Eq. (36) for the ExB energy U_E, by Eq. (61) for the thermal free energy U_n, and by Eq. (41) for the enstrophy W. The spectral range k is defined as the set of all Fourier pairs (k_x, k_y) for which the resulting $k_\perp$ is within $0.5(2\pi/L)$ of k, noting that $k_\perp^2 = k_x^2 + k_y^2$. The symmetry about $k' = k$ is obvious, and so the contours are drawn only where they are positive. Activity above the line $k' = k$ indicates an inverse transfer tendency, while the activity for a direct transfer appears below the line. The quantitative level of each transfer is given by its maximum value, denoted M in the figures. A local cascade is a successive transfer between spectral ranges corresponding to scales within a factor of two of each other, and indeed this is where we find the activity. As argued above, the transfer is clearly an inverse cascade in the ExB energy (energy associated with ϕ, or U_E), and a direct cascade in the mean squared vorticity (fluid enstrophy) and the thermal free energy (energy associated with p, or U_n).

The above introductory results show that the basic properties of the underlying hydrodynamic turbulence are present generally. Most notably, both cascade tendencies, dual for the ExB energy and enstrophy, and direct for the passive scalar quantity, are still active when the turbulence is severely forced by drive and dissipation. It is useful to demonstrate that the simultaneous presence of the nonlinear transfer dynamics for ϕ and p is present for all values of D except the extreme adiabatic limit of $D \to \infty$ at which $\phi = p$ and hence $\mathbf{v}_E \cdot \nabla p = 0$. In Fig. 7 we find the transfer dynamics for $D = 0.01$, 0.03, 0.3, and 1.0, from top to bottom. We find the same basic form in all cases, with the hydrodynamic regime (small D) showing the strongest dynamics along with the strongest turbulence (a wider range of the spectrum is hydrodynamic). For all cases shown the direct transfer in U_n is stronger than the inverse transfer in U_E. Larger values of D are necessary to find the true adiabatic limit. In general, this type of turbulence is either more nonadiabatic than in the $D > 1.0$ cases or it is too weak to be of consequence. We therefore always find that in cases of interest that both Euler fluid and passive scalar nonlinear dynamics is simultaneously present in gradient driven turbulence in magnetised plasmas, and that the nonlinear transfer in all the conserved quantities satisfies the property of a local cascade.

SUMMARY

Turbulence is a phenomenon involving mostly conservative interactions among a large number of statistically independent degrees of freedom. It involves nonlinear interactions among various accessible states of the disturbances, rather than quasilinear interactions between disturbances and background (profile) quantities. For low frequency turbulence in fluids and magnetised plasmas the principal nonlinearities are quadratic, with the result that the interactions are three wave triplets each following their own transfer dynamics. The number of such triplets, and the number of triplets of which each degree of freedom is a member, is arbitrarily large, so that each triplet is statistically independent of any other.

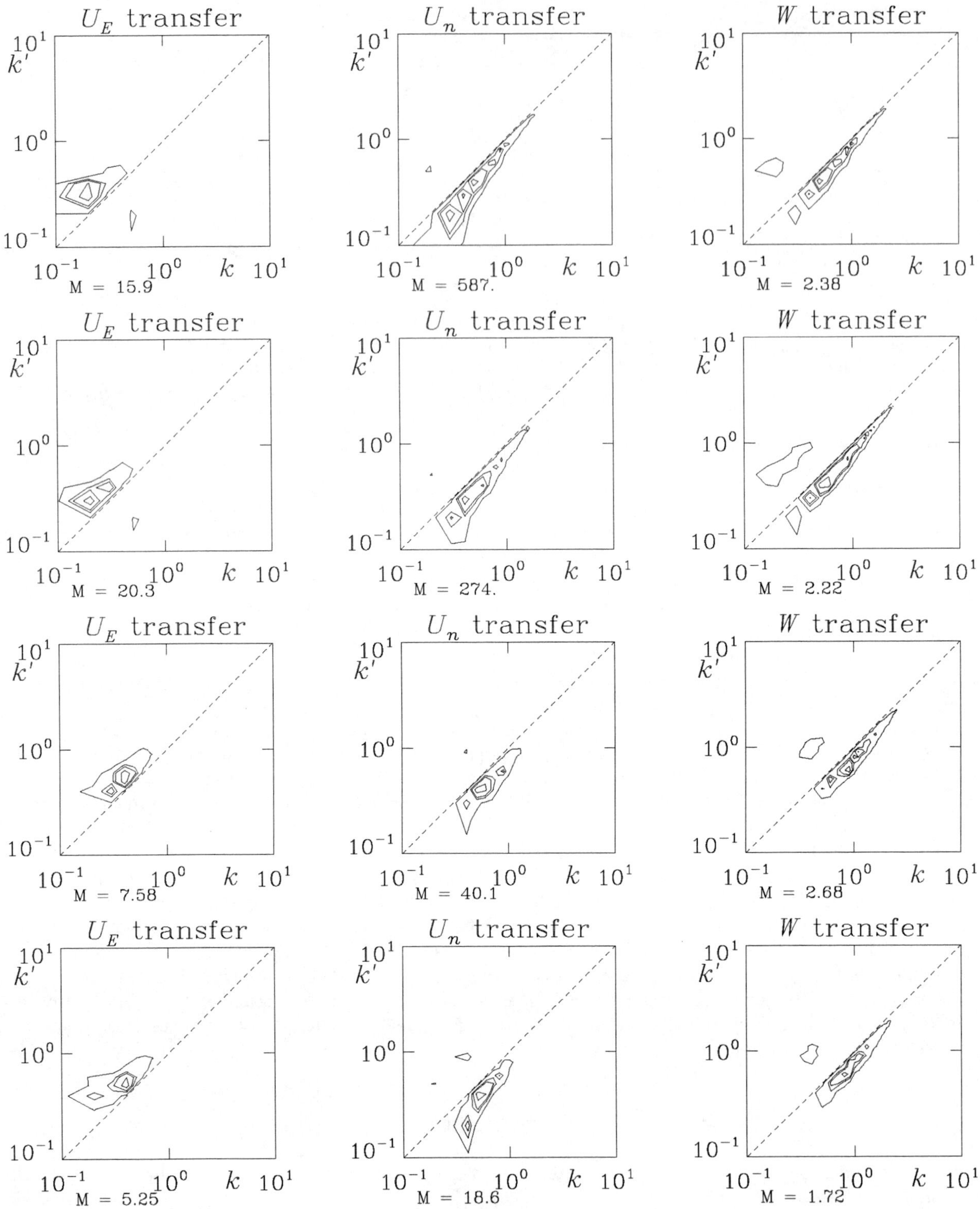

FIGURE 7. Energy and enstrophy transfer in the dissipative coupling model, from top to bottom with $D = 0.01$, 0.03, 0.3, and 1.0, respectively. Labelling is as in Fig. 6. These cases serve to show that the same basic transfer dynamics is present for all $D < 1.0$, with only the levels changing.

Three wave interactions transfer energy mostly locally in the spectrum of scales of motion. For most nonlinearities the tendency is to transfer free energy from larger scales towards smaller scales, where there are more accessible states. A spectrally local transfer chain is called a cascade, termed direct for transfer to smaller scales and inverse for transfer to larger scales. A two dimensional, incompressible fluid such as the ExB velocity exhibits a dual cascade due to the simultaneous conservation of energy and enstrophy (mean squared vorticity for an homogeneous fluid), with the enstrophy going to smaller scales and the energy to larger. This dual cascade is a basic consequence of the three wave constraint on the vorticity nonlinearity.

Dissipative coupling reaches limits in which it can be treated as hydrodynamic, with no correlation between the transporting eddies and the transported quantities, or adiabatic, in whose extreme limit the eddy stream function and transported quantity are constrained to be equal. Conventional fluid turbulence in the presence of a thermal gradient is hydrodynamic, and always develops motion at the very largest scales due to the dual cascade. When dissipative coupling is important the dominant scales of motion occur in a range whose lower limit is the drift scale, ρ_s, at which the dynamics loses the self similarity enjoyed by purely hydrodynamic turbulence. In deeply adiabatic dynamics, the direct cascade through the transported quantity is shut off, leaving the vorticity nonlinearity to act alone. Dissipative drift wave turbulence is always somewhere between these two limits, and in all regimes of interest the basic transfer dynamics of each nonlinearity retains its character.

FURTHER READING

Basic textbooks on fluid turbulence include those by U. Frisch, *Turbulence: the Legacy of A.N. Kolomogorov*, (Cambridge, 1995), M. Lesieur, *Turbulence in Fluids*, (3rd ed, Kluwer Academic Publishers, 1997), and A. S. Monin and A. M. Yaglom, *Statistical Fluid Mechanics: Mechanics of Turbulence*, Vol 2 (MIT Press, Cambridge, USA 1975). The standard cascade model is the one by A. N. Komolgorov, in *C. R. (Dokl.) Akad. Sci. URSS* **30** (1941) 301 [*J. Fluid Mech.* **13** (1962) 82], covered in all the basic texts. The dual cascade property of two dimensional turbulence was shown and explained by R. H. Kraichnan, in *Phys. Fl.* **10** (1967) 1417.

MHD turbulence *per se* is covered in a few of the texts on MHD proper, though many of these are mostly concerned with global equilibrium and stability. Recommended in general and for its chapters on MHD turbulence and reconnection in particular is *Nonlinear Magnetohydrodynamics*, by D. Biskamp (Cambridge, 1993).

On gradient driven turbulence in magnetised plasmas the best sources are still papers in the scientific literature, since most of the texts still suffer greatly from an over-reliance upon linear instabilities and their specific properties. The adiabatic and hydrodynamic limits for drift wave turbulence arise from the earliest standard models, developed by A. Hasegawa and H. Mima, in *Phys. Rev. Lett.* **39** (1977) 205 and *Phys. Fl.* **21** (1978) 87, and by A. Hasegawa and M. Wakatani, in *Phys. Rev. Lett.* **50** (1983) 682, and M. Wakatani and A. Hasegawa, in *Phys. Fl.* **27** (1984) 611, respectively. Statistical equilibrium models to investigate the cascade properties analytically were developed by D. Montgomery and L. Turner, in *Phys. Fl.* **23** (1980) 264 for the Hasegawa-Mima model, and by F. Y. Gang, B. D. Scott, and P. H. Diamond, in *Phys. Fl. B* **1** (1989) 1331 for the Hasegawa-Wakatani model, both of which are known by those names. A general review of the self organising property of flows with dual cascades is given by A. Hasegawa, in *Adv. Phys.* **34** (1985) 1. The energy transfer diagnostics presented herein are derived from their use in hydrodynamics by A. J. Domaradzki, in *Phys. Fl.* **31** (1988) 2747, and applied to the dissipative coupling model in the papers by S. Camargo, D. Biskamp, and B. Scott, in *Phys. Plasmas* **2** (1995) 48, and *Phys. Plasmas* **3** (1996) 3912.

Nonlinear Evolution of Pressure Gradient Driven Modes and Anomalous Transport in Plasmas

Satoshi Hamaguchi

Center for Atomic and Molecular Technologies, Graduate School of Engineering, Osaka University
2-1 Yamadaoka, Suita, Osaka 565-0871, Japan.

Abstract. Pressure gradients, or more widely, density gradients and/or temperature gradients can be free energy sources of various instabilities in a magnetically confined high-temperature plasmas. In the framework of magnetohydrodynamics (MHD), such an instability is essentially the same as Rayleigh Taylor instability that leads to Rayleigh Bénard convection of a charge-neutral fluid in a gravitational field. In this article, nonlinear evolution of such instabilities and transition mechanisms to fully developed turbulence are discussed.

Keywords: MHD, Resistive Interchange Instability, Rayleigh-Taylor Instability, Rayleigh Bénard Convection,
PACS: 52.30.Cv, 52.35.Py, 52.55.Tn

INTRODUCTION

In the presence of a strong and almost uniform magnetic field (such as the toroidal field in a tokamak), the plasma spreads rapidly along the magnetic field and becomes nearly uniform in the direction of each magnetic field line. This is, as is well known, due to the fact that gyrating motion of electrons and ions around each magnetic field line almost limits their transverse motion across the field lines in the fast time scale. On the other hand, in the slow time scale, since the plasma is almost uniform along each field line, it looks as if the dynamics of plasmas is almost limited to two dimensional, i.e., only across the field lines. The presence of the strong magnetic field, in this way, sometimes makes a problem of magnetohydrodynamics (MHD) a little easier to solve, to the lowest order approximation, than its counterpart of neutral fluid dynamics, the solution to which often requires fully three dimensional analyses.

In fluid dynamics, however, low dimensional (e.g., two dimensional) analyses for various fluid instabilities have been well developed. Due to the relative ease of analyses and also possibility of obtaining the exact analytical solutions to the problem, such low dimensional analyses often offer great insight into physics underlying complex phenomena observed in real fluids.

In this article, we consider an instability of a magnetized plasma driven by its own pressure gradient in the framework of MHD. It will be shown that the pressure driven MHD instability is similar to an instability of a charge neutral fluid in a gravitational field under some particular conditions. Such an analogy to neutral fluid dynamics may help one to better understand the physical mechanism of the MHD instability in a more intuitive manner despite the fact that MHD equations are far more complex than Navier-Stokes equations for neutral fluids. We shall also extend the linear stability analyses to a weakly nonlinear regime and see how such instabilities can possibly evolve into fully developed turbulence.

In the next section, we briefly summarize Rayleigh-Bénard convection [1,2] of a neutral fluid in a gravitation field. Rayleigh-Bénard convection is a motion of a fluid arising from Rayleigh-Taylor instability [1,2], which drives a lighter fluid near the heated bottom to move upward and replace heavier fluid on the top. In the third section, we introduce a pressure driven MHD instability and compare it with Rayleigh-Bénard convection of a neutral fluid. In the final section, we consider nonlinear evolution of the pressure driven instability and how turbulence can be generated from such instabilities.

CP1013, *Turbulent Transport in Fusion Plasmas, First ITER International Summer School*
edited by S. Benkadda

RAYLEIGH-BÉNARD CONVECTION

The dynamical equations for a charge neutral fluid are given by the following Navier-Stokes equations [1].

$$\frac{\partial \rho}{\partial t}+\nabla\bullet(\rho\mathbf{v})=0 \tag{1}$$

$$\rho\left(\frac{\partial \mathbf{v}}{\partial t}+\mathbf{v}\bullet\nabla\mathbf{v}\right)=-\nabla p-\rho g\hat{\mathbf{z}}+\nu\Delta\mathbf{v} \tag{2}$$

$$\frac{\partial T}{\partial t}+\mathbf{v}\bullet\nabla T+(\gamma-1)T\nabla\bullet\mathbf{v}=\kappa\Delta T \tag{3}$$

$$p=p(\rho,T) \tag{4}$$

Here we have assumed that the fluid is in a uniform gravitational field in the negative z direction. The notation follows convention for standard fluid dynamics; ρ, $\mathbf{v}$, p, and T represent the mass density, velocity, pressure, and temperature of the fluid. The parameters ν, κ, γ, and g are the viscosity, heat conductivity, ratio of specific heats, and gravitational acceleration. The unit vector in the vertical direction (i.e., z direction) is denoted by $\hat{\mathbf{z}}$. Equation (4) represents the equation of state.

The geometry we consider here is given in Fig. 1. The fluid is confined in the region $0\leq z\leq d$ and its temperature is maintained at $T=T_0$ and T_1 at $z=0$ and d planes. If the bottom temperature T_0 is higher than the top temperature T_1, the fluid heated at the bottom becomes expanded and lighter, moves upward due to the buoyancy force, and thus the Rayleigh-Bénard convection occurs.

It is customary to use Boussinesq approximation [3] to analyze Rayleigh-Bénard convection. Under Boussinesq approximation, Eqs. (1)-(4) may be written as

$$\nabla\bullet\mathbf{v}=0 \tag{5}$$

$$\rho_0\left(\frac{\partial \mathbf{v}}{\partial t}+\mathbf{v}\bullet\nabla\mathbf{v}\right)=-\nabla p-\rho g\hat{\mathbf{z}}+\nu\Delta\mathbf{v} \tag{6}$$

$$\rho(T)=\rho_0\left[1-\alpha(T-T_0)\right] \tag{7}$$

$$\frac{\partial T}{\partial t}+\mathbf{v}\bullet\nabla T=\kappa\Delta T \tag{8}$$

Here it is assumed that the fluid is essentially incompressible except that weak temperature dependence of the density is taken into account.

Equilibrium solutions are those with $\mathbf{v}$=0 and $\partial/\partial t=0$ in the above equations. Here all equilibrium quantities are assumed to depend only on z. Writing an equilibrium pressure, density, and temperature as $\bar{p}$, $\bar{\rho}$, and $\bar{T}$, we assume that, in equilibrium, the temperature depends linearly on z as

$$\bar{T}=T_0-\beta z\ .$$

Here the temperature gradient is given by $\beta=(T_0-T_1)/d$. Then we have equilibrium solutions

$$\bar{p}=p_0-\rho_0 gz\left(1+\frac{\alpha\beta}{2}z\right) \quad \text{and} \quad \bar{\rho}=\rho_0(1+\alpha\beta z).$$

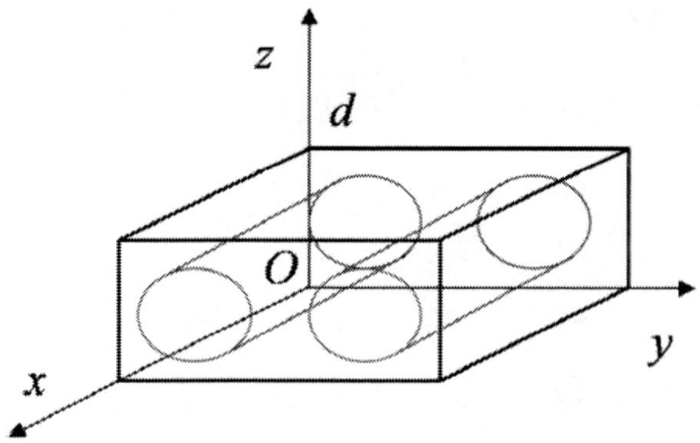

FIGURE 1. Geometry for Rayleigh-Bénard convection.

To consider infinitesimal motion of the fluid from the equilibrium, we set $\theta = T - \overline{T}(z)$ and $\tilde{p} = p - \overline{p}(z)$. Then the equations for the perturbed quantities $\mathbf{v}$, $\tilde{p}$ and θ derived from Eqs. (5)-(8) are then given by

$$\nabla \bullet \mathbf{v} = 0$$
$$\frac{\partial \mathbf{v}}{\partial t} + \mathbf{v} \bullet \nabla \mathbf{v} = -\frac{\nabla \tilde{p}}{\rho_0} + g\alpha\theta\hat{\mathbf{z}} + \nu\Delta\mathbf{v}$$
$$\frac{\partial \theta}{\partial t} + \mathbf{v} \bullet \nabla \theta = \beta w + \kappa\Delta\theta$$
$$w = \mathbf{v} \bullet \hat{\mathbf{z}} \quad ,$$

where we have written $\nu/\rho_0 \to \nu$. The boundary conditions for θ are $\theta(t,x,y,0) = \theta(t,x,y,d) = 0$.

With the following transformation

$$\mathbf{v} \to \frac{\nu}{d}\mathbf{v}\,,\ \theta \to \frac{\beta d \nu}{\kappa}\theta\,,\ \tilde{p} \to \frac{\rho_0 \nu^2}{d^2} p\,,\ t \to \frac{d^2}{\nu} t\,,\ \text{and}\ \ (x,y,z) \to d(x,y,z)\,,$$

we further reduce the number of parameters and obtain

$$\frac{\partial \mathbf{v}}{\partial t} + \mathbf{v} \bullet \nabla \mathbf{v} = -\nabla p + R\theta\hat{\mathbf{z}} + \Delta\mathbf{v} \tag{9}$$
$$\frac{\partial \theta}{\partial t} + \mathbf{v} \bullet \nabla \theta = \frac{1}{P} w + \frac{1}{P}\Delta\theta \tag{10}$$
$$\nabla \bullet \mathbf{v} = 0 \tag{11}$$

where the parameters R and P represent Rayleigh and Prandtl numbers defined by $R = g\alpha\beta d^4/\kappa\nu$ and $P = \nu/\kappa$.

Since Rayleigh-Bénard convection cells of general interest have three dimensional structures, Eqs. (9)-(10) have been widely analyzed in three dimensional space [1,2]. However, the goal of this section is to point out the similarity between Rayleigh-Bénard convection and pressure gradient MHD instability under strong equilibrium magnetic field, we now further simplify the equations, limiting the dynamics of the fluid in the *yz* plane. For such a flow, one can introduce the stream function ϕ as

$$\mathbf{v} = \nabla\phi(y,z) \times \hat{\mathbf{x}} = \left(0, \frac{\partial \phi}{\partial z}, -\frac{\partial \phi}{\partial y}\right).$$

Then Eqs. (9)-(11) can be further reduced to a pair of scalar equations

$$\frac{d}{dt}\Delta\phi = -R\frac{\partial\theta}{\partial y} + \Delta^2\phi \tag{12}$$

$$\frac{d\theta}{dt} = -\frac{1}{P}\frac{\partial\phi}{\partial y} + \frac{1}{P}\Delta\theta \tag{13}$$

with

$$\frac{d}{dt} = \frac{\partial}{\partial t} + \{\phi,\ \} \text{ with } \{f,g\} = (\nabla g \times \nabla f)\bullet\hat{\mathbf{z}} = \frac{\partial f}{\partial x}\frac{\partial g}{\partial y} - \frac{\partial g}{\partial x}\frac{\partial f}{\partial y}$$

As we shall see later, Eqs. (12) and (13) are similar to the equations of a pressure gradient driven MHD instability called resistive interchange instability. In this article, we do not further pursue the analyses of Rayleigh-Bénard convection. Instead we proceed to discuss the MHD instabilities.

RESISTIVE INTERCHANGE MODE

Now we turn to resistive MHD equations, which are given by

$$\frac{\partial\rho}{\partial t} + \nabla\bullet(\rho\mathbf{v}) = 0 \tag{14}$$

$$\rho\left(\frac{\partial\mathbf{v}}{\partial t} + \mathbf{v}\bullet\nabla\mathbf{v}\right) = -\nabla p + \mathbf{j}\times\mathbf{B} + \nu\Delta\mathbf{v} \tag{15}$$

$$\frac{\partial p}{\partial t} + \mathbf{v}\bullet\nabla p + \gamma p\nabla\bullet\mathbf{v} = \kappa\Delta p \tag{16}$$

$$\frac{\partial\mathbf{B}}{\partial t} = \nabla\times(\mathbf{v}\times\mathbf{B} - \eta\mathbf{j}) \tag{17}$$

$$\mu_0\mathbf{j} = \nabla\times\mathbf{B} \qquad \nabla\bullet\mathbf{B} = 0 \qquad p = p(\rho, T) \tag{18}$$

Fluid quantities are represented by the same symbols as those used in Eqs. (1)-(4). Electromagnetic quantities are also given by the standard notation: $\mathbf{B}$, $\mathbf{j}$, η, and μ_0 represent the magnetic field (magnetic flux density), current density, resistivity, and permeability of vacuum. In this article, we focus on instabilities driven by a pressure gradient. Assuming that the modes are localized along mean magnetic field lines, we apply the scaling of the resistive fast interchange mode of Ref. [4]. Denoting the mean quantities by subscript 0 and perturbed quantities by subscript 1, one may write the perturbed magnetic field $\mathbf{B}_1$, current $\mathbf{j}_1$, velocity $\mathbf{v}_1$, and pressure p_1 as

$$\mathbf{B}_1 = \nabla_\perp A\times\mathbf{b} - p_1\mathbf{b}$$

$$\mathbf{j}_1 = -(\nabla_\perp^2 A)\mathbf{b} - \nabla_\perp p_1\times\mathbf{b}$$

$$\mathbf{v}_1 = \nabla_\perp\phi\times\mathbf{b} - v\mathbf{b}$$

where

$$\nabla_\perp \equiv \nabla - \mathbf{b}(\mathbf{B}_0\nabla\bullet) \qquad \text{with} \qquad \mathbf{b} = \mathbf{B}_0/|\mathbf{B}_0|^2.$$

Here the scalar dependent variables A and ϕ have been introduced. The parallel velocity is denoted by $-\nu$. In cylindrical geometry with r, θ, and z denoting the radial distance, poloidal angle, and the longitudinal distance, respectively, we introduce new variables [5,6,7]

$$x = |\sigma|^{1/2}\left[(r - r_0)/r_0\right] \tag{19}$$

$$y = (B_\theta / r_0 B)|\sigma|^{1/2}\left[z - \mu(r)\theta\right] \tag{20}$$

$$\tilde{\theta} = |\sigma|\theta \tag{21}$$

$$\tau = \left(B_\theta |\sigma| / r_0 \sqrt{\rho_0}\right)t \tag{22}$$

$$\mu(r) = \frac{rB_z}{B_\theta} \quad \sigma = \frac{B_\theta}{B}\frac{d\mu}{dr}$$

with $B = |\mathbf{B}_0|$. The radius r_0 indicating a specific position in the vicinity of which we consider the growth of instabilities. The azimuthal (poloidal) and longitudinal components of the mean magnetic field $\mathbf{B}_0$ are denoted by B_θ and B_z. In Eqs. (19)-(22), all mean quantities are evaluated at $r = r_0$. Note that the direction of variable $\tilde{\theta}$ (with fixed x and y) is the direction along the equilibrium magnetic field line $\mathbf{B}_0(r)$. Equations (14)-(17) are then reduced to

$$\frac{d\tilde{A}}{d\tau} = \frac{\partial\tilde{\phi}}{\partial\tilde{\theta}} + \frac{1}{R}\Delta_\perp\tilde{A} \tag{23}$$

$$\frac{d}{d\tau}\Delta_\perp\tilde{\phi} = \frac{\partial}{\partial\tilde{\theta}}\Delta_\perp\tilde{A} + \left\{\tilde{A}, \Delta_\perp\tilde{A}\right\} - \frac{\partial\tilde{p}}{\partial y} + M\Delta_\perp^2\tilde{\phi} \tag{24}$$

$$\frac{d\tilde{p}}{d\tau} = -D\frac{\partial\tilde{\phi}}{\partial y} + \chi\Delta_\perp\tilde{p} \tag{25}$$

with $\tilde{A} = (1/r_0 B_\theta B)A$, $\tilde{\phi} = \left(\sqrt{\rho_0}/r_0 B_\theta B\right)\phi$, and $\tilde{p} = \left(2/B^2|\sigma|^{3/2}\right)p_1$. The parameters are given by

$$D = -\frac{2rp_0'}{B^2\sigma^2}, \quad R = \frac{r_0 B_\theta}{\eta\sqrt{\rho_0}}, \quad M = \frac{\nu}{r_0 B_\theta\sqrt{\rho_0}} \quad \text{and} \quad \chi = \frac{\kappa\sqrt{\rho_0}}{r_0 B_\theta},$$

with the prime denoting the derivative in r evaluated at $r = r_0$. The differential operators are

$$\frac{d}{d\tau} = \frac{\partial}{\partial\tau} + \{\phi, \ \},$$

$$\{f, g\} = (\nabla g \times \nabla f)\cdot\frac{\mathbf{B}_0}{B} = \frac{\partial f}{\partial x}\frac{\partial g}{\partial y} - \frac{\partial g}{\partial x}\frac{\partial f}{\partial y},$$

and

$$\Delta_\perp = \left(\frac{\partial}{\partial x} - \frac{\sigma}{|\sigma|}\tilde{\theta}\frac{\partial}{\partial y}\right)^2 + \frac{\partial^2}{\partial y^2}.$$

Assuming that instability is highly localized at the rational surface at $x=0$, we set the domain on which the solutions to Eqs. (23)-(25) as $-L_x/2 \le x \le L_x/2$, $-L_y/2 \le y \le L_y/2$, and $0 \le z \le L_z$, with some positive numbers L_x, L_y, and L_z.

In the electrostatic limit, one can set $d\tilde{A}/d\tau = 0$, and therefore $\partial\tilde{\phi}/\partial\tilde{\theta} = -(1/R)\Delta_\perp \tilde{A}$ from Eq. (3-15), and also ignore the nonlinear term $\{\tilde{A}, \Delta_\perp \tilde{A}\}$. Then Eqs. (23)-(25) become

$$\frac{d}{d\tau}\Delta_\perp \tilde{\phi} = -R\frac{\partial^2}{\partial\tilde{\theta}^2}\tilde{\phi} - \frac{\partial\tilde{p}}{\partial y} + M\Delta_\perp^2 \tilde{\phi} \tag{26}$$

$$\frac{d\tilde{p}}{d\tau} = -D\frac{\partial\tilde{\phi}}{\partial y} + \chi\Delta_\perp \tilde{p} \tag{27}$$

If there is no magnetic shear and all perturbed quantities are uniform along the field line, i.e., $\partial/\partial\tilde{\theta} = 0$, then the set of electrostatic Eqs. (26) and (27) becomes exactly the same as Eqs. (12) and (13). In this case, physical mechanism is also essentially the same except that the role of gravity in Eq. (6) is played by part of the $\mathbf{J} \times \mathbf{B}$ force.

Now we return to Eqs. (23)-(25) for electromagnetic resistive interchange modes. In this article, it is assumed that the ideal stability criterion $D \le 1/4$ (Suydam's criterion [8]) is satisfied and therefore instabilities arise only from the finite diffusivity.

If the mean pressure gradient p_0' becomes larger than the critical value p_{0L}' (the marginal stability limit), (or, in terms of the dimensionless parameter, $D \ge D_L$ with D_L being D at $p_0' = p_{0L}'$), the equilibrium specified by $\mathbf{B}_0$, p_0, ρ_0 and $\mathbf{v}_0 = 0$ become unstable and modes on the rational surface grow at $r = r_0$. If the pressure gradient is only slightly larger than the marginal stability limit, the unstable modes saturate at small amplitudes and form steady convection cells. Such convective flows can transport heat, in addition to the heat transport by ordinary conduction. The heat (or pressure) transported in the radial direction by such flows can be expressed as

$$\langle \tilde{p}\tilde{v}_r \rangle \propto \left\langle p\frac{\partial\phi}{\partial y} \right\rangle$$

where the average is defined by

$$\langle f \rangle = \frac{1}{L_x L_y L_z}\int dxdydz\, f(x,y,z,t)$$

with the spatial integral being taken over the entire domain $L_x \times L_y \times L_z$. The ratio of total heat transport (including that by convection) to the thermal conduction in the radial direction may then be expressed by the Nusselt number

$$Nu = 1 + \frac{\langle pv_r \rangle}{\kappa p_0''} \tag{28}$$

Figure 3(a) shows the time evolution of the Nusselt number evaluated from the solutions to Eqs. (23)-(25) when the pressure gradient parameter D = 0.2 [7]. Contours of the function ϕ, which is the stream function for the velocity perpendicular to the equilibrium magnetic field $\mathbf{B}_0$, are plotted in Fig. 3(b), which shows that steady state convection cells are formed on the rational surface x = 0. Although the results shown here are two dimensional solutions, full three dimensional simulations also provide similar results, as will be explained more in detail in the following section. This is what is known as quasi-linear saturation, where through nonlinear interaction, the $\tilde{\theta}$ and z independent component of a pressure perturbation is generated in such a way that the total pressure profile near the rational surface becomes flat, eliminating the free energy source for further growth of all perturbations.

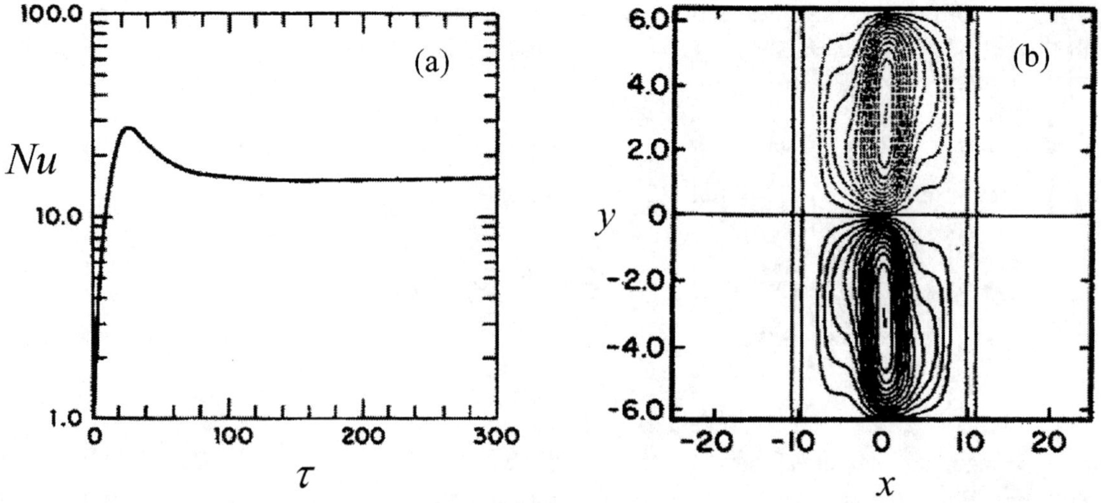

FIGURE 2. Formation of convection cells by a resistive interchange mode obtained from numerical simulations of Eqs. (23)-(25). (a) Time evolution of Nusselt number Nu and (b) contours of the stream function ϕ at τ =300. The parameters used for the simulations are D=0.2, R=M=1, and χ =0.01. [7]

The saturation level depends on the equilibrium pressure gradient p_0' . If the initial pressure gradient is that for the marginal stability limit (i.e., $D = D_L$), then there is no perturbation growth and therefore the saturation level is zero. If the initial pressure gradient is larger than the marginal stability limit, i.e., $D > D_L$, then the saturation level is essentially becomes a function of the difference $D - D_L$ if the difference is sufficiently small. In the following session, we examine the level of saturation as a function of the given equilibrium pressure gradient p_0' , i.e., the pressure gradient parameter D.

NONLINEAR ANALYSIS AND BIFURCATION

Now we evaluate the level of saturation by asymptotic expansion. Suppose that the magnitude of perturbed quantities is proportional to ε defined by

$$\varepsilon^2 = \frac{1}{2}\left\langle \left|\nabla_\perp \tilde{\phi}\right|^2 \right\rangle$$

$$\tilde{A} = \varepsilon \hat{A} \quad \tilde{\phi} = \varepsilon \hat{\phi} \quad \tilde{p} = \varepsilon \hat{p} \quad .$$

When the system is close to the marginally stable state, the parameter ε may be considered as a small parameter, with which all the physical variables may be expanded by Taylor series. We also assume that the system remains to be two dimensional as long as the saturation level is small and consider nonlinear interactions among the same helicity (i.e., modes arising on the same rational surface).

To make the analysis easier to handle, we transform the independent variables as $y + \tilde{\theta}x \to y$ and $\tilde{\theta} \to z$, so that we can transform $\partial/\partial\tilde{\theta} \to x\partial/\partial y + \partial/\partial z$ and $\Delta_\perp \to \partial^2/\partial x^2 + \partial^2/\partial y^2$. In the new coordinates (x,y,z), the independent variables x, y, and z represent the radial distance (as before), the distance in the poloidal (i.e., azimuthal) direction, and the distance in the longitudinal direction at a specific mode rational surface in the plasma cylinder, respectively. The equilibrium magnetic field $\mathbf{B}_0$ is in the z direction only at $x = 0$ in the new coordinate system and has a non-normal angle relative to the xy plane if $x \neq 0$. Especially in two dimensions, we may set

$$\frac{\partial}{\partial\tilde{\theta}}=\frac{\partial}{\partial z}+x\frac{\partial}{\partial y} \quad\rightarrow\quad x\frac{\partial}{\partial y}.$$

At saturation ($\partial/\partial\tau=0$), Eqs. (23)-(25) may be written with the new independent variables above as

$$\varepsilon\left\{\hat{\phi},\hat{A}\right\}=\tilde{x}\frac{\partial\hat{\phi}}{\partial\tilde{y}}+\frac{1}{R}\Delta_{\perp}\hat{A} \tag{29}$$

$$\varepsilon\left\{\hat{\phi},\Delta_{\perp}\hat{\phi}\right\}=\tilde{x}\frac{\partial}{\partial\tilde{y}}\Delta_{\perp}\hat{A}+\varepsilon\left\{\hat{A},\Delta_{\perp}\hat{A}\right\}-\frac{\partial\hat{p}}{\partial y}+M\Delta_{\perp}^{2}\hat{\phi} \tag{30}$$

$$\varepsilon\left\{\hat{\phi},\hat{p}\right\}=-D(\varepsilon)\frac{\partial\hat{\phi}}{\partial y}+\chi\Delta_{\perp}\hat{p} \tag{31}$$

Here we have explicitly stated that the amplitude ε and D are related, i.e., $D=D(\varepsilon)$. Equations (23)-(25) [and also Eqs. (29)-(31)] are nonlinear equations that always admit the trivial solutions $\hat{A}=\hat{\phi}=\hat{p}=0$. These trivial solutions are equilibrium solutions, which may or may not be stable. From linear analysis, we know that if $D < D_L$, the trivial solutions are stable in the sense that any perturbation given to the system will decay to zero exponentially. At $D = D_L$, Eqs. (29)-(31) admit non-trivial (i.e., nonzero) solutions $\hat{A}$, $\hat{\phi}$, and $\hat{p}$ with $\varepsilon=0$. Now we assume that, for a given D larger than the marginal stability limit D_L, there is a set of non-trivial solutions with nonzero ε. Such solutions may be considered as equilibrium solutions to the original nonlinear Eqs. (23)-(25) branched out from the trivial solutions that have become unstable. These equilibrium solutions with $\varepsilon\neq 0$ are called bifurcated solutions. Since the saturation level ε of the bifurcated solutions is controlled by the value of $D=D(\varepsilon)$ in the case of Eqs. (29)-(31) (if the other parameters are fixed) , D is called the bifurcation parameter of the system. This is similar to what we saw for Rayleigh-Bénard convection in fluid dynamics in the earlier section.

To solve Eqs. (29)-(31) perturbatively, we further expand the dependent variables, together with the parameter D, in ε in the following manner.

$$\hat{A}=\sum_{n=0}^{\infty}\hat{A}_n\varepsilon^n,\quad \hat{\phi}=\sum_{n=0}^{\infty}\hat{\phi}_n\varepsilon^n,\quad \hat{p}=\sum_{n=0}^{\infty}\hat{p}_n\varepsilon^n,\quad D=D_L+\sum_{n=0}^{\infty}D_n\varepsilon^n$$

By substituting these variables into Eqs. (29)-(31), we obtain the following equations to the lowest order.

$$\tilde{x}\frac{\partial\hat{\phi}_0}{\partial\tilde{y}}+\frac{1}{R}\Delta_{\perp}\hat{A}_0=0$$

$$\tilde{x}\frac{\partial}{\partial\tilde{y}}\Delta_{\perp}\hat{A}_0-\frac{\partial\hat{p}_0}{\partial y}+M\Delta_{\perp}^{2}\hat{\phi}_0=0$$

$$-D_L\frac{\partial\hat{\phi}_0}{\partial y}+\chi\Delta_{\perp}\hat{p}_0=0$$

The solutions to these equations represent nothing more than the marginal stability solutions. By eliminating A_0 from the above expressions, we can write

$$L\begin{pmatrix}\hat{\phi}_0\\ \hat{p}_0\end{pmatrix}=\begin{pmatrix}0\\ 0\end{pmatrix} \tag{32}$$

with the differential operator

$$L \equiv \begin{pmatrix} -R\tilde{x}^2 \partial^2 / \partial\tilde{y}^2 + M\Delta_\perp^2, & -\partial/\partial y \\ \partial/\partial\tilde{y}, & -(\chi/D_L)\Delta_\perp \end{pmatrix} \tag{33}$$

Note that the operator L is self-adjoint, i.e., $L = L^+$ (acing on a class of functions that satisfy reasonable boundary conditions.) The condition that the operator L is singular determines the value of D_L.

Due to the periodic boundary conditions in y, one may consider the normal mode solution in y for the sake of simplicity and assume that the solutions of Eq. (32) are given in the forms of

$$\hat{p}_0 = \hat{p}_{01}(x)\cos k\tilde{y} \quad \text{and} \quad \hat{\phi}_0 = \hat{\phi}_{01}(x)\sin k\tilde{y} \tag{34}$$

Then Eqs. (32) and (33) may be also simplified as

$$L_k \begin{pmatrix} \hat{\phi}_{01} \\ \hat{p}_{01} \end{pmatrix} = \begin{pmatrix} 0 \\ 0 \end{pmatrix} \quad \text{with} \quad L_k = \begin{pmatrix} Rk^2\tilde{x}^2 + M\left(\partial_x^2 - k^2\right)^2 & k \\ k & -(\chi/D_L)\left(\partial_x^2 - k^2\right) \end{pmatrix}. \tag{35}$$

Note that the operator L_k is self-adjoint ($L_k = L_k^+$) and singular.

To determine D_1, we proceed to the first order equations obtained from Eqs. (29)-(31), i.e.,

$$L\begin{pmatrix} \hat{\phi}_1 \\ \hat{p}_1 \end{pmatrix} = \begin{pmatrix} \left\{\hat{\phi}_0, \Delta_\perp \hat{\phi}_0\right\} - R\tilde{x}\left\{\hat{\phi}_0, \hat{A}_0\right\} - \left\{\hat{A}_0, \Delta_\perp \hat{A}_0\right\} \\ -\left(\left\{\hat{\phi}_0, \hat{p}_0\right\} + D_1 \partial\hat{\phi}_0/\partial y\right)/D_L \end{pmatrix} \tag{36}$$

$$\left\langle \nabla_\perp \hat{\phi}_0 \bullet \nabla_\perp \hat{\phi}_1 \right\rangle = 0$$

$$\Delta_\perp \hat{A}_1 = -R\tilde{x}\,\partial\hat{\phi}_1/\partial y + R\left\{\hat{\phi}_0, \hat{A}_0\right\}$$

For the zero-th order (i.e., marginally stable) solutions of Eq. (34), the first order solutions can be also expressed in Fourier series in y as $\hat{A}_1 = \sum_m \hat{A}_{1m} \cos mk\tilde{y}$, $\hat{\phi}_1 = \sum_m \hat{\phi}_{1m} \sin mk\tilde{y}$, and $\hat{p}_1 = \sum_m \hat{p}_{1m} \cos mk\tilde{y}$. It is clear that nontrivial solutions can exist only for the terms with $m = 0$, 1, and 2.

Especially for $m = 1$, Eq. (36) becomes

$$L_k \begin{pmatrix} \hat{\phi}_{11} \\ \hat{p}_{11} \end{pmatrix} = \begin{pmatrix} 0 \\ -k\, D_1 / D_L \hat{\phi}_{01}(\tilde{x}) \end{pmatrix} \tag{37}$$

with L_k being defined in Eq. (35). Since the operator L_k is singular, for Eq. (37) to be solved, its right-hand side (i.e., inhomogeneous term) has to satisfy the solvability condition [9]; it must be orthogonal to all solutions $\mathbf{u}$ of the homogeneous equation for the adjoint operator L_k^+, i.e., $L_k^+\mathbf{u} = \mathbf{0}$. Since L_k is self-adjoint ($L_k = L_k^+$) and the zero-th order solution $\mathbf{v} = (\hat{\phi}_{01}\sin k\tilde{y}, \hat{p}_{01}\cos k\tilde{y})^t$ (where the superscript t denotes the transpose of the vector, as usual) satisfies $L_k\mathbf{v} = \mathbf{0}$, the parameter D_1 needs to satisfy the condition

$$-k\frac{D_1}{D_L}\left\langle \hat{p}_{01}\hat{\phi}_{01}\right\rangle = 0 \quad , \tag{38}$$

which is derived from the condition that the inner product of the right-hand side of Eq. (37) and the zero-th order solution **v** be zero. This is a necessary condition for Eq. (37) to be solved. A mathematical proof for sufficiency is not discussed here, but the interested reader is referred to, e.g., Ref. [9], for general theory for Fredholm operators and the Fredholm alternative.

From the second equation of Eq. (35), i.e.,

$$k\hat{\phi}_{01} - \frac{\chi}{D_L}\left(\hat{p}''_{01} - k^2\hat{p}_{01}\right) = 0$$

one can easily show that

$$k\left\langle \hat{\phi}_{01}\hat{p}_{01}\right\rangle = -\frac{\chi}{D_L}\left(\left\langle \hat{p}'^{\,2}_{01}\right\rangle + k^2\left\langle \hat{p}^2_{01}\right\rangle\right) \tag{39}$$

This relation indicates that the term $\left\langle \hat{\phi}_{01}\hat{p}_{01}\right\rangle$ is nonzero for nonzero function $\hat{p}_{01}$. Therefore, from Eqs. (38) and (39), we obtain D_1=0.

Similarly one can proceed to the second order equations to determine D_2. We refer the reader to Ref. [7] for details. The parameter D_2 is related to the zero-th and first order solutions and its value, which can be numerically evaluated from these solutions, has been shown to be nonzero [7]. Therefore we may write

$$D = D_L + D_2\varepsilon^2 + \cdots$$

or, to the lowest order, the saturation level ε and bifurcation parameter D are related as

$$\varepsilon^2 = \frac{D - D_L}{D_2} \quad .$$

The heat flux (pressure flux) by fluid motion (i.e., convection) in the radial direction is then evaluated as

$$\left\langle \tilde{p}\tilde{v}_r\right\rangle \propto \left\langle p\frac{\partial\phi}{\partial y}\right\rangle = k\left\langle \hat{p}_{01}\hat{\phi}_{01}\right\rangle \frac{D - D_L}{D_2} \quad . \tag{40}$$

The convective hear flux (anomalous heat flux) is thus shown to depend linearly on the pressure gradient parameter D.

Figure 4 shows the Nusselt number *Nu* [defined in Eq. (28)] for saturated convections. The empty circles of Fig. 4 are the values of *Nu* evaluated from two dimensional (i.e., *z* independent) nonlinear numerical solutions to Eqs. (23)-(25) at steady state. As the pressure gradient parameter D increases from the marginal stability limit D_L, it is shown that the Nusselt number increases monotonically, indicating that time independent solutions to Eqs. (23)-(25) [i.e., solutions to Eqs. (29)-(31)] indeed exist in the range of D shown here. The solid line indicated as "theory" in Fig. 4 represent the Nusselt number evaluated from the formula (40), where the proportional constant has been evaluated from the zero-th and first order solutions $\hat{A}_n$, $\hat{\phi}_n$, and $\hat{p}_n$ with n=0 and 1 obtained from the perturbation analysis presented in this section. It is seen that the perturbation analysis presented here provides a good approximation to the value of anomalous heat transport when D is close to D_L.

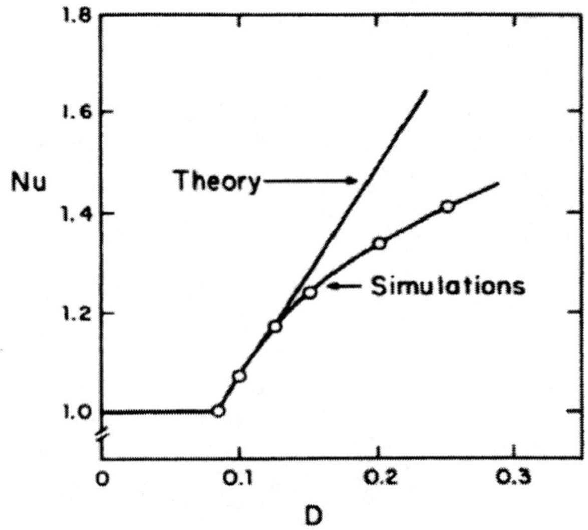

FIGURE 4. Values of Nusselt number *Nu* (empty circles) as a function of the pressure gradient parameter *D*, which are obtained from numerical solutions to Eqs. (23)-(25). The solid line indicated as "theory" is the *Nu* obtained from the perturbation analysis, given in Eq. (40), with the proportional constant also evaluated from the perturbation analyses. The perturbation analysis is shown to provide a good approximation of *Nu* when the parameter *D* is close to the marginal stability limit D_L. The parameters used for the simulations are R=M=1, and χ =0.1. [7]

The numerical simulation used to obtain Fig.4 is two dimensional, where the solutions are assumed to be independent of *z*. For three dimensional simulations, the solutions may be expanded in Fourier series in *y* and *z* as $\exp i(mk_y y + nk_z z)$ with the parameters *m* and *n* indicating the mode numbers. The ratio *m/n* is called helicity.

Note that the *z* dependence appears only in $\partial/\partial\tilde{\theta}$ in Eqs. (23)-(25). For the (*m,n*) Fourier mode solution, it may be written as

$$\frac{\partial}{\partial\tilde{\theta}} = i\left(xmk_y + nk_z\right) = i\left(x + \frac{nk_z}{mk_y}\right)mk_y$$

Therefore by shifting $x \rightarrow x - nk_z/mk_y$, the equations become exactly the same as those for the Fourier mode (*m*,0). This means that the Fourier mode solution of (*m,n*) is exactly the same as that of (*m*,0) except for their locations; the location of the (*m,n*) mode rational surface is at $x = -nk_z/mk_y$. From this argument, it is also clear that all the modes with the same helicity *m/n* are located on the same rational surface.

To study how the three dimensional solutions to Eqs. (23)-(25) evolve, we have solved them numerically. When the pressure gradient parameter *D* is slightly larger than the marginal stability limit D_L, the same convection cells as those we saw in two dimensional simulations appear and reach steady state on all possible rational surface positions specified by *n*, as shown in Fig. 5(a).

As D is increased further, the saturation level of each convection cell increases. If convection cells on different rational surfaces are sufficiently separated and do not interact, then the dependence of the saturation level on parameter *D* is the same as that in the case of two dimensional simulations. However, at a certain level of saturation, the interaction of neighboring convection cells through the nonlinear terms of Eqs. (23)-(25) can no longer be ignored.

For example, the (*m,n*) = (1,0) mode and (*m,n*) = (1,1) mode create the (*m,n*) = (2,1) mode through nonlinear interaction. The (*m,n*) = (2,1) mode is located at $x = -k_z/2k_y$, i.e., just in the middle of the locations of the two neighboring modes (1,1) and (1,0). When *D* becomes larger than a certain value D_{2L}, the (*m,n*) = (2,1) mode becomes linearly unstable and the (*m,n*) = (2,1) modes created by nonlinear interaction grow and they saturate by

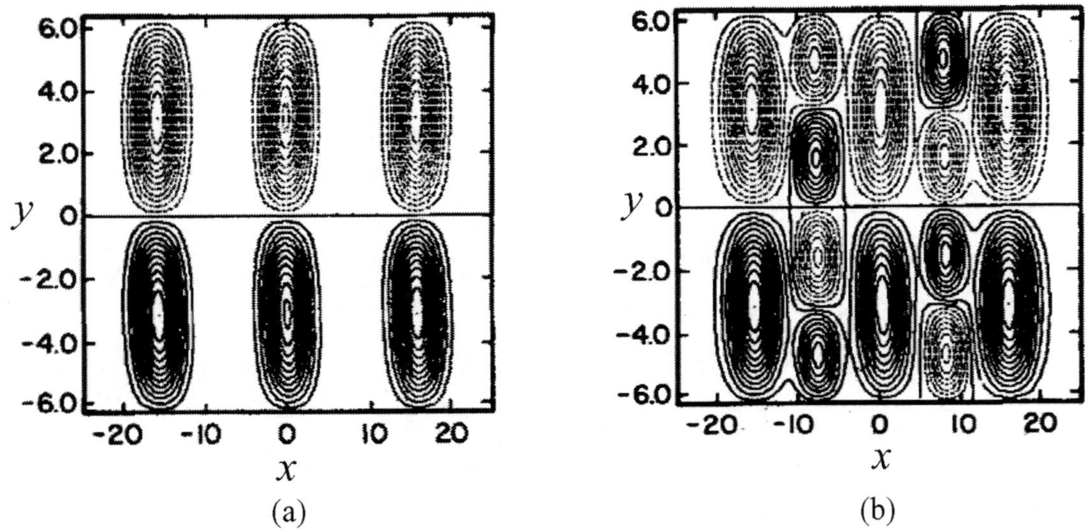

FIGURE 5. Contours of the stream function ϕ at steady state. The parameters used for the simulations are (a) D=0.15 and (b) D= 0.2, with R=M=1, and χ =0.06. [7]

their own quasilinear effects. This is called the secondary bifurcation. Such convection cells are shown in Fig. 5(b). For the parameters used in the simulations shown in Fig. 5, $D_{2L} \simeq 0.18$.

The two dimensional solutions to Eqs. (23)-(25) with single helicity modes and three dimensional solutions of multi helicity modes provide the same Nusselt number value in the parameter range $D_L \leq D \leq D_{2L}$ since only independent steady convection cells of the m = 1 modes contribute to convective heat transport. However, as mentioned above, the Nusselt number evaluated from three dimensional simulation results exhibits a clear secondary bifurcation at $D \geq D_{2L}$, where the m = 2 mode becomes linearly unstable whereas the Nusselt number evaluated from two dimensional simulation results show no drastic change as D exceeds D_{2L}, as shown in Fig. 6. We note here that the sample simulation results shown in Figs. 5 and 6 are under highly diffusive conditions, so that the bifurcations can be clearly seen. For real plasmas, the diffusivity is much lower, so the Nusselt number is typically much larger and higher-order bifurcations are much less clear. .

As the bifurcation parameter D is further increased, ternary and quaternary (and so on) bifurcations also should occur although it is often difficult, if not impossible, to observe clearly such higher-order bifurcations since convection cells produced by higher-order bifurcations (some of them are often oscillatory in time, rather than time independent) quickly mix with one another through nonlinear interactions. It is generally believed that such successive bifurcations of a fluid (or plasma) equilibrium represent the transition to turbulence.

SUMMARY

In this article we have seen that the mathematical structure of resistive interchange mode (resistive g-mode) of a plasma is essentially the same as that of the instability leading to Rayleigh-Bénard convection in a neutral fluid. In both cases, when there is a free energy source for instability (i.e., the pressure gradient for the former and the temperature gradient for the latter), convection cells can grow and saturate at some magnitudes as long as the free energy fed to the system is low. As the free energy provided from the source becomes larger, the system exhibits successive bifurcations, which may lead to the formation of turbulence. Although general theory of turbulence is far beyond the scope of this article, it has been shown that the primary, and sometimes secondary, bifurcations may be studied with some mathematical rigor and those for resistive interchange modes have been discussed here. It has been also shown that the convective heat transport scales as in Eq. (40) when the pressure gradient parameter D is close to the marginal stability limit D_L.

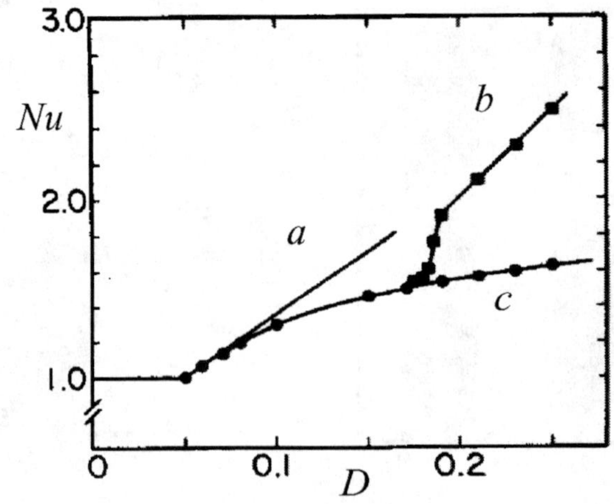

FIGURE 6. The Nusselt numbers as functions of the pressure gradient parameter D obtained from (a) two dimensional solutions of a single helicity (denoted by filled circles), and (c) three dimensional solutions of multi helicities. The parameters used for the simulations are R=M=1, and χ =0.06. [7]

In the perturbation analysis presented in this article, we have assumed that the convection cells have reached steady state and sought time-independent solutions only. In reality, such modes grow in time before reaching the steady state. The analysis presented in this article can be extended to the study of nonlinear time evolution of the solutions by including time dependence of expansion parameter $\varepsilon = \varepsilon(t)$ in

$$\hat{A} = \sum_{n=0}^{\infty} \hat{A}_n \varepsilon^n, \quad \hat{\phi} = \sum_{n=0}^{\infty} \hat{\phi}_n \varepsilon^n, \quad \hat{p} = \sum_{n=0}^{\infty} \hat{p}_n \varepsilon^n .$$

Then one can obtain a Landau type equation [1,10] for time dependence of the amplitude

$$d_0 \frac{d\varepsilon}{dt} + (D - D_L) d_1 \varepsilon + d_2 \varepsilon^3 = 0$$

with some parameters d_0, d_1, and d_2. Details may be found in Refs. [11,12].

REFERENCES

1. P.G. Drazin and W. H. Reid, *Hydrodynamic stability*, Cambridge: Cambridge University Press, 1981, pp 59-62.
2. S. Chandrasekhar, *Hydrodynamic and Hydromagnetic Stability,* New York. Dover, 1981, pp. 9-145.
3. J. Boussinesq, *Théorie analytique de la chaleur*, Paris: Gauthier-Villars, 1903.
4. B. Coppi, J.M. Greene, and J.L. Johnson, Nuclear Fusion **6**, 101-117 (1966).
5. A. Bhattacharjee and E. Hameiri, Phys. Rev. Lett. **57**, 206-209 (1986).
6. E. Hameiri and A. Bhattacharjee, Phys. Fluids **30**, 1743-1755 (1987).
7. S. Hamaguchi, Phys. Fluids B**1**, 1416-1430 (1989).
8. B. R. Suydam, in *Proceedings of the Second International Conference on Peaceful Uses of Atomic Energy* (United Nations, Geneva, 1985), **31**, p.157.
9. B. Friedman, *Principles and Techniques of Applied Mathematics*, New York, John Wiley & Sons, 1966.
10. L. D. Landau and E. M. Lifshitz, *Fluid Mechanics,2nd. Ed.*, London: Pergamon Press, 1987.
11. N. Nakajima, Phys. Fluids B**2**, 1170-1183 (1990).
12. N. Nakajima and S. Hamaguchi, Phys. Fluids B**2**, 1184-1189 (1990).

Tearing Modes in Tokamaks

R. B. White

Plasma Physics Laboratory, Princeton University, P.O.Box 451, Princeton, New Jersey 08543

Abstract. This lecture gives a basic introduction to magnetic £elds, magnetic surface destruction, toroidal equilibrium and tearing modes in a tokamak, including the linear and nonlinear development of these modes and their modi£cation by current drive and bootstrap current, and sawtooth oscillations and disruptions.

Keywords: equilibrium, tearing, magnetic coordinates, sawteeth, disruptions
PACS: 52.55.Fa 52.55.Dy

MAGNETIC CONFIGURATION

Because of the divergence free nature of the magnetic £eld, the simplest topological con£guration it can assume with no £eld lines exiting from a £xed volume is toroidal. Since particle trajectories, to lowest order in gyro radius, follow £eld lines, this is also the simplest con£guration which can give complete con£nement within this approximation.

In addition, because of the very high particle velocity along the £eld lines, the £eld must, at least approximately, possess magnetic surfaces, *i.e.* the £eld lines trace out a nested set of toroidal surfaces, to prevent rapid particle transport from the inside of the torus to the exterior, and thus allow the existence of pressure and temperature gradients.

Hamiltonian character of the magnetic £eld line trajectories

To describe a toroidal magnetic con£guration it is convenient to use coordinates de£ned by the £eld itself. Introduce general coordinates ρ, labeling the ¤ux surfaces, θ, a poloidal angle, and ζ a toroidal angle. Write $\vec{B} = \nabla \times \vec{A}$, since $\nabla \cdot \vec{B} = 0$. Write the vector potential in terms of the coordinates ρ, θ, ζ with $\vec{r}(\rho, \theta, \zeta)$

$$\vec{A} = A_\rho \nabla\rho + A_\theta \nabla\theta + A_\zeta \nabla\zeta. \tag{1}$$

Then, de£ne $\partial_\rho G = A_\rho$. Since $\nabla G = \partial_\rho G \nabla\rho + \partial_\theta G \nabla\theta + \partial_\zeta G \nabla\zeta$ we have

$$\vec{A} = \nabla G + (A_\theta - \partial_\theta G)\nabla\theta + (A_\zeta - \partial_\zeta G)\nabla\zeta \tag{2}$$

or $\vec{A} \equiv \nabla G + \psi\nabla\theta - \psi_p \nabla\zeta$ and thus

$$\vec{B} = \nabla\psi \times \nabla\theta - \nabla\psi_p \times \nabla\zeta. \tag{3}$$

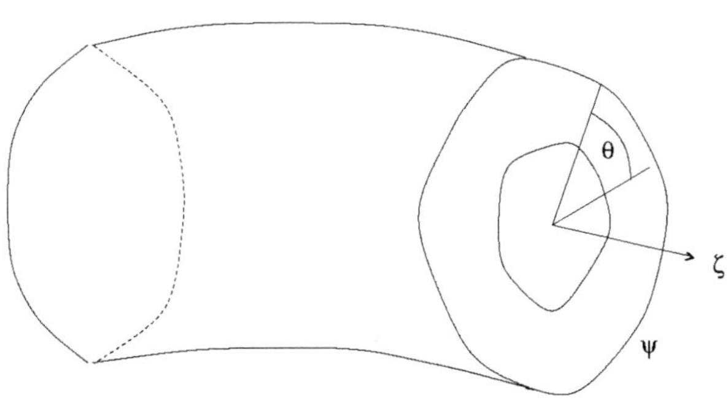

FIGURE 1. General Coordinates.

CP1013, ***Turbulent Transport in Fusion Plasmas, First ITER International Summer School***
edited by S. Benkadda

Now take ψ to be the ¤ux surface coordinate, with Jacobian $\mathcal{J}^{-1} = \nabla\zeta\cdot(\nabla\psi\times\nabla\theta)$. The coordinate system is shown in Fig. 1. The functions ψ and ψ_p are readily identi£ed. The toroidal ¤ux is $(1/2\pi)\int(\vec{B}\cdot\nabla\zeta)\mathcal{J}\,d\zeta d\theta d\psi = 2\pi\psi$, and the poloidal ¤ux is $(1/2\pi)\int(\vec{B}\cdot\nabla\theta)\mathcal{J}\,d\zeta d\theta d\psi = 2\pi\psi_p$, where the integration extends from the magnetic axis to the ¤ux surface ψ.

Using this form for $\vec{B}$ we can easily show that the £eld line trajectories are Hamiltonian in nature. The magnetic £eld lines are de£ned by $d\psi/d\zeta = \vec{B}\cdot\nabla\psi/\vec{B}\cdot\nabla\zeta, \quad d\theta/d\zeta = \vec{B}\cdot\nabla\theta/\vec{B}\cdot\nabla\zeta$ or

$$\frac{d\psi}{d\zeta} = -\frac{\nabla\psi\cdot(\nabla\psi_p\times\nabla\zeta)}{(\nabla\psi\times\nabla\theta)\cdot\nabla\zeta}, \qquad \frac{d\theta}{d\zeta} = -\frac{\nabla\theta\cdot(\nabla\psi_p\times\nabla\zeta)}{(\nabla\psi\times\nabla\theta)\cdot\nabla\zeta}. \tag{4}$$

But the gradient of ψ_p is given by $\nabla\psi_p = \partial_\psi\psi_p\nabla\psi + \partial_\theta\psi_p\nabla\theta + \partial_\zeta\psi_p\nabla\zeta$. Thus

$$\frac{d\psi}{d\zeta} = -\partial_\theta\psi_p \qquad \frac{d\theta}{d\zeta} = \partial_\psi\psi_p. \tag{5}$$

This is of Hamiltonian form with $\psi_p(\psi,\theta,\zeta)$ the Hamiltonian, ψ the momentum, θ the coordinate, and ζ the time.

Surface destruction

A two dimensional surface de£ned by a function $f(\vec{r}) = constant$ is said to be a magnetic surface if at any point the magnetic £eld lies within the surface, ie. $\vec{B}\cdot\nabla f = 0$. According to a theorem by Kolmogorov, small perturbation of a £eld possessing nested surfaces leaves well de£ned magnetic surfaces existing everywhere except in a small volume proportional to the square root of the perturbation, where the £eld assumes an island topology with regions of stochastic character.

The simplest way to guarantee the existence of approximate surfaces is to take the con£nement device to be a small perturbation of a system in which $\vec{B}\cdot\nabla\psi = \vec{B}\cdot\nabla\psi_p = 0$ so that $\psi_p = \psi_p(\psi)$. Shown in Fig. 2 is a poloidal surface, (£xed θ), and a toroidal surface (£xed ζ), each bounded by the magnetic axis and a ¤ux surface de£ned by ψ.

To illustrate the way in which nested toroidal magnetic surfaces are destroyed, introduce a small perturbation of the poloidal ¤ux function, Fourier decompose it in the variables θ, ζ, and consider a single harmonic

$$\psi_p = \int\frac{d\psi}{q} + V cos(n\zeta - m\theta), \tag{6}$$

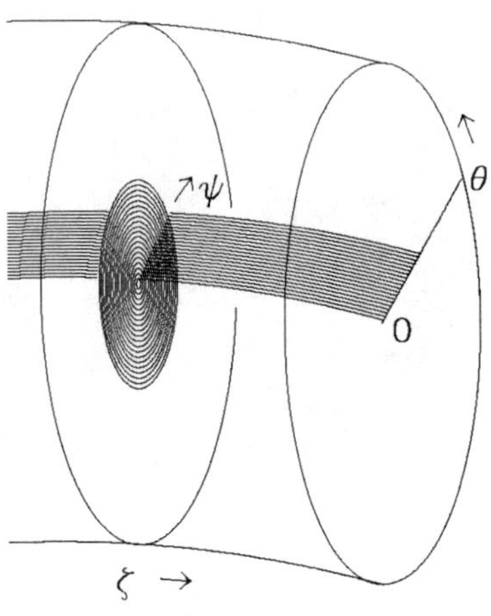

FIGURE 2. Flux surfaces

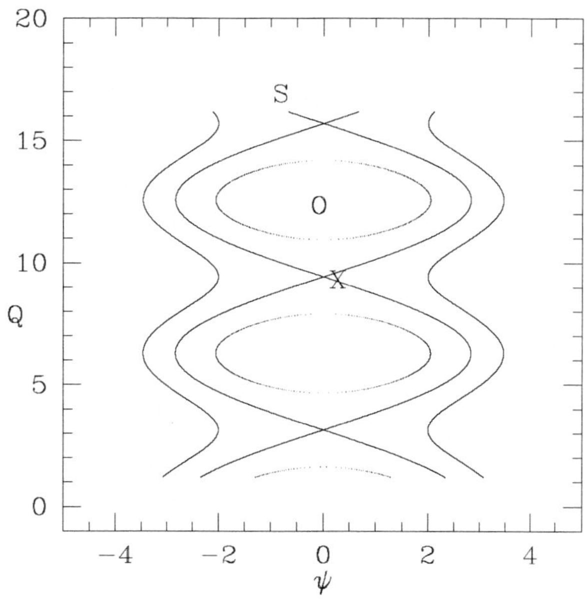

FIGURE 3. Island geometry

and the equations for the £eld lines become, from Eq. 5

$$\frac{d\psi}{d\zeta} = -Vmsin(n\zeta - m\theta), \qquad \frac{d\theta}{d\zeta} = \frac{1}{q(\psi)}. \tag{7}$$

The safety factor $q(\psi)$ de£nes the £eld helicity in the θ, ζ variables on the unperturbed surface ψ . Note from Eq. 3 that $\vec{B} \cdot \nabla\psi = mVsin(n\zeta - m\theta)/\mathcal{J}$, *i.e.* the perturbation V introduces a component of $\vec{B}$ directed across the original ¤ux surfaces. If $q(\psi)$ is not close to the rational m/n, ψ is oscillatory and $q(\psi)$ can be approximated as constant, giving $\theta = \zeta/q + \theta_0$ and $\psi - \psi_0 = Vcos(n\zeta - m\theta)/(n - m/q)$. The ¤ux surfaces are distorted but they remain topologically nested surfaces. However if $nq - m \simeq 0$ the resonant denominator creates large excursions in ψ and this solution is not valid. In this case expand $q(\psi)$ about ψ_0, with $q(\psi_0) = m/n$, and introduce the variable $Q = n\zeta - m\theta$. We then £nd

$$dQ = \left(\frac{mq'}{q^2}\right)(\psi - \psi_0)d\zeta, \qquad d\psi = -mVsin(Q)d\zeta \tag{8}$$

where $q' = dq/d\psi$, evaluated at $\psi = \psi_0$. Integrating these equations we £nd

$$(\psi - \psi_0)^2 = \frac{q^2V}{q'}[cos(Q) + k] \tag{9}$$

where the constant k is determined by the initial position. Note that one cannot derive these equations by setting the Hamiltonian equal to a constant because this is a "time" dependent Hamiltonian.

The points in the ψ, θ plane at $\zeta = 0$ formed by successive transits of a £eld line are called a Poincaré plot. If q is rational the Poincaré plot will consist only of a £nite set of the points on the surface given by Eq. 9, but if q is irrational the whole surface will be covered. The points given by $cos(Q) = 1$, $k = -1$, and those with $cos(Q) = -1$, $k = 1$ are periodic points of period m. Expanding the cosine function near $cos(Q) = +1$ we £nd

$$(\psi - \psi_0)^2 = \frac{q^2V}{q'}(1 - Q^2/2 + k) \tag{10}$$

i.e. for $Vq' > 0$ the points lie on elliptic manifolds about the periodic points. They form either a continuous curve or discrete points depending on the value of q. Similarly, near the points $cos(Q) = -1$ the points lie on hyperbolic

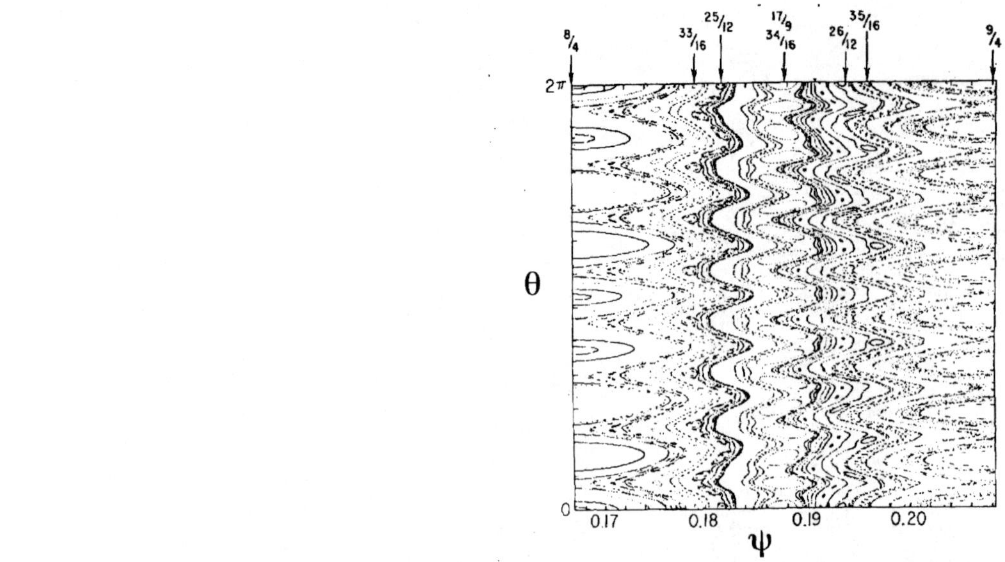

FIGURE 4. Magnetic £eld structure due to perturbations.

manifolds. Thus there is a chain of elliptic and hyperbolic points at $\psi = \psi_0$. This island chain is separated from the topologically toroidal surfaces by a separatrix which passes through the X-points. The separatrix is described by choosing the integration constant $k = 1$, giving islands with width

$$\delta\psi = 4\left(\frac{Vq^2}{q'}\right)^{1/2}. \tag{11}$$

In Fig. 3 are shown the resulting ¤ux surfaces for a single island chain. Note however that this plot is simpli£ed in that the separatrix is a simple curve only for the case when the Hamiltonian is exactly that of a pendulum. Any additional perturbations, certainly present here because of the approximations made, turn the separatrix into a stochastic band which grows in width with the strength of the perturbation.

In Fig. 4 is shown an example of magnetic surface destruction due to perturbations δ_{nm} for $m/n = 8/4, 9/4$. The higher order islands are due to nonlinear interaction of these two modes. Modes m,n and m',n' couple to produce a mode at $m+m', n+n'$, etcetera. This Fibonacci series means islands of decreasing size exist at rational surfaces to arbitrarily large m,n.

TOKAMAK EQUILIBRIA

Consider an axisymmetric £eld con£guration with the topology of nested toroidal ¤ux surfaces with label ψ_p. Since $\vec{B}$ is orthogonal to $\nabla\psi_p$ use the Clebsch representation

$$\vec{B} = \nabla\psi_p \times \nabla V \tag{12}$$

and choose $V = q(\psi_p)\theta - \zeta + \lambda(\psi_p, \theta, \zeta)$ where θ and ζ are arbitrary poloidal and toroidal angles, respectively, the only requirement being that they must increase monotonically from 0 to 2π as one moves around a ¤ux surface.

Choose $2\pi\psi_p$ to be the poloidal ¤ux inside the magnetic surface. The element of surface area in a poloidal section is $d\vec{S}_p = \mathcal{J}_p \nabla\theta d\zeta d\psi_p$, giving (see Fig. 2)

$$2\pi\psi_p = \int \vec{B}\cdot\nabla\theta\,\mathcal{J}_p d\psi_p d\zeta = \int d\psi_p d\zeta + \int_{\zeta_0}^{\zeta_0+2\pi} \frac{\partial\lambda}{\partial\zeta} d\psi_p d\zeta \tag{13}$$

for any ζ_0, with the Jacobian given by $\mathcal{J}_p^{-1} = \nabla\psi_p \cdot \nabla\theta \times \nabla\zeta$, and thus λ is periodic in ζ.

We then have the **contravariant** representation for $\vec{B}$

$$\vec{B} = (q + \lambda'_\theta)\nabla\psi_p \times \nabla\theta + (1 - \lambda'_\zeta)\nabla\zeta \times \nabla\psi_p \tag{14}$$

Now choose $2\pi\psi = 2\pi\int_0^{\psi_p} q(\psi_p)d\psi_p$ to be the toroidal flux inside the surface. The element of surface area in a toroidal section is $d\vec{S}_t = \mathscr{J}_p\nabla\zeta d\theta d\psi_p$, giving

$$2\pi\psi = \int \vec{B}\cdot\nabla\zeta\mathscr{J}_p d\psi_p d\theta = \int q d\psi_p d\theta + \int_{\theta_0}^{\theta_0+2\pi} \frac{\partial\lambda}{\partial\theta} d\psi_p d\theta \tag{15}$$

for any θ_0, and thus λ is periodic in θ and $d\psi/d\psi_p = q$.

It is also useful to note the vector potential associated with $\vec{B}$, with $\vec{B} = \nabla\times\vec{A}$,

$$\vec{A} = \psi\nabla\theta + \psi_p\nabla\zeta - \lambda\nabla\psi_p. \tag{16}$$

Straight field line coordinates are often used for equilibrium and stability studies. Replace $\zeta - \lambda$ with ζ in Eq. 12, equivalent to setting λ equal to zero. The straight field line toroidal angle differs from the usual cylindrical coordinate only by a periodic oscillation.

Force balance

Now we make use of the equilibrium property of $\vec{B}$. We assume a scalar pressure equilibrium, $p = p(\psi)$. From the equilibrium condition $\vec{j}\times\vec{B} = \nabla p$ with $p = p(\psi)$ it follows that $\vec{j}\cdot\nabla\psi = 0$. Write $\vec{j} = \nabla\psi\times\nabla W$, and choose $W = \bar{I}'(\psi)\theta + \bar{g}'(\psi)\zeta + \alpha(\psi,\theta)$. Then

$$\vec{j} = (\bar{I}'(\psi) + \alpha'_\theta)\nabla\psi\times\nabla\theta + \bar{g}'(\psi)\nabla\psi\times\nabla\zeta \tag{17}$$

with primes indicating derivatives with respect to the subscripted variable. Choose $2\pi\bar{I}$ to be the toroidal current inside ψ, and $2\pi\bar{g}$ to be the poloidal current outside ψ. Then as is the case with λ, α is periodic in θ,ζ. Then $\nabla\times\vec{B} = \vec{j}$ gives the **covariant** representation for $\vec{B}$

$$\vec{B} = g\nabla\zeta + I\nabla\theta + \delta\nabla\psi \tag{18}$$

with $< g >= \bar{g}$, $< I >= \bar{I}$ and the brackets indicate averaging over θ and ζ and we also find in the case of axisymmetry $g = g(\psi)$.

We now restrict ourselves to an axisymmetric equilibrium, by requiring that $\partial_\zeta f = 0$ for all equilibrium functions f. The ignorable coordinate ζ is related to the cylindrical coordinate ϕ through $\zeta = \phi - \nu(\psi,\theta)$, and the straight field line condition $q = \vec{B}\cdot\nabla\zeta/\vec{B}\cdot\nabla\theta$ gives, using $\vec{B}\cdot\nabla\zeta = \vec{B}\cdot\nabla\phi - \partial_\theta\nu\vec{B}\cdot\nabla\theta$, and evaluating $\vec{B}\cdot\nabla\phi = g/X^2$, and $\vec{B}\cdot\nabla\theta = 1/(\mathscr{J}q)$,

$$\frac{\partial\nu}{\partial\theta} = \frac{gq\mathscr{J}}{X^2} - q. \tag{19}$$

Consider the force balance equation in the ϕ,ψ,θ system, giving

$$\mathscr{J}(j^\theta B^\phi - j^\phi B^\theta) = p'. \tag{20}$$

Again using the contravariant representation, Eq. 14 gives

$$\vec{j}\cdot\nabla\phi = \nabla\cdot(\vec{B}\times\nabla\phi) = \nabla\cdot[(\nabla\zeta\times\nabla\psi_p)\times\nabla\phi] = \nabla\cdot[(\nabla\phi\cdot\nabla\zeta)\nabla\psi_p] \tag{21}$$

which gives simply $\vec{j}\cdot\nabla\phi = \nabla\cdot\nabla\psi_p/X^2$, and using the covariant representation, Eq. 18 we find $\vec{j}\cdot\nabla\theta = \nabla\cdot(\vec{B}\times\nabla\theta) = -g'/\mathscr{J}$.

Force balance becomes, upon substituting these expressions into Eq. 20 and changing to the variable ψ_p

$$\nabla\cdot\frac{\nabla\boldsymbol{\psi}_\mathbf{p}}{\mathbf{X^2}} + \mathbf{p}' + \frac{\mathbf{gg}'}{\mathbf{X^2}} = \mathbf{0} \tag{22}$$

where now primes indicate derivatives with respect to ψ_p. This is the Grad Shafranov equation. It must be solved with boundary conditions given by a conducting wall or external coils.

The physical coordinate system chosen is shown in Fig. 5. Note that the toroidal current is in the direction of positive ϕ (ζ), the field points in the direction of positive ζ and θ with the functions g and I positive and that ψ,θ,ζ form a right handed system. Not all authors use the same coordinates, so one must beware of sign changes.

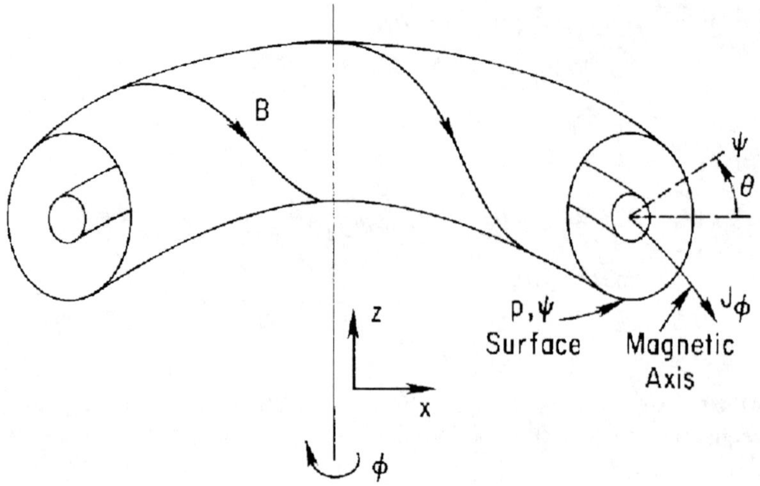

FIGURE 5. Physical cylindrical coordinates X, Z, ϕ.

Large aspect ratio

For very large aspect ratio, with a circular conducting wall, and at low $\beta = 2p/B^2$, toroidal effects are negligible and the equilibrium is approximately cylindrical. This approximation allows many analytic calculations that cannot be carried out in general geometry, and still give much qualitative insight into the behavior of a tokamak. The magnetic £eld, normalized to its value on axis, has the covariant form

$$\vec{B} = \nabla\phi + \frac{r^2}{q(r)}\nabla\theta \tag{23}$$

with the major radius $R = 1$, and $r \ll 1$. $\nabla\phi = \hat{\phi}$, $\nabla\theta = \hat{\theta}/r$. The contravariant form is

$$\vec{B} = \frac{r}{q(r)}\nabla\phi \times \nabla r \;+\; r\nabla r \times \nabla\theta. \tag{24}$$

The toroidal ¤ux function is r^2 and the poloidal ¤ux function is $\int r dr/q$.

Express £eld perturbations in Fourier harmonics of the form $f(r)exp(i(m\theta - n\phi))$. The wave vector is thus $\vec{k} = \nabla(m\theta - n\phi)$. The helicity, m/n, is preserved in nonlinear products of terms with the same helicity. Introduce the helicity vector $\vec{h} = \nabla\phi + nr^2\nabla\theta/m$ and the helicity variable $\tau = \theta - (n/m)\phi$ with $\vec{h}\cdot\nabla\tau = \vec{h}\cdot\vec{k} = 0$. It is convenient to introduce the helical ¤ux

$$\psi_h = \int r dr \vec{B}\cdot\nabla\tau = \psi_p - \left(\frac{n}{m}\right)\frac{r^2}{2}. \tag{25}$$

The magnetic £eld is given by

$$\vec{B} = \vec{h} + \nabla\phi \times \nabla\psi_h \tag{26}$$

and ψ_h is a magnetic surface function, $\vec{B}\cdot\nabla\psi_h = 0$ provided that $\psi_h = \psi_h(r,\tau)$. In Fig. 6 is shown a typical equilibrium helical ¤ux function, with slope equal to zero at the rational surface where $q = m/n$.

We £nd for the current

$$\vec{j} = \left(\nabla_\perp^2\psi_h + \frac{2n}{m}\right)\nabla\phi - \partial_\phi\nabla_\perp\psi_h. \tag{27}$$

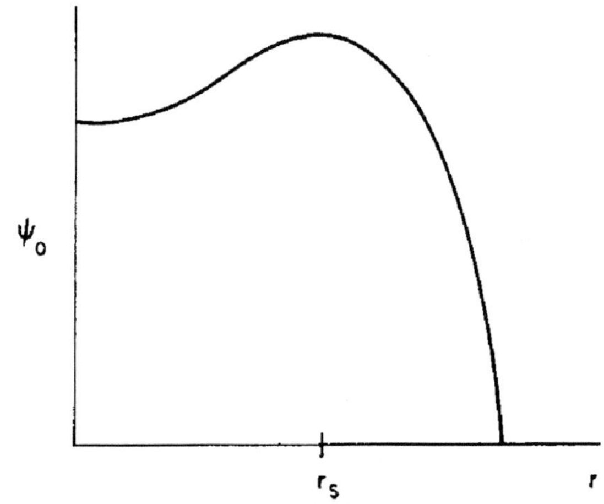

FIGURE 6. Equilibrium helical flux.

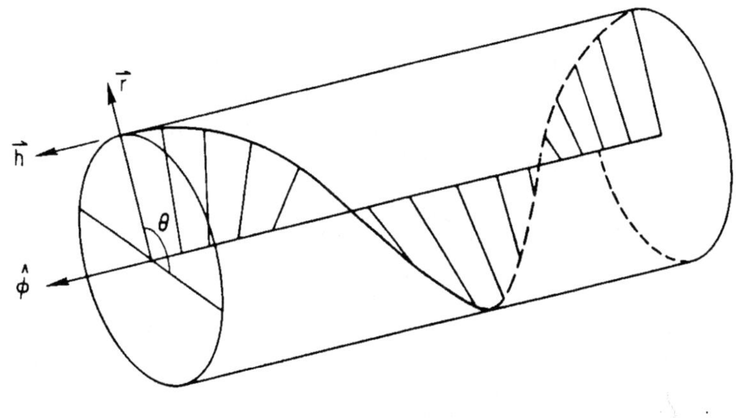

FIGURE 7. Helical Surface.

TEARING MODES

The forces driving the tearing mode are large compared to toroidal effects, and this mode can be examined in cylindrical geometry. Introduce a perturbation of a single helicity through $\psi = \psi_0(r) + \psi_1(r)e^{i(m\theta - n\phi)}$. Thus the perturbation wave vector is $\vec{k} = \nabla(m\theta - n\phi)$ and $\vec{k}\cdot\vec{B}_0 = (m/r)\psi_0'$.

The perturbation ψ_1 is simply related to the ideal displacement, $\vec{\xi}$. To lowest order in $\varepsilon = r/R$, *i.e.* the cylindrical geometry tokamak approximation, $\vec{\xi}$ is given by $\xi_r = \xi, \xi_\theta = id(r\xi)/(mdr), \xi_\phi = 0$, and thus

$$\vec{\xi}\times\vec{B} = \left[\hat{r}\frac{id(r\xi)}{mdr} - \hat{\theta}\xi + \hat{\phi}\xi\left(\psi_0' + \frac{nr}{m}\right)\right]e^{i(m\theta - n\phi)}. \tag{28}$$

Also, from Eq. 26, to lowest order the perturbed field is given by $\delta\vec{B} = -\nabla\times(\psi_1\hat{\phi})$.

Thus writing $-\psi_1\hat{\phi} = \vec{\xi}\times\vec{B} + \nabla G$ with G the difference in gauge between the two representations we find $G = -(i/m)r\xi e^{i(m\theta - n\phi)}$ and $\psi_1(r) = \xi\psi_0'$, or $\psi_1(r) = \xi r(1/q - n/m)$. Thus for finite ξ, $\psi_1(r)$ vanishes at the rational surface.

Consider the helical ribbon defined by the vector $\vec{h} = \hat{\phi} + nr\hat{\theta}/m$, as shown in Fig. 7. At the rational surface, $\psi_0' = 0$, the $\vec{B}$ field lies in the surface defined by $\vec{h}$. The vectors $\vec{k}, \vec{h}$, and $\vec{r}$ are mutually orthogonal, and in the coordinate system

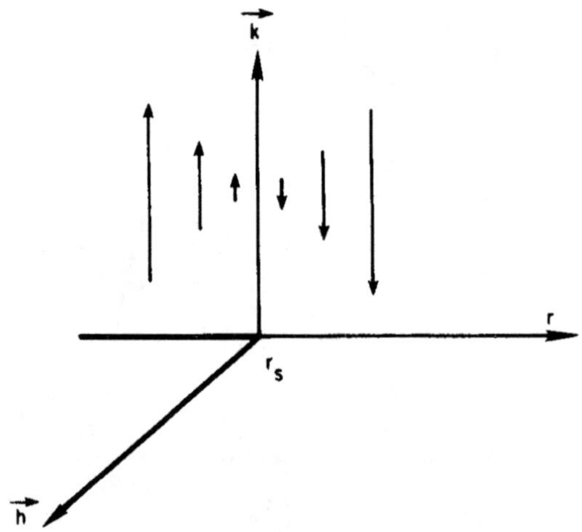

FIGURE 8. Magnetic £eld near $r = r_s$.

de£ned by them the component of $\vec{B}$ in the direction of $\vec{k}$ vanishes at $r = r_s$ as shown in Fig. 8.

The surfaces of constant $\psi = \psi_0 + \psi_1(r)cos(m\theta - n\phi)$ at £xed ϕ de£ne magnetic islands at the surface where ψ_0' vanishes, with width $w = 4(-\psi_1/\psi_0'')^{1/2}$ as shown in Fig. 9. In the exterior region (away from the island) the equations of ideal magnetohydrodyanics (MHD) are valid and ψ_1 can be found by minimizing the potential energy[1] δW. Let $\chi = \xi F = \psi_1/r$, with $F = (1/q - n/m)$ and carry out the variation of δW using χ. Integrating by parts we £nd

$$\delta W = -\frac{\pi r^3}{2}\chi \frac{d\chi}{dr}\bigg|_-^+$$
$$+\frac{\pi}{2}\int dr\left\{-\chi\frac{d}{dr}\left(r^3\frac{d\chi}{dr}\right) + \left[(m^2-1)r + \frac{1}{F}\frac{d}{dr}\left(r^3\frac{dF}{dr}\right)\right]\chi^2\right\}. \tag{29}$$

Thus the Euler-Lagrange equation is given by the vanishing of the integrand, and

$$\delta W_{min} = -\frac{\pi}{2} r^3 \chi \frac{d\chi}{dr}\bigg|_-^+. \tag{30}$$

Substituting $\psi_1 = r\chi$ and using the fact that ψ_1 is continuous across the resistive layer, but not necessarily $d\psi_1/dr$, we £nd

$$\delta W_{min} = -\frac{\pi}{2} r_s \psi_1^2 \Delta', \qquad \Delta' = \frac{1}{\psi_1}\frac{d\psi_1}{dr}\bigg|_-^+ \tag{31}$$

and Δ' is the discontinuity of ψ_1' across the tearing layer, and is a measure of the magnetic energy to be gained by a perturbation producing magnetic islands at $r = r_s$. Substituting ψ_1 into the integrand and using $F'' = j_0'/r - 3F'/r$ we £nd the Euler-Lagrange equation

$$\nabla_\perp^2 \psi_1 = \frac{dj_0}{d\psi_0}\psi_1 \tag{32}$$

where $dj_0/d\psi_0 = dj_0/(\psi_0' dr)$ and $\nabla_\perp$ is the gradient in the r, θ plane.

The solution of this equation for ψ_1 in the exterior region determines Δ'. The energy differential driving the mode and proportional to Δ' is a function of the whole current pro£le, not a quantity determined locally at the rational surface.

Solutions ψ_1 behave as r^m for small r, vanish at the conducting wall, and have discontinuities in $d\psi_1/dr$ at $r = r_s$. The $m = 1$ solution is rapidly changing in the vicinity of $r = r_s$, but for $m > 1$ ψ_1 is essentially constant near $r = r_s$.

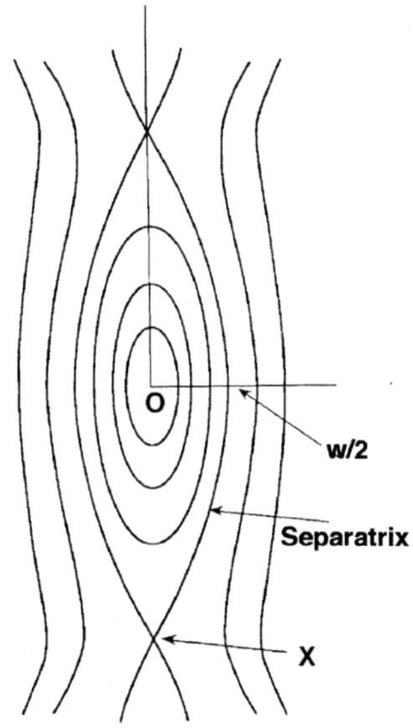

FIGURE 9. Magnetic island.

Now consider the full resistive equations, which are valid in the ideal domain as well as in the tearing layer. The poloidal ¤ux and the helical ¤ux satisfy very similar equations, as expected from the simple way they are related. First write $\vec{B}$ in terms of the poloidal ¤ux, $\vec{B} = \nabla\phi - \nabla \times \psi_p \nabla\phi$ and substitute it into Faraday's Law and Ohm's Law, giving

$$\nabla \times [\partial_t \psi_p \nabla\phi] = \nabla \times (\eta \vec{j} - \vec{v} \times \vec{B}), \qquad \partial_t \psi_p \nabla\phi = \eta \vec{j} - \vec{v} \times \vec{B} \tag{33}$$

except for a gauge term ∇G, which can be taken to be zero

Dot Eq. 33 with $\vec{h}$ and substitute Eq. 26, $\vec{B} = \vec{h} + \nabla\phi \times \nabla_\perp \psi_h$ and Eq. 27 for $\vec{j}$. Using also $\psi_h = \psi_h(r,\tau)$ so that $\vec{h} \cdot \nabla\psi_h = 0$ we £nd

$$\partial_t \psi_h + \vec{v} \cdot \nabla_\perp \psi_h = \eta j_\phi, \qquad j_\phi = \nabla_\perp^2 \psi_h + \frac{2n}{m}. \tag{34}$$

The velocity is governed by the equation of motion $\rho d\vec{v}/dt = \vec{j} \times \vec{B} - \nabla p$ where $d/dt = \partial_t + \vec{v} \cdot \nabla$ is the convective derivative.

The large aspect ratio approximation, $R/r \gg 1$ and tokamak ordering with $\beta_\theta = 2p/B_\theta^2 \simeq 1$, $q \simeq 1$ can be used to give a consistent ordering in the parameter $\varepsilon = r/R$. The result is that B_ϕ, ρ, $\nabla_\perp$, ∂_ϕ are of order 1, and B_θ, B_r, j_ϕ, $v_\perp$, ∂_t are of order ε, while δB_ϕ, p, $j_\perp$, v_ϕ are of order ε^2 and $\nabla \cdot \vec{v}_\perp$ is of order ε^3. In cylindrical approximation $\nabla \cdot \vec{v}_\perp \simeq 0$ which implies that the velocity can be represented by a stream function U with $\vec{v} = \hat{\phi} \times \nabla_\perp U$ where $\perp$ refers to the r,θ plane. Operating with $\hat{\phi} \cdot \nabla \times$ we £nd

$$\frac{d}{dt}\rho \nabla^2 U = \hat{\phi} \cdot \nabla \times (\vec{j} \times \vec{B}). \tag{35}$$

We now use only the helical ¤ux ψ_h and omit the subscript. Substituting the expression for $\vec{v}$ into Eq. 34, and the expression for $\vec{B}$ into the equation for U, we have two coupled nonlinear differential equations for ψ, U

$$\frac{\partial \psi}{\partial t} + \hat{\phi} \cdot (\nabla_\perp U \times \nabla_\perp \psi) = \eta \left(\nabla_\perp^2 \psi + \frac{2n}{m} \right) \tag{36}$$

$$\frac{d}{dt}\rho \nabla_\perp^2 U = \hat{\phi} \cdot [\nabla_\perp \psi \times \nabla_\perp (\nabla_\perp^2 \psi)]. \tag{37}$$

These equations give the full nonlinear evolution of a single helicity mode. To study the linear behavior, introduce the perturbations,

$$\psi = \psi_0(r) + \psi_1(r)cos(m\theta - n\phi), \qquad U = \frac{r\gamma\xi(r)}{m}sin(m\theta - n\phi). \tag{38}$$

Substituting these expressions then results in two coupled second order ordinary differential equations for ξ, ψ_1

$$\psi_1 - \psi_0'\xi = \frac{r_s^2}{\gamma\tau_R}\nabla_\perp^2\psi_1, \qquad \gamma^2\tau_A^2\nabla_\perp^2 r\xi = \frac{m^2\psi_0'}{r}\left(\frac{dj_0}{d\psi_0}\psi_1 - \nabla_\perp^2\psi_1\right) \tag{39}$$

where now $\nabla_\perp^2 = \frac{1}{r}\frac{d}{dr}r\frac{d}{dr} - \frac{m^2}{r^2}$ and $\tau_A = R\sqrt{\rho}/B_z$ is the characteristic Alfvèn time for these equations, equal to the poloidal Alfvèn time divided by q, and $\tau_R = r_s^2/\eta$ is the resistive time given by the rational surface.

Outside the tearing layer the operator $r_s^2\nabla_\perp^2$ is of order unity and using $\gamma\tau_R \gg 1$, $\gamma\tau_A \ll 1$ we £nd the solutions in the external region satisfy

$$\nabla_\perp^2\psi_1 = \frac{dj_0}{d\psi_0}\psi_1, \qquad \psi_1 = \xi\psi_0' \tag{40}$$

which is Eq. 32, *i.e.* these solutions are those derived by minimizing the ideal δW.

Near the tearing layer the linear eigenfunctions have the form

$$\psi_1(x) = \begin{cases} \psi_0''\xi_0 x\theta(-x) & m = 1 \\ \psi_1 & m \geq 2 \end{cases} \tag{41}$$

$$\xi(x) = \begin{cases} \xi_0\theta(-x) & m = 1 \\ \frac{\psi_1}{\psi_0'' x} & m \geq 2 \end{cases} \tag{42}$$

where $x = r - r_s$, r_s is the rational surface; $q(r_s) = m/n$.

To complete the analysis, boundary layer solutions must be obtained in the interior region, and matched to the exterior ideal solutions. In the interior region, $|x| \ll r_s$ and ψ_1' is rapidly changing. Thus we can take $\nabla_\perp^2\psi_1 \simeq \psi_1''$, and $\psi_0' = -snx/m$ with $s = r_s q'/q$. We then £nd

$$\psi_1(x) + \frac{snx\xi(x)}{m} = \frac{r_s^2\psi_1''(x)}{\gamma\tau_R}, \qquad \gamma^2\tau_A^2\xi''(x) = \frac{smnx}{r^2}\psi_1''(x). \tag{43}$$

The nature of the boundary layer is found using a Kruskal-Newton diagram [2, 3, 4, 5]. For a given linear homogeneous differential equation of the form

$$\sum_{q,n,m} C_{q,m,n}\varepsilon^q x^m \frac{d^n y}{dx^n} = 0, \tag{44}$$

construct a Kruskal-Newton diagram with terms given by $x^p\varepsilon^q$. The layer is at $x = 0$ so count powers of x as well as powers of dx giving $p = m - n$. The function y and the numbers $C_{q,m,n}$ are all assumed to be of order one. Each term in the sum can be represented as a point in the p, q plane. All points above a given line represent terms smaller than points on the line, and all points below a line represent terms larger than those on the line.

To £nd all possible combinations of dominant balance for a given equation, £nd all possible placements of a line so that it includes two or more terms of the equation, with all other points lying above the line. Graphically this may be understood as bringing the line up from below until it makes contact with a point, and then rotating it one way or the other until it makes contact with a second point. The scaling of x is quickly determined by the slope of the line in the diagram. If the terms de£ning the line are given by $x^p\varepsilon^q$ and $x^r\varepsilon^s$ then balancing them gives the scaling $x \sim \varepsilon^{(s-q)/(p-r)}$, the power of ε being minus the slope of the line. Such a plot is known as a Kruskal-Newton diagram. It was £rst used by Newton[3] in considering in£nitesimal displacements for the development of differential calculus and subsequently further developed by Kruskal[4].

THE TEARING MODE , $m \neq 1$

Because of the quite different form of the solutions for $m = 1$, it must be treated as a special case. First examine $m \neq 1$. From the nature of the exterior solution for $m \neq 1$ we can approximate $\psi_1(x)$ as constant. Substituting we then obtain a single second order differential equation

$$\psi_1 + \frac{snx\xi(x)}{m} = \frac{r_s^4\gamma^2\tau_A^2\xi''(x)}{smnx\gamma\tau_R}. \tag{45}$$

The Kruskal Newton diagram is shown in Fig. 10, where the point a is the ξ'' term, b is the ξ term, and c the constant ψ_1, and $\gamma^2\tau_A^2/(\gamma\tau_R) \sim \varepsilon^3$. Other neglected terms by taking $\nabla_\perp^2\psi_1 \simeq \psi_1''$ and $\psi_0' \simeq -snx/m$ are given by dots appearing above the line joining a and b.

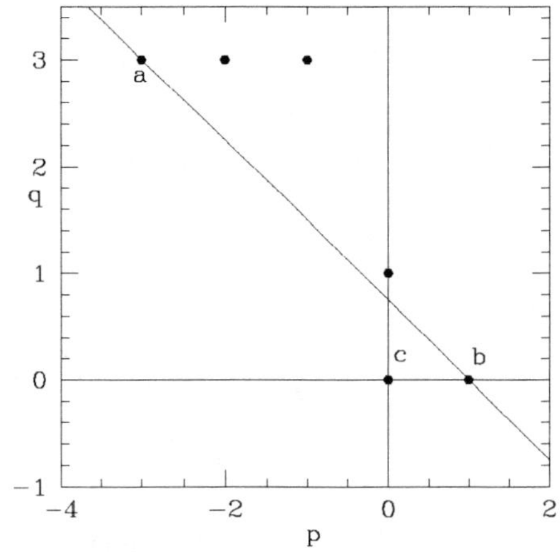

FIGURE 10. Kruskal Newton diagram for $m \geq 2$

We now use Kruskal's asymptotic principle of maximal balance according to which "the most informative ordering is that which simplifies the least, maintaining a maximal set of comparable terms"[4]. We obtain this by renormalizing ξ through $\xi = m\psi_1(\gamma\tau_R/(n^2s^2\gamma^2\tau_A^2))^{1/4}\chi(z)/r_s$ to bring points a,b,c to lie on the same line. The slope of the line gives the scaling of the layer as $x \sim (\gamma^2\tau_A^2/\gamma\tau_R)^{1/4}$, and we introduce the independent variable z through $x = (\gamma^2\tau_A^2/\gamma\tau_R n^2s^2)^{1/4}r_s z$, giving

$$\chi'' - z^2\chi = z, \qquad \chi = -\frac{z}{2}\int_0^1 d\mu e^{-z^2\mu/2}(1-\mu^2)^{-1/4}. \tag{46}$$

For the necessary matching to the exterior solution we have

$$\frac{r^2\gamma^2\tau_A^2}{msn\psi_1(r_s)}\int_{-\infty}^{\infty}\frac{\xi''}{x}dx = \Delta'. \tag{47}$$

The matching condition becomes

$$\gamma^{5/4}\tau_A^{5/4}S^{3/4}(ns)^{-1/2}\int_{-\infty}^{\infty}\frac{dz}{z}\chi'' = \Delta' r, \tag{48}$$

with $S = \tau_R/\tau_A$ the ratio of resistive to Alfvèn times, typically greater than 10^6 in a tokamak, and $\gamma\tau_R \gg 1$ and $\gamma\tau_A \ll 1$, as assumed.

To evaluate the integral $I = \int_{-\infty}^{\infty}(dz/z)\chi''$, substitute χ and integrate over z, giving

$$I = \sqrt{\pi/2}\int_0^1 d\mu\frac{\mu^{1/2}}{(1-\mu^2)^{1/4}} = \pi\Gamma(3/4)/\Gamma(1/4) \tag{49}$$

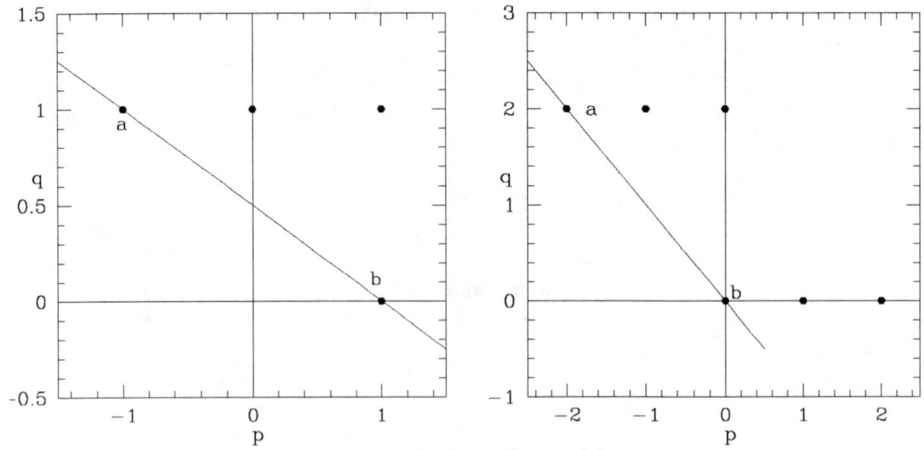

FIGURE 11. Kruskal Newton diagrams for $M = 1$

Thus

$$\gamma\tau_A = \left(\frac{\Gamma(1/4)\Delta' r}{\pi\Gamma(3/4)}\right)^{4/5} S^{-3/5}(ns)^{2/5}. \tag{50}$$

We thus £nd $\gamma \sim \tau_R^{-3/5}\tau_H^{-2/5}$, where the hydromagnetic time $\tau_H = \tau_A/(sn)$ is the time for a shear Alfvèn wave to cross the plasma.

THE $m = 1$ TEARING MODE

For the $m = 1$ mode recall that the exterior displacement ξ is to lowest order a step function, and ψ_1 is also very rapidly changing at the rational surface. Thus the constant ψ approximation used to obtain Eq. 46 is not valid. Return to the coupled Eqs. 43 for the interior region and substitute $\psi_1 = xZ$ giving equations in order one quantities

$$xZ + \frac{snx\xi(x)}{m} = \frac{r_s^2(2Z' + xZ'')}{\gamma\tau_R}, \qquad \gamma^2\tau_A^2\xi''(x) = \frac{smn}{r^2}Z \tag{51}$$

The Kruskal-Newton diagrams for these equations are shown in Fig. 11. In the left plot $\varepsilon = 1/(\gamma\tau_R)$ and point a is the derivative term, point b the xZ and $x\xi$ terms. This plot gives a layer scaling of $x \sim 1/\sqrt{\gamma\tau_R}$. In the right plot point a is the ξ'' term, and point b the Z term. This plot gives a layer scaling of $x \sim \gamma\tau_A$. But of course these equations are coupled and there is only one layer, and equating these we £nd $\gamma \sim \tau_A^{-2/3}\tau_R^{-1/3}$.

Introducing the layer variable $z = (\gamma\tau_R)^{1/2}x/r_s$ we £nd an exact solution which satis£es the boundary conditions $\psi_1 \to 0, x \to +\infty$ and $\psi_1 \to -\xi_0 snx, x \to -\infty$,

$$\xi = \frac{\xi_0}{\sqrt{2\pi}}\int_z^\infty e^{-z^2/2}dz, \qquad \psi_1 = \frac{\xi_0 r_s}{\sqrt{2\pi}}\frac{sn}{(\gamma\tau_R)^{1/2}}\left[e^{-z^2/2} - z\int_z^\infty e^{-z^2/2}dz\right], \tag{52}$$

and $\gamma = \tau_R^{-1/3}\tau_H^{-2/3}$, where again $\tau_H = \tau_A/(sn)$, the hydromagnetic time, is the time for a shear Alfvèn wave to cross the plasma. This growth rate is much larger than the growth rate for $m \geq 2$, and further the $m = 1$ mode is unstable to all pro£les with $q_0 < 1$. The very different behavior is due to existence of the ideal $m = 1$ mode, which causes a steepening of the £eld gradients near the rational surface and drives the reconnection.

NONLINEAR TEARING MODE ANALYSIS

We are interested in the domain near the rational surface, but outside the tearing layer, and in magnetic islands which are large compared to the tearing layer, which is the width at which nonlinear effects begin to play a role. In this domain the helical flux can be approximated as constant for modes with $m \geq 2$ and as linear in the distance from the rational surface for $m = 1$. These approximations suffice to carry out a nonlinear analysis which is valid when the island width is large compared to the tearing layer, but small compared to the system size. We will present here simple heuristic derivations which give good approximate results. The tearing layer width is given by $w_t = (Sns)^{-1/3} r_s$ for the $m = 1$ mode and $w_t = (Sns)^{-2/5} r_s$ for $m \geq 2$.

In this region the linear eigenfunctions have the form given by Eq.42. The relation between ψ_1 and ξ is given by the ideal MHD solutions. For $m \geq 2$, ψ_1 is approximately constant near $r = r_s$, and $\xi = const$ near $r = r_s$ follows from the $m = 1$ solution. For $m > 2$ the perturbation produces a magnetic island structure with width $w = 4(-\psi_1/\psi_0'')^{1/2}$ and within the constant ψ approximation the X-point and O-point are located at the tearing layer $r = r_s$. This is not the case for the $m = 1$ mode. In this case surfaces of constant ψ are given by

$$\psi = \psi_0'' x^2/2 + \psi_0'' \xi_0 x(1 - \theta^2/2) \tag{53}$$

for small x, θ. The point $\nabla\psi = 0$ is located at $x = -\xi_0$, $\theta = 0$. Expanding ψ about this point we find

$$\psi = const + \psi_0''(x + \xi_0)^2/2 + \psi_0'' \xi_0^2 \theta^2/2 + \ldots \quad . \tag{54}$$

Thus this is an O-point and the island width is given by $w = 2\xi_0$, which scales directly as the magnitude of ψ_1, not its square root. The flux surface describing the interior separatrix is given by $x - \xi_0(1 + cos\theta)$. Recall that the ideal solution is $\xi = constant$, thus the surfaces inside r_s consist of displaced circles. The X-point occurs at the tangent point of the circles $r = r_s$ and the displaced circle of radius $r_s - \xi_0$, as shown in Fig. 12.

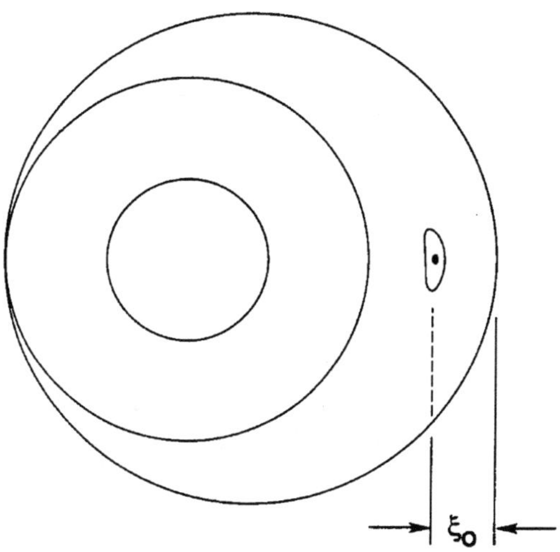

FIGURE 12. Magnetic surfaces for a large $m = 1$ island.

To obtain the quasilinear evolution of the mode for $m \geq 2$ the convection term is eliminated by averaging over a flux surface. Carry out a flux surface average $\oint dl/\nabla\psi$. The convection term becomes $\oint dl\vec{v} \cdot \hat{n}$ where $\hat{n} = \nabla\psi/|\nabla\psi|$ is the normal to the flux surface. By incompressibility this integral is zero. Thus

$$\langle \partial_t \psi \rangle = \langle \eta(j - j_d) \rangle \tag{55}$$

where j_d is driven current and $\langle f \rangle = \oint (dl/\nabla\psi) f / \oint (dl/\nabla\psi)$.

Evaluated at the O-point, where $cos(m\theta) = 1$ and at the X point where $cos(m\theta) = -1$ the flux surface averaged equation for $\partial_t \psi$ gives

$$\dot{\psi}_0(r_0) + \dot{\psi}_1(r_0) = \eta[j_0 + j_1(r_0) - j_d(r_0)], \qquad \dot{\psi}_0(r_x) - \dot{\psi}_1(r_x) = \eta[j_0 - j_1(r_x) - j_d(r_x)]. \tag{56}$$

For $m \geq 2$ the X and O points are located on the same flux surface, simplifying these evaluations. The current is not negligible, and must be evaluated at the X and O-points, which lie within the tearing layer. By the constant ψ approximation, the exterior solutions can be used for ψ_1 in this region. Using the narrowness of the island,

$w \ll 1$, $\nabla_\perp^2 \psi_1 = j_1$, and the form of ψ_1, we £nd $j(r_0) = j_0(r_0) + [\Delta'(w)\psi_1/w]cosm\theta$ where $\Delta'(w) = [\psi_1'(w/2) - \psi_1'(-w/2)]/\psi_1(0)$, and similarly for $j(r_x)$. We then £nd, on subtracting X and O-point values and using $r_0 \simeq r_s$,

$$\dot{\psi}_1 = \frac{\eta(r_s)\Delta'(w)\psi_1}{w} - \eta j_d(r_0) + \eta j_d(r_x). \tag{57}$$

describing the nonlinear saturation of the mode and the effect of current drive. Substituting $\psi_1' = -\psi_0'' w^2/16$ we £nd

$$\dot{w} = \frac{\eta\Delta'(w)}{2} \tag{58}$$

plus term given by j_d, giving the usual algebraic growth of the width. Saturation occurs approximately when $\Delta'(w) = 0$, which is simply interpreted that the magnetic energy of the island state has reached a local minimum.

Now consider the case of $m = 1$. The $m = 1$ mode analysis differs because of the inward shift of the O-point due to ψ_1 being linear in x. Because of the form of ψ_1, the $m = 0$ part contributes signi£cantly through the inertial $\vec{v} \cdot \nabla\psi$ term. No saturation can be derived. This is what is observed in resistive MHD simulations, the mode continues at approximately the linear growth rate far into the nonlinear regime. Thus the inertia continues to play a role in the evolution of the $m = 1$ mode in the nonlinear regime, with plasma ¤ow entering into the force balance equation. This means that the plasma does not evolve through a succession of MHD equilibria, as it does for modes with $m \geq 2$.

In addition, the fact that the $m = 1$ tearing mode does not enter a simple nonlinear regime means that the growth rate is subject to other physical effects, including electron inertia and the generation of other modes. This mode is thus still under intense investigation.

SAWTEETH

The sawtooth oscillation was £rst observed on tokamaks with the introduction of soft X-ray detectors for the measurement of internal temperatures. The internal temperature was observed to repeatedly slowly rise to a peak value and then suddenly crash. Kadomtsev[6] realized that this process could be realized by a complete reconnection of ¤ux due to an $m = 1$ tearing mode. The ¤ux evolution during the reconnection is shown in Fig. 13. The $m = 1$ mode consists of a rigid displacement of the plasma inside the $q = 1$ surface, ξ is a constant. The original center of the plasma moves over to the q=1 surface and is shifted outside this surface, with the island O-point taking over as the new plasma center. The evolution of the helical ¤ux pro£le, the current pro£le and the q pro£le are also shown in Fig. 13. Although this process gives a reasonable heuristic explanation of the sawtooth crash, the time scale of the crash, the extent of stochasticity, and the sawtooth period are not completely understood. By redistributing the current inside the $q = 1$ surface the sawtooth effectively prevents sustained operation with $q < 1$. Furthermore repeated sawtoothing ¤atten the

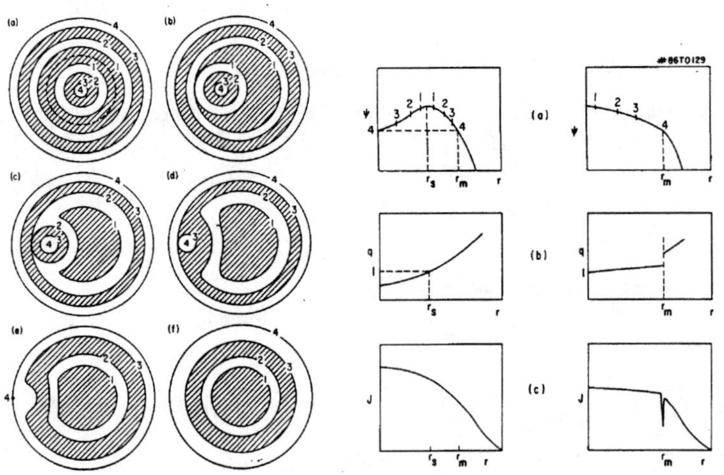

FIGURE 13. Helical ¤ux, q pro£le, and current pro£le before and after a Kadomtsev reconnection.

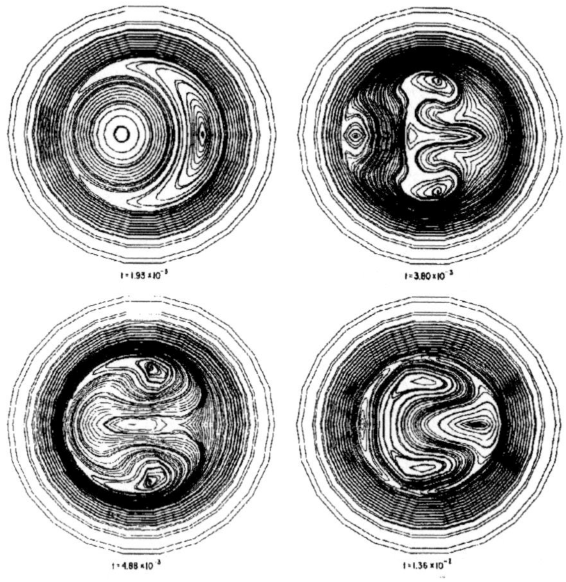

FIGURE 14. A sawtooth crash simulation

current pro£le, making the discharge unbstable to the development of a large $m = 2$ mode. In Fig. 14 is shown the result of a computer simulation of a sawtooth crash, giving results very similar to the heuristic scenario of Kadomtsev.

DISRUPTION

Although there are many variations, a plasma disruption is thought to be primarily due to the development of a large scale $m = 2$ island. A large $m = 2$ island can nonlinearly destabilize other islands leading to overlap and large scale stochastic £elds. The saturation width of an island depends on the current pro£le through Δ'. If the current pro£le is fairly ¤at, the saturation width is large. This scenario is very probable in an extended discharge, because repeated sawtooth oscillations tend to lead to a ¤at current pro£le. In this case, as the $m = 2$ island grows, it ¤attens the pro£le locally and causes steepening nearby. This steepening can destabilize nearby rational surfaces which were stable in the original equilibrium. An example of this is shown in Fig. 15. The $m = 2$ island causes a nonlinear destabilization of the 3/2 and the 5/3 islands. At the end of the simulation a large part of the plasma cross section has become stochastic, causing rapid and large scale loss of particles. Disruption of a reactor size tokamak can lead to severe wall damage and must be avoided if possible.

BOOTSTRAP CURRENT

The collisional trapping and detrapping of particles in the banana regime gives rise to a net toroidal current which is called the bootstrap current. It is a net toroidal current in a collisional plasma. It results from a small perturbation of the particle distribution from that of uniform pitch, and is due to toroidal curvature and the existence of banana orbits.

Consider a ¤ux surface ψ, and the two associated ion banana orbits with opposite velocities at $\theta = 0$ on the ¤ux surface of their common origin, as shown in Fig. 16. The inside banana is co-moving at its initial point, *ie* the ion velocity is parallel to the plasma current, and the outside banana is counter moving. Because of the density gradient, on this ¤ux surface there will be more co-moving particles than the counter-moving ones by a factor $-(r/R)^{1/2}(dn/dr)W_b$, with W_b the banana width. The trapped particles carry no net current however, because they do not move toroidally except for the toroidal drift, which is a smaller effect than the bootstrap current. The bootstrap current is due to the associated asymmetry in the co- and counter-moving passing particles necessary to maintain steady

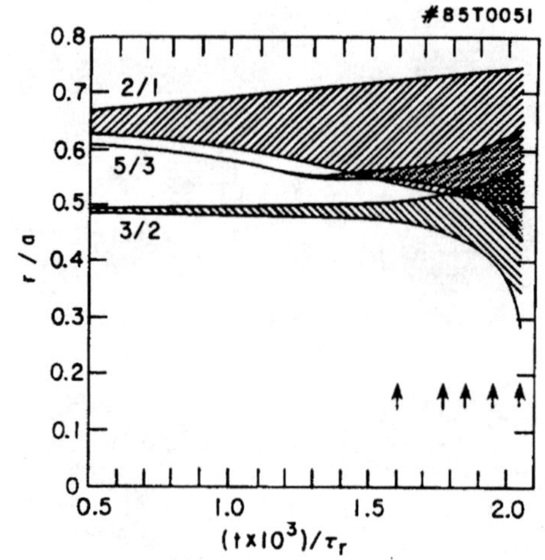

FIGURE 15. Evolution of large m=2 island

state in the presence of trapping and detrapping of the co- and counter-moving populations through small pitch angle scattering. Collisional relaxation of the nonisotropic ¤ow of passing particles produces a net transfer of particles from the more populated inner banana orbits of Fig. 16 to the outer one. For equilibrium the co-moving passing population must be greater than the counter-moving by an amount $-(dn/dr)W_b$. Substituting $W_b = (R/r)^{1/2}q\rho$, multiplying by the average velocity of the passing particles, $\rho/2$ and using the expression for the pressure, $p = n\rho^2/2$, and also noting that we have caluclated the ¤ux of ions, giving negative current, we £nd for the magnitude of the bootstrap current

$$j_{BS} = \left(\frac{r}{R}\right)^{1/2} \frac{1}{B_\theta} \frac{\partial p}{\partial r}. \tag{59}$$

and the current direction is so as to add to the equilibrium toroidal current. This expression is for a given pressure, independent of mass. Electrons produce a ¤ow in the opposite direction, and thus a current also given by Eq. 59.

The resulting passing particle density is modi£ed as shown in Fig. 16. Thus the bootstrap current is driven at the expense of net outward particle diffusion, and is not completely without a price. With no outward ¤ow there can be no

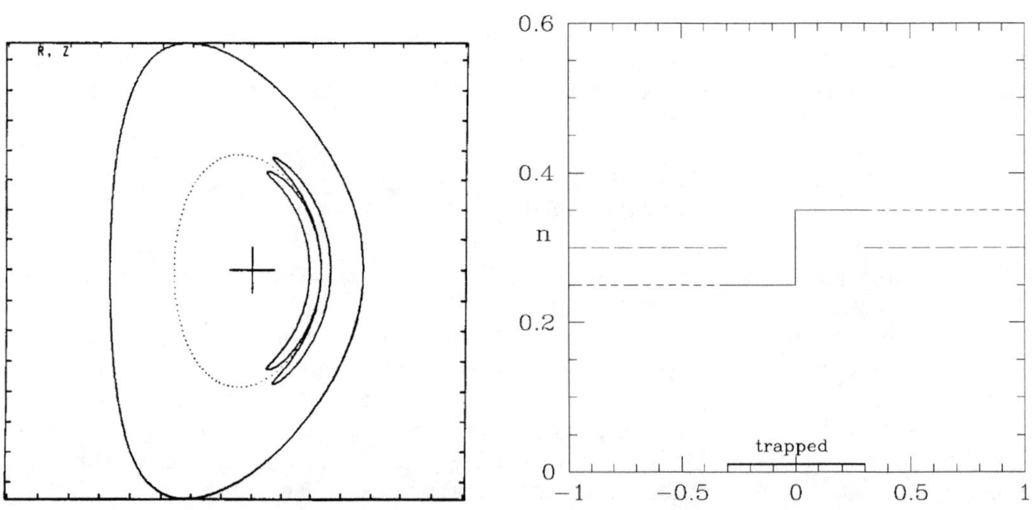

FIGURE 16. Banana orbits originating at $\theta = 0$ with opposite pitches, and schematic plasma density at the ¤ux surface. Dashed line is collisionless passing particle uniform pitch distribution, short dashed line is the collisional modi£cation of it.

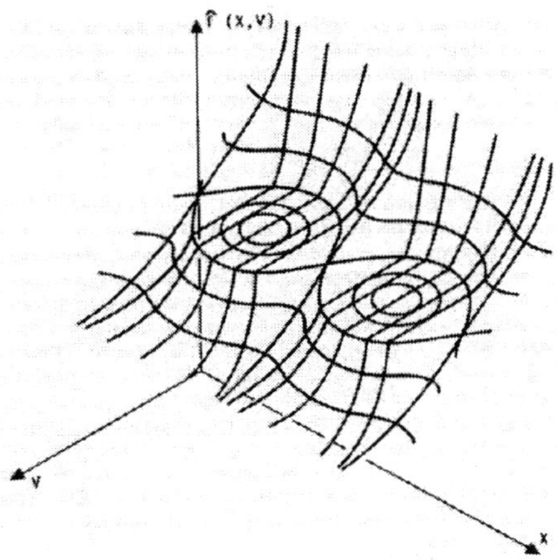

FIGURE 17. Schematic plasma density near an island, showing resulting large density gradient near X-point and flattening near O-point

bootstrap current.

Bootstrap current is especially important because of the possibility of producing almost all of the current profile for a tokamak, and thus making steady state operation possible. This is much easier to do if the desired current profile has very little current on axis, or equivalently very high q on axis. This is called an advanced tokamak configuration, and is of special interest because of the very low levels of transport discovered to exist in configurations with very small or even negative shear.

NEOCLASSICAL TEARING

Small magnetic islands can always occur due to fluctuations in the magnetic field. Once such an island occurs, it modifies the local pressure and density gradient. In the vicinity of the island X-point particles freely pass from inner to outer flux surfaces, whereas the density is locally flattened by an O-point.

In Fig. 17 is shown the modification of the plasma density in the vicinity of an island. Note that a large density gradient is present at the island X-point, but there is no gradient at the O-point. By Eq. 57 this is destabilizing for island growth through the production of bootstrap current. This effect is called neoclassical tearing.

ACKNOWLEDGMENTS

This work was partially supported by the U.S. Department of Energy Grant DE-FG03-94ER54271. The author is indebted to Leonid Zakharov and Allen Boozer for many useful discussions. The Kruskal-Newton analysis was developed with Thomas Fischaleck.

REFERENCES

1. White, R. B., *The Theory of Toroidally Confined Plasmas*, Imperial College Press (2006). See this book for many other references on tearing modes, equilibria, and other topics related to tokamaks.
2. White, R. B., *Asymptotic Analysis of Differential Equations*, Imperial College Press (2005).
3. Isaac Newton, *Methods of series and fluxions*, in *The Mathematical Papers of Isaac Newton* edited by D. T. Whiteside, Volume III 1670-1673, pp 50-71, Cambridge University Press, 1969
4. Kruskal M., Asymptotology, in "Mathematical Models in Physical Sciences" (proceedings of conference of that name at University of Notre Dame, Apr 15–17, 1962), edited by S Drobot and P A Viebrock, Prentice-Hall, Englewood Cliffs NJ, 17–47, with discussion 47–48 (1963).
5. T. Fischaleck and R. B. White, to be published.
6. Kadomtsev, B.B., Fiz. Plazmy 1, 710 (1975) [Sov. J. Plasma Phys. 1, 389 (1975)]

Transport of parallel momentum by collisionless drift wave turbulence

P. H. Diamond*, C. J. McDevitt*, Ö. E. Gürcan*, T. S. Hahm† and V. Naulin**

*Center for Astrophysics and Space Sciences, University of California, San Diego, 9500 Gilman Dr., La Jolla, CA92093-0424
†Princeton University, Princeton Plasma Physics Laboratory, Princeton, NJ 08543-0451 USA
**Danish Tech. Univ., Riso Natl. Lab., EURATOM Assoc., DK-4000 Roskilde, Denmark

Abstract. This paper presents a novel, unified approach to the theory of turbulent transport of parallel momentum by collisionless drift waves. The physics of resonant and non-resonant off-diagonal contributions to the momentum flux is emphasized, and collisionless momentum exchange between waves and particles is accounted for. Two related momentum conservation theorems are derived. These relate the resonant particle momentum flux, the wave momentum flux and the refractive force. A perturbative calculation, in the spirit of Chapman-Enskog theory, is used to obtain the wave momentum flux, which contributes significantly to the residual stress. A general equation for mean $k_{\|}$ ($\langle k_{\|}\rangle$) is derived and used to develop a generalized theory of symmetry breaking. The resonant particle momentum flux is calculated, and pinch and residual stress effects are identified. The implications of the theory for intrinsic rotation and momentum transport bifurcations are discussed.

Keywords: Impurities, turbulence, tokamaks, transport
PACS: 52.25.Vy,52.35.Ra,51.20.+d

INTRODUCTION

Momentum transport - in particular, the turbulent transport of toroidal momentum has long been of interest to researchers in magnetic fusion and tokamak dynamics. The earliest investigations of toroidal momentum confinement were motivated by the desire to understand neutral beam injection performance and to probe the basic physics of confinement processes by comparing thermal and momentum diffusivities, χ_i, χ_ϕ, etc [1]. Subsequently, the dual realizations that electric field shear is a key control parameter for enhanced confinement (via turbulence suppression [2]) and that toroidal rotation 'fed back' on electric field shear via radial force balance sparked [3] further interest in toroidal momentum transport and its interplay with confinement [4]. This growth in interest was accompanied by the developing realization that both self-generated (via Reynolds stress [5]) and externally driven flows contributed to electric field shear v'_E. Key experiments included the discovery of the VH-mode [6], in which toroidal velocity shear regulated itself and other confinement channels via v'_E, the discovery that a 'pinch' or inward convection in the flux of toroidal momentum was required to model momentum profile structure [7] (much as a particle pinch is required to explain density profile structure), and studies of the fluctuation-driven parallel Reynolds stress $\langle \tilde{v}_r \tilde{v}_{\|}\rangle$ and its relation to local flow profile evolution [8].

Interest in toroidal momentum transport was further stimulated by the discovery of 'spontaneous' or 'intrinsic' rotation [9] and the realization that such intrinsic rotation may be required to suppress resistive wall modes in ITER [10], where NBI is of limited utility and high cost. Spontaneous rotation refers to the observation that tokamak plasmas appear to rotate toroidally, at quite healthy velocities, in the absence of any apparent toroidal momentum input. The effort to understand spontaneous rotation has two thrusts, namely:

i.) an effort to understand non-diffusive, off-diagonal contributions to the momentum flux which may account for radially inward transport of momentum and resulting rotation profile peaking on axis.

ii.) an effort to understand the role of edge plasma dynamics in the onset of spontaneous rotation. In particular, several observations of the rotation evolution suggest that the profile of spontaneous rotation evolves inward, from the separatrix, so the edge may constitute a 'dynamic' boundary condition or (possibly) a source.

As part of (ii.), there has been intense interest in uncovering a possible fluctuation-driven momentum pinch or inward stress and relating it to the overall profile self-organization structure and energetics of the plasma [11]. Efforts in this

CP1013, *Turbulent Transport in Fusion Plasmas, First ITER International Summer School*
edited by S. Benkadda

vein have discovered an inward convective pinch by perturbation experiments [12] utilizing ripple loss with pulsed NBI, and have linked spontaneous rotation to the off-set rotation values deduced from scans of rotation vs. torque [13]. As part of (ii.), one striking observation which has emerged is that the Rice/ITPA database scaling $\Delta v_\phi(0) \sim \Delta w_P/I_P$ is most clearly manifested in H-mode plasmas [14], suggesting that an edge effect may be fundamental to its origin. Other edge plasma studies suggest a link of spontaneous rotation in L-mode to scrape-off-layer flows (SOL), which can be manipulated by moving the divertor magnetic null-point location [15]. This has lead to the speculation that the SOL flows may "spin-up" the core, but care must be taken with this interpretation of the data. Another striking feature of these experiments is the apparent disconnect between L-mode and H-mode spontaneous rotation phenomenology, suggesting that there may be an intimate link between transition dynamics (involving v'_E, for example) and spontaneous rotation. Studies of the relation between edge pedestal structure and intrinsic rotation velocity scaling could be relevant here, but are conspicuous by their absence. Overall, a picture is emerging which suggests that in H-mode plasmas, intrinsic rotation builds up near the edge and is then re-distributed inward, thus producing a peaked rotation profile. In addition, interesting studies on the TCV tokamak have also uncovered a new class of momentum transport bifurcations which occur in the core plasma [16]. In a somewhat related vein, recent studies on the JT60-U tokamak also suggest that there is a departure from a simple diffusion-convection model of momentum transport in regions of steep ∇P located in the discharge core [17]. Generally, however, rather little is known about the interesting possibility of intrinsic rotation in torque-free, ITB plasmas. Finally, virtually none of these studies take any serious account of the possibility of anomalous transport of *poloidal* momentum, as observed in JET [18]. Poloidal rotation may be an especially important player at the edge or during an L→H transition.

Theoretical approaches to the problem of turbulent transport of toroidal momentum and intrinsic rotation have focused on attempts to calculate the various elements of the momentum flux [19]. Various works include calculations of the momentum diffusivity χ_ϕ [20], calculations of the momentum convection velocity V in various models [21, 22, 23, 24], and calculations of the residual stress driven by fluctuations and ∇P, which can act as a *local* anomalous momentum source [25]. Most of the calculations of off-diagonal flux elements involve some assumption of a mechanism for broken $k_\parallel$ spectral symmetry, since $\langle \tilde{v}_r \tilde{v}_\parallel \rangle \sim \langle \nabla_y \tilde{\phi} \nabla_\parallel \tilde{\phi} \rangle$ requires $\langle k_\parallel \rangle_{eff} \neq 0$ for a significant non-diffusive component. In one case, such symmetry breaking occurred via an interesting interplay of curvature coupling and ballooning structure [23]. In scenarios involving the electric field shear, progress was facilitated by drawing upon previous results for the effect of shear on turbulence and transport [26, 27, 28, 29]. Other approaches have invoked the effects of blobs or other coherent structures [30]. Most of the calculations implemented so far have been extremely simple and based on fluid models. Even the few kinetic calculations have not treated the response of both resonant and non-resonant particles and have not addressed parallel acceleration effects. However, the general structure of a kinetic model has been discussed, to some extent, in Ref. [31].

This paper presents a novel unified approach to the theory of turbulent transport of parallel (i.e. ~toroidal) momentum by collisionless drift waves. There are two principal motivations for yet another new look at this very old problem. These are:

i.) the need to identify and understand off-diagonal contributions to the turbulent momentum flux in collisionless drift wave turbulence. Such off-diagonal terms are not limited to convection (i.e. a "pinch"), but may also involve a residual stress, i.e. an element of the momentum flux not proportional to $\langle v_\phi \rangle$ or $\partial \langle v_\phi \rangle / \partial r$. Also in a collisionless plasma, off-diagonal terms also can conflate momentum *transport* with momentum *transfer* between resonant particles and waves.

ii.) The need to address (i.) within a theoretical framework which accounts for wave-particle momentum exchange and the somewhat related problem of distinguishing between resonant and non-resonant transport [32]. Confronting the latter is crucial for understanding momentum transport in a stationary plasma, where the quasi-linear theory of non-resonant transport is no longer meaningful, but where significant momentum transport may occur via waves.

The phenomenology which drives the interest in off-diagonal contributions to the momentum flux, such as a possible inward convection (i.e. momentum "pinch") and a possible ∇P-driven residual stress, has been discussed above. The considerations which drive motivation (ii.) become apparent upon comparison of the simple physics of turbulent momentum transport to the more familiar phenomena of turbulent transport of particles and energy. Unlike density, momentum can be exchanged between waves and particles via resonant interaction with the underlying turbulence. Unlike the thermal energy density, which *always* exceeds the wave energy density (i.e. $\varepsilon_\omega \ll nT$) since $\{\tilde{n}/n, e\tilde{\Phi}/T\} \ll 1$, the non-resonant particle momentum density of the fluctuations is in fact *equal* to the wave momentum density for electrostatic drift wave turbulence, since the field momentum density of the latter is negligible. This momentum equality has obvious implications for the onset of intrinsic rotation. Hence while transport theories may often neglect wave-particle energy exchange (though electron-ion energy coupling via fluctuations has been studied), and wave energy transport, they cannot ignore wave-particle momentum exchange or transport of wave momentum. Thus,

i.) wave-particle momentum *transfer* is necessarily a key element of the momentum transport process, since "wave momentum" is equal to non-resonant particle momentum and can be exchanged with resonant particles.
ii.) the distinction between resonant and non-resonant transport processes is critical, since the former is mediated primarily by *particles* and likely is dominant near marginality in stiff profile regimes, while the latter is mediated by *waves* and likely is dominant in regimes of strong turbulence. A complete theory must address *both* processes.

Since non-resonant transport is mediated primarily by waves, it seems appropriate to describe it in terms of wave dynamics, quantities, populations etc. The strong coupling between momentum transport and momentum transfer is readily appreciated upon a moments consideration of the mean field (i.e. quasi-linear) evolution equation for the mean distribution function $\langle f \rangle$. This may easily be derived by averaging the drift/gyro-kinetic equation

$$\frac{\partial f}{\partial t} + v_{\|}\nabla_{\|} f - \frac{c}{B}\nabla\Phi \times \hat{\mathbf{b}} \cdot \nabla f + \frac{e}{m}E_{\|}\frac{\partial f}{\partial v_{\|}} = 0$$

to obtain:

$$\frac{\partial \langle f \rangle}{\partial t} = \frac{1}{r}\frac{\partial}{\partial r}\left[r\left\{ D_{r,r}\frac{\partial}{\partial r}\langle f \rangle + D_{r,v_{\|}}\frac{\partial \langle f \rangle}{\partial v_{\|}} \right\}\right] + \frac{\partial}{\partial v_{\|}}\left\{ D_{v_{\|},r}\frac{\partial \langle f \rangle}{\partial r} + D_{v_{\|},v_{\|}}\frac{\partial \langle f \rangle}{\partial v_{\|}} \right\}$$

Here $D_{r,r}$ accounts for the familiar cross-field, scattering and transport process, and $D_{v_{\|},v_{\|}}$ is simply the stochastic acceleration along the mean magnetic field. More interesting indeed, however, are the cross-terms $D_{r,v_{\|}}$ and $D_{v_{\|},r}$ which represent correlated radial scattering and parallel acceleration. Within the quasi-linear approximation, these are:

$$D_{r,v_{\|}} \approx \sum_{\mathbf{k}} c_s^2 \rho_s^2 k_\theta k_{\|} \left| \frac{e\tilde{\Phi}}{T_e} \right|^2 \Omega_i R\left(\omega - k_{\|} v_{\|}\right)$$

where $R\left(\omega - k_{\|}v_{\|}\right)$ is the response function. $D_{r,v_{\|}}$ is striking in that:

i.) it describes, within kinetic theory, precisely the process of correlated radial scattering and parallel acceleration which is required to generate a sheared parallel flow, as suggested by the phenomenon of intrinsic rotation.
ii.) it is explicitly proportional to the correlation $\langle \nabla_\theta \Phi \nabla_{\|} \Phi \rangle$, as is the Reynolds stress $\langle \tilde{v}_{r_{E\times B}} \tilde{v}_{\|} \rangle$, which is responsible for regulating radial transport of parallel momentum. Note that the sign of this correlator, and of $D_{r,v_{\|}}$ can be either positive or negative, and so can produce a net inward or outward momentum flux.
iii.) wave-particle resonance provide a direct way for obtaining for $D_{r,v_{\|}} \neq 0$. In particular, in contrast to speculations advanced in the course of previous studies, neither electric field shear nor toroidicity are necessarily required for off-diagonal momentum flux contributions.

A second motivation for this work is the long standing question of how to reconcile non-resonant transport with stationary turbulence. This question is often overlooked by transport models. Many important drift instabilities, such as the ion temperature gradient driven (ITG) mode, can have a strongly non-resonant, fluid-like character - particularly in regimes far from marginal stability. This observation naturally motivates the question of how to calculate the stationary nonlinear counterpart to this intrinsically non-stationary process. The traditional procedure of replacing the factor of $|\gamma_k|$ (or, incorrectly, γ_k) in the quasilinear theory with a decorrelation rate ignores the fact that for electrostatic turbulence, non-resonant particle dynamics are closely linked to wave dynamics, and so should be described in terms of the wave population, and other quantities characteristic of wave dynamics. In particular, consideration of momentum transport leads us to realize that since non-resonant particle momentum is the same as the wave momentum, then transport of non-resonant particle momentum is most naturally described as a process of transport of wave momentum. Another key issue here is the physics of the irreversibility which underlies the wave momentum transport. *Thus, a major focus of this paper is the formulation of the calculation of the stationary state counterpart of the non-resonant particle momentum flux in terms of the properties and dynamics of the wave population density.* As we will see, this formulation bares certain similarities to the structure of the theory of radiation hydrodynamics in the limit of large optical depth [33]. The approach and results may also be of interest in the context of other problems involving mesoscale dynamics, most notably that of turbulence spreading [34, 35, 36, 37].

In this paper, then, we present a general theory of toroidal momentum transport by collisionless drift wave turbulence, with special emphasis on the physics of both non-resonant and resonant off-diagonal contributions to the momentum flux. Starting from the fundamental gyro-kinetic equation, we derive a momentum theorem (MT-I) relating momentum evolution to the resonant particle momentum flux and the wave momentum flux. We extend this theorem, derived using a wave-energy balance relation (i.e. generalized Poynting theorem), to one which includes the effect of forces arising from the recoil on the plasma dielectric medium due to wave refraction (i.e. bending of wave rays must exert a force on the plasma!). This more general momentum theorem (MT-II) is derived using the full wave-kinetic equation. We then discuss the varieties of stationary momentum balance, and the possible interplay between the resonant particle momentum flux, the wave momentum flux and the refractive force, along with the significance of these for intrinsic rotation. After this, we undertake a perturbative calculation, in the spirit of Chapman-Enskog theory, to relate the wave momentum flux to the structure of the mean macroscopic wave quanta density distribution in both **x** and **k**. The spatial and wave vector gradient dependence of this flux arise from spatial propagation and shear flow-induced wave wind-up. A generalized equation for the evolution of the mean parallel wave vector $\langle k_{\parallel} \rangle$ is derived from the wave kinetic equation, and used to quantify the possible scenarios for competition between a number of possible and relevant symmetry breaking mechanisms. Of course, the resonant particle momentum transport is calculated as well, and shown to consist of the sum of a diffusive flux, a convective flux (a pinch, i.e. inward convection, in certain cases) and a residual stress contribution. The interplay among these is strongly sensitive to the mechanism of symmetry breaking. Finally, we consider the "bottom line" implications of the theory for intrinsic rotation and momentum transport bifurcations in tokamaks. In particular, we propose a novel class of momentum transport bifurcations and discuss the relation of these to the more general phenomenon of intrinsic rotation in H-mode plasmas.

The remainder of this paper is organized as follows. In Section II, we prove two momentum conservation theorems at the level of mean field theory. The first theorem is a special, albeit simpler and more accessible case of the second. Section III presents a calculation of the radial flux of parallel wave momentum. A related approach allows the derivation of a general evolution equation for the net parallel quasiparticle momentum $\left(\int d\mathbf{k} k_{\parallel} N\right)$, which enables the quantitative comparison of different, competing symmetry breaking mechanisms. Section IV presents the calculation of the resonant particle momentum flux. In Section V, we discuss the theoretical results in the light of known aspects of momentum transport phenomenology. Special attention is devoted to theoretical predictions of various types of momentum transport bifurcations. This section is useful for those who are mainly interested in applications rather than development of the theory. Section VI presents a discussion and conclusions.

MEAN FIELD THEORY AND CONSERVATION OF MOMENTUM IN COLLISIONLESS DRIFT WAVE TURBULENCE

In this section we discuss conservation of momentum and the structure of the momentum budget relation in the mean field theory of drift wave turbulence. We assume a simple geometry r, θ, ϕ and leave extensions to the torus to a future paper. Two theorems of conservation of momentum (MT-I, MT-II) are proved. The latter (MT-II) constitutes a generalization of the former (MT-I). Special emphasis is placed on delineating the roles of resonant particles and waves in momentum balance and transport, as well as on the resolution of several questions related to how non-resonant diffusion effects are manifested in a stationary state.

A brute force approach to the problem of momentum balance and conservation might proceed as follows. Starting from the drift kinetic equation (i.e. for $k_{\perp}\rho_i \ll 1$) for ions in collisionless electrostatic turbulence:

$$\frac{\partial f}{\partial t} + v_{\parallel}\nabla_{\parallel} f - \frac{c}{B}\nabla\Phi \times \hat{\mathbf{b}}\cdot\nabla f + \frac{e}{m}E_{\parallel}\frac{\partial f}{\partial v_{\parallel}} = 0 \tag{1a}$$

the mean field equation follows as

$$\frac{\partial \langle f\rangle}{\partial t} + \frac{1}{r}\frac{\partial}{\partial r}\left[r\left\langle \tilde{v}_r \tilde{f}\right\rangle\right] + \frac{\partial}{\partial v_{\parallel}}\left\langle \frac{e}{m}\tilde{E}_{\parallel}\tilde{f}\right\rangle = 0 \tag{1b}$$

so that the usual quasi-linear calculation gives:

$$\frac{\partial \langle f\rangle}{\partial t} = \frac{1}{r}\frac{\partial}{\partial r}\left[r\left\{D_{r,r}\frac{\partial}{\partial r}\langle f\rangle + D_{r,v_{\parallel}}\frac{\partial \langle f\rangle}{\partial v_{\parallel}}\right\}\right] + \frac{\partial}{\partial v_{\parallel}}\left\{D_{v_{\parallel},r}\frac{\partial \langle f\rangle}{\partial r} + D_{v_{\parallel},v_{\parallel}}\frac{\partial \langle f\rangle}{\partial v_{\parallel}}\right\} \tag{2a}$$

where

$$D_{r,r} = \rho_s c_s \sum_{\mathbf{k}} \rho_s^2 k_\theta^2 \left| \frac{e\widetilde{\Phi}}{T_e} \right|^2 \Omega_i R\left(\omega - k_\parallel v_\parallel\right) \tag{2b}$$

$$D_{r,v_\parallel} = D_{v_\parallel,r} = c_s^2 \sum_{\mathbf{k}} \rho_s^2 k_\theta k_\parallel \left| \frac{e\widetilde{\Phi}}{T_e} \right|^2 \Omega_i R\left(\omega - k_\parallel v_\parallel\right) \tag{2c}$$

$$D_{v_\parallel,v_\parallel} = c_s^2 \Omega_i \sum_{\mathbf{k}} \rho_s^2 k_\parallel^2 \left| \frac{e\widetilde{\Phi}}{T_e} \right|^2 \Omega_i R\left(\omega - k_\parallel v_\parallel\right) \tag{2d}$$

are the coefficients corresponding to radial spatial scattering by stochastic $E \times B$ drifts ($D_{r,r}$), parallel stochastic acceleration as in the 1D Vlasov plasma ($D_{v_\parallel,v_\parallel}$), and correlated parallel acceleration and radial scattering ($D_{r,v_\parallel}$), which is related to the generation of a radially sheared parallel flow. Since plasma momentum $\langle P_\parallel \rangle = m_i \int dv_\parallel v_\parallel \langle f \rangle$ is carried by ions, taking the $v_\parallel$ moment of Eqn. (2a) says that parallel momentum evolves according to:

$$\begin{aligned} \frac{\partial}{\partial t} \langle P_\parallel \rangle = & \frac{m_i}{r} \frac{\partial}{\partial r} \left[r \int dv_{\parallel i} v_{\parallel i} \left(D_{r,r} \frac{\partial \langle f \rangle_i}{\partial r} + D_{r,v_\parallel} \frac{\partial \langle f \rangle_i}{\partial v_\parallel} \right) \right] \\ & + m_i \int dv_{\parallel i} v_{\parallel i} \left[D_{v_\parallel,r} \frac{\partial \langle f \rangle_i}{\partial r} + D_{v_\parallel,v_\parallel} \frac{\partial \langle f \rangle_i}{\partial v_\parallel} \right] \end{aligned} \tag{3}$$

We see, then, that Eqn. (3) has the form:

$$\frac{\partial \langle P_\parallel \rangle}{\partial t} = -\frac{1}{r} \frac{\partial}{\partial r} \left[r \Pi_{r,\parallel} \right] + S_\parallel \tag{4}$$

and states that momentum evolves via *transport* by turbulence induced stress ($\Pi_{r,\parallel}$) and via a local "source or sink", due to wave-particle scattering (which we term momentum *transfer*). Note that the definitions of $\Pi_{r,\parallel}$ and $S_\parallel$ in terms of fundamental microscopic quantities are clear from Eqns. (3,4).

The result of Eqn. (4) is somewhat unsettling on account of the appearance of an explicit local fluctuation driven source $S_\parallel$ for $\partial \langle f \rangle / \partial t$. Recall that the aim here is to determine the off-diagonal momentum *flux* (i.e. transport), in order to understand possible inward convection and stresses. The presence of a local, fluctuation induced *source* significantly complicates this task. One does not know how significant this effect is - though since $D_{r,v_\parallel} \sim D_{v_\parallel,r}$, $S_\parallel$ is clearly not negligible - and neither does one know the fate of momentum coupled to/from the particles to the fluctuations. One thing that *is* clear, however, is that the mean field contributions to $\langle f \rangle$ evolution from parallel acceleration ($e/m \left\langle \widetilde{E}_\parallel \partial \widetilde{f} / \partial v_\parallel \right\rangle$) should not be neglected *a priori*. Another is that a natural way to formulate the theory is in terms of resonant particles and waves, since these conserve momentum between them.

The physical content of $S_\parallel$ in Eqn. (4) is best resolved by re-considering the basic energy and momentum conservation theorems for the quasi-linear theory of the 1D Vlasov plasma. For that well-studied system, we know that energy balance may be stated in either of two equivalent ways : as conservation of the sum of resonant particle (ε_{RP}) and wave (ε_W) energy densities, i.e.

$$\frac{d}{dt} \left(\varepsilon_{RP} + \varepsilon_W \right) = 0 \tag{5a}$$

or as conservation of the sum of particle and electrostatic field energy density, i.e.

$$\frac{d}{dt} \left(\varepsilon_P + \varepsilon_F \right) = 0\,. \tag{5b}$$

The equivalence of these two results follows from the fact that $\varepsilon_W = \varepsilon_F + \varepsilon_{NRP}$ - i.e. waves (collective modes) are supported both by non-resonant particles and fields. We observe that Eqn. (5a) can be alternatively re-cast in terms of the quasi-particle density N (wave quanta density or - usually action density) by noting that $\varepsilon_W = N(\mathbf{k}, \mathbf{x}, t)\, \omega_{\mathbf{k}}$, so that the sum of resonant particle energy and quasi-particle energy is conserved [32].

Similarly, for momentum, (in the absence of external sources) we can write:

$$\frac{d}{dt} \left(P_{RP} + P_W \right) = 0 \tag{6a}$$

and

$$\frac{d}{dt}(P_P + P_F) = 0 \,. \tag{6b}$$

Here P_{RP} is the resonant particle momentum density, P_W is the wave momentum density (sometimes referred to as the pseudo-momentum density), and $P_P = P_{RP} + P_{NRP}$ is the total particle momentum density. Equation (6b) follows from (6a) since the field momentum P_F for electrostatic turbulence is necessarily zero ($\widetilde{E} \times \widetilde{B}/4\pi c^2 = 0$, since $\widetilde{B} = 0$). *Thus the increment of non-resonant particle momentum density is necessarily equal to the wave momentum density*, i.e.

$$\delta P_{NRP} = P_W \,, \tag{6c}$$

where $\delta P_{NRP} = P_{NRP} - P_{NRP}^{(0)}$ is the momentum density associated with the *fluctuation driven* non-resonant particle flow, and $P_{NRP}^{(0)}$ is the zeroth order non-resonant particle momentum density, associated with the background (mean) flow. Equation (6c) is a simple, albeit important, identity which is applicable to *any* system of electrostatic turbulence, including electrostatic drift wave turbulence.

In this paper, we focus on electrostatic drift wave driven transport. Thus, turning to momentum conservation in 3D electrostatic drift wave turbulence, it seems incontrovertible that resonant particle momentum density will be conserved against wave momentum density. In contrast to the "textbook" 1D Vlasov problem, however, the radial fluxes of both quantities will enter, as well. Thus, we can expect a conservation relation of the form:

$$\frac{\partial}{\partial t}\langle P_\parallel\rangle_{RP} + \frac{1}{r}\frac{\partial}{\partial r}\left[r\left\langle \widetilde{v}_E \widetilde{P}_\parallel\right\rangle_{RP}\right] + \frac{\partial}{\partial t}\langle P_\parallel\rangle_w + \frac{1}{r}\frac{\partial}{\partial r}\left[r\Pi^w_{r,\parallel}\right] = 0 \tag{7}$$

where $\left\langle \widetilde{v}_E \widetilde{P}_\parallel\right\rangle$ is the radial $E \times B$ flux of resonant particle parallel momentum and $\Pi^w_{r,\parallel}$ is the radial flux of parallel wave momentum. In a quasi-particle or eikonal formulation of wave dynamics, $\left\langle \Pi^w_{r,\parallel}\right\rangle = \int dk_\parallel v_{gr} k_\parallel \langle N\rangle$, where v_{gr} is the radial group velocity of the waves [38]. Since, however, $P_W = \delta P_{NRP}$, Eqn. (7) immediately undergoes an important simplification to

$$\frac{\partial}{\partial t}\langle \delta P_\parallel\rangle + \frac{1}{r}\frac{\partial}{\partial r}\left[r\left(\left\langle \widetilde{v}_{Er}\widetilde{P}_\parallel\right\rangle_{RP} + \left\langle \Pi^w_{r,\parallel}\right\rangle\right)\right] = 0 \tag{8}$$

At this point, a careful discussion of how $\langle \delta P_\parallel\rangle$ (the mean momentum increment driven by the fluctuations) relates to the evolution of the background momentum density is necessary. The total momentum density $\langle P_\parallel\rangle$ is defined as the sum of the momenta of the mean flow and that of the fluctuations. Thus, we have

$$\langle P_\parallel\rangle = P_{\parallel,0} + \langle \delta P_\parallel\rangle \tag{9a}$$

where

$$\frac{d\langle P_\parallel\rangle}{dt} = S \tag{9b}$$

i.e. we find it useful to introduce an external momentum density source $S(r,t)$ (i.e. such as NBI-driven momentum input). Thus,

$$\frac{\partial}{\partial t}\left(P_{\parallel,0} + \langle \delta P_\parallel\rangle\right) + \frac{1}{r}\frac{\partial}{\partial r}\left(r\left[\left\langle \Pi^{NRP}_{r,\parallel}\right\rangle_0 + \left\langle \widetilde{v}_{Er}\widetilde{P}_\parallel\right\rangle_{RP} + \left\langle \Pi^w_{r,\parallel}\right\rangle\right]\right) = S \tag{9c}$$

follows as the dynamical evolution equation for total momentum. Here $\langle \Pi_{NR}\rangle_0$ is the zeroth order (non-resonant) momentum flux which corresponds to the non-resonant quasi-linear flux plus the neoclassical flux (i.e. the latter being the only surviving contributor to the flux in the absence of fluctuations). The quasi-linear part is $O\left(\left|e\widetilde{\Phi}/T_e\right|^2\right)$ and is proportional to the magnitude of the linear growth rate, $|\gamma_k|$. This differs from the wave momentum flux in a subtle way. In this regard, and by way of contrast, note that $\left\langle \Pi^w_{r,\parallel}\right\rangle$ in contrast, need not be limited to a second order effect in fluctuation level, and can involve "spreading" of the momentum of fluctuations, and other nonlinear processes. Physically, $\left\langle \Pi^{NRP}_{r,\parallel}\right\rangle_0$ is a necessary part of the description of the momentum density evolution in time, though it asymptotes to the (usually negligible) collisional level in the case of stationary turbulence.

Regarding temporal evolution, it is useful here to discuss the distinctions between "stationary mean momentum profile", "stationary turbulence" and "intrinsic rotation". Starting from Eqn. (9c), stationarity of the mean momentum profile requires $\partial_t \left[P_{\parallel,0} + \langle \delta P_\parallel \rangle\right] = 0$, so the momentum is steady on *transport* time scales. This then gives the usual balance between momentum input and transport, i.e.

$$\frac{1}{r}\frac{\partial}{\partial r}\left(r\left[\left\langle \Pi^{NRP}_{r,\parallel}\right\rangle_0 + \left\langle \widetilde{v}_{Er}\widetilde{P}_\parallel\right\rangle_{RP} + \left\langle \Pi^{w}_{r,\parallel}\right\rangle\right]\right) = S \tag{10a}$$

so the mean profile is determined by

$$\left\langle \Pi^{NRP}_{r,\parallel}\right\rangle_0 + \left\langle \widetilde{v}_{Er}\widetilde{P}_\parallel\right\rangle_{RP} + \left\langle \Pi^{w}_{r,\parallel}\right\rangle = \frac{1}{r}\int r' dr' S\left(r'\right) \tag{10b}$$

and the profile boundary conditions. In addition, it is necessary to parametrize $\left\langle \Pi^{w}_{r,\parallel}\right\rangle$ in terms of plasma parameter profiles. Generally this need not be a purely local relation. Note that Eqn. (10b) allows the possibility that the turbulence can be at least weakly non-stationary, though the profile is not globally evolving. For stationary turbulence, we have $\left\langle \Pi^{NRP}_{r,\parallel}\right\rangle_0 \to \langle \Pi_{r,\parallel}\rangle^C$ where $\langle \Pi_{r,\parallel}\rangle^C$ is the collisional momentum flux, so that Eqn. (10b) becomes:

$$\langle \Pi_{r,\parallel}\rangle^C + \left\langle \widetilde{v}_{Er}\widetilde{P}_\parallel\right\rangle_{RP} + \left\langle \Pi^{w}_{r,\parallel}\right\rangle = \frac{1}{r}\int r' dr' S\left(r'\right) \tag{10c}$$

which is the usual stationary state (for both mean field and turbulence) criterion. Note that $\left\langle \Pi^{w}_{r,\parallel}\right\rangle$ accounts for the non-resonant transport in this stationary state, and that this transport is purely *wave transport*. Equation (10c) indicates that the most important results of this paper are the momentum theorems for the evolution of $\langle \delta P_\parallel \rangle$ (i.e. MT-I, MT-II, proved in this Section) and the actual calculations of $\left\langle \widetilde{v}_{Er}\widetilde{P}_\parallel\right\rangle_{RP}$ and $\left\langle \Pi^{w}_{r,\parallel}\right\rangle$ (discussed in Sections III and IV respectively). Finally, the case of "intrinsic rotation" for stationary turbulence corresponds to one where $S \to 0$, i.e. where:

$$\langle \Pi_{r,\parallel}\rangle^C + \left\langle \widetilde{v}_{Er}\widetilde{P}_\parallel\right\rangle_{RP} + \left\langle \Pi^{w}_{r,\parallel}\right\rangle = 0 \tag{10d}$$

and which has non-trivial profile solutions in the absence of momentum input. Note that the boundary conditions for the solution of Eqn (10d) will have a significant impact on the answer to this question. The interesting possibility of $\langle S \rangle = 0$ but $\widetilde{S} \neq 0$, i.e. a momentum noise source due to avalanches etc. [39, 40], will be discussed in a future paper. In particular, (for $\langle \widetilde{S} \rangle = 0$) it would be interesting to compare the intrinsic rotation profile for $\widetilde{S} = 0$ with that calculated for small but finite $\widetilde{S}$.

Equation (8) constitutes the first momentum conservation theorem (MT-I) to be presented in this study. Two salient features of Eqn. (8) are:

i.) the somewhat ill-defined non-resonant particle momentum flux has been eliminated in favor of the wave-momentum flux $\left\langle \Pi^{w}_{r,\parallel}\right\rangle$, which is well defined for stationary turbulence. The irreversibility which underpins $\left\langle \Pi^{w}_{r,\parallel}\right\rangle$ in a stationary state must necessarily originate in nonlinear wave-wave and wave-particle interactions.

ii.) the ambiguous anomalous source $S_\parallel$ which appears in Eqn. (3) is replaced by $-\nabla \cdot \left\langle \hat{\mathbf{r}}\Pi^{w}_{r,\parallel}\right\rangle$, which suggests that the apparent "source or sink" due to wave-particle interaction must ultimately dispose of momentum by coupling it to radially propagating waves. This suggests that momentum transport is best described in terms of the combined flux of resonant particles and waves, since these two populations conserve momentum between them. This leads us naturally to think in terms of a picture of resonant particles and quasi-particles.

Points (i.) and (ii.) go a long way toward resolving the questions inherent in Eqn. (3). The challenge, now, is to *prove* MT-I, as expressed by Eqn. (8). We now turn to this task.

Momentum theorem I is most directly proved by relating evolution of net resonant particle momentum to the sum of the resonant particle momentum flux *and* the net force exerted on the plasma by the resonant particles. The latter can then be linked to wave-momentum, yielding a net momentum conservation theorem. The evolution equation for the *total resonant particle momentum* is derived by adding the parallel velocity moments of the individual resonant electron and resonant ion drift kinetic equations:

$$\frac{\partial}{\partial t}\langle P_\parallel \rangle_{RP} + \frac{1}{r}\frac{\partial}{\partial r}\left[r\left\langle \widetilde{v}_{Er}\widetilde{P}_\parallel\right\rangle_{RP}\right] = e\left\langle \widetilde{E}_\parallel\left(\widetilde{n}_i - \widetilde{n}_e\right)\right\rangle_{RP} \tag{11}$$

The right hand side of Eqn. (11) does not vanish on account of quasi-neutrality, since $\widetilde{n}_i = \widetilde{n}_e$ is not equivalent to $\widetilde{n}_{iR} = \widetilde{n}_{eR}$, i.e. *resonant* ions need not be neutralized by resonant electrons (see Appendix A for the explicit evaluation of this term). It is convenient to express the density perturbations in terms of the susceptibilities $\chi_{i,e}(k, \omega_k)$ via

$$\widetilde{n}_{(i,e)\mathbf{k},\omega_{\mathbf{k}}} = \frac{n_0 e}{T_e} \widetilde{\Phi}_{\mathbf{k},\omega_{\mathbf{k}}} \chi_{(i,e)}(\mathbf{k}, \omega_{\mathbf{k}})$$

Note that $\chi_{(i,e)}$ refers to a response function, not a thermal diffusivity! Recalling that $\chi_{(i,e)}(\mathbf{k}, \omega_{\mathbf{k}})$ for resonant particles is imaginary, we thus obtain

$$e\left\langle \widetilde{E}_{\|} (\widetilde{n}_i - \widetilde{n}_e) \right\rangle = -n_0 T_e \sum_{\mathbf{k}} k_{\|} \left(\mathrm{Im}\chi_i(\mathbf{k}, \omega_{\mathbf{k}}) - \mathrm{Im}\chi_e(\mathbf{k}, \omega_{\mathbf{k}})\right) \left| \frac{e\widetilde{\Phi}_{\mathbf{k}}}{T_e} \right|^2$$

This relates the net force density acting on resonant particles to their dielectric response functions. It is now convenient to recall the quasi-linear wave-energy ("Poynting") theorem for drift waves:

$$\frac{\partial}{\partial t} \varepsilon_{\omega_{\mathbf{k}}} + \nabla \cdot \mathbf{S}_{\mathbf{k}} + Q_{\mathbf{k}} = 0 \tag{12}$$

i.e. that wave energy density $\varepsilon_{\omega_{\mathbf{k}}}$ evolves via divergence of the wave energy density flux $\mathbf{S}_{\mathbf{k}}$ and local dissipation $Q_{\mathbf{k}}$. Here, $\varepsilon_{\omega_{\mathbf{k}}}$, $\mathbf{S}_{\mathbf{k}}$ and $Q_{\mathbf{k}}$ are given, respectively, by:

$$\varepsilon_{\omega_{\mathbf{k}}} = n_0 T_e \omega_{\mathbf{k}} \left(\frac{\partial \chi}{\partial \omega}\right)_{\omega_{\mathbf{k}}} \left| \frac{e\widetilde{\Phi}}{T_e} \right|^2 \tag{13a}$$

$$\mathbf{S}_{\mathbf{k}} = -n_0 T_e \omega_{\mathbf{k}} \left(\frac{\partial \chi}{\partial \mathbf{k}}\right)_{\omega_{\mathbf{k}}} \left| \frac{e\widetilde{\Phi}}{T_e} \right|^2 \tag{13b}$$

$$Q_{\mathbf{k}} = -n_0 T_e \omega_{\mathbf{k}} \left(\mathrm{Im}[\chi]\right)_{\omega_{\mathbf{k}}} \left| \frac{e\widetilde{\Phi}}{T_e} \right|^2 \tag{13c}$$

where

$$\chi(\mathbf{k}, \omega) = \chi_i(\mathbf{k}, \omega) - \chi_e(\mathbf{k}, \omega) \tag{13d}$$

Equation (12) is a straightforward adaptation of a more general result for waves in a dielectric medium to the special case of drift waves, which we consider here. The result of Eqns. (12,13) is proved in Refs. [41, 42]. Using Eqns. (12,13), we immediately note that

$$e\left\langle \widetilde{E}_{\|} (\widetilde{n}_i - \widetilde{n}_e) \right\rangle_R = \sum_{\mathbf{k}} \frac{k_{\|}}{\omega_{\mathbf{k}}} Q_{\mathbf{k}} = -\sum_{\mathbf{k}} \frac{k_{\|}}{\omega_{\mathbf{k}}} \left(\frac{\partial}{\partial t} \varepsilon_{\omega_{\mathbf{k}}} + \nabla \cdot \mathbf{S}_{\mathbf{k}} \right)$$

thus directly relating the dielectric medium's force on resonant particles to the evolution of wave energy! It is no surprise to see the proportionality between resonant density imbalance $(\widetilde{n}_{iR} - \widetilde{n}_{eR})$ and $\nabla \cdot \mathbf{S}$ (the divergence of the wave energy density flux) emerge in this theory, since the latter includes radial wave propagation, which originates in the polarization charge contribution to quasi-neutrality, and which is required to neutralize $\widetilde{n}_{iR} - \widetilde{n}_{eR}$. Since the spatio-temporal evolutions of $\varepsilon_{\omega_{\mathbf{k}}}$ and $\mathbf{S}_{\mathbf{k}}$ are, by construction, 'slow' - i.e. corresponding to envelope scales - we have $\varepsilon_{\omega}^{-1} \partial \varepsilon_{\omega} / \partial t \ll \omega_{\mathbf{k}}$ and $|S|^{-1} |\nabla_r S| < |k|$, so

$$e\left\langle \widetilde{E}_{\|} (\widetilde{n}_i - \widetilde{n}_e) \right\rangle_{RP} = -\sum_{\mathbf{k}} \left(\frac{\partial}{\partial t} \langle P_{\|} \rangle_W + \frac{1}{r} \frac{\partial}{\partial r} \left[r \left\langle \Pi^{w}_{r,\|} \right\rangle \right] \right) \tag{14}$$

Since $P^w_{\|} = (k_{\|}/\omega_{\mathbf{k}}) \varepsilon_{\omega}$ and $(k_{\|}/\omega_{\mathbf{k}}) \mathbf{S} = \Pi^w$, consistent with the simple geometry assumed here, only radial variation of the envelope is considered. To complete the proof of (MT-I), we need only combine Eqn. (14) with equation (11), and note that $\langle P_\parallel \rangle_{RP} + \langle P_\parallel \rangle_W = \langle \delta P_\parallel \rangle$, since wave momentum and non-resonant particle momentum are identical in the frame of the background plasma. These steps yield:

$$\frac{\partial \langle \delta P_\parallel \rangle}{\partial t} + \frac{1}{r}\frac{\partial}{\partial r}\left[r \left(\left\langle \widetilde{v}_{Er} \widetilde{P}_\parallel \right\rangle_{RP} + \int d\mathbf{k} v_{gr} P^{w}_{\parallel \mathbf{k}} \right) \right] = 0 \tag{15}$$

which proves the first momentum theorem. Here $P^{w}_{\parallel \mathbf{k}}$ is the parallel wave momentum density for the wave with the wave vector $\mathbf{k}$, and radial group velocity v_{gr}. Note that Eqn. (15) eliminates the local force in (11), in favor of the mean radial flux of parallel wave momentum $\left\langle \Pi^{w}_{r,\parallel} \right\rangle$. More generally, Eqn. (15) relates the total, fluctuation-induced parallel force on the plasma to the sum of the fluxes of resonant particle and wave momentum. Of course, given the disparity between electron and ion inertia ($m_e \ll m_i$), $\langle p_\parallel \rangle$ is carried primarily by ions. However we emphasize that no limiting assumptions concerning the electron response (i.e. "adiabatic" or "i-δ") were made in the course of this derivation.

The alert reader may notice that the result of Eqn. (15) (i.e. MT-I) does *not* account for the force exerted on the plasma due to wave refraction. In particular, given sheared flows and other profile variations, drift wave trajectories will bend and refract, so the wave packet momentum will exhibit a concomitant change in direction [43, 44]. Of course, since the waves and dielectric media are (strongly) coupled, the process of wave refraction will induce a recoil force on the plasma. This *refractive force* will then necessarily change the plasma momentum, and thus should enter the momentum budget [45, 46, 47]. A proper treatment of the refractive force requires an extension of MT-I. We note in passing that the force which drives the amplification of zonal flow shear is just the radial component of the refractive force, and in that case, the net power expended by the refractive force is precisely the rate of growth of zonal flow energy. The basic physics and energetics of the refractive force are discussed further in Appendix B. Here we focus on the derivation of a momentum conservation theorem which accounts for refraction effects (i.e. MT-II).

In order to derive MT-II, it is advantageous to start from the wave kinetic equation, which effectively constitutes a Boltzmann equation for the quasi-particle or wave quanta density $N(\mathbf{x},\mathbf{k},t)$, Here, in the context of simple drift wave models, such as extended Hawegawa-Mima models, the $(\mathbf{x},\mathbf{k})$ phase space density which satisfies the wave kinetic equation is closely related to the potential enstrophy density $\Omega = \left(1 + k_\perp^2 \rho_s^2\right)^2 \left| e\hat{\phi}_{\mathbf{k}}/T \right|^2$, since $N(\mathbf{x},\mathbf{k},t) = \left(1 + k_\perp^2 \rho_s^2\right)^2 I_w$, where the Wigner distribution function $I_w = \int d\mathbf{q} e^{i\mathbf{q}\cdot\mathbf{x}} \langle \hat{\phi}_{\mathbf{k}+\mathbf{q}} \hat{\phi}_{-\mathbf{k}} \rangle$. The interpretation of Ω as a 'quanta density' follows from the intuition that Ω may be viewed as a vortex or 'roton' density [48, 49, 50, 51]. The wave kinetic equation is:

$$\frac{\partial N}{\partial t} + \nabla \cdot [(\mathbf{v}_g + \mathbf{V})N] + \nabla_{\mathbf{k}} \cdot \left[-\frac{\partial}{\partial \mathbf{x}} (\omega + \mathbf{k}\cdot\mathbf{V})N \right] = C(N) \tag{16}$$

where the collision integral may be decomposed as

$$C(N) = C_{w-w}(N) + C_{w-p}(N)\ .$$

Here $C_{w-w}(N)$ corresponds to resonant nonlinear wave-wave interactions which do not conserve N, while $C_{w-p}(N)$ corresponds to resonant linear and nonlinear wave-particle interactions. Thus, for (generalized) three wave interaction processes arising from quadratic nonlinearity, we can expect $C_{w-w}(N)$ to have the form:

$$C_{w-w}(N) = \sum_{\mathbf{k}'} \left\{ C_1(\mathbf{k}', \mathbf{k}-\mathbf{k}') N_{\mathbf{k}'} N_{\mathbf{k}-\mathbf{k}'} - C_2(\mathbf{k},\mathbf{k}') N_{\mathbf{k}'} N_{\mathbf{k}} \right\} \delta(\omega_{\mathbf{k}-\mathbf{k}'} + \omega_{\mathbf{k}'} - \omega_{\mathbf{k}})$$

where conservation of energy and momentum by the microscopic interaction processes require that

$$\int dk \begin{pmatrix} \mathbf{k} \\ \omega_{\mathbf{k}} \end{pmatrix} C_{w-w}(N) = 0$$

independent of the detailed structure of $C_{w-w}(N)$. Note that the Manley-Rowe relations [52] state that while the microscopic interactions do conserve energy and momentum, they do *not* conserve quasi-particle number (apart from the important exception of induced diffusion [53]). $C_{w-p}(N)$, which represents wave-particle interaction, is necessarily dissipative, since quasi-particles exchange energy and momentum with particles via linear and nonlinear resonant interaction. The lowest order contributions to the latter occurs, for example, via the familiar beat wave resonance associated with nonlinear Landau damping (i.e. when $\omega + \omega' = \left(k_\parallel + k'_\parallel\right) v_\parallel$).

In order to derive MT-II, we proceed from the $k_\parallel$ "moment" of the wave kinetic equation, which gives:

$$\frac{\partial}{\partial t}\left\langle P_{\parallel}^{w}\right\rangle + \frac{1}{r}\frac{\partial}{\partial r}\left[r\left\langle \Pi_{r,\parallel}^{w}\right\rangle\right] - \int d\mathbf{k}\left\{\left\langle \frac{dk_{\parallel}}{dt}N\right\rangle + \frac{\partial k_{\parallel}}{\partial k_r}\left\langle \frac{dk_r}{dt}N\right\rangle\right\}$$
$$= \left\langle \int d\mathbf{k}k_{\parallel}\left(C_{w-w}(N) + C_{w-p}(N)\right)\right\rangle . \quad (17)$$

Equation (17) describes the evolution of the mean parallel wave momentum $\left\langle P_{\parallel}^{w}\right\rangle$. Note that since the basic wave-wave interactions conserve momentum, we can immediately write:

$$\int d\mathbf{k}k_{\parallel}C_{w-w}(N) = 0 \quad (18)$$

i.e. wave-wave "collisions" conserve net wave momentum as a consequence of the **k** matching or selection rules. In practice, a concrete demonstration of Eqn. (18) for a specific case requires writing $\int d\mathbf{k}k_{\parallel}C_{w-w}(N)$ as a double integral $\int d\mathbf{k}\int d\mathbf{k}'$ symmetrizing **k** and **k**′ , etc. Detailed examples of such calculations may be found in Ref. [54]. The result of Eqn. (18) still applies to the special case where one element of the wave triad is a zonal flow, though in that case the parallel momentum selection rule $k_{\parallel} = k_{\parallel}' + k_{\parallel}''$ degenerates to the identity $k_{\parallel} = k_{\parallel}'$, since zonal flows have essentially no variation along the field line. Stochastic straining by zonal flows conserves N, as well.

Since the wave-wave interaction process conserves momentum, the mean parallel wave momentum density equation may be simplified as:

$$\frac{\partial}{\partial t}\left\langle P_{\parallel}^{w}\right\rangle + \frac{1}{r}\frac{\partial}{\partial r}\left(r\Pi_{r,\parallel}^{w}\right) - \left\langle f_{\parallel}^{ref}\right\rangle = \int d\mathbf{k}\left\langle k_{\parallel}C_{w-p}(N)\right\rangle . \quad (19a)$$

Here, we see that wave momentum density evolves due to the combined actions of the mean radial flux of parallel wave momentum $\left\langle \Pi_{r,\parallel}^{w}\right\rangle$, the parallel component of the mean refractive force (allowing for magnetic shear - i.e. $\partial k_{\parallel}/\partial k_r \neq 0$)

$$\left\langle f_{\parallel}^{ref}\right\rangle = \int d\mathbf{k}\left\{\left\langle \frac{dk_{\parallel}}{dt}N\right\rangle + \frac{\partial k_{\parallel}}{\partial k_r}\left\langle \frac{dk_r}{dt}N\right\rangle\right\} \quad (19b)$$

and the net dissipation of mean wave momentum by *all* resonant wave-particle interactions, i.e.

$$\left(\frac{d\left\langle P_{\parallel}\right\rangle}{dt}\right)_{diss} = \int d\mathbf{k}\left\langle k_{\parallel}C_{w-p}(N)\right\rangle \quad (19c)$$

However, since the wave momentum density dissipated by resonant interaction with the particles is simply the net rate of change of the *resonant particle* momentum density, we immediately have:

$$\left(\frac{d\left\langle P_{\parallel}\right\rangle}{dt}\right)_{diss} = \int d\mathbf{k}\left\langle k_{\parallel}C_{w-p}(N)\right\rangle$$
$$= -\frac{d}{dt}\left\langle P_{\parallel}\right\rangle_R$$
$$= -\frac{\partial}{\partial t}\left\langle P_{\parallel}\right\rangle_R - \frac{1}{r}\frac{\partial}{\partial r}\left[r\left\langle \widetilde{v}_{Er}\widetilde{P}_{\parallel}\right\rangle\right] . \quad (20)$$

Note that the contribution from the divergence of the resonant particle momentum flux appears here because:

i.) it necessarily enters the time rate of change of the *local* resonant particle momentum density

ii.) it is required for a treatment of wave-particle interactions that is consistent with that applied to the dissipation term.

Indeed, point (i.) above follows from the fact that wave particle interaction in drift wave turbulence scatters particles in radius at the same time it accelerates particles along the field. Thus $C_{w-p}(N)$ and $\left\langle \widetilde{v}_{Er}\widetilde{P}_{\parallel}\right\rangle_R$ must be treated symmetrically and consistently - order by order in perturbation theory - as *each* is an element of the *total* wave particle interaction process. In principle, both $d\left\langle P_{\|}\right\rangle_{diss}/dt$ and $\left\langle \widetilde{v}_{Er}\widetilde{P}_{\|}\right\rangle_R$ contain contributions from *all* even orders in perturbation theory, and so may be written in the generic form [55]

$$\begin{pmatrix} \left\langle \widetilde{v}_{Er}\widetilde{P}_{\|}\right\rangle \\ d\left\langle P_{\|}\right\rangle_{diss}/dt \end{pmatrix} = \sum_n \left\{ \sum_{\mathbf{k}_1}\sum_{\mathbf{k}_2}\cdots\sum_{\mathbf{k}_n} C_n(\mathbf{k}_1,\mathbf{k}_2,\cdots,\mathbf{k}_n)\left\langle \widetilde{\Phi}^2\right\rangle_{\mathbf{k}_1}\left\langle \widetilde{\Phi}^2\right\rangle_{\mathbf{k}_2}\cdots\left\langle \widetilde{\Phi}^2\right\rangle_{\mathbf{k}_n} \right.$$
$$\left. \times\delta\left(\omega_1+\omega_2+\cdots+\omega_n - v_{\|}\left[k_{1\|}+k_{2\|}+\cdots k_{n\|}\right]\right)\right\} . \tag{21}$$

Here $n=1$ corresponds to the quasi-linear piece, $n=2$ to the contribution from nonlinear Landau damping or Compton scattering, etc. Thus, the irreversibility underpinning the $n=1$ contribution is stochasticity due to overlap of primary wave-particle resonances, that governing $n=2$ is stochasticity due to overlap of *beat wave* resonances, etc. Since here we are concerned only with the most basic mean field theory analysis, we limit $\sum_n$ to only the $n=1$ term and so, in the vein of quasi-linear theory, retain only the primary resonance effects in the calculation of $\left\langle \widetilde{v}_{Er}\widetilde{P}_{\|}\right\rangle$.

Proceeding by combining Eqns. (19,20) and again utilizing Eqn. (6c) to combine wave and resonant particle momentum finally leads us to the second or extended momentum conservation theorem (MT-II), which is:

$$\frac{d\left\langle \delta P_{\|}\right\rangle}{dt}+\frac{1}{r}\frac{\partial}{\partial r}\left[r\left\langle \widetilde{v}_{Er}\widetilde{P}_{\|}\right\rangle + r\int d\mathbf{k}\left\langle (v_{gr}+V_r)k_{\|}N\right\rangle\right]$$
$$= \int d\mathbf{k}\left\{\left\langle\frac{dk_{\|}}{dt}\right\rangle\langle N\rangle + \frac{\partial k_{\|}}{\partial k_r}\left\langle\frac{dk_r}{dt}\right\rangle\langle N\rangle + \left\langle\left(\frac{d}{dt}\widetilde{k}_{\|}\right)\widetilde{N}\right\rangle\right\} \tag{22}$$

Here we have written the mean refractive force as the product of the means plus the mean of the product of fluctuations. MT-II, as given by Eqn. (22), states that the total parallel momentum density of the particles evolves by radial transport of resonant particle momentum, radial transport of parallel wave momentum due to magnetic shear (i.e. which renders $\partial k_{\|}/\partial k_r \neq 0$)and by the force exerted on the plasma due to refraction of waves [i.e. as given in the RHS of Eqn (22)]. Note that the refractive force density is a *local* momentum drive, and is not related to the divergence of a stress tensor, i.e. it represents a local force density exerted on the plasma, and has not been considered in previous studies. Clearly, MT-II is more general and inclusive than MT-I. Note also that if refractive force effects are ignored, the predictions of MT-I and MT-II are identical with regard to macroscopics (i.e. stresses).

At this point, it is appropriate to discuss the physics of the refractive force in some depth. As mentioned above, the refractive force is a simple and straightforward consequence of the fact that bending of wave ray trajectories in a dielectric medium (e.g. plasma) necessarily induces a recoil force on that medium. The refractive force is not *a priori* dependent upon dissipation of wave energy, wave breaking, nonlinear transfer or any other "strongly nonlinear" processes. Note that the refractive force consists of the sum of two pieces, a mean or coherent piece ($\sim \left\langle dk_{\|}/dt\right\rangle\langle N\rangle + \left(\partial k_{\|}/\partial k_r\right)\left\langle dk_r/dt\right\rangle\langle N\rangle$) and a stochastic piece [$\sim \left\langle\left(d\widetilde{k}_{\|}/dt\right)\widetilde{N}\right\rangle$]. The coherent refractive force $f^{ref}_{\|,coh}$ is given by

$$f^{ref}_{\|,coh} = -\int d\mathbf{k}\left\{\left\langle\nabla_{\|}(\omega+\mathbf{k}\cdot\mathbf{V})\right\rangle\langle N\rangle + \frac{\partial k_{\|}}{\partial k_r}k_\theta\left\langle v_E\right\rangle'\langle N\rangle\right\}$$

In practice the first term will occur as a result of poloidal asymmetry in mean plasma flows or the parameters which determine the mean wave frequency, whereas the second term will be relevant in the presence of poloidal flow shear in sheared magnetic geometry. Also note that for the second term, since $\partial k_{\|}/\partial k_r \sim k_\theta$, the overall term $\sim k_\theta^2$, so no additional poloidal symmetry breaking is required. The stochastic component of the refractive force may be calculated using a quasi-linear approximation to the mean product $\left\langle\left(d\widetilde{k}_{\|}/dt\right)\widetilde{N}\right\rangle$ and the linear response for $\widetilde{N}$. Thus,

$$\widetilde{N}_{\mathbf{q},\Omega} = -R(\Omega-\mathbf{q}\cdot\mathbf{v}_g)\left(\widetilde{\omega}+\mathbf{k}\cdot\widetilde{\mathbf{V}}\right)_{\mathbf{q}}\mathbf{q}\cdot\frac{\partial\langle N\rangle}{\partial\mathbf{k}} \tag{23a}$$

where

$$R(\Omega-\mathbf{q}\cdot\mathbf{v}_g) = \frac{|\gamma_{\mathbf{k}}|}{(\Omega-\mathbf{q}\cdot\mathbf{v}_g)^2+|\gamma_{\mathbf{k}}|^2} \tag{23b}$$

is the strain field wave packet resonance function. Here $\mathbf{q}$ and Ω are the wave vector and frequency of the strain field, respectively. Thus

$$f_{\parallel,stoch}^{ref} = -\hat{\mathbf{b}} \cdot \overleftrightarrow{D_{\mathbf{k}}} \cdot \frac{\partial \langle N \rangle}{\partial \mathbf{k}} \tag{23c}$$

where $\overleftrightarrow{D_{\mathbf{k}}}$ is the $\mathbf{k}$-space diffusion tensor.

$$\overleftrightarrow{D_{\mathbf{k}}} = \sum_{\mathbf{q}} \mathbf{q}\mathbf{q} \left| \left(\tilde{\omega} + \mathbf{k} \cdot \tilde{\mathbf{V}} \right)_{\mathbf{q}} \right|^2 R(\Omega - \mathbf{q} \cdot \mathbf{v}_g) \tag{23d}$$

and $\hat{\mathbf{b}} \cdot \overleftrightarrow{D_{\mathbf{k}}}$ is given by

$$\hat{\mathbf{b}} \cdot \overleftrightarrow{D_{\mathbf{k}}} = \sum_{\mathbf{q}} q_{\parallel}\mathbf{q} \left| \left(\tilde{\omega} + \mathbf{k} \cdot \tilde{\mathbf{V}} \right)_{\mathbf{q}} \right|^2 R(\Omega - \mathbf{q} \cdot \mathbf{v}_g) \tag{23e}$$

Equation (23e) indicates that the correlations between $q_{\parallel}$ and the other components of the strain field wave vector are critical to determining various components of the net refractive force and that the parallel (i.e. poloidal) variation of the strained field is the ultimate origin of the parallel stochastic refractive force. This observation immediately suggests that geodesic acoustic modes (GAMs) are natural candidates for a physical origin of the refractive force, since on account of their poloidal variation, they produce a strain field with non-zero $q_{\parallel}$ and q_{θ}. All these considerations suggest a simpler and more explicit version of (MT-II), which is just:

$$\begin{aligned} &\frac{d\langle \delta P_{\parallel} \rangle}{dt} + \frac{1}{r}\frac{\partial}{\partial r}\left[r\left\langle \tilde{v}_{Er}\tilde{P}_{\parallel} \right\rangle + r\int d\mathbf{k} \langle (v_{gr} + V_r) k_{\parallel} N \rangle \right] \\ &= -\int d\mathbf{k} \left\{ \nabla_{\parallel} \langle \omega + \mathbf{k} \cdot \mathbf{V} \rangle \langle N \rangle + \frac{\partial k_{\parallel}}{\partial k_r} k_{\theta} \langle v_E \rangle' \langle N \rangle + \hat{\mathbf{b}} \cdot \overleftrightarrow{D_{\mathbf{k}}} \cdot \frac{\partial \langle N \rangle}{\partial \mathbf{k}} \right\} . \end{aligned} \tag{24}$$

We now discuss the general structure of the momentum balance relation and its implications for intrinsic rotation. More generally, to address phenomenon of intrinsic rotation demands an affirmative answer to the question of "can the turbulence sustain a stationary, non-trivial (i.e. peaked) profile of toroidal rotation in the absence of external sources?". For intrinsic rotation at an arbitrary radius a, this requires that a non-trivial profile result from the stationary solution to Eqn. (24), within $r = a$. Imposing the standard boundary condition that fluxes vanish on axis then leads to the momentum stationarity condition:

$$\begin{aligned} \left(r\left\langle \tilde{v}_{Er}\tilde{P}_{\parallel} \right\rangle_R + r\left\langle \Pi_{r,\parallel}^{w} \right\rangle \right)\Big|_a = &-\textstyle\int_0^a r' dr' \int d\mathbf{k} \left\{ \langle \nabla_{\parallel} (\omega + \mathbf{k} \cdot \mathbf{V}) \rangle \langle N \rangle + \frac{\partial k_{\parallel}}{\partial k_r} k_{\theta} \langle v_E \rangle' \langle N \rangle \right. \\ &\left. + \hat{\mathbf{b}} \cdot \overleftrightarrow{D_{\mathbf{k}}} \cdot \frac{\partial \langle N \rangle}{\partial \mathbf{k}} \right\} , \end{aligned} \tag{25}$$

i.e. that the resonant particle momentum flux at $r = a$ plus the wave momentum flux at $r = a$ balance the refractive force at $r = a$. The resonant particle momentum flux is calculated in detail in Section IV. We can, however, expect it to have a structure of the general form:

$$\left\langle \tilde{v}_{Er}\tilde{P}_{\parallel} \right\rangle_R = -D_{\parallel}\frac{\partial}{\partial r}\langle P_{\parallel} \rangle + (\Pi_{r,\parallel})_R^{ND} \tag{26a}$$

where the non-diffusive piece due to resonant particles $(\Pi_{r,\parallel})_R^{ND}$ may be further decomposed into convective and residual stress components, i.e.

$$(\Pi_{r,\parallel})_R^{ND} = V_r \langle P_{\parallel} \rangle + (\Pi_{r,\parallel})_R^{residual} . \tag{26b}$$

Note that further information or assumptions concerning the specific dynamical model are required to separate the particle flux and flow transport (stress) contributions to $(\Pi_{r,\parallel})_R^{ND}$. In general, however, sustaining a non-trivial (positive) flow profile requires at least one of:

$$(\Pi_{r,\parallel})_R^{ND} < 0 \tag{27a}$$

i.e. either inward resonant convection of momentum ($V_r < 0$) or an inward resonant residual stress $\left(\Pi_{r,\|}\right)_R^{residual} < 0$, or

$$\left(\Pi_{r,\|}\right)_R^w < 0 \tag{27b}$$

i.e. either an inflow of positive, or an outflow of negative, wave momentum, or

$$\int_0^a f_{ref} dr \neq 0$$

i.e a net refractive force.

It is interesting to note that the microscopic irreversibility which underpins $\left\langle \widetilde{v}_{Er} \widetilde{P}_\| \right\rangle$ is stochasticity of resonant particles, while that which underlies $\left(\Pi_{r,\|}\right)_R^w$ is overlap of wave resonances or drift wave ray chaos resulting from interaction with zonal flow or GAM shears. Ray chaos is also the underpinning of the stochastic refractive force.

Further progress clearly requires concrete calculation of $\left\langle \Pi_{r,\|}^w \right\rangle$ - the mean wave momentum flux, and of $\left\langle \widetilde{v}_{Er} \widetilde{P}_\| \right\rangle$ - the resonant particle momentum flux. Loosely speaking, one can expect the former to be more important for fluid-like instabilities and/or in regimes of strong turbulence, while the latter to be more important in kinetic regimes near marginality, i.e. for stiff profiles. We expect the refractive force to be significant only when poloidal symmetry is strongly broken, or when electric field shear is strong. In practice, this is likely to occur only at the plasma edge or in transport barriers. Given the special nature and complexity of the edge plasma transport and barrier dynamics, we postpone further discussion of the effects of the refractive force on rotation to a future paper.

CALCULATING THE WAVE MOMENTUM FLUX - A CHAPMAN-ENSKOG APPROACH

In this section, we present the concrete calculation of $\left\langle \Pi_{r,\|}^w \right\rangle$ - the mean radial flux of parallel wave momentum. This is one of the two principal components of the total momentum flux, and is most important in regimes of strong, hydrodynamic-like turbulence.

The key physical idea which enables us to systematically calculate $\left\langle \Pi_{r,\|}^w \right\rangle = \int d\mathbf{k} \left\langle v_{gr} k_\| N \right\rangle$ is that waves propagate while being scattered by nonlinear interaction. Hence we develop an approach similar to that used to describe quasiparticle transport processes in solids and Fermi liquids [56]. Hence, it is useful to construct a kinetic theory of the wave population density N in the phase space of $\mathbf{k}$ and $\mathbf{x}$. To this end, we first list and order the basic time and space scales. Drift wave packets propagate radially at $v_{gr} = -2k_\theta k_r \rho_s^2 v_* / \left(1 + k_\perp^2 \rho_s^2\right)^2$, where $v_* = \rho_s c_s / L_n$ is the diamagnetic velocity. Thus a drift wave packet will:

i.) propagate one radial wavelength in

$$\left(\tau_{prop}\right)_\mathbf{k} \sim \left(k_r v_{gr}\right)^{-1} \sim -\frac{\left(1 + k_\perp^2 \rho_s^2\right)^2}{2k_\theta k_r^2 \rho_s^2 v_*} \gtrsim \omega_\mathbf{k}^{-1}$$

ii.) scatter or decay via nonlinear interaction in

$$\left(\tau_{decay}\right)_\mathbf{k} \sim \left(k_\perp^2 D\right)^{-1} \sim \omega_\mathbf{k}^{-1}$$

where the equalities follow from using the "mixing length" prediction of fluctuation levels, but should be regarded as crude estimates, only. Zonal flow shearing will yield a roughly comparable decay time. It is useful to note that where fluctuation levels are comparable to the mixing-length level, $\left(\tau_{prop}\right)_\mathbf{k} \sim \left(\tau_{decay}\right)_\mathbf{k}$ follows directly from $\widetilde{v} \sim v_*$. For smaller fluctuation levels, i.e. ($\widetilde{v} < v_*$), then $\left(\tau_{prop}\right)_\mathbf{k} \ll \left(\tau_{decay}\right)_\mathbf{k}$, so wave transport is the faster process.

iii.) propagate one fluctuation intensity gradient scale length distance (L_I, where $L_I^{-1} \equiv \langle N \rangle^{-1} \partial \langle N \rangle / \partial r$) in

$$\left(\tau_{prop}\right)_{L_I} \sim \left(k_r L_I\right) \left(\tau_{prop}\right)_\mathbf{k} .$$

These time scales are summarized in Table 1. Hence, for the "typical" regime where $k_r L_I \gg 1$, we are justified in

TABLE 1. Time scales for wave population response

time scale	definition
oscillation time	$\tau \sim \omega_{\mathbf{k}}^{-1}$
self-propagation (time to propagate λ_r)	$(\tau_{prop})_{\mathbf{k}} \sim (k_r v_{gr})^{-1}$
decay time (time of nonlinear scattering)	$\tau_{decay} \sim (k_\perp^2 D)^{-1}$
intensity gradient scale	$(\tau_{prop})_{L_I} \sim L_I/v_{gr} \sim$
propagation time	$(k_r L_I)\, \tau_{prop\mathbf{k}}$

applying the ordering

$$(\tau_{prop})_{\mathbf{k}} \sim (\tau_{decay})_{\mathbf{k}} \ll (\tau_{prop})_{L_I} \,.$$

This implies that a drift wave packet will be nonlinearly scattered much more rapidly than it will transit one intensity profile scale length. Thus, drift wave packets can be thought of as having "short mean free path" or, in the language of radiation hydrodynamics, as satisfying the conditions for a regime of large optical depth [33]. Central to this ordering is the assumption of a clear separation of scales between the radial wavelength (k_r^{-1}) and/or correlation length Δr_c on the one hand, and the intensity profile scale length L_I on the other. L_I may be loosely estimated as $L_\perp$ (i.e. a typical profile scale length), since the turbulence intensity profile is supported by the heat and particle fluxes which produce the profiles of plasma temperature, density etc. Alternatively put, since $T(r)$ decays with radius for central deposition, $\langle N(r)\rangle$ must *increase* with radius - see Fig. 1. This is both independently plausible and also consistent with the empirical finding that the fluctuation intensity profile is almost always observed to *increase* with radius [57]. One possible exception to the assumed $L_I k_r \gg 1$ ordering would be the instance of a fully developed strong transport barrier.

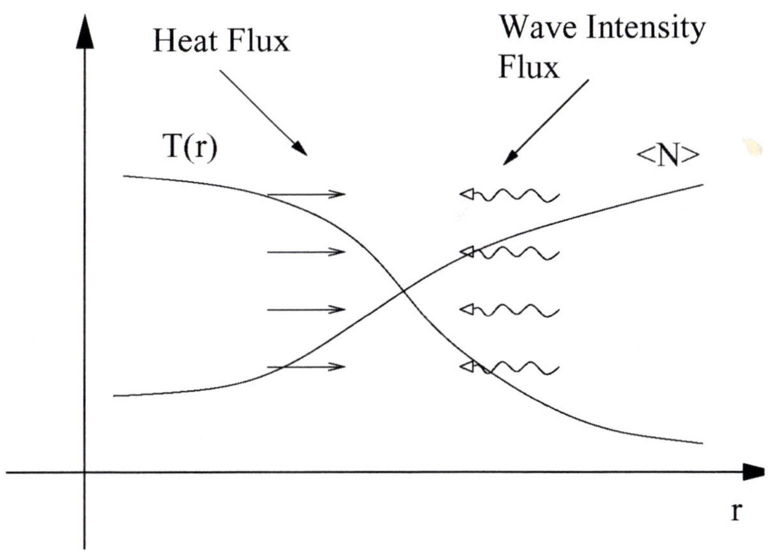

FIGURE 1. Cartoon of radial profile of temperature and turbulence intensity. Solid arrows denote outward flux of heat, whereas the squiggles indicate an inward flux of wave intensity.

Given time and space scale ordering discussed above, we can formulate a Chapman-Enskog expansion approach to the calculation of $\Pi^w_{r,\|}$. The basic idea is to write

$$N = \langle N(\mathbf{x}, \mathbf{k}, t)\rangle + \delta N$$

where $\langle N\rangle$ is the slowly varying "mean" wave population density, and δN is the perturbation to it induced by gradients of $\langle N\rangle$ in radius and $\mathbf{k}$. Since, δN will be shown to be proportional to $\partial\langle N\rangle/\partial r$ and $\partial\langle N\rangle/\partial k_r$, it ultimately produces the net flux of wave momentum. As usual, δN is calculated perturbatively, exploiting the time scale ordering.

Proceeding, we hereafter ignore the mean radial convection, and write the wave kinetic equation in the form:

$$\frac{\partial N}{\partial t} + \mathbf{v}_g \cdot \nabla N - \frac{\partial}{\partial \mathbf{x}}(\mathbf{k}\cdot\mathbf{V})\frac{\partial}{\partial \mathbf{k}} N = C(N) \tag{28a}$$

where

$$N = \langle N \rangle + \delta N\,. \tag{28b}$$

We assume the "collisions" (i.e. the nonlinear wave-wave and wave-particle scattering events) underpinning $C(N)$ occur with collision frequency $\nu_{T\mathbf{k}} \sim \Delta\omega_{\mathbf{k}} \sim 1/\left(\tau_{decay}\right)_{\mathbf{k}}$, (where $\Delta\omega_{\mathbf{k}}$ is the decorrelation rate for scale $\mathbf{k}$), and that $v_{gr}/L_I \ll \Delta\omega_{\mathbf{k}}$. More precisely, $\nu_{T\mathbf{k}}$ is the characteristic rate of wave-wave energy transfer or wave particle scattering, as represented by $C(N)$. Here $C(N)$ includes both C_{w-w} and C_{w-p} and all relevant orders (i.e. both linear and non-linear) of each type of scattering process. Thus, to lowest order, Eqn. (28a) is just

$$C(N) = 0 \tag{29a}$$

which formally defines the mean $\langle N \rangle$ as:

$$N^{(0)} = \langle N \rangle\,. \tag{29b}$$

The restrictions and caveats concerning the structure of $\langle N \rangle$ require a brief discussion. In practice, the mean $\langle N \rangle$ is defined by the balance of linear growth with coupling to dissipation via non-linear wave-wave and wave-particle interaction. Here "wave-wave interaction" includes the possibility of zonal shearing, and ν_T is the characteristic rate for nonlinear relaxation to this mean state. Obviously, $\nu_{T\mathbf{k}} \sim \Delta\omega_{\mathbf{k}}$, the turbulence decorrelation rate. In general, we can expect $\langle N \rangle$ to have a power law structure, in between limits set by low-k (i.e. flow damping) and high-k (i.e. Landau damping and collisions) dissipation. Thus, there may be some limitation on *which* higher order moments of $\langle N \rangle$ actually exist without sensitive dependence upon dissipation. We defer detailed discussion of this subtle question to a future publication. $\langle N \rangle$ must be symmetric in k_r, since there should be no intrinsically preferred direction of wave packet propagation or the condition of stationarity of $\langle N \rangle$ would be violated. Thus, $\langle N \rangle$ makes no contribution to $\left\langle \Pi^{w}_{r,\parallel} \right\rangle$. Since $N = \langle N \rangle + \delta N$, to next order we have

$$\frac{\partial}{\partial t}\delta N + \mathbf{v}_g \cdot \nabla \delta N - \frac{\partial}{\partial r}\left(k_\theta \langle v_E \rangle\right)\frac{\partial}{\partial k_r}\delta N - \nu_T \delta N = -v_{gr}\frac{\partial \langle N \rangle}{\partial r} + \frac{\partial}{\partial r}\left(k_\theta \langle v_E \rangle\right)\frac{\partial \langle N \rangle}{\partial k_r} \tag{30a}$$

with $k_\theta \to m/r$, where m is the poloidal mode number. Here we have assumed that the mean flow is predominantly a sheared $E \times B$ poloidal flow, so

$$\mathbf{k}\cdot\mathbf{V} = k_\theta \langle v_E \rangle\,. \tag{30b}$$

Also, we have employed a Krook approximation to $C(N)$, i.e.

$$C(N_{\mathbf{k}}) \cong -\nu_{T\mathbf{k}}\left(N_{\mathbf{k}} - \langle N \rangle\right) = -\nu_{T\mathbf{k}}\delta N_{\mathbf{k}}\,. \tag{30c}$$

This states that the effect of wave-wave or wave-particle collisions on N is simply to drive it to relax to the mean $\langle N \rangle$. Recall that the Manley-Rowe relations state that $C(N)$ need not conserve quanta number so a Krook model will suffice. Then, in the limit of strong collisionality, where $\nu_T > v_{gr}/L_I$, Ω, $k_\theta \langle v_E \rangle' / k_r$, etc., we have

$$\delta N = -\frac{v_{gr}}{\nu_{T_{\mathbf{k}}}}\frac{\partial \langle N \rangle}{\partial r} + \frac{1}{\nu_{T_{\mathbf{k}}}}\frac{\partial}{\partial r}\left(k_\theta \langle v_E \rangle\right)\frac{\partial \langle N \rangle}{\partial k_r}\,. \tag{31}$$

Here δN, the excursion from the uniform "equilibrium" distribution is driven by both $\partial \langle N \rangle / \partial r$ and $\partial \langle N \rangle / \partial k_r$ - i.e. the quanta density profile gradients in both *radius* and *radial wave-number*. Substituting δN from Eqn. (31) into $\left\langle \Pi^{w}_{r,\parallel} \right\rangle$ then gives

$$\begin{aligned}\left\langle \Pi^{w}_{r,\parallel} \right\rangle &= \int d\mathbf{k}\, v_{gr} k_\parallel \left\{ -\frac{v_{gr}}{\nu_{T_{\mathbf{k}}}}\frac{\partial \langle N \rangle}{\partial r} + \frac{k_\theta}{\nu_{T_{\mathbf{k}}}}\frac{\partial \langle v_E \rangle}{\partial r}\frac{\partial \langle N \rangle}{\partial k_r} \right\} \\ &= -\int d\mathbf{k}\, k_\parallel \left\{ \frac{v_{gr}^2}{\nu_{T_{\mathbf{k}}}}\frac{\partial \langle N \rangle}{\partial r} + \frac{\partial}{\partial k_r}\left(\frac{v_{gr}}{\nu_{T_{\mathbf{k}}}}\right)\langle v_E \rangle' k_\theta \langle N \rangle \right\}\end{aligned}$$

where we have integrated by parts to re-write the second term on the RHS. It is apparent that the wave momentum flux consists of two pieces. The first is a Fickian flux, driven by $\partial \langle N \rangle / \partial r$ with effective diffusivity $D_{rad} \sim v_{gr}^2 / \nu_{T\mathbf{k}}$. The second is a flow shear driven "mobility", proportional to the divergence in k_r of the wave packet group velocity. The second term arises from the combined effects of shearing, which tends to increase k_r by straining, and the presence of a population density profile in k_r, i.e. $\partial \langle N \rangle / \partial k_r$. Thus $\left\langle \Pi^w_{r,\|} \right\rangle$ may be written in the approximate diffusion-convection form:

$$\left\langle \Pi^w_{r,\|} \right\rangle = -D_w \frac{\partial}{\partial r} \left\langle P^w_\| \right\rangle + V_w \left\langle P^w_\| \right\rangle \tag{32a}$$

where the wave momentum diffusivity is

$$D_w \cong \frac{\int d\mathbf{k} \left(\frac{v_{gr}^2}{\nu_{T_\mathbf{k}}} \right) \langle N \rangle}{\int d\mathbf{k} \langle N \rangle} \tag{32b}$$

and the convection velocity is

$$V_w = -\frac{\int d\mathbf{k} \left(\frac{\partial}{\partial k_r} \left(\frac{v_{gr}}{\nu_{T_\mathbf{k}}} \right) k_\theta \langle v_E \rangle' \right) \langle N \rangle}{\int d\mathbf{k} \langle N \rangle} \tag{32c}$$

Of course, the first term on the RHS of Eqn. (32a) gives the diffusive flux, while the second term drives the convective flux. Given the identity between wave-momentum density and non-resonant particle momentum density, $\left\langle \Pi^w_{r,\|} \right\rangle$ directly implies a corresponding flux of non-resonant particle momentum. However since the convention is to define transport coefficients such as diffusion, convection etc. in terms of how they act on moments of the particle distribution function, $\left\langle \Pi^w_{r,\|} \right\rangle$ ultimately is classified as a contribution to the residual stress part of the radial flux of parallel momentum. Finally, we note that, at this stage of the calculation the origin of symmetry breaking (i.e. the origin of $\langle k_\| \rangle = \int d\mathbf{k} k_\| \langle N \rangle / \int d\mathbf{k} \langle N \rangle \neq 0$) has not yet been addressed.

The physics of the first $\partial \langle N \rangle / \partial r$-driven contribution to $\left\langle \Pi^w_{r,\|} \right\rangle$ is straightforward, and describes diffusion of wave momentum due to propagation and scattering of the wave packets which carry the momentum [33]. As shown in Fig. 2, since $v_{gr} \sim k_r$ (as $\chi \sim -k_\perp^2 \rho_s^2$ for drift waves), a wave packet originating at a particular radius r will propagate a distance $\ell_{mfp} \sim v_{gr} \tau_c$ before undergoing absorption and re-emission as part of the nonlinear wave-wave interaction process. This immediately implies the existence of a diffusive flux of wave momentum $\left\langle \Pi^w_{r,\|} \right\rangle_D = -D_{rad} \partial \langle P_\| \rangle / \partial r$ with $D_{rad} \sim v_{gr} \ell_{mfp} \sim v_{gr}^2 / \nu_T$. It is important to realize that this flux is due to wave radiation (albeit in the short mean free path limit) rather than turbulent mixing. Interestingly, since, for "typical" mixing length type fluctuation levels, $v_{gr} \sim v_*$, $\nu_T \sim \omega_\mathbf{k} \sim c_S / L_\perp$, we have $D_{rad} \sim D_{GB}$ so that the radiative diffusivity D_{rad} is similar to the familiar GyroBohm turbulent diffusivity! This, together with the presence of a heat flux driving and the fact that $nT \gg \varepsilon_\omega$, explains why the edge peaked intensity gradient does not simply collapse via wave radiation. It also explains the long-standing difficulty in disentangling wave transport processes [38] from eddy transport processes, since transport of both heat and wave intensity will occur at a similar rate. One interesting consequence of this wave quanta diffusion process is that since the turbulent intensity field is always peaked toward the edge, there will always be a concomitant *inward* diffusive flux of wave quanta. This implies a spreading of edge turbulence inward into the core by radiative diffusion, as originally predicted by B. B. Kadomtsev [58]. Though parallel momentum density and wave momentum density are, of course, different moments of $\langle N \rangle$, a similar inward diffusive flux of wave momentum from the edge can also be expected. Indeed, given that the parallel wave momentum flux corresponds to a portion of the parallel Reynolds stress $\langle \tilde{v}_r \tilde{v}_\| \rangle$ and that stress induced change in momentum necessarily involves an intensity gradient, it is no surprise that $\left\langle \Pi^w_{r,\|} \right\rangle$ contains an intensity gradient dependent contribution. This is significant since several symmetry breaking mechanisms may be at work in the edge plasma [15], so such inward quanta diffusion would transmit the broken symmetry, as well.

The physics of the second, $\partial \langle N \rangle / \partial k_r$ driven contribution to $\left\langle \Pi^w_{r,\|} \right\rangle$ (from which the convective wave momentum flux originates) is due to the distortion in N induced by shearing, which tends to increase k_r. Formally, $\delta N \sim \tau_{c\mathbf{k}} (\partial k_r / \partial t) (\partial \langle N \rangle / \partial k_r) \sim \tau_{c\mathbf{k}} k_\theta \langle v_E \rangle' \partial \langle N \rangle / \partial k_r$, so $\left\langle \Pi^w_{r,\|} \right\rangle \sim \int d\mathbf{k} k_\| v_{gr} \delta N$ is just the flux of wave momentum which occurs during the change in population-density induced by a shearing event. The key element here is k_r dependence - i.e. shearing alters k_r, and since both v_{gr} and $\langle N \rangle$ are functions of k_r, a flux of wave momentum density results. Thus,

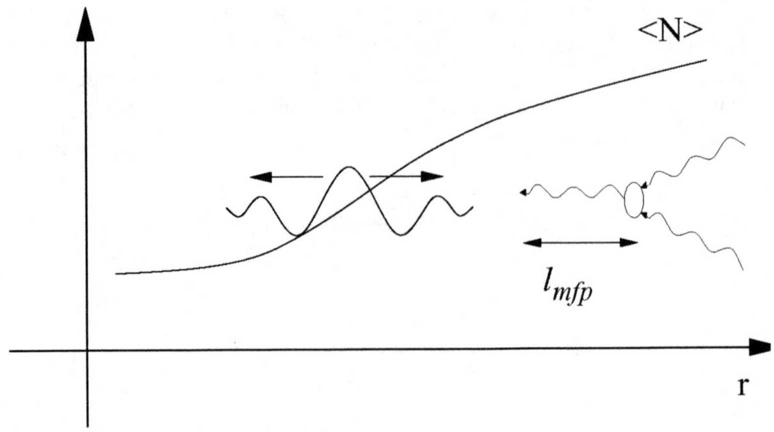

FIGURE 2. Cartoon of radial profile of turbulence intensity. Wave propagation direction equally likely, however inhomogeneity in turbulence intensity provides means of net flux of wave intensity. Here we consider regimes in which l_{mfp} is small in comparison to the profile variation.

one way to understand this convective momentum flux is as a process of wave "wind-up" due to shear, as shown in Fig. 3. Note, however that this effect is conceptually distinct from its better known counterpart related to wind up of

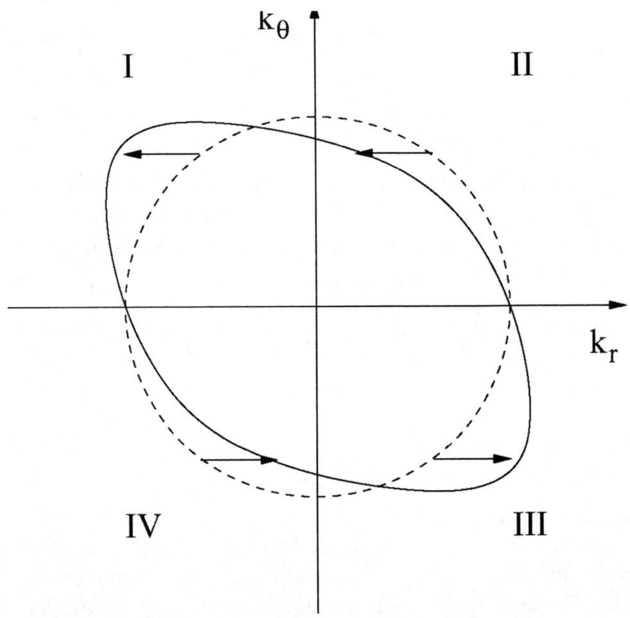

FIGURE 3. Contours of N_k in the $k_\theta - k_r$ plane. The Broken line corresponds to an initial isotropic case, the solid line to the spectrum after being refracted by mean shear flow. It's apparent that the wave population in regions I and III increases while regions II and IV decrease.

$k_\parallel$ with k_r, which we discuss later in the context of symmetry breaking. Alternatively, an integration by parts gives

$$\left\langle \Pi^{w}_{r,\parallel} \right\rangle_{convective} \cong \int d\mathbf{k} \left[\frac{\partial}{\partial k_r} \left(\frac{v_{gr}}{v_{T_\mathbf{k}}} \right) \right] k_\theta \left\langle v_E \right\rangle' \left\langle N \right\rangle ,$$

thus linking the convective flux to the k_r dependence of v_{gr}. This suggests that when shearing alters k_r, the variation in wave packet propagation speed (i.e. the "compressibility" of v_{gr} with respect to k_r) will then induce a net wave momentum flux. Note that since $v_{gr} \sim k_\theta$, $\left\langle \Pi^w_{r,\parallel} \right\rangle_{convective}$ is *even* in k_θ, so no additional poloidal symmetry breaking (i.e. other than that driven by $\langle v_E \rangle'$) is required in order to induce a flux. Finally, it is useful to observe that since $v_{gr} \sim v_*$, $\left\langle \Pi^w_{r,\parallel} \right\rangle' \sim v_* \langle v_E \rangle'$, so the off-diagonal wave-momentum flux is determined by the *product* of the electric field shear and the mode propagation velocity (v_{*e} or v_{*i}). Thus a sudden change or bifurcation in the convective momentum flux can be induced by either

i.) an abrupt change in $\langle v_E \rangle'$, as in an electric field shear induced transport bifurcation, as observed for example in L-H mode transition and ITB formation.

ii.) a change in mode propagation direction, such as would occur along with a change in the mode population composition. One example of this might be a change from predominance of v_{*e}-direction propagating waves (i.e. electron drift waves) to v_{*i}-direction propagating waves (i.e. ion drift waves or ITG modes) as the density n exceeds the threshold for excitation of ITG modes at saturation of energy confinement in OH plasmas.

Note that momentum transport bifurcations originating with a change in V_w thus can occur either in connection with a change in confinement (i.e. $\langle v_E \rangle'$ increases, thus suppressing turbulence and forming a transport barrier) or as a change in momentum transport at roughly constant thermal energy content. We also comment that shearing and wave refraction can also induce a convective flux of $\langle N \rangle$, as well as $\langle P_\parallel \rangle$, and so may play a role in the dynamics of turbulence spreading. Finally, we note that the *total* momentum flux should include the contribution from convection by radial flows (i.e. due to streamers, for example) and so is

$$\left\langle \Pi^w_{r,\parallel} \right\rangle = \int d\mathbf{k} k_\parallel \left\{ \langle V_r \rangle \langle N \rangle - \frac{v_{gr}^2}{v_{T_\mathbf{k}}} \frac{\partial \langle N \rangle}{\partial r} + v_{gr} \frac{k_\theta}{v_{T_\mathbf{k}}} \langle v_E \rangle' \frac{\partial \langle N \rangle}{\partial k_r} \right\} \tag{33}$$

All the results given up till now have been calculated for the regime of strong scattering, where Ω, v_{gr}/L_I, $k_\theta \langle v_E \rangle' / k_r \ll v_{T_\mathbf{k}}$. However, in regimes of weaker turbulence and/or sharper intensity or electric field gradients, the inequality above may break down, so that $\ell_{mfp} > L_I$, etc. This forces us to confront the analogue of flux limited radiative transfer. In that case, and considering the spatial flux for example, the stationary, gradient driven perturbation would follow from

$$(\mathbf{v}_g \cdot \nabla + v_{T_\mathbf{k}}) \delta N = -v_{gr} \frac{\partial \langle N \rangle}{\partial r} \tag{34a}$$

so the response δN would no longer be local, but instead take the integral form

$$\delta N = -\int_{r_a}^{r} dr' K(r,r',t)\, v_{gr} \frac{\partial \langle N(r') \rangle}{\partial r} \tag{34b}$$

where $K(r,r',t)$ is a non-local influence kernel of width ℓ_{mfp}, which is obtained by inverting the LHS of Eqn. (34a) [59]. Continuing in this vein, a crude approximation would eventually yield a tractable form for D_{rad} such as:

$$D_{rad} \cong v_{gr}^2 \left(\frac{v_T}{(v_{gr}/L_I)^2 + v_T^2} \right). \tag{34c}$$

Thus for smooth intensity profiles and strong scattering where $v_T > v_{gr}/L_I$ etc., $D_{rad} \sim v_{gr}^2 / v_T$, as before. On the other hand, for steep profiles and/or weak scattering (with consequent long ℓ_{mfp}),

$$D_{rad} \cong \frac{v_{gr}^2 v_T}{(v_{gr}/L_I)^2} \cong v_T L_I^2 , \tag{34d}$$

as in the case of flux limited transport [33]. Similar approximate forms which ensure proper behavior in regimes of long packet mean free path, weak scattering, strong shearing or fast variation can easily be derived. In general, in such cases the response is non-local in space and/or time, so δN is given by a response integral. In the limit of very long packet mean free path, a coherent wave-packet dynamics approach, using integration along rays (in the spirit of Longuett-Higgins and Stewart [60], and later Mattor and Diamond [61] in the MFE context) becomes advantageous.

Up till now, we have not addressed the origin of symmetry breaking - i.e. *the key question of what sets the net mean* $k_\parallel$ (i.e. $\langle k_\parallel \rangle$), which appears so prominently in the wave-momentum flux. Indeed, $\langle k_\parallel \rangle$ also plays an important role in the resonant particle momentum flux, particularly in the residual stress. One advantage of the general framework we present here is that it allows us to address and compare *several* possible symmetry breaking mechanisms and quantitatively explore the competition between them. In contrast, all previous works have postulated a single symmetry breaking mechanism *a priori*.

The approach to $\langle k_\parallel \rangle$ is best formulated by considering the mean field wave kinetic equation for $\langle N \rangle$, which is:

$$\frac{\partial \langle N \rangle}{\partial t} + \frac{1}{r}\frac{\partial}{\partial r}\left[r\langle (v_{gr}+V_r)N\rangle\right] - \frac{\partial}{\partial k_r}\left[k_\theta \langle v_E\rangle' \langle N\rangle\right] - \frac{\partial}{\partial k_\parallel}\left(k_\parallel \langle v_\parallel\rangle' \langle N\rangle\right)$$
$$= \nabla_{\mathbf{k}} \cdot \left[\overleftrightarrow{\mathbf{D}}_{\mathbf{k}} \cdot \nabla_{\mathbf{k}} \langle N\rangle\right] + 2\gamma_{\mathbf{k}} \langle N\rangle + \langle C_{NL}(N)\rangle \tag{35}$$

where the first term on the RHS accounts for random straining, the second for linear wave growth and the third for nonlinear wave-wave scattering. Then, taking $\int d\mathbf{k}k_\parallel$ to obtain a general evolution equation for $\langle k_\parallel \rangle$ yields.

$$\frac{\partial}{\partial t}\langle k_\parallel\rangle = -\frac{1}{r}\frac{\partial}{\partial r}\left[r\left\langle \Pi^w_{r,\parallel}\right\rangle\right] - \int d\mathbf{k}\left(\frac{\partial k_\parallel}{\partial k_r}\right) k_\theta \langle v_E\rangle' \langle N\rangle$$
$$- \int d\mathbf{k}\hat{\mathbf{b}} \cdot \overleftrightarrow{\mathbf{D}}_{\mathbf{k}} \cdot \nabla_{\mathbf{k}}\langle N\rangle + 2\int d\mathbf{k}k_\parallel \gamma_{\mathbf{k}}\langle N\rangle - \gamma_{NL}\langle k_\parallel\rangle\ . \tag{36a}$$

Here the last term on the RHS reflects the *net decay* of wave momentum by coupling to dissipation via nonlinear interaction of all sort. Heuristically γ_{NL} should be thought of as a nonlinear decay rate for parallel wave momentum. Clearly γ_{NL} is comparable to v_T but not necessarily precisely *equal* to it. Thus, at steady state, $\langle k_\parallel \rangle$ satisfies:

$$\gamma_{NL}\langle k_\parallel\rangle = -\frac{1}{r}\frac{\partial}{\partial r}\left[r\left\langle \Pi^w_{r,\parallel}\right\rangle\right] - \int d\mathbf{k}\left(\frac{\partial k_\parallel}{\partial k_r}\right) k_\theta \langle v_E\rangle' \langle N\rangle$$
$$- \int d\mathbf{k}\hat{\mathbf{b}} \cdot \overleftrightarrow{\mathbf{D}}_{\mathbf{k}} \cdot \nabla_{\mathbf{k}}\langle N\rangle + 2\int d\mathbf{k}k_\parallel \gamma_{\mathbf{k}}\langle N\rangle \tag{36b}$$

so the net local wave momentum density is determined by the competition between nonlinear decay and

i.) the local inflow or outflow of wave momentum density by transport, given by $-\nabla_r \cdot \left\langle \Pi^w_{r,\parallel}\right\rangle$. For example this process would account for growth of local wave momentum density in the core plasma as a result of influx from an edge or SOL momentum source, as depicted in Fig. 4.

ii.) the enhancement of $\langle k_\parallel \rangle$ via the synergistic interaction of $\mathbf{E}\times\mathbf{B}$ shearing ($\langle v_E\rangle'$) and magnetic shear (required for $\partial k_\parallel/\partial k_r \neq 0$). This effect, represented by $-\int d\mathbf{k}\left(\partial k_\parallel/\partial k_r\right) k_\theta \langle v_E\rangle' \langle N\rangle$, captures the now well-known process of $E\times B$ shear-generated shifting of the spectrum off the resonant surface, which has been studied by many authors [27, 62]. In the eikonal theory formulation given here, that process appears naturally as one whereby $k_\parallel$ "winds up" in response to an increase in k_r induced by mean velocity shear. Magnetic shear (i.e. $\partial k_\parallel/\partial k_r \neq 0$) is absolutely essential to the viability of this process. Note also that since $\partial k_\parallel/\partial k_r \sim k_\theta$, the overall term $\sim k_\theta^2$, so no additional poloidal symmetry breaking is required.

iii.) the input of wave momentum by the stochastic refractive force. This mechanism is driven by fluctuating mesoscale flows with parallel structure induced by poloidal asymmetry. Geodesic acoustic modes (GAMs) are thus a natural candidate. Hence, symmetry breaking by the GAM-induced refractive force is a possible wave momentum source in the plasma edge, where GAM activity is strong. The relation between GAM activity and intrinsic rotation has not been explored and merits further study.

iv.) asymmetry in growth. This effect captures the possibility that one sign of $k_\parallel$ may be preferred, dynamically. Such a phenomenon can occur when parallel velocity shear is present. This mechanism has been advocated by Coppi [63].

The symmetry-breaking mechanisms are summarized as follows:

i.) Radial flux of wave momentum
   - possible influx of momentum from scrape of layer

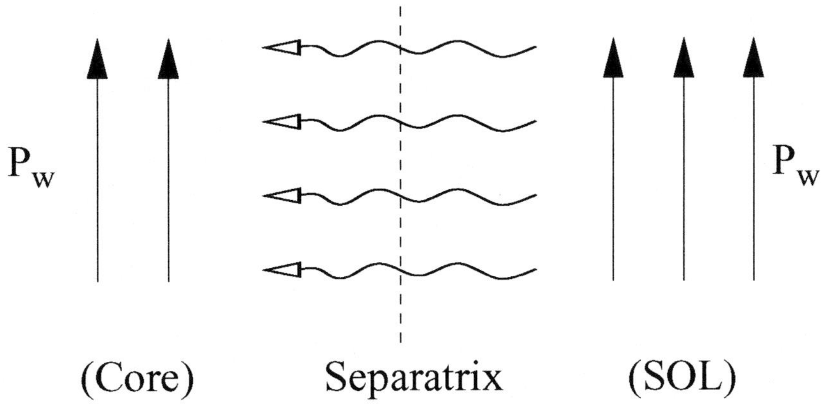

FIGURE 4. Sketch of influx of wave momentum from SOL.

- potentially critical in edge region, close to symmetry breaking

ii.) $\mathbf{E} \times \mathbf{B}$ shearing
- requires magnetic shear, i.e. $\partial k_{\parallel}/\partial k_r \neq 0$
- critical in or near barrier regions, either pedestal or ITB, but not limited to these

iii.) Growth asymmetry $[\gamma(+k_{\parallel}) \neq \gamma(-k_{\parallel})]$
- may enter via parallel velocity shear

iv.) Refraction by GAMs
- likely significant near edge
- largely unexplored

A virtue of this general approach to the question of symmetry breaking is that it allows quantitative comparisons between competing mechanisms. For example, by comparing the first and second terms on the RHS of Eqn. (36b), one could determine the strength of the electric field shear required for the "shearing" mechanism of local symmetry breaking to exceed the influx of net wave momentum from neighboring regions, such as the edge. Comparisons such as these will be very useful in understanding the dynamics of momentum transport bifurcation, changes in intrinsic rotation and the relation of these two phenomena to changes in confinement.

This section has presented a rather lengthy reformulation of the quasi-linear theory of momentum transport by non-resonant particles in terms of wave-momentum transport. The skeptical (and tired!) reader is no doubt wondering "just what have we gained by all this?". The answer to this very reasonable question is threefold, namely:

i.) since non-resonant transport is fundamentally a wave process (i.e. associated with particle "sloshing"), it should, in principle, be formulated in terms of wave dynamics. This approach, in the vein of radiative transfer and radiation hydrodynamics, accomplishes precisely that! Ultimately, non-resonant transport is related to wave population gradients in both $\mathbf{x}$ and $\mathbf{k}$, with radiative transport coefficients and correlation times set by wave-wave interaction. A novel off-diagonal momentum flux contribution driven by $\langle v_E \rangle'$ was identified, as well.

ii.) the origin of the irreversibility which underpins the stationary state counterpart of the non-resonant quasi-linear flux is clearly identified.

iii.) a systematic way to assess the relative strength and importance of several possible competing symmetry breaking mechanisms has been developed.

Finally we note that these ideas, approaches and methods may also be useful in the study of non-locality phenomena in turbulent transport. In particular, they suggest one route toward a systematic theory of turbulence spreading.

CALCULATING THE RESONANT PARTICLE MOMENTUM FLUX

In this section, we present the calculation of the parallel momentum flux carried by resonant particles. In contrast to the calculation of the non-resonant portion, calculating the resonant particle momentum flux is relatively straightforward and follows from a direct application of methods familiar from quasi-linear theory.

The quasi-linear gyrokinetic equation for the evolution of the resonant particle (i.e. ion) momentum is just

$$\frac{\partial}{\partial t}\langle P_\parallel\rangle_R+\frac{1}{r}\frac{\partial}{\partial r}\left[r\left\langle\Pi^R_{r,\parallel}\right\rangle\right]=\int d^3v\dot{v}_\parallel\tilde{f}^R_i \tag{37a}$$

where the resonant particle momentum flux is

$$\Pi^R_{r,\parallel}=\left\langle\tilde{v}_{Er}\tilde{P}_{\parallel i}\right\rangle=m_in_0\int d^3vv_\parallel\frac{dr}{dt}\tilde{f}^R_i \tag{37b}$$

and $\tilde{f}^R_i$ is the resonant, linear ion response. As shown in Section II, the RHS corresponds to momentum exchange between waves and particles and cancels identically with its counterpart involving $C_{w-p}(N)$ in the $k_\parallel$ moment of the wave kinetic equation. Hence, the RHS is hereafter neglected, and we need only focus on $\left\langle\Pi^R_{r,\parallel}\right\rangle$. As usual in quasi-linear theory, we simply plug the resonant linear response $\tilde{f}^R_i$ into $\left\langle\Pi^R_{r,\parallel}\right\rangle$ to calculate the flux. The linear response $\tilde{f}^R_{i,\mathbf{k}}$ is:

$$\tilde{f}^R_{i,\mathbf{k}}=\frac{-J_0(\lambda)}{\omega_k-\omega_{EB}-k_\parallel v_\parallel+i\varepsilon}c_s^2\left\{\frac{k_\theta}{\omega_{ci}}\frac{\partial\langle F_i\rangle}{\partial x}+k_\parallel\frac{\partial\langle F_i\rangle}{\partial v_\parallel}\right\}\frac{e\tilde{\phi}_\mathbf{k}}{T_e} \tag{38}$$

where $\lambda\equiv k_\perp\rho_\perp$ and $\omega_{EB}\equiv k_\theta\langle v_E\rangle$. After considerable (albeit straightforward) algebra we obtain the resonant ion momentum flux

$$\left\langle\Pi^R_{r,\parallel}\right\rangle=n_0m_i\left[-\chi_\phi\frac{\partial}{\partial r}\langle v_\parallel\rangle+V_r\langle v_\parallel\rangle+S\right] \tag{39a}$$

where for a shifted Maxwellian $\langle f\rangle$,

$$\chi_\phi=\sqrt{2\pi}\sum_\mathbf{k}(k_\theta\rho_i)^2\frac{v_{ti}}{|k_\parallel|}\Omega^2\Gamma_0(b)e^{-\Omega^2}\left|\frac{e\tilde{\Phi}_\mathbf{k}}{T_i}\right|^2 \tag{39b}$$

$$V_r=\sqrt{\frac{\pi}{2}}\sum_\mathbf{k}(k_\theta\rho_i)^2\frac{v_{ti}}{|k_\parallel|}\frac{1}{L_n}\Gamma_0(b)\left|\frac{e\tilde{\Phi}_\mathbf{k}}{T_i}\right|^2e^{-\Omega^2}\left\{1-\frac{\eta_i}{\eta_i^{crit}}-\frac{\bar{\omega}_\mathbf{k}}{\omega_{*i}}-\eta_i\Omega^2\right\} \tag{39c}$$

and

$$S=\sqrt{\pi}\sum_\mathbf{k}(k_\theta\rho_i)^2\frac{v_{ti}^2}{L_n}\frac{1}{|k_\parallel|}\Omega\Gamma_0(b)\left|\frac{e\tilde{\Phi}_\mathbf{k}}{T_i}\right|^2e^{-\Omega^2}\left\{1-\frac{\eta_i}{\eta_i^{crit}}-\frac{\bar{\omega}_\mathbf{k}}{\omega_{*i}}-\eta_i\Omega^2\right\}. \tag{39d}$$

Here the notation is standard, so $\eta_i^{crit}=2\left[1+2b(1-I_1/I_0)\right]^{-1}$, $\Omega=\bar{\omega}_\mathbf{k}/\sqrt{2}k_\parallel v_{ti}$, $b=k_\perp^2\rho_i^2$, I_0 and I_1 are the modified Bessel functions., $v_{*e,i}$ are the electron and ion diamagnetic velocities, $\bar{\omega}_k$ is the Doppler shifted wave frequency defined by $\omega_\mathbf{k}-\omega_{EB}\equiv\bar{\omega}_\mathbf{k}+k_\parallel\langle v_\parallel\rangle$, and $\omega_{*e,i}=k_\theta v_{*e,i}$, $\tau=T_e/T_i$, $\Gamma_0=I_0e^{-b}$ etc.

Some discussion of the transport coefficients χ_ϕ, V_r and S is appropriate at this point. Predictably, $\chi_\phi\sim\chi_i$ but $\chi_\phi\neq\chi_i$, on account of the structure of resonant coupling to the spectrum of wave phase velocities. χ_ϕ (and obviously, the entire resonant particle flux) decays rapidly [$\sim exp\left(-\Omega^2\right)$] for non-resonant particles, and may be written as $\chi_\phi \sim \left\langle \tilde{v}_E^2 \right\rangle \tau_{ac}$ where τ_{ac} is set by the dispersion in the distribution of the ion transit rate. Of course, $\chi_\phi \sim D_{GB}$ for "mixing length estimate" fluctuation levels. The convection velocity V_r is rather sensitive and model dependent. V_r is inward (corresponding to a pinch) for ITG modes in the resonant regime near threshold ($|\Omega| \lesssim 1$). For regimes far from threshold ($\Omega \gg 1$) V_r is necessarily outward, however as noted above, the magnitude of this term is then negligible. Near marginality, but for electron drift waves, the sign of the convective term depends sensitively on η_i, and requires a quantitative description of the micro-turbulence spectrum. Note that the convective term scales as $V_r/\chi_\phi \sim 1/\left(\Omega^2 L_\perp\right)$, a profile scale length. Thus, the resonant particle pinch complements the non-resonant turbulent convection (TurCo) pinch [23], derived for toroidal geometry. The latter is non-resonant (and so must represent wave transport), is inward for electron drift waves, can attain either sign (depending on plasma parameters) for ITG modes and has $V_r/\chi_\phi \sim 1/R$ [i.e. $O(\varepsilon)$ smaller], but is insensitive to the resonance function. In other words, if the percentage weighting of resonant particles is lower than ε, TurCo provides the main convection effect. On the other hand, near marginal stability, where the percentage of resonant particles is high, the resonant particle pinch is the primary cause of convection. Furthermore, near marginality $|\bar{\omega}_\mathbf{k} \ll \omega_{*i}|$ and $\left|\Omega^2\right| \lesssim 1$, so that

$$\frac{V_r}{\chi_\phi} \approx \frac{1}{\Omega^2}\frac{1}{L_n}\left\{1 - \frac{\eta_i}{\eta_i^{crit}}\right\},$$

so that for unstable ITG, the ∇T_i driven pinch is *inward* in rough agreement with Ref. [21], but the ∇n driven pinch is *outward*, opposite to the non-resonant pinch predicted in Ref. [22]. Considering now the third term in the resonant momentum flux given by Eqn. (39a), since $S/V_r \approx \Omega v_{ti}$, S must vanish in the absence of symmetry breaking (i.e. $S \to 0$ as $\left\langle k_\parallel \right\rangle \to 0$). As is also clear, for regimes of finite symmetry breaking, the sign of the residual stress follows identical rules to that of the resonant particle convection term, but multiplied by sgn $\left(\left\langle k_\parallel \right\rangle\right)$, which is determined by Eqn. (36a). Also, it is readily apparent that the residual stress is a much broader concept than that encountered while considering $\mathbf{E}\times\mathbf{B}$ shearing effects on momentum transport. More generally, it is clear that S emerges naturally while constructing the quasi-linear theory of momentum transport in the presence of symmetry breaking ($\left\langle k_\parallel \right\rangle \neq 0$).

Here we note that the above result is in agreement with a similar calculation of the resonant particle flux performed in Ref. [21]. However, we emphasize that while the resonant particle flux contains a host of tantalizing non-diffusive terms, neglecting the wave momentum flux - as was done in Ref. [21] - is in general not justified since the relative magnitude of the resonant particle flux in comparison to the non-diffusive wave momentum flux is strongly parameter dependent. Note that the results in Ref. [21] don't exhibit an apparent mode-dependency which is obvious from the standard calculations using the gyrokinetic equation. We also note that the basic scaling $V_r/\chi_\phi \sim 1/\Omega^2\left(\Omega^2 L_\perp\right)$ is suggestive of an underlying physical mechanism which does not depend on toroidicity (i.e. exists in simpler geometry), leaving some doubts on the scaling with respect to L_n proposed in Ref. [22]. Furthermore, the critical role of wave momentum is highlighted when one considers that by neglecting this component, the RHS of Eqn. (37a) (which corresponds to wave-particle interactions, and is generally non-zero) is not cancelled by its counterpart in the wave momentum equation. Thus, any description of momentum transport purely in terms of resonant particle evolution leaves an unaccounted for sink/source of momentum in the mean momentum evolution equation! Finally, it should be noted that this analysis has not addressed the nature of particle transport or the non-adiabatic electron response. Considering these will surely introduce additional momentum transport effects related to the interplay of particle and momentum transport. Such cross coupling is discussed further in Refs. [25, 24].

IMPLICATIONS FOR MOMENTUM TRANSPORT PHENOMENOLOGY

Till now, we have focused on the general structure of the theory of momentum transport. Here we discuss the implications for and applications of, the theory to several aspects of the phenomenology of toroidal momentum transport in tokamaks. We proceed by first summarizing the structure of the predicted momentum flux, first discussing its implication for quasi-stationary momentum transport phenomena and then for *momentum transport bifurcation* phenomena. Detailed modeling of specific cases is left to future papers. Here, rather, we explore the general scope of potential applications which are possible within this theory.

The principal results obtained so far are the stationary state values of the radial flux of parallel momentum $\left\langle \Pi_{r,\parallel}^w \right\rangle$, the evolution equation for $\left\langle k_\parallel \right\rangle$ (which links symmetry breaking to dynamics), and the resonant particle parallel momentum flux $\left\langle \widetilde{v}_{Er}\widetilde{P}_{\parallel}\right\rangle_R$. These are respectively:

$$\left\langle \Pi^w_{r,\parallel}\right\rangle = \int d\mathbf{k}k_\parallel \left\{ \langle V_r\rangle N - \frac{v_{gr}^2}{v_{T_\mathbf{k}}}\frac{\partial\langle N\rangle}{\partial r} + v_{gr}\frac{k_\theta}{v_{T_\mathbf{k}}}\langle v_E\rangle' \frac{\partial\langle N\rangle}{\partial k_r}\right\} \tag{40a}$$

$$\frac{\partial}{\partial t}\langle k_\parallel\rangle + \gamma_{NL}\langle k_\parallel\rangle = -\frac{1}{r}\frac{\partial}{\partial r}\left[r\left\langle \Pi^w_{r,\parallel}\right\rangle\right] - \int d\mathbf{k}\left(\frac{\partial k_\parallel}{\partial k_r}\right)k_\theta\langle v_E\rangle'\langle N\rangle - \int d\mathbf{k}\hat{\mathbf{b}}\cdot\overleftrightarrow{\mathbf{D}}_\mathbf{k}\cdot\nabla_\mathbf{k}\langle N\rangle \tag{40b}$$

$$\left\langle \widetilde{v}_{Er}\widetilde{P}_{\parallel}\right\rangle_R = n_0 m_i\left[-\chi^R_\phi\frac{\partial}{\partial r}\langle v_\parallel\rangle + V^R_r\langle v_\parallel\rangle + S_R\right] \tag{40c}$$

where

$$\chi^R_\phi = \sqrt{2\pi}\sum_k (k_\theta\rho_i)^2\frac{v_{ti}}{|k_\parallel|}\Omega^2\Gamma_0(b)e^{-\Omega^2}\left|\frac{e\Phi_\mathbf{k}}{T_i}\right|^2 \tag{40d}$$

$$V_r = \sqrt{\frac{\pi}{2}}\sum_\mathbf{k}(k_\theta\rho_i)^2\frac{v_{ti}}{|k_\parallel|}\frac{1}{L_n}\Gamma_0(b)\left|\frac{e\widetilde{\Phi}_\mathbf{k}}{T_i}\right|^2 e^{-\Omega^2}\left\{1-\frac{\eta_i}{\eta_i^{crit}}-\frac{\bar{\omega}_\mathbf{k}}{\omega_{*i}}-\eta_i\Omega^2\right\} \tag{40e}$$

$$S = \sqrt{\pi}\sum_\mathbf{k}(k_\theta\rho_i)^2\frac{v_{ti}^2}{L_n}\frac{1}{|k_\parallel|}\Omega\Gamma_0(b)\left|\frac{e\widetilde{\Phi}_\mathbf{k}}{T_i}\right|^2 e^{-\Omega^2}\left\{1-\frac{\eta_i}{\eta_i^{crit}}-\frac{\bar{\omega}_\mathbf{k}}{\omega_{*i}}-\eta_i\Omega^2\right\} \tag{40f}$$

The total turbulent momentum flux is conventionally expressed for purposes of data analysis and modeling in the form

$$\left\langle \Pi^T_{r,\parallel}\right\rangle = n_0 m\left[-\chi_\phi\frac{\partial}{\partial r}\langle v_\phi\rangle + V_r\langle v_\phi\rangle + S\right] \tag{41}$$

where χ_ϕ is the diffusivity (i.e. turbulent viscosity accounting for the radial scattering of toroidal flow), V_r is the radial convection velocity (i.e. pinch, if inward) and S is the residual stress, which arises since waves and particles can exchange momentum. S has no counterpart in the particle flux. Loosely speaking, anomalous electron-ion energy coupling might be thought of as a process analogous to S in the energy balance, though residual stress accounts for ion-wave momentum exchange, rather than ion-electron momentum exchange. Both anomalous thermal transfer and residual stress (related to anomalous momentum transfer) are processes mediated by waves. Pending possible re-expression of $\left\langle \Pi^w_{r,\parallel}\right\rangle$ in terms of $\langle v_\parallel\rangle$, Eqn (40) implies that:

$$\chi_\phi = \chi^R_\phi$$

$$V_r = V^R_r$$

$$S = S_R + \left\langle \Pi^w_{r,\parallel}\right\rangle$$

so the momentum diffusivity and the convection velocity are carried primarily by resonant particles, while the residual stress is supported by both resonant particles and waves. Note that the residual stress is usually neglected in traditional "D and V models", but can indeed be the dominant off-diagonal contribution to momentum transport, particularly in regimes of strong turbulence away from marginality. Surely the most basic message of this section is the need to address the residual stress contribution to the momentum flux.

We first discuss the implications of the theory for transport of momentum in quasi-stationary states. Obviously, $\chi_\phi \sim \chi_i$ but $\chi_\phi \neq \chi_i$, in accord with all experimental results. Inward convection (i.e. $V_r < 0$, a momentum pinch) due to resonant transport is predicted for profiles which are "stiff" to ITG turbulence (i.e. near marginal). Here $V/\chi_\phi \sim 1/L_\perp$, but we must emphasize that both the magnitude and the sign of V_r are *very* sensitive to parameter variations and must be studied quantitatively in order to meaningfully apply the theory to perturbative transport experiments, such as those of Yoshida, *et. al.* on JT-60U [12]. We again note that the resonant ion pinch discussed here is complementary to the non-resonant TurCo pinch discussed in Ref. [23]. It is very important, however, to note that the residual stress S can drive inward non-diffusive momentum transport in addition to V_r. Both resonant and non-resonant particles can contribute to S. A nearly universal contribution to S is the radiative diffusive flux of wave momentum ($\sim -D_{rad}\partial\left\langle P_{\parallel}^{w}\right\rangle/\partial r$), which is *inward*, since fluctuation intensity tends to peak toward the outside (i.e. $\langle N\rangle$ and $\langle P_{\parallel}\rangle$ increase while $T(r)$ and $n(r)$ etc. decrease). Also, the edge region is replete with potentially viable symmetry breaking mechanisms, so $\langle k_{\parallel} \neq 0\rangle$ there. Thus, inward diffusion of wave momentum is clearly a robust candidate mechanism to support residual stress contributions to the off-diagonal momentum flux. However, we hasten to add that electric field shear may also contribute to the inward momentum flux via S, as should be apparent from Eqn (40a). Indeed note that $\langle v_E\rangle'$ can drive a non-diffusive momentum flux even if electric field shear is not the principal symmetry-breaking mechanism which determines $\langle k_{\parallel}\rangle$. This is due to the fact that flux and symmetry breaking of $\langle k_{\parallel}\rangle$ are distinct in this theory, though both operate via $\langle v_E\rangle'$.

Turning to the interesting phenomenon of momentum transport bifurcations, we focus primarily on the wave momentum flux, since this is the most relevant to strong turbulence regimes from which bifurcations initiate, and since $\left\langle \Pi_{r,\parallel}^{w}\right\rangle$ is clearly sensitive to electric field shear, which is a key bifurcation order parameter. We remind the reader that momentum transport bifurcations related to a change in the structure of the momentum flux have been observed both at the edge and in the core of tokamak plasmas. In general, it is clear from Eqns. (40a) and (40b) that both electric field shear and fluctuation profile affects are important in regulating the local wave momentum flux. Electric field shear enters both as a symmetry breaking mechanism, which along with magnetic shear renders $\langle k_{\parallel}\rangle \neq 0$. $\langle v_E\rangle'$ also acts as a direct driver of $\left\langle \Pi_{r,\parallel}^{w}\right\rangle$ with the multiplicative factor $k_{\theta}\partial v_{gr}/\partial k_r$, which is mode phase velocity dependent. Of course, electric field shear also regulates fluctuation levels. Thus, we see that a generic scaling is $\left\langle \Pi_{r,\parallel}^{w}\right\rangle \sim \hat{s}\left(\langle v_E\rangle'\right)^{\alpha} v_{ph}\langle I\rangle$, where $\alpha \sim 1 \rightarrow 2$. Here I is the fluctuation intensity level, $I(r)$. The fluctuation profile structure, often associated with the edge, is also very relevant. Radiative diffusion of wave momentum $\left(\sim -D_{rad}\partial\langle P_{\parallel}\rangle/\partial r\right)$ can produce an influx of wave momentum from the edge into the core. Similarly, $-\nabla\cdot\left\langle \Pi_{r,\parallel}^{w}\right\rangle$ can also act to produce or enhance symmetry breaking (i.e. $\langle k_{\parallel}\rangle$) and wave momentum transport via an inward wave momentum flux. Since these intensity gradient driven fluxes compete with the electric field shear driven convective flux of wave momentum to regulate the overall structure of $\left\langle \Pi_{r,\parallel}^{w}\right\rangle$, any changes in the balance between the competitors may trigger possible momentum transport bifurcations.

Given the rich structure of $\left\langle \Pi_{r,\parallel}^{w}\right\rangle$, one can identify at least two scenarios for momentum transport bifurcations. These are:

i.) a change in $\left\langle \Pi_{r,\parallel}^{w}\right\rangle$ from an intensity diffusion dominated state $\left(\Pi_{r,\parallel}^{w} \sim -D_{rad}\partial\langle P_{\parallel}\rangle/\partial r\right)$ to a $\langle v_E\rangle'$-driven convection dominated state $\left(\Pi_{r,\parallel}^{w} \sim (k_{\theta}\partial v_{gr}/\partial k_r)\langle v_E\rangle'\langle P_{\parallel}\rangle/v_T\right)$.

ii.) a change in a convection dominated state of $\left\langle \Pi_{r,\parallel}^{w}\right\rangle$ due to a change in the product $v_{ph}\langle v_E\rangle'$. This could occur either via an electric field bifurcation *or* via a change in v_{ph}, the turbulence phase velocity.

Type (i.) qualitatively resembles an $L \rightarrow H$ transition, which occurs at the edge and is associated with an increase in electric field shear so that $V_w\langle N\rangle$ increases while $-D_{rad}\partial\langle N\rangle/\partial r$ drops. In this scenario, $\langle k_{\parallel}\rangle$ is set by $\langle v_E\rangle'$, as well, strengthening the dependence on $\langle v_E\rangle'$. Type (i.) bifurcations seem consistent with the observed strong correlation between scaling trends in intrinsic rotation and the $L \rightarrow H$ transition. Type (ii.) bifurcations qualitatively resemble internal momentum transport bifurcations, such as occur with ITBs and as that observed recently on the TCV tokamak [16]. Such internal bifurcations can occur via an increase in local $\langle v_E\rangle'$ in the core, thus increasing both convective wave momentum flux and symmetry breaking. This scenario is relevant to ITB-formation, where changes in the momentum flux would be accompanied by improvements in confinement. However, the theory suggests that internal momentum transport bifurcations can also occur via changes in v_{ph}, thereby reversing the direction of wave momentum convection. This scenario is especially relevant to the core plasma, where intensity gradients are weak, and to instances where the change in momentum transport is *not* accompanied by a noticeable change in confinement. This is the case in a recently reported, interesting experiment on TCV, during which core rotation 'flipped' sign from counter to co-current direction, as the density was increased in ohmic plasmas (N.B. similar phenomena have been observed on Alcator C-Mod [64]). No significant change in thermal transport was observed. We speculate that the observed momentum transport bifurcation could be explained as a change in v_{ph} from v_{*e} direction to v_{*i} direction, while the ambient fluctuation population changes from predominantly electron drift waves to predominantly ITG modes, as density increases while approaching the OH-saturation regime. Indeed, excitation of ITG turbulence has long been thought to explain the saturation of energy confinement in high density ohmic plasmas. Detailed, quantitative modelling of this fascinating momentum transport phenomenon using the theory presented here is ongoing, and will be discussed in a future publication. One interesting question which is stimulated by the TCV results is that while the experiment clearly exhibits hysteresis phenomena in density, in that co-direction rotation persists when density is lowered below OH-saturation levels, the theory suggests a *smooth* evolution with v_* as it changes from v_{*i} to v_{*e}. Put another way, the theory seems to predict a second order transition while the data suggests a first order transition. We speculate that a possible resolution of this dilemma is 'turbulence spreading' or, turbulent overshoot of the ITG population into the electron drift wave population. Turbulence intensity fronts are well known to exhibit 'effective inertia' [65], thus enabling them to continue to advance when local drive is lowered. We suggest that the dynamics may be described by two coupled nonlinear diffusion equations for the electron and ion drift wave quanta density. In the past, such simple nonlinear diffusion equations have had some success in modelling the non-local dynamics of turbulence spreading. Constructing a credible theoretical model of the two interacting turbulence populations and then using that model to explain the observed density hysteresis constitute *very* significant challenges to our ongoing studies of TCV phenomenology. Finally, we comment that a theoretical understanding of the implications of reversed or weakly negative magnetic shear for the $\langle v_E \rangle'$-driven convective wave momentum flux would be of great interest in elucidating intrinsic rotation in ITB plasmas. This will be pursued in future works.

CONCLUSIONS

In this paper, we have presented a general theory of parallel momentum transport by collisionless electrostatic drift wave turbulence. The principal results of this paper are:

i.) the proof of a general momentum conservation theorem, which relates total mean momentum evolution to the resonant particle momentum flux, the wave momentum flux and the refractive force exerted by flow structures, has been derived. This theorem accounts for, and resolves all ambiguities related to the interaction between momentum *transport* by turbulence and momentum *transfer* by wave-particle interaction. The physics of the fluctuation-dependent momentum source/sink which appears in the naive quasilinear theory is clarified.

ii.) the calculation of the wave momentum flux valid for stationary turbulence. The theoretical formulation which we present yields a general expression for the radial flux of parallel wave momentum, as given by Eqn. (40a). The wave momentum flux contains both radiative diffusive flux and a novel $\langle v_E \rangle'$-driven convective flux. The wave momentum flux contributes to the total residual stress - i.e. that part of the momentum flux not explicitly proportional to $\langle v_\phi \rangle$ or $\partial \langle v_\phi \rangle / \partial r$. All the usual ambiguities associated with calculating non-resonant parallel momentum transport have been resolved.

iii.) the calculation of the resonant particle momentum flux, as given by Eqns. (40c)-(40f). The resonant momentum flux consists of a diffusive flux (with $\chi_\phi \sim \chi_i$ but $\chi_\phi \neq \chi_i$), a novel convective flux which may be inward in certain regimes, and a novel residual stress piece, driven by ∇T_i and explicitly dependent upon the symmetry breaking mechanism (i.e. the origin of $\langle k_\parallel \rangle \neq 0$, see (v.) below).

iv.) the calculation of the refractive force, which is shown to depend sensitively on the structure of secondary parallel flows, such as those produced by GAMs, and on radial electric field shear.

v.) a general expression for the evolution of $\langle k_\parallel \rangle$, which allows evaluation and comparison of competing symmetry breaking mechanisms, as shown in Eqn. (40b).

vi.) the application of the theory to outline two possible scenarios for momentum transport bifurcations. Both of these involve a change in the momentum flux balance, and arise from evolution of the convective component of the wave momentum flux, proportional to $v_{ph} \langle v_E \rangle'$. In one case, $\langle v_E \rangle'$ rises, as in a conventional electric field shear bifurcation, leading to overall improvement in confinement. In the other, v_{ph} changes direction, due to a change in the propagation direction of the underlying drift waves, but $\langle v_E \rangle'$ and confinement do not change. The second scenario may be relevant to the phenomena recently observed in the core of the TCV tokamak.

The theory presented here has several significant limitations. First, the model is purely electrostatic, so field momentum transport is not addressed. In the case of Alfvenic turbulence, such as that driven by Alfven eigenmodes, the field momentum density can be expected to be comparable to the non-resonant particle momentum density, thus complicating the wave momentum budget, and possibly introducing an Abraham force on the plasma [66]. Second, we have not addressed the edge boundary condition on the rotation profile, beyond entertaining the possibility of a possible influx of wave momentum from the SOL. This boundary condition is quite likely non-trivial, as SOL flows are significant, they can exert a stress on the core plasma, and can modify wave propagation and wave momentum in the SOL plasma [15]. Third, the Krook model employed in Section III needs improvement. This leads one to confront the interesting theoretical question of how nonlinear interaction and transfer processes driven by cascading or modulational instability respond to perturbations, such as those induced by population gradients. An improved model of 'turbulent collisions' is required, and an approach using a modified Leith model holds promise [67, 68]. Finally, the results in this paper indicate that the calculation of wave momentum transport cannot really be separated from the calculation of turbulence spreading and other non-locality phenomena which influence turbulence envelope evolution[69, 70].

While this paper is theoretical, it does offer several ideas and results of potential interest to experimentalists. First it presents a *complete* calculation and discussion of the structure and physics of the momentum flux (i.e. resonant particle + wave) and identifies (quantitatively) the residual stress as a key contributor to momentum transport. The residual stress is ignored in traditional "D and V models", and requires dual perturbation experiments to elucidate it. Second, we present an improved calculation of a resonant particle momentum pinch, which may be relevant to collisionless core plasmas near ITG marginality. This pinch has $V_r/\chi_\phi \sim 1/L_T$. Third, we identify two possible momentum transport bifurcation scenarios. Well resolved, localized fluctuation measurements could be used to test the momentum transport bifurcation scenario based on a change in fluctuation v_{ph}. Experiments which address intrinsic rotation evolution in ITB/plasmas with steepen ∇P would be of interest, too. Fourth, we identify mean electric field shear and GAMs as likely agents for transmitting the refractive force, and suggest future studies of the relation between these quantities and intrinsic rotation. Fifth, we offer a framework within which to compare and contrast the strength and relevance of competing turbulence symmetry breaking mechanisms.

Future work will focus on extensions of the theory and on applications to specific experiments.

ACKNOWLEDGMENTS

The authors would like to extend special thanks to Sergey Nazarenko for insightful explanation of the Leith model and its relevance to turbulent collisions and to Xavier Garbet, K. Itoh and S.-I Itoh for discussions and collaboration on the subjects of momentum transport and nonlocality in turbulence. We also thank (in alphabetical order) C. Bourdelle, C.S. Chang, G. Falchetto, Ph. Ghendrih F. Hinton, Z. Lin, M. Malkov, Y. Sarazin and the participants in the 2007 Festival du Theorie for stimulating discussions of the theory and (also in alphabetical order) K. Burrell, P. deVries, B. Duval, C. Hidalgo, Y. Kamada, S. Kaye, V. Parail, J. E. Rice, W. Solomon and M. Yoshida for discussions of the experimental data. This research was supported by DOE Grant No. FG02-04ER 54738 (UCSD) and DOE Grant No. DE-AC02-76-CHO-3073 (PPPL). P.D. would also like to acknowledge support from the MEXT project on Structure Formation in Plasma and the hospitality of Kyushu University, where part of this work was performed.

EXPLICIT DERIVATION OF WAVE ACTION DENSITY

In this appendix we derive an explicit expression for the wave action density from the gyrokinetic equation, and relate it to momentum exchange between resonant particles and waves. The induced plasma response to an external electrostatic perturbation may be written as

$$\delta F_{k,\omega}^{\mathrm{ind}} = \frac{J_0\left(k_\perp \rho_\perp^{(s)}\right)}{\omega - \omega_{EB} - k_\parallel v_\parallel + i\varepsilon}\left\{\frac{c}{B}\left(\hat{\mathbf{b}} \times \mathbf{k}\right)\cdot\frac{\partial \langle F_s\rangle}{\partial \mathbf{x}} - \frac{q_s}{m_s}k_\parallel\frac{\partial\langle F_s\rangle}{\partial v_\parallel}\right\}\delta\phi_{k,\omega}^{\mathrm{tot}} \equiv L_{k,\omega}\delta\phi_{k,\omega}^{\mathrm{tot}}, \tag{42}$$

where $\delta\phi^{\mathrm{tot}} = \delta\phi^{\mathrm{ind}} + \delta\phi^{\mathrm{ext}}$. Considering the gyrokinetic Poisson equation, an equation for the induced response may be written as

$$k_\perp^2\delta\phi_{k,\omega}^{\mathrm{ind}} = 4\pi\sum_s q_s\int d^3\bar{v}\left\{J_0(\lambda)\,\delta F_{k,\omega}^{\mathrm{ind}} + \left[J_0^2(\lambda) - 1\right]\frac{q_s\delta\phi_{k,\omega}^{\mathrm{ind}}}{T_s}\langle F_s\rangle\right\}, \tag{43}$$

where it's understood that $\delta F_{k,\omega}^{\mathrm{ind}}$ is species dependent. Eqn. (43) may be simplified via the following notation

$$\varepsilon_{\perp}(k)\, k_{\perp}^2\, \delta\phi_{k,\omega}^{\mathrm{ind}} = 4\pi \sum_s q_s \int d^3v' J_0(\lambda)\, \delta F_{k,\omega}^{\mathrm{ind}}, \tag{44}$$

where $\varepsilon_{\perp}(k) \equiv 1 + \sum_s \left(k_{Ds}^2/k_{\perp}^2\right)[1 - I_0(b_s)\exp(-b_s)]$. This permittivity may be profitably understood to correspond to a "gyrokinetic vacuum" [71].

After substitution of Eqn. (42) into Eqn. (44), the induced response of the electrostatic field can be written in terms of the external perturbations, yielding:

$$(1 + \chi_{k,\omega})\, \varepsilon_{\perp}(k)\, k_{\perp}^2\, \delta\phi_{k,\omega}^{\mathrm{ind}} = 4\pi \sum_s q_s \int d^3v' J_0\left(k_{\perp}\rho_{\perp}^{(s)}\right) L_{k,\omega}\, \delta\phi_{k,\omega}^{\mathrm{ext}}. \tag{45}$$

Defining the susceptibility by

$$\chi_{k,\omega} = -4\pi \sum_s \frac{q_s}{\varepsilon_{\perp}(k)\, k_{\perp}^2} \int d^3\bar{v} \frac{J_0^2\left(k_{\perp}\rho_{\perp}^{(s)}\right)}{\omega - \omega_{EB} - k_{\parallel}v_{\parallel} + i\varepsilon} \left\{ \frac{c}{B}(\hat{\mathbf{b}} \times \mathbf{k}) \cdot \nabla \langle F_s \rangle - \frac{q_s}{m_s} k_{\parallel} \frac{\partial \langle F_s \rangle}{\partial v_{\parallel}} \right\}, \tag{46}$$

and the dispersion relationship is defined as

$$0 = D_{k,\omega} \equiv 1 + \chi_{k,\omega}. \tag{47}$$

The growth rate may be approximated as

$$\gamma_k = \left. \frac{-\mathrm{Im} D_{k,\omega}}{\partial D_{k,\omega}/\partial\omega} \right|_{\omega=\omega_k}. \tag{48}$$

Finally, the rate at which momentum is exchanged between waves and resonant particles is given by the RHS of Eqn. 37a, and may be rewritten in terms of the susceptibility as:

$$S^R = \frac{1}{4\pi} \int d\mathbf{k} k_{\parallel} \mathrm{Im}\chi_{k,\omega_k}\, \varepsilon_{\perp}(k)\, k_{\perp}^2 \left|\delta\phi_k\right|^2, \tag{49}$$

utilizing Eqns. (48) and (47), this may be simplified, yielding:

$$S^R = -2 \int d\mathbf{k} k_{\parallel} \gamma_k N_k, \tag{50}$$

where N_k is the wave action density, and is defined as

$$N_k \equiv \frac{1}{8\pi} \left. \frac{\partial D_{k,\omega}}{\partial \omega} \right|_{\omega=\omega_k} \varepsilon_{\perp}(k)\, k_{\perp}^2 \left|\delta\phi_k\right|^2. \tag{51}$$

PHYSICS OF THE REFRACTIVE FORCE

This appendix presents a brief discussion of the basic physics of the refractive force. A pedagogical tactic employed here is to illustrate aspects of the refractive force in the context of concepts and results familiar from the theory of zonal flow generation.

Put most generally, the refractive force is simply the rate of change of wave momentum density induced by wave packet refraction. Since the evolution of wave momentum density is described by the k-moment of the wave kinetic equation, we have:

$$\frac{\partial}{\partial t}\mathbf{P}_w + \nabla \cdot \overleftrightarrow{\Pi}_w = \mathbf{f}_{refr} + \left(\frac{dP_w}{dt}\right)_{diss} \tag{52}$$

where:

$$\overleftrightarrow{\Pi}_w = \int d\mathbf{k}\mathbf{k}\,(\mathbf{V} + \mathbf{v}_g)\, N \tag{53a}$$

is the total wave momentum density flux,

$$\left(\frac{dP_w}{dt}\right)_{diss} = \int d\mathbf{k}\mathbf{k} C(N) = \int d\mathbf{k}\mathbf{k}\,(C_{w-w}(N) + C_{w-p}(N)) \tag{53b}$$

is the total wave momentum dissipated by quasi-particle interactions (N.B. in practice, such momentum dissipation almost always occurs via linear and non-linear wave-particle momentum transfer), and

$$\mathbf{f}_{refr} = \int d\mathbf{k}\left(-\frac{\partial}{\partial \mathbf{x}}(\omega + \mathbf{k}\cdot\mathbf{V})N\right) \tag{53c}$$

is the refractive force. As noted in the text for fluctuating strain fields, the wave population density may be further decomposed into a mean and fluctuating piece (i.e. $N = \langle N\rangle + \widetilde{N}$), so the mean field refractive force may be written as:

$$\langle \mathbf{f}_{refr}\rangle = -\int d\mathbf{k}\left\{\left(\frac{\partial}{\partial \mathbf{x}}\langle \omega + \mathbf{k}\cdot\mathbf{V}\rangle\right)\langle N\rangle + \overleftrightarrow{\mathbf{D}}_{\mathbf{k}}\cdot\frac{\partial\langle N\rangle}{\partial \mathbf{k}}\right\} \tag{54a}$$

where

$$\overleftrightarrow{\mathbf{D}}_{\mathbf{k}} = \sum_{\mathbf{q}} \mathbf{q}\mathbf{q}\left|\left(\widetilde{\omega} + \mathbf{k}\cdot\widetilde{\mathbf{V}}\right)_q\right|^2 R(\Omega - \mathbf{q}\cdot\mathbf{v}_{gr}) \tag{54b}$$

is the k-space diffusion tensor for stochastic refraction.

The interpretation of $\langle \mathbf{f}_{refr}\rangle$ as a force density follows from the definition $\partial \mathbf{P}_w/\partial t = \langle f_{refr}\rangle + \cdots$ and the fact that $dP_{NR}/dt = dP_W/dt$. Moreover, $\langle f_{refr}\rangle$ is a local source/sink of momentum, and not a divergence of a flux. The physics of the refractive force is simple - refraction tilts wave packet trajectories, thus changing their momentum. For example the familiar case of a mean shear flow $\langle v_y(x)\rangle$, the refractive force is just

$$\langle \mathbf{f}_{refr}\rangle = \langle f_{ref}\rangle \hat{\mathbf{x}} = -\frac{\partial v_y}{\partial x}\int d\mathbf{k}k_y\langle N\rangle \ . \tag{55a}$$

This corresponds to the familiar cartoon explanation of "zonal flow formation", where "tilting" of wave packets and eddies by coherent shear flows, which diverts $\hat{\mathbf{x}}$-direction momentum into $\hat{\mathbf{y}}$-direction momentum and so applies a force (or stress) on the wave packet, leading to shear flow amplification. Similarly for stochastic (i.e. diffusive) refraction by a spectrum of zonal flows (with $\mathbf{q} = q\hat{\mathbf{x}}$ only),

$$\langle \mathbf{f}_{refr}\rangle = \langle f_{refr}\rangle \hat{\mathbf{x}} = -\int d\mathbf{k}D_{k_x,k_x}\frac{\partial\langle N\rangle}{\partial k_x} \tag{55b}$$

so the refractive force is just the integrated flux in $\mathbf{k}$ produced by random shearing. Here $D_{k_x,k_x} = \hat{\mathbf{x}}\cdot\overleftrightarrow{\mathbf{D}}_{\mathbf{k}}\cdot\hat{\mathbf{x}}$. Proceeding heuristically, we can immediately note that the power expended by the wave packet against the refractive force is just

$$P = \mathbf{f}\cdot\mathbf{V} \approx \int d\mathbf{k}v_{gr}f_r = -\int d\mathbf{k}v_{gr}D_{k_x,k_x}\frac{\partial\langle N\rangle}{\partial k_x} \tag{56}$$

This result may be recovered systematically from wave-kinetics by taking the energy moment of the wave kinetic equation [72], i.e.

$$\frac{\partial \varepsilon_\omega}{\partial t} = -\int d\mathbf{k}\omega_{\mathbf{k}}\frac{\partial}{\partial \mathbf{k}}\cdot\left[\frac{d\mathbf{k}}{dt}N\right] = \int d\mathbf{k}\mathbf{v}_g\cdot\frac{d\mathbf{k}}{dt}N$$

so for stochastic zonal flow shears

$$\frac{\partial \varepsilon_\omega}{\partial t} = -\int d\mathbf{k}v_{gr}D_{k_x,k_x}\frac{\partial\langle N\rangle}{\partial k_x}$$

in agreement with the heuristic argument. Note that here, N is the wave action density. Interestingly, using $N = \Omega$ along with the appropriate corresponding relation $\varepsilon_\omega = N/\left(1 + k_\perp^2\rho_s^2\right)$ yields an identical result for P. Which is the macroscopic observable (i.e. power expended by waves). Since the total energy of waves and zonal flows is conserved up to dissipation, i.e. $d\left(\varepsilon_{Wave} + \varepsilon_{ZF}\right)/dt = 0$ [43], this establishes that the refractive force of random shearing is in fact *precisely* what does the work to drive zonal flows against collisional dissipation, and that the refractive force may be viewed as the effective body force which drives the zonal flow, thus connecting the refractive force to familiar concepts. Note that a parallel wave group velocity is required to energize a parallel mean flow via the parallel refractive force, just as a radial group velocity is required to energize zonal flows via the radial refractive (i.e. "eddy tilting") force.

Finally, we comment that since the refractive force enters the balance of wave quanta density N, wave quanta momentum density $\mathbf{P}_w = \mathbf{k}N$ and wave quanta energy density $\varepsilon_w = \omega N$ it necessarily must play a role in the dynamics of spreading or propagation of wave intensity $I \sim \int d\mathbf{k}N$, wave momentum density $\mathbf{P}_w = \int d\mathbf{k}\mathbf{k}N$ and wave energy density $\varepsilon_w = \int d\mathbf{k}\omega N$. It is not clear, however, that the transport of all moments of N will be affected equally. Clearly, the "radiation hydrodynamics" of turbulence spreading and nonlocality phenomena is a promising and interesting topic for future research.

REFERENCES

1. S. D. Scott, P. H. Diamond, R. J. Fonck, et al., Phys. Rev. Lett. **64**, 531 (1990).
2. H. Biglari, P. H. Diamond, and P. W. Terry, Phys. Fluids B **2**, 1 (1990).
3. R. J. Groebner, K. H. Burrell, and R. P. Seraydarian, Phys. Rev. Lett. **64**, 3015 (1990).
4. K. H. Burrell, Phys. Plasmas **4**, 1499 (1997), ISSN 1070-664X.
5. P. H. Diamond and Y. B. Kim, Phys. Fluids B **3**, 1626 (1991), ISSN 0899-8221.
6. K. H. Burrell, E. J. Doyle, and P. Gohil, *et al.*, Phys. Plasmas **1**, 1536 (1994).
7. K. Ida et al., Phys. Rev. Lett. **86**, 3040 (2001).
8. B. Gonçalves, C. Hidalgo, M. A. Pedrosa, R. O. Orozco, E. Sanchez, and C. Silva, Phys. Rev. Lett. **96**, 145001 (2006).
9. J. E. Rice, M. Greenwald, I. H. Hutchinson, E. S. Marmar, Y. Takase, S. M. Wolfe, and F. Bombarda, Nucl. Fusion **38**, 75 (1998).
10. J. E. Rice, A. Ince-Cushman, J. S. deGrassie, et al., 21th IAEA Fusion Energy Conference, Chengdu, China (International Atomic Energy Agency, Vienna, 2006), pp. EX/p3–12.
11. J. S. deGrassie, J. E. Rice, K. H. Burrell, R. J. Groebner, and W. M. Solomon, Physics of Plasmas **14**, 056115 (2007), URL `http://link.aip.org/link/?PHP/14/056115/1`.
12. M. Yoshida, Y. Koide, H. Takenaga, H. Urano, N. Oyama, K. Kamiya, Y. Sakamoto, and Y. Kamada, Plasma Phys. Controlled Fusion **48**, 1673 (2006), ISSN 0741-3335.
13. W. Solomon et al., *Momentum confinement at low torque*, submitted to Plasma Phys. Controlled Fusion (2007).
14. J. E. Rice, E. S. Marmar, P. T. Bonoli, R. S. Granetz, M. J. Greenwald, A. E. Hubbard, J. W. Hughes, I. H. Hutchinson, J. H. Irby, B. Labombard, et al., FUSION SCIENCE AND TECHNOLOGY **51**, 288 (2007), ISSN 1536-1055.
15. B. LaBombard, J. E. Rice, A. E. Hubbard, et al., Nucl. Fusion **44**, 1047 (2004).
16. A. Bortolon, B. P. Duval, A. Pochelon, and A. Scarabosio, Phys. Rev. Lett. **97**, 235003 (2006).
17. M. Yoshida, Y. Koide, H. Takenaga, et al., in *Proc. of the 21th IAEA Fusion Energy Conference, Chengdu, China* (International Atomic Energy Agency, Vienna, 2006), pp. IAEA/EX–P3–22.
18. K. Crombe, Y. Andrew, M. Brix, C. Giroud, S. Hacquin, N. C. Hawkes, A. Murari, M. F. F. Nave, J. Ongena, V. Parail, et al., Phys. Rev. Lett. **95**, 155003 (2005), ISSN 0031-9007.
19. S. I. Itoh, Phys. Fluids B **4**, 796 (1992).
20. N. Mattor and P. H. Diamond, Phys. Fluids **31**, 1180 (1988).
21. K. C. Shaing, Phys. Plasmas **8**, 193 (2001).
22. A. G. Peeters, C. Angioni, and D. Strintzi, Phys. Rev. Lett. **98**, 265003 (2007).
23. T. S. Hahm, P. H. Diamond, O. D. Gurcan, and G. Rewoldt, Physics of Plasmas **14**, 072302 (2007).
24. Ö. D. Gürcan, P. H. Diamond, and T. S. Hahm, *Turbulent equipartition and homogenization of angular momentum*, submitted to Phys. Rev. Lett. (2007).
25. Ö. D. Gürcan, P. H. Diamond, T. S. Hahm, and R. Singh, Phys. Plasmas **14**, 042306 (2007).
26. Y. B. Kim, P. H. Diamond, Biglari, and P. W. Terry, Phys. Fluids B **2**, 2143 (1990).
27. R. R. Dominguez and G. M. Staebler, Phys. Fluids B **5**, 3876 (1993).
28. P. H. Diamond, V. B. Lebedev, Y. M. Liang, A. V. Gruzinov, I. Gruzinova, and M. Medvedev, in *Proc. of the 15th IAEA Fusion Energy Conference, Seville, Spain* (International Atomic Energy Agency, Vienna, 1994), pp. IAEA–CN–60/D–2–II–6.
29. X. Garbet, Y. Sarazin, P. Ghendrih, S. Benkadda, P. Beyer, C. Figarella, and I. Voitsekhovitch, Phys. Plasmas **9**, 3893 (2002).
30. J. R. Myra, J. Boedo, B. Coppi, D. A. D'Ippolito, S. I. Krasheninnikov, B. P. LeBlanc, M. Lontano, R. Maqueda, D. A. Russell, D. P. Stotler, et al., in *Proc. of the 21th IAEA Fusion Energy Conference, Chengdu, China* (International Atomic Energy Agency, Vienna, 2006), pp. IAEA/TH–P6–21.
31. L. Chen, Journal of Geophysical Research **104**, 2421 (1999).
32. R. Z. Sagdeev and A. A. Galeev, *Nonlinear Plasma Theory* (New York, W. A. Benjamin, 1969).
33. D. Mihalas and B. Weibel-Mihalas, *Foundations of Radiation Hydrodynamics* (Oxford University Press, 1984).
34. X. Garbet, L. Laurent, A. Samain, and J. Chinardet, Nuclear Fusion **34**, 963 (1994).
35. Z. Lin and T. S. Hahm, Phys. Plasmas **11**, 1099 (2004).
36. T. S. Hahm, P. H. Diamond, Z. Lin, K. Itoh, and S. I. Itoh, Plasma Phys. Controlled Fusion **46**, A323 (2004), ISSN 0741-3335.
37. Ö. D. Gürcan, P. H. Diamond, and T. S. Hahm, Phys. Rev. Lett. **97**, 024502 (2006).
38. M. N. Rosenbluth and C. S. Liu, Phys. Fluids **19**, 815 (1976).
39. P. H. Diamond and T. S. Hahm, Phys. Plasmas **2**, 3640 (1995).
40. B. A. Carreras, D. Newman, V. E. Lynch, and P. H. Diamond, Phys. Plasmas **3**, 2903 (1996).
41. P. H. Diamond and Y.-B. Kim, Phys. Fluids B **3**, 1626 (1991).
42. H. L. Berk and K. Molvig, Phys. Fluids **26**, 1385 (1983).
43. P. H. Diamond, S. I. Itoh, K. Itoh, and T. S. Hahm, Plasma Phys. Controlled Fusion **47**, R35 (2005), ISSN 0741-3335.
44. P. H. Diamond, M. N. Rosenbluth, F. L. Hinton, M. Malkov, J. Fleischer, and A. Smolyakov, in *Proc. of the 17th IAEA Fusion Energy Conference, Yokohama, Japan*, 17th IAEA Fusion Energy Conference, Yokohama, Japan (International Atomic Energy Agency, Vienna, 1998), pp. IAEA–CN–69/TH3/1.
45. F. P. Bretherton, J. Fluid Mech. **36**, 785 (1969).
46. S. V. Nazarenko, N. J. Zabusky, and T. Scheidegger, Phys. Fluids **7**, 2407 (1995).
47. O. Buhler and M. E. McIntyre, J. Fluid Mech. **492**, 207 (2003).
48. G. B. Whitham, *Linear and Nonlinear Waves* (Wiley Interscience, New York, 1974).

49. A. M. Balk, S. V. Nazarenko, and V. E. Zakharov, Phys. Lett. A. **146**, 217 (1990).
50. A. I. Smolyakov and P. H. Diamond, Phys. Plasmas **6**, 4410,4413 (1999).
51. B. Dubrulle and S. Nazarenko, Physica D **110**, 123 (1997).
52. J. R. Pierce, *Travelling Wave Tubes* (D. Van Nostrand Company Inc., New York, 1950).
53. C. H. McComas and F. P. Bretherton, J. Geo. Res. **82**, 1367 (1977).
54. V. E. Zakharov, V. S. Lvov, and G. Falkovich, *Kolmogorov Spectra of Turbulence I: Wave Turbulence* (Springer-Verlag, 1992).
55. W. M. Manheimer and T. H. Dupree, Phys. Fluids **11**, 2709 (1968).
56. E. M. Lifshitz and L. P. Pitaevskii, *Physical Kinetics* (Butterworth-Heinemann Ltd, 1981).
57. C. P. Ritz, H. Lin, T. L. Rhodes, and A. J. Wootton, Phys. Rev. Lett. **65**, 2543 (1990).
58. B. B. Kadomtsev, Plasma Phys. Controlled fusion **34**, 1931 (1992).
59. E. A. Spiegel, Ap. J. **138**, 216 (1963).
60. M. S. Longuet-Higgins and R. W. Stewart, Deep-Sea Research **11**, 529 (1964).
61. N. Mattor and P. H. Diamond, Phys. Rev. Lett. **72**, 486 (1994).
62. J. Q. Dong, W. H. R. D. Bengtson, and G. X. Li, Phys. Plasmas. **1**, 3250 (1994).
63. B. Coppi, Nucl. Fusion **42**, 1 (2002).
64. J. E. Rice (2006), private communication.
65. J. D. Murray, *Mathematical Biology* (Springer-Werlag Berlin Heidelberg, 1993), 2nd ed.
66. L. D. Landau, E. M. Lifshitz, and L. P. Pitaevskii, *Electrodynamics of Continuous Media* (Pergamon Press Ltd, Oxford, England, 1981).
67. C. E. Leith, Phys. Fluids **11**, 671 (1968).
68. C. Connaughton and S. Nazarenko, Phys. Rev. Lett. **92**, 044501 (2004).
69. X. Garbet, Y. Sarazin, F. Imbeaux, P. Ghendrih, C. Bourdelle, O. D. Gurcan, and P. H. Diamond, *Front propagation and critical gradient transport models* (2007).
70. V. Naulin, A. H. Nielsen, and J. J. Rasmussen, Phys. Plasmas **12**, 122306 (2005).
71. J. Krommes, Phys. Fluids B **5**, 1066 (1993).
72. P. H. Diamond, M. N. Rosenbluth, E. Sanchez, et al., Phys. Rev. Lett. **84**, 4842 (2000).

Physics of Zonal Flows

K. Itoh[1], S.-I. Itoh[2], P. H. Diamond[3], A. Fujisawa[1], Y. Nagashima[2], M. Yagi[2], A. Fukuyama[4], T. S. Hahm[5], K. Hallatschek[6], G. R. Tynan[3]

[1]*National Institute for Fusion Science, Toki 509-5292, Japan*
[2]*Research Institute for Applied Mechanics, Kyushu University, Kasuga 816-8580, Japan*
[3]*University of California at San Diego, La Jolla, CA 92093-0319 USA*
[4]*Department of Nuclear Engineering, Kyoto University, Kyoto 606-8501, Japan*
[5]*Princeton Plasma Physics Laboratory, P.O. Box 451, Princeton, NJ 08543-0451 USA*
[6]*Max-Planck-Institut für Plasmaphysik, 85748 Garching, Germany*

Abstract. Zonal flows, by which we mean azimuthally symmetric band-like shear flows, are ubiquitous phenomena in nature and the laboratory. It is now widely recognized that zonal flows are a key constituent in virtually all cases and regimes of drift wave turbulence, indeed, so much so that this classic problem is now frequently referred to as "drift wave-zonal flow turbulence." In this lecture note, we present new viewpoints and unifying concepts which facilitate understanding of zonal flow physics, via theory, computation and their confrontation with the results of laboratory experiment. Special emphasis is placed on identifying avenues for further progress. We briefly survey issues such as (i) mechanism of zonal flows excitation, (ii) back interaction on turbulence, (iii) saturation mechanism of zonal flows, (iv) energy partition between fluctuations and flows, (v) turbulent transport coefficient dressed by zonal flows, and (vi) experimental efforts to verify these fundamental processes.

Keywords: zonal flows, plasma turbulence, drift waves, turbulent transport, anomalous diffusion, structure formation, transport barrier, experimental validation

PACS: 52.35.Ra, 52.35.Kt, 52.25.Fi, 52.35.Qz

I INTRODUCTION

Zonal flows in toroidal plasmas are, as a zero-th order description, constant flows on magnetic surface, being independent of poloidal and toroidal angles, but the sign of which oscillates in radius. They are damped

CP1013, ***Turbulent Transport in Fusion Plasmas, First ITER International Summer School***
edited by S. Benkadda

modes in linear theory, but are driven by background microscopic turbulence, which are induced by plasma inhomogeneities. The gradient-driven microscopic instabilities (such as drift wave instabilities) generate turbulent-driven transport in magnetically-confined plasmas. The partition of energy from microscopic turbulence to linearly-stable zonal flows thus reduces the level of microscopic fluctuations and turbulent transport. The influence of zonal flows has been a central issue in the physics of confinement. See, for a review, [1], where early related work (such as [2-11]) are also surveyed. Figure 1 illustrates the fluctuation pattern, which coexists with zonal flows, in a direct nonlinear simulation [12].

Recent focused attention on zonal flows has two principal reasons. The first is their critical importance on the reality of the design of fusion experimental reactor. The plasma confinement time has the fundamental impacts on the design. The fulfillment of the condition of the plasma ignition (or required power amplification factor Q) demands the minimum of confinement time, which, following to the scaling law, the minimum size of the device. At the same time, the larger size of the device (as a consequence of poorer confinement) demands more challenging conditions for the device (such as the heat load per unit area) [13]. Zonal flows, which have a possibility to reduce the turbulent transport, are no doubt a key issue in fusion research [1, 14]. In addition, the better *understanding of confinement* is as important as better confinement itself for the confidence of the project such as ITER. The second motivation of the research of zonal flows is that the issue is a prototypical example in the fundamental problem of structure formation in nature. The study of zonal flows on experimental devices gives a chance to investigate physics such as zonal flows and dynamo magnetic field in solar and celestial plasmas.

Representing the recent extensive work on this important subject, there are a number of reviews on the subjects [1, 15-21]. Details are left for extensive reviews; we shall introduce essential elements in physics of zonal flows. In this article, we briefly survey issues such as (i) mechanism of zonal flows excitation, (ii) back interaction on turbulence, (iii) saturation mechanism of zonal flows, (iv) energy partition between fluctuations and flows, (v) turbulent transport coefficient dressed by zonal flows, and (vi) experimental efforts to verify these fundamental processes.

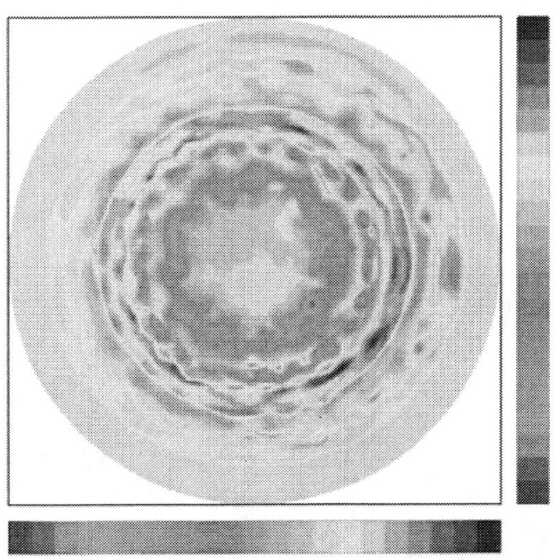

FIGURE 1: Snapshot contour of temperature fluctuation, which coexist with zonal flows.

II. THEORETICAL MODELS

2.1 Short introduction of linear properties of the zonal flows

In the slab plasma, zonal flows are in the direction of the diamagnetic drift velocity. In toroidal plasmas, however, the flow is associated with return flow along the field line, so that the zonal flows take a form of helical flow.

Figure 2 shows a magnetic surface, magnetic field line and $E \times B$ drift velocity. Owing to the toroidal effect, the magnetic field line deviates from the geodesic line (which connects two points on curved surface with minimum arc length). As a result of the geodesic curvature, the $E \times B$ drift velocities (at neighbouring points) are no longer parallel, so that the divergence of the $E \times B$ drift does not necessarily vanish. This divergence can be compensated in two ways.

One way is that this divergence is cancelled by the parallel flow [22, 23]

$$V_{//} = -2q \cos\theta \, V_{\theta} , \tag{1a}$$

where q is the safety factor. The net helical flow is shown in Fig.2. There is no pressure difference on the magnetic surface, and the flow remains to be static one,

$$\omega = 0 \ . \tag{1b}$$

The divergence of the momentum leads to the return flow. Analogous phenomena are the Pfirsch-Schlüter current, which is generated to compensate the diamagnetic current on the toroidal magnetic surface.

The other path to resolve this divergence is to induce a density perturbation (with up-down anti-symmetry), which causes temporal oscillation of the flow with frequency

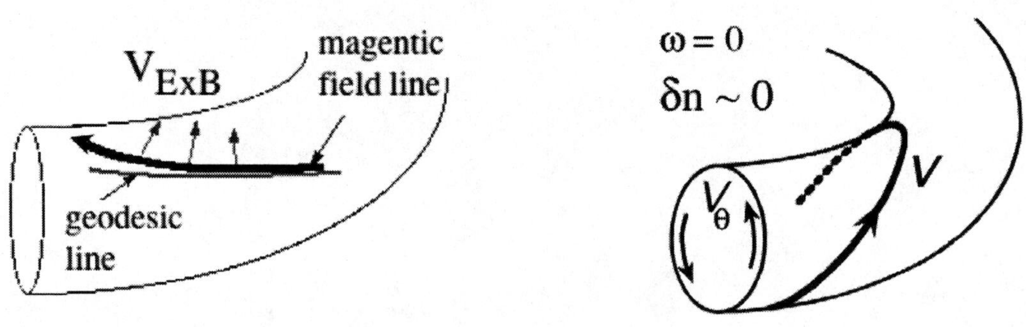

FIGURE 2: Magnetic field line and $E \times B$ drift. The zonal flows are helical flows in toroidal plasmas

$$\omega_{\mathrm{GAM}} = \sqrt{2}\, c_s / R \quad \text{(1c)}$$

This is called geodesic acoustic modes (GAMs) [2, 24]. This mode has poloidal and toroidal mode numbers $(m, n) = (1, 0)$ and the phase velocity along the field line is given by $\sqrt{2}\, q c_s$. When this phase velocity is much faster than the ion thermal velocity $V_{\mathrm{Th},\,i}$ (i.e., either $q > 1$ or $T_e >> T_i$), the ion Landau damping of GAMs are small. (Note that the m=2 component in the GAMs have substantial influence on the damping rate [25].) The importance of GAMs near tokamak edge has been demonstrated by direct simulations (see, e.g., [26] for the survey of this issue.) Mode patterns of low frequency zonal flow and GAMs are summarized in Table 1. The low frequency one is simply called zonal flows here.

Table 1: Linear mode structures of zonal flow and GAMs

	Zonal flow	**Geodesic Acoustic Modes**
poloidal velocity	$V_\theta(r)$	$V_\theta(r)$
toroidal velocity	$-2q \cos\theta\, V_\theta(r)$	$q^{-1} \cos\theta\, V_\theta(r)$
density perturbation	$\simeq 0$	$\tilde{n}/n = -\sqrt{2}\, q_r \rho_s \sin\theta\, e\tilde{\phi}_{\mathrm{GAM}}/T$
oscillation frequency	$\omega = 0$	$\omega = \omega_{\mathrm{GAM}}$

2.2 Mechanism of Zonal Flows Excitation

There are several paths to understand the mechanism of the zonal flow excitation by microscopic waves: the argument on tilt of vortex, envelope formalism and modulational instability (e.g., [27]). Detailed explanations are found in literature [1]. Since the impact of zonal flows on turbulent transport may occupy the interest of readers here, discussion based on the conservation of the drift wave actions [28,29] is explained here.

A basic assumption there is the separability of scales of microscopic fluctuations and zonal flows. When spatio-temporal scales of zonal flows are longer (slower) that those of micro fluctuations, one can applies the Manley-Law relation in the interaction of zonal flows and fluctuations. In the interaction where two modes with $(\omega_1, \boldsymbol{k}_1)$ and $(\omega_2, \boldsymbol{k}_2)$ couple to generate the third one with $(\omega_3, \boldsymbol{k}_3)$, the total energy and action are conserved. (An action spectrum N_k and energy spectrum W_k satisfies the relation $W_k = \omega_k N_k$.) When one of the three is the zonal flow with $\omega \simeq 0$ and the other two are drift waves, one concludes that the action of drift waves is conserved. The drift wave action density is given as $N_k = \left(1 + k_\perp^2 \rho_s^2\right)^2 |\tilde{\phi}_k|^2$, where k is the wave number, $\tilde{\phi}_k$ is the amplitude
of drift wave (electrostatic potential) and ρ_s is the ion cyclotron radius at ion sound speed. The set of basic equations then consists of the conservation equation of drift wave action density in the presence of zonal flows

$$\frac{\partial}{\partial t} N_k + \frac{\partial \omega_k}{\partial \boldsymbol{k}} \cdot \frac{\partial N_k}{\partial \boldsymbol{x}} - \frac{\partial \omega_k}{\partial \boldsymbol{x}} \cdot \frac{\partial N_k}{\partial \boldsymbol{k}} = \Gamma_k N_k , \tag{2a}$$

and the evolution of zonal flow in the presence of drift waves

$$\frac{\partial}{\partial t} V_Z = \frac{\partial}{\partial r} \frac{c^2}{B^2} \int \mathrm{d}^2 k \frac{k_\theta k_r}{\left(1 + k_\perp^2 \rho_s^2\right)^2} N_k - \gamma_{damp} V_Z , \tag{2b}$$

where V_Z is the poloidal velocity of zonal flow, γ_{damp} is the linear damping rate of zonal flow (by, e.g., collisional damping process), $\partial \omega_k / \partial \boldsymbol{k} = v_{g\,r}$ denotes the group velocity of drift waves, $\partial \omega_k / \partial \boldsymbol{x}$ is the gradient of Doppler shift frequency by zonal flows, and Γ_k is the growth rate of drift waves including linear as well as nonlinear processes. (It is often modelled as $\Gamma_k = \gamma_{\mathrm{L}} - \Delta\omega_k$ by using the linear growth rate γ_{L} and nonlinear decorrelation rate $\Delta\omega_k$.)

When zonal flows has the form like $\exp\left(i q_r x - i \Omega t\right)$, Eqs.(2a) and (2b) provide

$$\frac{\partial}{\partial t} V_Z = - D_{rr} \frac{\partial^2}{\partial r^2} V_Z - \gamma_{\mathrm{damp}} V_Z \tag{3a}$$

$$D_{rr} = \frac{1}{B^2} \int \mathrm{d}^2 k \frac{i}{\Omega - q_r v_{g\,r} + i \Delta\omega_k} \frac{k_\theta^2 k_r}{\left(1 + k_\perp^2 \rho_s^2\right)^2} \frac{\partial N_k}{\partial k_r} . \tag{3b}$$

When the autocorrelation time of drift wave is short and $\Delta\omega_k$ dominates the denominator in the integrand of Eq.(3b), the growth rate of the zonal flow $\gamma_Z = \mathrm{Im}\ \Omega$ is given as

$$\gamma_Z = \frac{q_r^2}{\Delta\omega_k} \frac{1}{B^2} \int \mathrm{d}^2 k \frac{k_\theta^2 N_k}{\left(1 + k_\perp^2 \rho_s^2\right)^2} - \gamma_{\mathrm{damp}} \tag{4a}$$

In the limit where γ_Z dominates the denominator in the integrand of Eq.(3b), one has

$$\gamma_Z = q_r \frac{1}{B} \sqrt{\int \mathrm{d}^2 k \frac{k_\theta^2 N_k}{\left(1 + k_\perp^2 \rho_s^2\right)^2}} \tag{4b}$$

which recovers the limit of modulational instability of a plane drift wave.

Another mechanism was also pointed out. It has been known that the poloidal asymmetry of the mean radial flow of plasmas can drive poloidal flow [30, 31]. The dynamic shearing of fluctuations by GAMs causes the poloidal asymmetry of turbulent transport and thus generates the GAMs. This loop works as a destabilization mechanism of GAMs [32].

Equation (3b) shows that zonal flows grow when N_k is a decreasing function of k_r. In the case where $\Delta\omega_k$ dominates the denominator (often the case of experimental circumstance), the growth rate scales as $D_{rr}q_r^2$, so that the zonal flows with shorter wavelength have larger growth rate. When the radial wave number becomes too large, the dispersion effect $q_r v_{gr}$ in the denominator stops the excitation of zonal flows. (Some of the readers may wonder 'Then what happens to the turbulence viscosity if the growth rate scales as $D_{rr}q_r^2$? ' This issue is discussed later.)

2.3 Back Interaction on Turbulence

Along the line of Eq.(3), the back interaction of (driven) zonal flow on turbulence can be formulated. The term $\partial\omega_k/\partial x \cdot \partial N_k/\partial k$ indicates that the drift wave spectrum spread in the wave number space by zonal flows. Taking the terms up to the second order of V_Z, the quasi-linear evolution equation for mean wave action density $\langle N_k \rangle$ can be obtained as [33]

$$\frac{\partial}{\partial t}\langle N_k\rangle = \Gamma\langle N_k\rangle + \frac{\partial}{\partial k_r} D(k) \frac{\partial}{\partial k_r}\langle N_k\rangle \quad , \tag{5a}$$

$$D(k) = k_\theta^{(0)2} q_x^2 V_Z^2 \tau_{ac} \; . \tag{5b}$$

In the process of diffusion of the action in wave number space, the total wave action is conserved. However, the energy density of the drift waves is given by $\omega_k \langle N_k \rangle$. Thus the increment in k_r, which reduces the wave frequency, decreases total wave energy. (Of cause, this reduction of wave energy is transformed into the flow energy, which is the dissipated via damping processes.)

Combining these interactions between zonal flows and drift wave, a point-model can be constructed for the zonal flow intensity V_Z^2 and total drift wave energy W as

$$\frac{\partial}{\partial t} W = \gamma_L W - \alpha V_Z^2 W - \Delta\omega_2 W^2 \tag{6a}$$

$$\frac{\partial}{\partial t} V_Z^2 = \alpha W V_Z^2 - \gamma_{damp} V_Z^2 - \gamma_{NL}\left(V_Z^2\right) V^2 \tag{6b}$$

The term $\alpha W V^2$ in Eq.(6) denotes the energy exchange where α is a coupling coefficient. The nonlinear damping rate of drift wave is expressed as $\Delta\omega = \Delta\omega_2 W$, and the term $\gamma_{NL}(V_Z^2)$ symbolically represents the self-nonlinear saturation mechanism of zonal flow, which is explained in the following.

2.4 Saturation Mechanism of Zonal Flows

When the amplitude of zonal flow becomes large enough, the nonlinear mechanism of the zonal flow itself shows the influence on the evolution of the drift wave – zonal flow system.

One possible self-nonlinear mechanism is the tearing of zonal flow by, e.g., excitation of Kelvin-Helmholtz instability associated with the flow shear of a large-amplitude zonal flow [34-36]. The second is the trapping of drift wave packet in the trough of zonal flows [37]. This mechanism is effective when the autocorrelation time of drift wave becomes longer. The third is the nonlinear wave-kinetic process, which is an analogous process of nonlinear ion Landau damping in Vlasov system. The wave packet trapping and the nonlinear wave kinetic model are discussed here.

The wave packets of drift waves behaves in the trough of zonal flows following the trajectory

$$\frac{dx}{dt} = v_g \qquad \frac{d}{dt} k_x = k_y \frac{d}{dx} V_Z, \tag{7}$$

where x and y indicate the radial and poloidal directions, respectively. (Because the zonal flow is homogeneous in the poloidal direction, $k_y = k_{y0}$ hold.) The dispersion of drift waves gives the relation $v_{gr} \propto -k_x$, so that in the bottom of the trough, packets follow the harmonic oscillation with the bounce frequency

$$\omega_{bounce}{}^2 = \frac{2\rho_s^2 k_y q_r}{1 + \rho_s^2 k_y^2} \omega_{k0} \left.\frac{dV_Z}{dr}\right]_{max}. \tag{8}$$

Figure 3 illustrates the bounce motion of packets. When the turbulence is weak and the period of bounce motion is

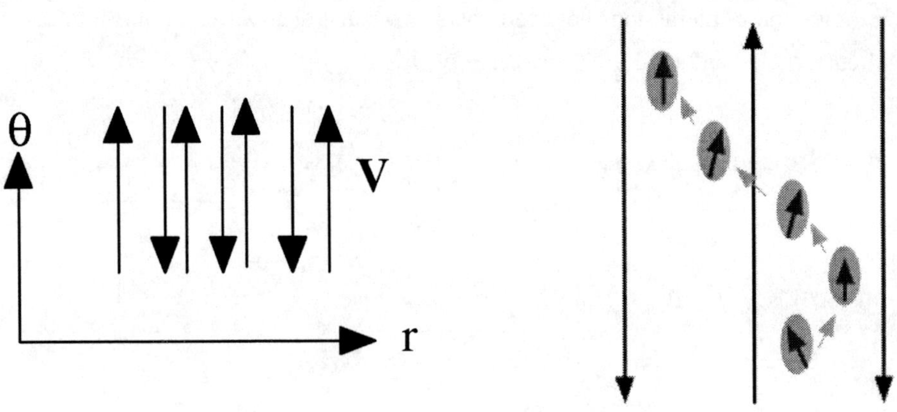

FIGURE 3 Zonal flows are illustrated (left). Motion of a drift wave packet in a trough of ZFs is schematically shown in

(right). Arrows on a packet denote the wave vector, indicating the modification of k_r.

shorter than the autocorrelation time of drift waves, wave packets are trapped in the flow structure, and the action density distribution is flattened. Thus, the growth of zonal flows stops.

In contrast, when the turbulence is strong and the period of bounce motion is longer than the autocorrelation time of drift waves, the flattening remains partial. The deformation of the mean action density leads to the saturation of the zonal flow growth. Expanding Eq.(2a) up to the third order of zonal flow vorticity U, and retaining up to the third order terms with respect to $U/\Delta\omega_k$, one obtains

$$\frac{\partial}{\partial t} U = -D_{rr}\left(\frac{\partial^2}{\partial r^2} U + K_0^{-2}\frac{\partial^4}{\partial r^4} U\right) + D_3\frac{\partial^2}{\partial r^2} U^3 - \gamma_{\mathrm{damp}} U \quad , \tag{9a}$$

with

$$D_3 = -\frac{1}{B^2}\int \mathrm{d}^2 k \frac{k_\theta^4 k_r}{\Delta\omega^3\left(1 + k_\perp^2\rho_s^2\right)^2}\frac{\partial^3 N_k}{\partial k_r^3} \quad . \tag{9b}$$

The third order term leads to the nonlinear saturation of zonal flows [38]. Renormalization of higher order nonlinear terms has also been reported in [39].

The equation (9a) is also useful in analyzing the selection rule of the radial scale length of zonal flows. The competition between zonal flows among different radial wavelengths has been analyzed [40]. In the phase of quasilinear growth (i.e., the third order term of $U/\Delta\omega_k$ is unimportant), the mode with $K_0/\sqrt{2}$ has the largest growth rate. In the later phase, the energy exchange between zonal flows with different wave numbers takes place. The radial scale length of $K_{\mathrm{m}} = K_0/2$ is chosen as a result of the nonlinear competition.

2.5 Energy Partition between Fluctuations and Flows

Stationary state of drift waves and zonal flows, which follows Eq.(6) is dictated by $\alpha W V^2$ and $\gamma_{\mathrm{NL}}\left(V^2\right) V^2$. In order to have an analytic insight, let us choose a simple model of the form $\gamma_{\mathrm{NL}}\left(V^2\right) = \alpha_2 V^2$. Stationary solution for Eq.(6) is given as, either the one without flow,

$$W = \frac{\gamma_{\mathrm{L}}}{\Delta\omega_2} \qquad V^2 = 0 \tag{10a}$$

for $\gamma_{\mathrm{L}} < \gamma_{\mathrm{damp}}\Delta\omega_2\alpha^{-1}$, or the solution with flow,

$$W = \frac{\gamma_{\mathrm{damp}}\alpha + \gamma_{\mathrm{L}}\alpha_2}{\alpha^2 + \Delta\omega_2\alpha_2} \qquad V^2 = \frac{\gamma_{\mathrm{L}}\alpha - \gamma_{\mathrm{damp}}\Delta\omega_2}{\alpha^2 + \Delta\omega_2\alpha_2} \tag{10b}$$

for the case of $\gamma_L > \gamma_{damp} \Delta\omega_2 \alpha^{-1}$. When the driving source of drift waves is weak, zonal flows are not excited and bare fluctuations are realized. When the parameter exceeds the threshold, the zonal flows are excited, and increment of fluctuation level becomes slower.

In the presence of zonal flow, Eq.(10b), the fluctuation intensity is given by $W = \gamma_{damp}/\alpha$, when the nonlinear damping rate of zonal flows $\gamma_{NL}(V^2)$ is smaller than the collisional damping rate γ_{damp}. The fluctuation level is independent on γ_L but is controlled by the damping rate of zonal flows γ_{damp}. In the opposite limit, $\gamma_{NL}(V^2) >> \gamma_{damp}$, both the flow energy and fluctuation energy are enhanced by increment of γ_L. The partition is given by the relation $V^2/W = \alpha/\alpha_2$.

Equation (10) is relevant when the nonlinear decorrelation rate of drift waves is large. Special care is necessary when the drift wave amplitude is low so that $\Delta\omega_k$ is small. A famous example is the Dimits shift problem [41]. (Solutions of initial value problems have shown that fluctuations are excited first but disappear in a longer time scale after converting the energy into zonal flows [42, 43].)

In such a case, the saturation of zonal flows takes place when trapping become effective, i.e., $\Delta\omega_k = \omega_{bounce} \propto \sqrt{V_Z}$. By use of the renormalized growth rate of zonal flow, the saturation condition for zonal flows in the absence of linear damping ($\gamma_{damp} = 0$) was derived as

$$\Delta\omega_k^2 = C k_\theta^2 \rho_s^2 U^2 - \omega_{bounce}^2 \tag{11}$$

where $C \sim 2$ is a numerical coefficient of order unity [39]. Note the relation $\omega_{bounce}{}^2 \propto \omega_k U$, as is given in Eq.(8). Figure 4 illustrates the energy partition between flow and fluctuations in the limit of $\gamma_{damp} = 0$. Even in the absence of fluctuations, zonal flows can have finite amplitude.

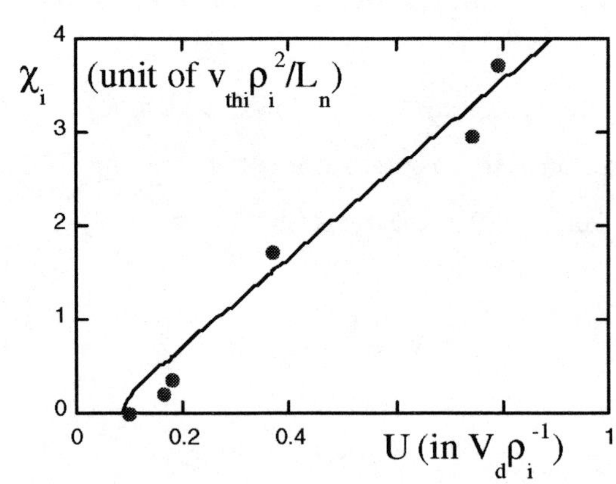

FIGURE 4 Comparison of the relations $\chi_i(U)$ for the steady state of ITG mode. Zonal flow vorticity is measured in a unit of $V_d \rho_i^{-1}$ and thermal conductivity is in $v_{thi}\rho_i^2 L_n^{-1}$. Theory (solid line) and DNS data (dots) quoted from [44].

2.6 Turbulent Transport Coefficient Dressed by Zonal Flows

When the damping rate of zonal flow is large, the turbulent transport coefficient is determined by bare microscopic fluctuations. In the case that the fluctuations are driven by linear instability, one has the thermal diffusivity

$$\chi_i = \frac{\gamma_L}{k_r^2} . \qquad (12a)$$

When the zonal flows are excited but are weak so that $\gamma_{\rm NL}\left(V^2\right) < \gamma_{\rm damp}$ holds, one has

$$\chi_i \sim \sqrt{\frac{\gamma_{\rm damp}}{\alpha}}\,\frac{\omega_*}{k_r^2} , \qquad (12b)$$

which is independent of $\gamma_{\rm L}$. In the limit of large $\gamma_{\rm L}$ so that $\gamma_{\rm NL}\left(V^2\right) >> \gamma_{\rm damp}$ holds, Eq.(11) describes the stationary state. In this case, the thermal diffusivity is given as

$$\chi_{\rm III} = \frac{C k_\theta^2 \rho_s^2 \omega_*}{2\alpha}\left(-1 + \sqrt{1 + \frac{4\alpha}{C k_\theta^2 \rho_s^2 \omega_*^2}\left(\gamma_{\rm L} - \gamma_{\rm L,c}\right)}\right)\frac{\omega_*}{k_r^2} , \qquad (12c)$$

where the boundary $\gamma_{\rm L,c}$ is given as $\gamma_{\rm L,c}/\omega_* = 2\sqrt{2}\,C^{-1} q_r^2 k_\perp^{-2}$. (See Fig.5 for schematic illustration.)

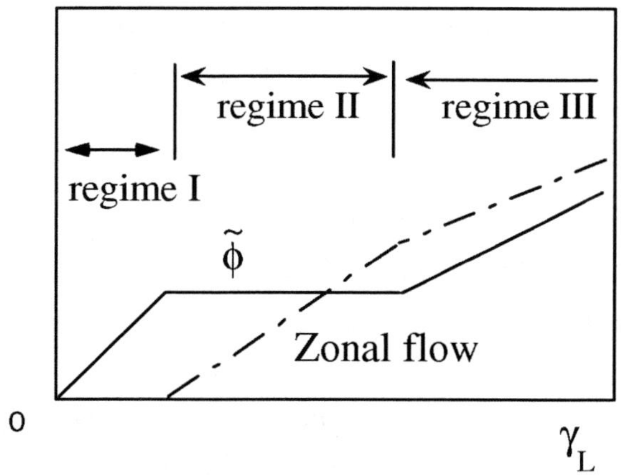

FIGURE 5: The dependence of the amplitude of drift wave fluctuations $\hat\phi$ (solid line) and zonal flow (chained line) vs. $\gamma_{\rm L}$ in the collisional case. (Here, $\gamma_{\rm damp}/\omega_*$ and α/ω_* are kept constant.) Three regions appear.

Now we discuss the turbulent shear viscosity. The mean poloidal flow $U_L(r)$ can be driven by drift waves via quasi-linear process. However, zonal flows with shorter radial wavelength are excited faster than the mean poloidal flow. Therefore, the growth of the mean poloidal flow must be calculated by taking into account of the screening effect by zonal flows. Putting $U(r) = U_Z(r) + U_L(r)$ (where $U_Z(r)$ is the dominant zonal flow) into Eq.(9a) and taking the saturation amplitude for zonal flows as $U_Z^2 = 3D_{rr}/4D_3$, the evolution equation for the mean poloidal flow is obtained as [45]

$$\frac{\partial}{\partial t} U_L - \frac{5}{4} D_{rr} \frac{\partial^2}{\partial r^2} U_L = 0 \quad . \tag{13}$$

This result shows that the long-wavelength perturbation is subject to turbulent viscous damping.

III. EXPERIMENTAL STUDIES

Stimulating developments have been made in experimental studies of zonal flows and GAMs. Articles [19, 20, 46], those in cluster articles [47-61] as well as [62-73], for examples, describe very rapid progress in this field of research. Contributions from various experiments including toroidal systems as well as linear devices, (in alphabetical order) ASDEX-U, CASTOR, CHS, CLM, CSDX, DIII-D, H1-heliac, HL-2A, HT-6M, HT-7, JFT-2M, JIPPT-IIU, T-10, TEXTOR, TEXT-U, TJ-II, TJ-K are summarized in Table 1 of [20], and further reports on LMD-U is found in [19].

3.1 Identification of Zonal Flow and GAMs

The most important progress was the experimental discovery of the zonal flow, which was performed on CHS device. The radial electric field is measured on CHS device at two different locations, which are separated in toroidal direction by (more than) 90 degree. Power spectrum of electric field fluctuation is quoted in Fig.6(a) together with the cross correlation between the electric fields at two different locations. When the two position of measurement are on a same magnetic surface, strong correlation are observed in the low frequency regime, showing the presence of the zonal flows. The two-point and two-time correlation of the low-frequency components of electric field is shown in Fig.6(b). The reference radius is chosen at r = 12 cm for one HIBP measurement, and the radius of other observation is varies. This result shows that the zonal flow has a radial wavelength of few cm and the autocorrelation time of a few ms. They are longer than those of microscopic fluctuations, confirming that zonal flows have meso-scales. Several methods without multipoint detection have also been invented to detect stationary zonal flows. For example, HT-7 tokamak (using a forked Langmuir probe) reported the existence of perturbation with a long-poloidal wavelength [66], the autocorrelation width technique was applied in the CASTOR tokamak [56] and the detection technique of Doppler frequency modulation due to zonal flow in the Colombia Linear Machine (CLM) [54].

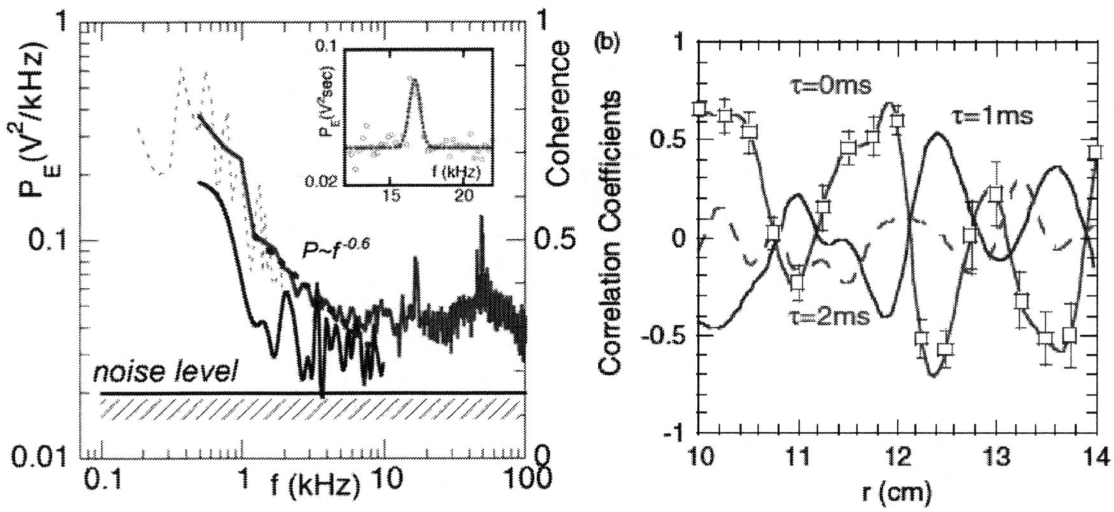

FIGURE 6: (*a*) Spatio-temporal structure of stationary zonal flows evaluated from the correlation function in CHS. Spatial correlation with the time lag of $\tau = 0$, 1 and 2 ms. The reference radius is chosen at r = 12 cm [20, 46].

More abundant reports have been given on the observation of the GAM oscillations in toroidal plasmas. Many tokamaks and toroidal plasmas have reported a sharp peak in the fluctuation spectrum. Examples of observations are quoted in Fig.7. Comparing the peak frequency with theoretical prediction, the peaks are attributed to the GAM oscillations. Assessment of the comparison between the observed peak and expected GAM frequency was reported in [20, 68]. Figure 8 is reproduced from [20]. The frequency of GAMs, Eq.(1c), is now widely validated.

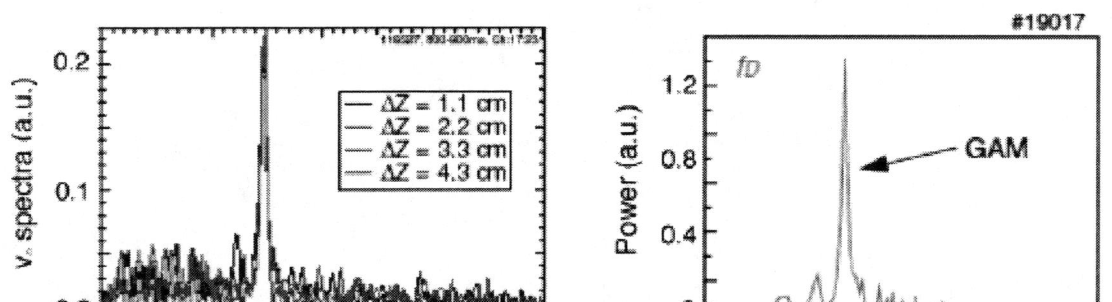

FIGURE 7: Examples of fluctuation spectra, which show the presence of GAMs [20]. Velocity fluctuation spectra measured with BES in DIII-D (left) [65]. Spectrum of Doppler frequency measured with DR in ASDEX-U (right) [68].

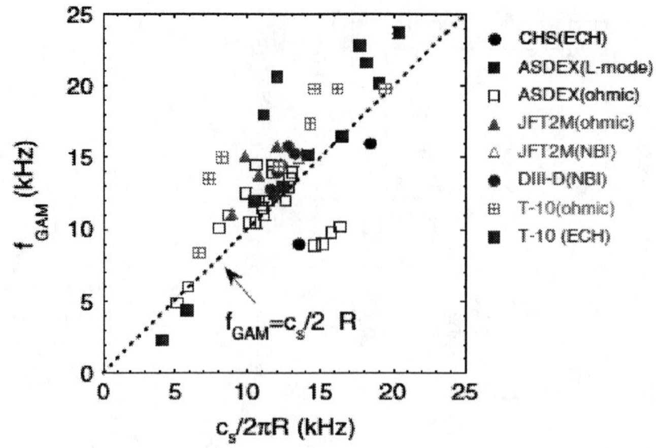

FIGURE 8: A comparison between the observed GAM frequency and theoretical prediction. Here, c_s and R are the ion sound velocity and the major radius, respectively. (Quoted from [20].)

The radial structure of GAMs has also been investigated. The observation in JFT-2M has reported the eigenmode characteristics of the GAM. That is, the mode is localized in ~ 5 cm inside the separatrix without a change in frequency, while the electron temperature increases significantly towards the core (see Figure 9(a)). In addition, the outward propagation characteristics of the GAM are deduced, as shown in Figure 9(b) [73]. It has been predicted by theory that GAMs have eigenmode structure owing to the temperature inhomogeneity, and can propagate in redial direction [74, 75]. The knowledge of property of radial eigenmode will be used in future advanced diagnostics. For instance, an application was made for the identification of the plasma surface [76], and the GAM spectroscopy has been proposed to study the measurement of ion density [77].

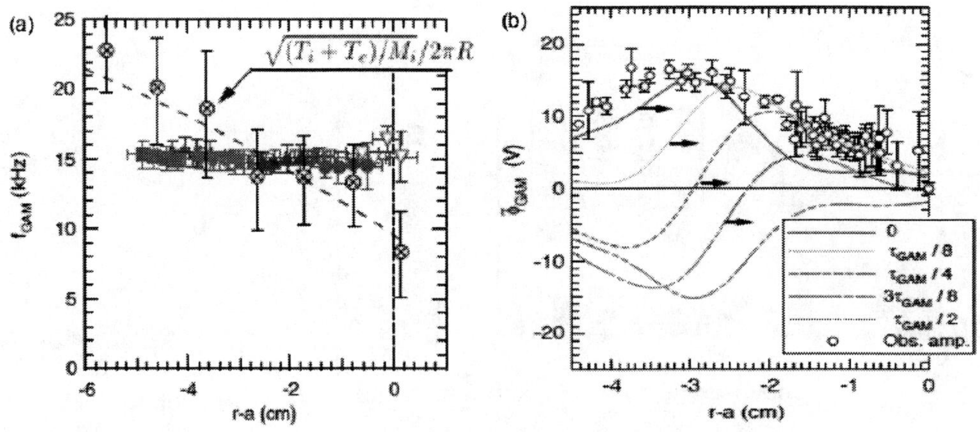

FIGURE 9: GAM structure in JFT-2M [73]. (*a*) GAM frequency as a function of radius. The circles show the GAM frequency expected from the ion temperature measured with charge exchange spectroscopy with an assumption of $Te = Ti$. The dashed line indicates a fitting line. (*b*) The temporal change in the GAM structure that indicates outward propagation.

3.2 Nonlinear Interactions among zonal flow and fluctuations

Fundamental issue in the study of the zonal flow – drift wave system is to quantify the nonlinear interaction among them. Identification of zonal flow requires the measurement of nonlinear coupling between zonal flow and fluctuations.

A systematic study has been developed on the coupling between GAMs and drift waves by use of the bicoherence analysis [43, 73, 78-80]. Figure 10 shows the bicoherence analysis for the edge fluctuation of JFT-2M plasma. Squared bicoherence is a measure that two fluctuation components (with frequencies ω_1 and ω_2) are nonlinearly couple with the third one (with frequency $\omega_3 = \omega_1 + \omega_2$). The component with the local GAM frequency has an unambiguous squared bicoherence for $\omega_1 + \omega_2 = \omega_{GAMs}$, which verifies the coupling with broadband microscopic fluctuations (in the range higher than 30 kHz) (Fig.10(a)). Figure 10(b) denotes the biphase. We observe that, when one of the three-coupled modes is the GAM component, the biphase only weakly depends on the frequency of the other modes. That is, any pair of two drift wave modes interacts with a common GAM component, so that the biphase takes almost constant values. In contrast, for combinations of three drift waves (i.e., all of ω_1, ω_2 and ω_3 belong to the range of background fluctuations), the biphase randomly changes by a choice of interacting pair.

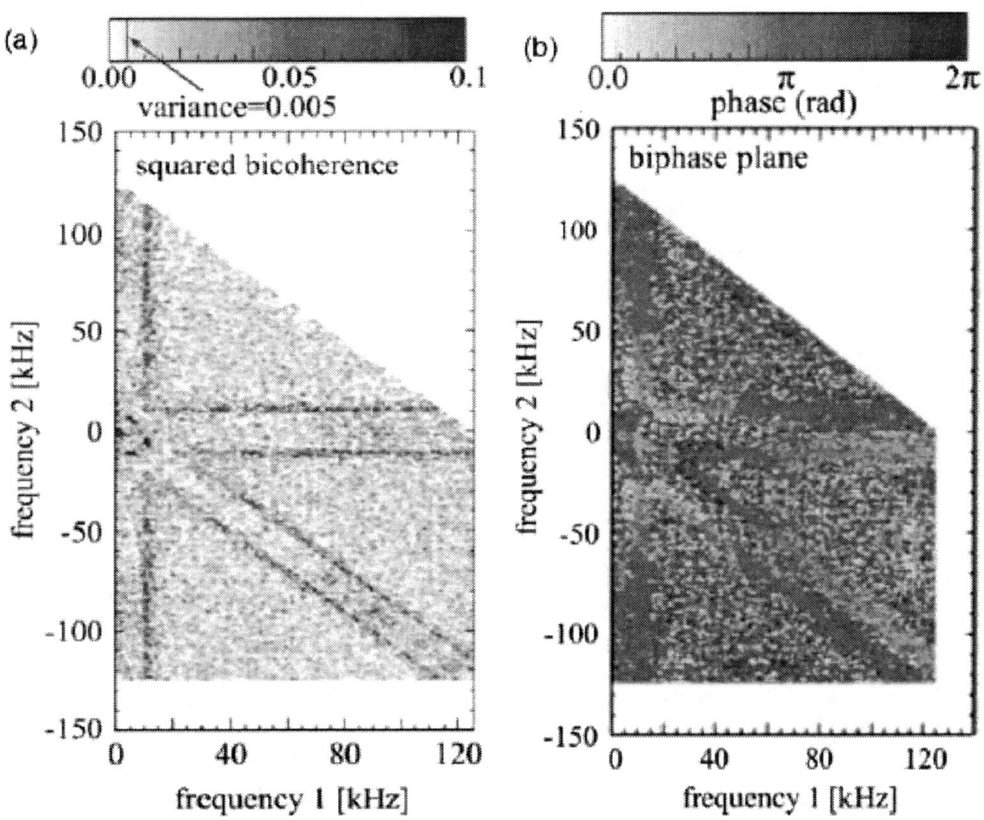

FIGURE 10: Bicoherence analysis to show the coupling between GAM and background turbulence in JFT-2M [47, 73]. (*a*) Bicoherence evaluated with potential fluctuation at the plasma edge and (*b*) corresponding biphase.

Absolute values of the bicoherence and biphase were compared with theoretical prediction [81] in the article [79]. It was found that observed value of bicoherence is in the same order of magnitude of theoretical predictions, providing quantitative confirmation that the observation of GAMs has been performed.

The other way to identify the nonlinear interaction between zonal flows and micro-fluctuations is to observe the amplitude modulation of the envelope of microscopic fluctuations. As is explained in the theory of zonal flow growth based on the modulational instability, the excitation of zonal flows is accompanied by the envelope modulation of background micro fluctuations [1]. This approach has been used in various experiments [47, 50, 73, 82].

The fluctuation-driven Reynolds stress is an essential issue in the physics of zonal flows. It has been measured in various devices. Fluctuation velocities (in the radial as well as poloidal directions, $\tilde{v}_r$ and $\tilde{v}_\theta$) are measured, and a time average stress $\Pi_{r\theta} = \langle \tilde{v}_r \tilde{v}_\theta \rangle$ has been evaluated. (It should be noted, because the measurement of $\langle \tilde{v}_r \tilde{v}_\theta \rangle$ is often limited to one point in space, the average over the magnetic surface is often evaluated by the time average.) Detailed comparison has been reported in basic plasma experiments on linear plasmas. By taking the radial derivative of $\Pi_{r\theta}$, the torque in the poloidal direction was estimated (Fig.11(a)). Using the neutral particle drag, calculated by use of the measured neutral profile, one can estimate the plasma flow velocity in the poloidal direction. This estimated flow was compared to the direct measurement of the flow. Reference [51] reports semi-quantitative agreement, showing that the turbulence-driven stress is the dominant element of the drive for the global plasma flow shear. (See Fig.11(b).)

As was explained in the Review [1], the generation of the shear by the convection motion driven by instabilities is associated with the tilt of the convection pattern. Such a tilt of the convection pattern has been observed in detail in linear plasma experiments [83, 84]. In toroidal plasmas, the tilt of convection instability has also been observed [85].

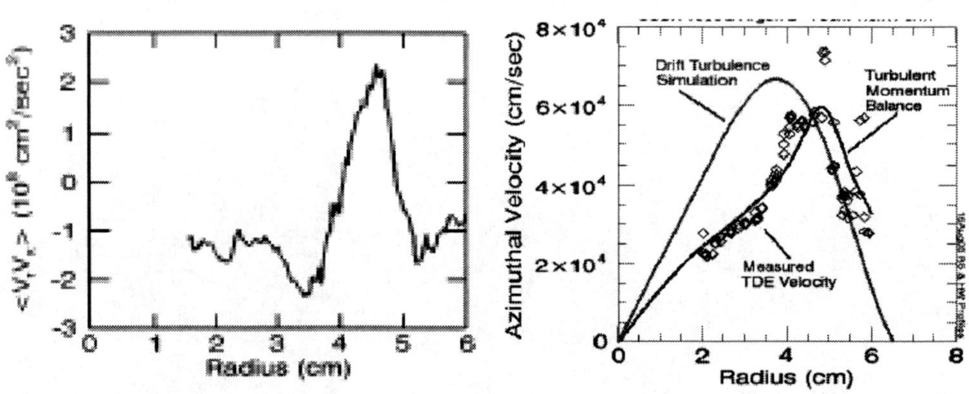

FIGURE 11: Radial profile of the turbulence-driven Reynolds stress (a) and poloidal flow (b) in the CSDX experiment. In (b), the plasma flow velocity is compared to the estimation given by balancing the turbulent Reynolds stress and drag on ions. Prediction of turbulence simulation is also illustrated. (Quoted from [51])

3.3 Suppression of Turbulent Transport by zonal flows

The identification of suppression of turbulence and turbulent transport by zonal flows is the focus of experimental studies of magnetic confinement.

Measurement of the effects of zonal flows on turbulent transport has been made on CHS [61]. The HIBP measurement provides local data of the electric field perturbation $\tilde{E}_\omega$, the density perturbation $\tilde{n}_\omega$, and the phase difference between them α_ω. The particle flux is measured as $\Gamma = \sum_\omega \Gamma_\omega$, $\Gamma_\omega = \frac{1}{B}\left\langle \tilde{E}_\omega \tilde{n}_\omega \cos\alpha_\omega \right\rangle$. It was found that Γ_ω (carried by high frequency fluctuations, $\omega > 40\mathrm{kHz}$) is modulated by the zonal flows. Γ_ω is high at the bottom of the zonal flow plot (where the ZF electric field is opposite to the mean DC radial electric field) and low at the peak of the zonal flow plot (where the ZF electric field is the same as the mean DC radial electric field) in the plasmas which fall in the electron-root branch. (See Fig.12.) This observation is consistent with theoretical modeling, which predicts that fluctuations tend to accumulate in the trough of zonal flows.

Observations have also been reported on the interaction of GAM oscillations and turbulent transport. A characteristic example is the HIBP measurement of the edge turbulence on the JFT-2M tokamak [73]. The particle flux $\Gamma = \sum_\omega \Gamma_\omega$ was evaluated based on HIBP data. It was found that Γ is modulated at the frequency of the observed GAM oscillations. The time series data $\Gamma(t)$ is analyzed, and its Fourier spectrum $\Gamma(\omega)$ was calculated. It is clear that the particle flux $\Gamma(t)$ is modulated at the GAM frequency.

The theoretical prediction of turbulent transport, which is dressed by zonal flows, has shown that the turbulent transport is strongly influenced by the damping rate of zonal flow [14]. This introduces an idea to control turbulent transport by modifying the damping rate of zonal flow. The test of this working hypothesis was made in the analysis of CHS plasma. The improvement of the core electron confinement has been reported for helical plasmas [86-89]. An electric field interface is induced owing to the nonlinear response of neoclassical particle flux against radial electric field [23, 90]. It has been known that the electron turbulent transport is reduced in the plasma

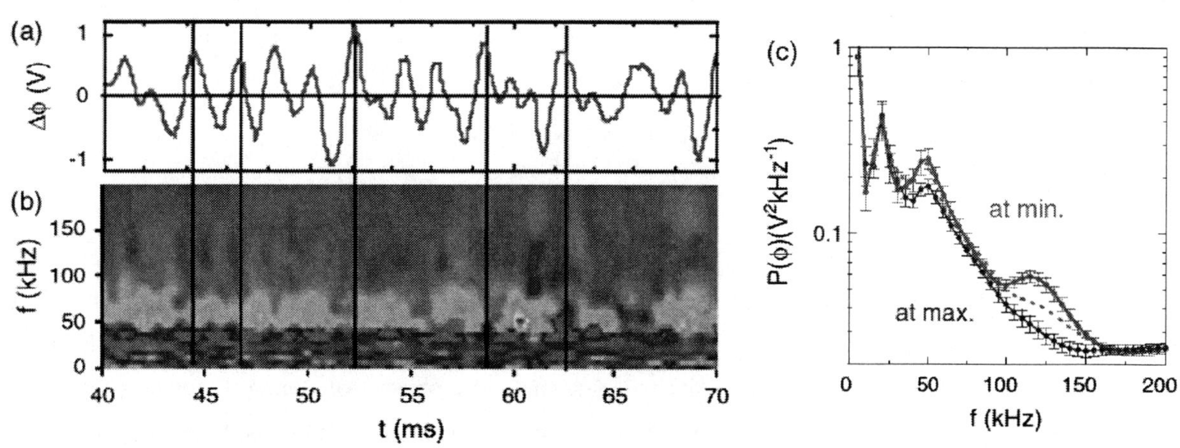

FIGURE 12: The modulation effect of stationary zonal flow on particle transport in CHS [20]. (*a*) Temporal evolution of stationary zonal flows and (*b*) an image plot of particle flux. (*c*) Conditional averages of potential fluctuation spectra in the time windows discriminated by the phase of zonal flow; i.e. maxima, zero and minima.

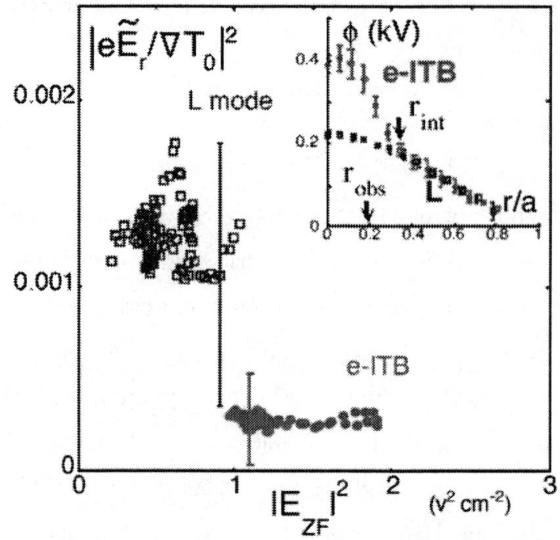

FIGURE 13 Simultaneously observed zonal flows and micro turbulence intensities during the e-ITB phase (solid circle, red) and the L-phase (open circle, blue). Vertical and horizontal axes indicate normalized fluctuation amplitude $|eE_r/\nabla T_0|^2$ and zonal flows amplitude $|E_{ZF}|$, respectively. The insert shows the profile of mean potential for e-ITB and L plasmas. Locations of interface and observation, r_{int} and r_{obsv}, are also indicated. [91]

core, where the electric field itself is long but its shear is not strong enough to suppress turbulence (see a review [89]). The role of zonal flows in the formation of an internal transport barrier in a toroidal helical plasma is analyzed [91]. The damping rate of zonal flow become smaller when the plasma state changes from the branch of a weak negative radial electric field to the strong positive one. Thus the turbulent transport is predicted to be reduced in the state with strong positive radial electric field, and induces a new transition of turbulent. The reduction of the turbulent transport in the core of e-ITB plasma, via the enhanced zonal flows, has been experimentally confirmed, as is shown in Fig.13. It is clearly demonstrated, theoretically and experimentally, that the damping rate of the zonal flows governs the global confinement of toroidal plasmas.

3.4 Zonal Fields

The electromagnetic counterpart for zonal flow is the zonal field [1]. The evolution of zonal fields can influence on the current profile or stability of tokamaks. In addition, the zonal fields are considered to be a mesoscale dynamo. This is a very good opportunity to study the physics of dynamo magnetic field [92, 93], through a laboratory measurement on the process where structured magnetic field is generated by microscopic turbulence. To this very important issue of modern physics, experimental evidence has recently been made [94]. Zonal fields which are in the poloidal direction and change the sign with radial wave length of few cm was discovered in CHS plasmas by use of dual HIBP system.

IV. SUMMARY

In conclusion, the study of zonal flow physics had led to a paradigm shift in our basic concept of "drift wave turbulence and transport." We now think of "drift wave - zonal flow turbulence" as a self-regulating system. This natural self-regulation can be important for magnetic fusion, both by defining a reduction in transport relative to gyroBohm and by its effect on transport barrier formation. Very rapid progress has been made these years concerning on the experimental studies of zonal flows. We can look forward with confidence to future discoveries of the role of secondary flows and fields in different areas of fusion physics. Physics of the drift wave – zonal flow system is a prototypical example of this evolution in understanding the turbulence and structure formation in plasmas.

There are several aspects of zonal flow physics where significant future study is required.

First, the rule governing how the system selects among the various saturation mechanisms must be derived and validated. For this, understanding of the interaction and competition between various secondary structures, such as zonal flows GAMs, and avalanches or streamers, must be improved. In addition, further development of statistical theory for the system of drift wave and zonal flows (for which readers are refereed to [29, 95-98] and review [1]) are necessary to address these issues.

Second, convincing experimental links between zonal flow excitation and enhancement of confinement need to be established. Challenge in this direction has many open issues. The factors limiting the efficiency of the control of zonal flows must be determined for future applications.

Through these researches, quantitative predictability (with understanding) must all be developed. Clearly, many interesting challenges and discoveries await researchers working on the subject of zonal flows.

ACKNOWLEDGEMENTS

This review has been performed with the help of numerous collaborators world wide, names of which are listed in [1]. In preparing this article, special thanks are to Dr. T. Ido for useful discussions and the authors of papers in *Plasma Phys. Contr. Fusion* cluster on "Experiments on zonal flows and turbulence". This work was partly supported by the Grant-in-Aid for Specially-Promoted Research of MEXT (16002005) and by the Grant-in-Aid for Scientific Research of JSPS (19360418) Japan, by the Collaboration Programs of NIFS (NIFS07KOAP017, NIFS06KDAD005) and of the RIAM of Kyushu University, by Asada Eiichi Research Foundation, by the U.S. DOE under Grant Nos. FG02-04ER54738, and by the U.S. DOE Contract No DE-AC02-76-CHO-3073. Principal author (KI) acknowledges Prof. S. Benkadda and Organizing Committee of ITER International Summer School for realizing this lecture and Prof. O. Motojima for encouragements.

REFERENCES

[1] P. H. Diamond, S.-I. Itoh, K. Itoh, T. S. Hahm: Plasma Phys. Contr. Fusion **47** R35 (2005)

[2] N. Winsor, J. L. Johnson, and J. M. Dawson, Phys. Fluids **11**, 2448 (1968)

[3] H. Okuda H and J. M Dawson: Phys. Fluids **16** 408 (1973)

[4] R. Z. Sagdeev, V. D. Shapiro, V. I Shevchenko: Sov. J. Plasma Phys. **4** 306 (1978) [Fiz. Plasmy **4** 551 (1978)]

[5] A. Hasegawa, et al., Phys. Fluids **22** 2122 (1979).

[6] A. Hasegawa and M. Wakatani: Phys. Rev. Lett. **59** 1581 (1987)

[7] A. Balk, *et al.,* Sov. Phys. JETP **71** 249 (1990).

[8] P. H. Diamond and Y. B. Kim: Phys. Fluids B**3** 1626 (1991)

[9] G. Hammet, et aal.: Plasma Phys. Control. Fusion **35** 973 (1993)

[10] R. Waltz, et al.: Phys. Plasmas **1** 2229 (1994)

[11] Z. Lin, et al.: Science **281** 1835 (1998).

[12] M. Yagi, et al.: Plasma Phys. Control. Fusion **48** A409 (2006)

[13] K. Itoh, S.-I. Itoh, A. Fukuyama: Fusion Engineering and Design **15** 297 (1992)

[14] M. N. Rosenbluth and F. L. Hinton, Phys. Rev. Lett. **80** 724 (1998).

[15] P. H. Diamond, S.-I. Itoh, K. Itoh and T. S. Hahm: NIFS-805 (2004).

[16] P. H. Diamond, et al. "*Overview of zonal flow physics*", *20th IAEA Conf. on Fusion Energy* (Portugal: IAEA, 2004) Paper OV 2-1

[17] K. Itoh, et al.: Phys. Plasmas **13** 055502 (2006)

[18] Cluster papers in Plasma Phys. Control. Fusion **48** No.4 (2006)

[19] S.-I. Itoh: J. Plasma Fusion Res. **83** 241 (2007) 241 (in Japanese)

[20] A. Fujisawa, et al.: Nucl. Fusion **47** S718 (2007)

[21] K. Hallatschek: Plasma Phys. Control. Fusion **49** B137 (2007)

[22] K. Itoh and S.-I. Itoh: Plasma Phys. Control. Fusion **38**, 1 (1996)

[23] K. Itoh, S.-I. Itoh and A. Fukuyama : *Transport and structural formation in plasmas* (IOP, England, 1999)

[24] K. Hallatscheck: Phys. Rev. Lett. **84** 5145 (2000).

[25] H. Sugama and T.-H. Watanabe: J. Plasma Phys. **72** 825 (2006)

[26] B. Scott: Plasma Phys. Control. Fusion **48** B277 (2006)

[27] Liu Chen, Lin Z and White R B: Phys. Plasmas **7** 3129 (2000)

[28] A. Smolyakov and P. H. Diamond: Phys. Plasmas **7** 1349 (2000)

[29] P. H. Diamond, *et al.*, Nuclear Fusion **41** (2001) 1067.

[30] T. Stringer: Phys. Rev. Lett. **22**, 770 (1969).

[31] A. B. Hassam and J. F. Drake: Phys. Fluids B **5**, 4022 (1993).

[32] K. Itoh, K. Hallatschek, S.-I. Itoh: Plasma Phys. Control. Fusion **47** 451 (2005)

[33] M. Malkov, et al.: Phys. Plasmas **9** 3996 (2001)

[34] H. Sanuki and J. Weiland: J. Plasma Phys. **23** 29 (1980)

[35] B. Rogers, et al.: Phys. Rev. Lett. **81** 4396 (1998)

[36] E. Kim and P. H. Diamond: Phys. Plasmas **9** 4530 (2002)

[37] P. K. Kaw, et al., Plasma Phys. Contr. Fusion **44** (2002) 51.

[38] K. Itoh, K. Hallatschek, S. Toda, H. Sanuki, and S.-I. Itoh: J. Phys. Soc. Jpn. **79** 2921 (2004)

[39] K. Itoh, K. Hallatschek, S.-I. Itoh, P. H. Diamond, S. Toda: Phys. Plasmas **12** 062303 (2005)

[40] S. Toda, K. Itoh, K. Hallatschek, M. Yagi and S. -I. Itoh: J. Phys. Soc. Jpn. **75** 104501 (2006)

[41] A. Dimits, et al.: Phys. Plasmas **7** 969 (2000)

[42] M. Malkov, et al., Phys. Plasmas **9** 5073 (2001)

[43] N. Bian, S. Benkadda, O. E. Garcia, J.-V. Paulsen, X. Garbet : Phys. Plasmas **10** 1382 (2003)

[44] K. Hallatschek: Phys. Rev. Lett. **93** 065001 (2004)

[45] K. Itoh, et al.: J. Phys. Soc. Jpn. **76** 084502 (2007)

[46] A. Fujisawa, et al.: Phys. Rev. Lett. **93** 165002 (2004)

[47] Y. Nagashima, et al.: Plasma Phys. Contr. Fusion **48** S1 (2006)

[48] M. G. Shats et al.: Plasma Phys. Contr. Fusion **48** S17 (2006)

[49] A. Fujisawa et al.: Plasma Phys. Contr. Fusion **48** S31 (2006)

[50] T. Ido et al.: Plasma Phys. Contr. Fusion **48** S41 (2006)

[51] G. R. Tynan et al.: Plasma Phys. Contr. Fusion **48** S51 (2006)

[52] M. Vergote et al.: Plasma Phys. Contr. Fusion **48** S75 (2006)

[53] A. V. Melnikov et al.: Plasma Phys. Contr. Fusion **48** S87 (2006)

[54] V. Sokolov et al.: Plasma Phys. Contr. Fusion **48** S111 (2006)

[55] G. R. McKee et al.: Plasma Phys. Contr. Fusion **48** S123 (2006)

[56] A. Bencze et al.: Plasma Phys. Contr. Fusion **48** S137 (2006)

[57] M. Hirsch, et al.: Plasma Phys. Contr. Fusion **48** S155 (2006)

[58] C. Hidalgo, et al.: Plasma Phys. Contr. Fusion **48** S169 (2006)

[59] Y. Hamada, et al.: Plasma Phys. Contr. Fusion **48** S177 (2006)

[60] N. Vianello, et al.: Plasma Phys. Contr. Fusion **48** S193 (2006)

[61] A. Fujisawa, et al.: Plasma Phys. Contr. Fusion **48** S205 (2006)

[62] C. Hidalgo, et al.: Phys. Rev. Lett. **83** 2203 (1999)

[63] Y. H. Xu et al.: 2000 Phys. Rev. Lett. **84** 3867 (2000)

[64] C. Lechte, S. Neidner. and U. Stroth: 2002 New J. Phys. **4** 34 (2002)

[65] G. R. Mckee, et al.: Plasma Phys. Control. Fusion **45** A477 (2003)

[66] G.S. Xu et al.: 2003 Phys. Rev. Lett. **91** (2003)

[67] P.M. Schoch, et al.: Rev. Sci. Instrum. **74** 1846 (2003)

[68] G.D. Conway, et al.: Plasma Phys. Control. Fusion **47** 1165 (2005)

[69] Y. Nagashima et al.: Phys. Rev. Lett. **95** 095002 (2005)

[70] Y. Hamada. et al 2005 Nucl. Fusion **45** 81 (2005)

[71] K. J. Zhao et al.: Phys. Rev. Lett. **96** 255004 (2006)

[72] A. Kramer-Flecken, et al.: 2006 Phys. Rev. Lett. **97** 045006 (2006)

[73] T. Ido, et al.: *Nucl. Fusion* **46** 512 (2006)

[74 K. Itoh, et al.: Plasma Fusion Res. **1** 037 (2006)

[75] M. Sasaki, et al.: "Radial eigenmodes of geodesic acoustic modes" Contrib. Plasma Phys. **76** (2007) in press

[76] Y. Nagashima, et al.: “Test of magnetic separatrix location by zonal flow measurements”, submitted to Plasma Phys. Contr. Fusion (2007).
[77] S.-I. Itoh, et al.: Plasma Phys. Control. Fusion **49** L7 (2007)
[78] Y. Nagashima, et al.: Plasma Fusion Res. **1** 041 (2006)
[79] Y. Nagashima, et al.: Rev. Sci. Instrum. **77** 045110 (2006)
[80] Y. Nagashima, et al.: Plasma Phys. Control. Fusion **49** 1611 (2007)
[81] K. Itoh et al.: Phys. Plasmas **12** 102301 (2005)
[82] H. Xia and M. G. Shats: Phys. Plasmas **11** 561 (2004)
[83] C. Schröder, O. Grulke, T. Klinger, V. Naulin: Phys. Plasmas **11** 4249 (2004)
[84] T. Yamada: presented at APS (2007)
[85] U. Stroth, presented at International workshop on structure formation and selection rule in turbulent plasmas (February 2005, RIAM Kyushu University).
[86] A. Fujisawa et al.: Phys. Rev. Lett. **82** 2669 (1999)
[87] U. Stroth, et al.: Phys. Rev. Lett. **86** 5910 (2001)
[88] K. Ida, et al.: Phys. Rev. Lett. **91** 085003 (2003)
[89] A. Fujisawa: Plasma Phys. Control. Fusion **45** R1 (2003).
[90] S. Toda and K. Itoh: Plasma Phys. Control. Fusion **43** 629 (2001)
[91] K. Itoh, et al.: Phys. Plasmas **14** 020702 (2007)
[92] A, Yoshizawa, S.-I. Itoh, K. Itoh, and N. Yokoi: Plasma Phys. Contr. Fusion, **46**, R25 (2004)
[93] P. H., Diamond, D. W. Hughes and E.-J. Kim: "Self-consistent mean field electrodynamics in two and three dimensions" in *The Fluid Mechanics of Astrophysics and Geophysics* eds. A. M. Soward, C. A. Jones, D. W. Hughes and N. O. Weiss, (CRC Press, London, 2004) Vol 12, 145.
[94] A. Fujisawa, et al.: Phys. Rev. Lett. **98** 165001 (2007)
[95] S.-I. Itoh and K. Itoh: J. Phys. Soc. Jpn. **69** 427 (2000)
[96] J. A. Krommes and Kim: Phys. Rev. E **62** 8508 (2000)
[97] S.-I. Itoh and K. Itoh: Plasma Phys. Contr. Fusion **43** 1055 (2001)
[98] R. Baleascu, Phys. Rev. E **68** 046409 (2003)

High Energy Particles in Tokamaks

R. B. White

Plasma Physics Laboratory, Princeton University, P.O.Box 451, Princeton, New Jersey 08543

Abstract. This lecture covers the derivation of guiding center equations in a tokamak, orbit classification, the effect of magnetic perturbations and ripple, the interaction of particles with magnetohydrodyanamic modes, including passing particle resonance, toroidal Alfvèn mode drive and saturation, the fishbone mode, and sawtooth stabilization.

Keywords: guiding center motion, tokamaks, magnetic perturbations, beam particle interaction, fishbone, sawtooth
PACS: 52.20.Dq, 52.25.Fi, 52.55.Fa, 52.40.Mj, 52.65.Cc

GUIDING CENTER MOTION

The Larmor radius of charged particles in a reactor scale plasma is small compared to the plasma dimensions and the radius of curvature of the magnetic field. Thus charged particle motion is well described using the drift approximation. The guiding center motion is determined by the mass, charge and energy of the particle and the field strength and configuration. These first four however, enter the equations in one combination only; the cyclotron radius. Thus there is actually only a single parameter characterizing a particle insofar as its guiding center motion is concerned, and it is convenient to use units in which this is apparent. Introducing units of time given by ω_0^{-1}, where $\omega_0 = eB/(mc)$ is the on axis cyclotron frequency, and units of distance given by the major radius R, the basic unit of energy becomes $m\omega_0^2 R^2$, which can also be written as $(mv^2/2)(2R^2/\rho^2)$, the gyro radius is $\rho = v/B \ll 1$, and the magnetic moment $\mu = v_\perp^2/(2B)$ is of order ρ^2. Particle motion both along and across the field lines is of order ρ but to leading order the cross field motion is the periodic cyclotron motion, and cross field drift is of order ρ^2.

Begin with the Lagrangian for the motion of a charged particle in an electromagnetic field [1, 2],

$$L = [\vec{A}(x,t) + \vec{v}] \cdot \dot{\vec{x}} - H(\vec{v},\vec{x}) \tag{1}$$

with $\vec{B} = \nabla \times \vec{A}$, and the Hamiltonian $H = \vec{v}^2/2 + \Phi(\vec{x},t)$, where $\Phi(\vec{x},t)$ is the electric potential. Putting the Lagrangian in the form $\sum p_i \dot{q}_i - H$ we recognize the canonical momentum for a charged particle in a electromagnetic field to be $\vec{p} = \vec{v} + \vec{A}$. For purposes of deriving equations of motion, the momentum variable $\vec{v}$ is considered independent of the velocity $\dot{\vec{x}}$, even though the equations of motion in the end give $\vec{v} = \dot{\vec{x}}$. Now explicitly separate the motion along the field from the cross field motion through

$$\vec{v} = v_\parallel \hat{b} + w\hat{c} \tag{2}$$

with $\hat{b} = \vec{B}/B$, and $\hat{c} = -sin\xi\hat{e}_1 - cos\xi\hat{e}_2$ with $\hat{e}_1$ and $\hat{e}_2$ unit vectors orthogonal to $\vec{B}$ and to each other chosen so that $\hat{e}_1 \times \hat{e}_2 = \hat{b}$, and ξ the gyro phase, and w the magnitude of the perpendicular velocity $\vec{v}_\perp$. Define the particle gyro center through

$$\vec{x} = \vec{X} + \frac{w\hat{a}}{B} \tag{3}$$

with $\hat{a} = cos\xi\hat{e}_1 - sin\xi\hat{e}_2$, so that $\hat{c} \times \hat{a} = \hat{b}$, and all quantities on the right are evaluated at the guiding center $\vec{X}$. The Lagrangian then becomes

$$L = [\vec{A}(\vec{x},t) + v_\parallel \hat{b} + w\hat{c}] \cdot \left[\dot{\vec{X}} + \frac{d}{dt}\left(\frac{w\hat{a}}{B}\right)\right] - H. \tag{4}$$

Now make an expansion in the small parameter w. We will assume that all spatial scales are large enough compared to w so that the time variations of the field viewed by the particle, as well as all explicit time variations of the field, are

CP1013, ***Turbulent Transport in Fusion Plasmas, First ITER International Summer School***
edited by S. Benkadda

slow compared to the gyration frequency $d\xi/dt$. Write

$$\vec{A}(x,t) \simeq \vec{A}(\vec{X},t) + \frac{w\hat{a}}{B} \cdot \nabla \vec{A}(\vec{X},t) \tag{5}$$

and substitute into the Lagrangian, giving

$$L = \left[\vec{A} + v_{\parallel}\hat{b} + w\hat{c}\right] \cdot \dot{\vec{X}} + v_{\parallel}\hat{b} \cdot \frac{d}{dt}\left(\frac{w\hat{a}}{B}\right) + w\hat{c} \cdot \frac{d}{dt}\left(\frac{w\hat{a}}{B}\right) +$$
$$\left(\frac{w\hat{a}}{B} \cdot \nabla\right)\vec{A} \cdot \frac{d}{dt}\left(\frac{w\hat{a}}{B}\right) + \vec{A} \cdot \frac{d}{dt}\left(\frac{w\hat{a}}{B}\right) + \left(\frac{w\hat{a}}{B} \cdot \nabla\right)\vec{A} \cdot \dot{\vec{X}} - H \tag{6}$$

where all $\vec{A}$ are now evaluated at $\vec{X},t$. Now add to the Lagrangian the perfect derivative dS/dt with $S = -(w/B)\hat{a} \cdot \vec{A}(\vec{X},t)$, and use the fact that $d\vec{A}/dt = \dot{\vec{X}} \cdot \nabla \vec{A} + \partial \vec{A}/\partial t$. Further note in taking time derivatives that $d\hat{a}/dt = \dot{\xi}\hat{c}$ plus terms small in the ordering of the time dependence of $\vec{B}$ compared to ξ. The Lagrangian then reduces to

$$L = \left[\vec{A} + v_{\parallel}\hat{b} + w\hat{c}\right] \cdot \dot{\vec{X}} + \frac{w^2\dot{\xi}}{B} - \frac{w}{B}\hat{a} \cdot \frac{\partial \vec{A}}{\partial t} +$$
$$\frac{w\dot{w}}{B^2}(\hat{a} \cdot \nabla)\vec{A} \cdot \hat{a} + \frac{w^2\dot{\xi}}{B^2}(\hat{a} \cdot \nabla)\vec{A} \cdot \hat{c} - H. \tag{7}$$

Now again add a perfect derivative to L, of the form dS/dt with $S = -(w^2/2B^2)(\hat{a} \cdot \nabla)\vec{A} \cdot \hat{a}$, and note that

$$(\hat{a} \cdot \nabla)\vec{A} \cdot \hat{c} - (\hat{c} \cdot \nabla)\vec{A} \cdot \hat{a} = -B \tag{8}$$

giving $L = \left[\vec{A} + v_{\parallel}\hat{b} + w\hat{c}\right] \cdot \dot{\vec{X}} + w^2\dot{\xi}/2B - w\hat{a} \cdot \partial_t \vec{A}/B - H$. Now average over the fast gyro motion time scale. The $w\hat{c}$ and the $\hat{a} \cdot \partial_t \vec{A}$ terms vanish, leaving to second order

$$L = \left[\vec{A} + v_{\parallel}\hat{b}\right] \cdot \dot{\vec{X}} + \frac{w^2\dot{\xi}}{2B} - H(v_{\parallel}, w, \vec{X}, t), \tag{9}$$

from which we conclude that $w^2/2B \equiv \mu$ is a constant of the motion, and that $\dot{\xi}$ is constant. No assumptions were made about the form of the magnetic field.

Write the Lagrangian in the form

$$L = (\vec{A} + \rho_{\parallel}\vec{B}) \cdot \vec{v} + \mu\dot{\xi} - H \tag{10}$$

with $\rho_{\parallel} = v_{\parallel}/B$ the normalized parallel velocity, $H = \rho_{\parallel}^2 B^2/2 + \mu B + \Phi$ the Hamiltonian and Φ the electric potential. Now introduce toroidal coordinates for the guiding center $\vec{X}$, with ψ_p the radial coordinate, and θ and ζ poloidal and toroidal coordinates, respectively. The field magnitude B and the potential are thus functions of ψ_p, θ and ζ.

We now use the straight field line representation (see **Tearing modes in tokamaks** by R. White, this lecture series). Use the contravariant representation for $\vec{B}$

$$\vec{B} = q\nabla\psi_p \times \nabla\theta + \nabla\zeta \times \nabla\psi_p \tag{11}$$

and the covariant representation for $\vec{B}$

$$\vec{B} = g\nabla\zeta + I\nabla\theta + \delta\nabla\psi. \tag{12}$$

Here $2\pi\psi_p$ is the poloidal flux and $2\pi\psi$ is the toroidal flux between the axis and the surface given by $\psi_p = constant$, $d\psi/d\psi_p = q(\psi_p)$, and $2\pi I$ is the toroidal current inside ψ, $2\pi g$ is the poloidal current outside ψ, and δ is a measure of the nonorthogonality of the coordinate system.

Rewrite Eq. 11 in the form $\vec{B} = \nabla \times (\psi\nabla\theta - \psi_p\nabla\zeta) = \nabla \times \vec{A}$ and substitute $\vec{A}$ and Eq. 12 for $\vec{B}$ into the Lagrangian $L(\rho_{\parallel}, \psi_p, \theta, \zeta, \dot{\rho}_{\parallel}, \dot{\psi}_p, \dot{\theta}, \dot{\zeta})$, giving

$$L = (\psi + \rho_{\parallel}I)\dot{\theta} + (\rho_{\parallel}g - \psi_p)\dot{\zeta} + \mu\dot{\xi} + \delta\rho_{\parallel}\dot{\psi}_p - H. \tag{13}$$

Lagrange's equations are

$$\frac{d}{dt}\frac{\partial L}{\partial \dot{q}} = \frac{\partial L}{\partial q}. \tag{14}$$

We then find

$$\begin{bmatrix} \dot{\rho}_\parallel \\ \dot{\psi}_p \\ \dot{\theta} \\ \dot{\zeta} \end{bmatrix} = \frac{1}{D}\begin{bmatrix} 0 & 0 & C & -F \\ 0 & 0 & -g & I \\ -C & g & 0 & -\delta \\ F & -I & \delta & 0 \end{bmatrix}\begin{bmatrix} \partial_{\rho_\parallel} H \\ \partial_{\psi_p} H \\ \partial_\theta H \\ \partial_\zeta H \end{bmatrix} \tag{15}$$

where $C = -1 + \rho_\parallel(g'_{\psi_p} - \delta'_\zeta)$, $F = q + \rho_\parallel(I'_{\psi_p} - \delta'_\theta)$ with denominator

$$D = gq + I + \rho_\parallel(gI'_{\psi_p} - Ig'_{\psi_p} - g\delta'_\theta + I\delta'_\zeta). \tag{16}$$

The equations become

$$\dot{\rho}_\parallel = -\frac{(1-\rho_\parallel g')}{D}\left[(\mu + \rho_\parallel^2 B)\frac{\partial B}{\partial \theta} + \frac{\partial \Phi}{\partial \theta}\right] \tag{17}$$

$$\dot{\psi}_p = -\frac{g}{D}\left[\mu + \rho_\parallel^2 B)\frac{\partial B}{\partial \theta} + \frac{\partial \Phi}{\partial \theta}\right] \tag{18}$$

$$\dot{\theta} = \frac{\rho_\parallel B^2}{D}(1-\rho_\parallel g') + \frac{g}{D}\left[(\mu + \rho_\parallel^2 B)\frac{\partial B}{\partial \psi_p} + \frac{\partial \Phi}{\partial \psi_p}\right] \tag{19}$$

$$\begin{aligned} \dot{\zeta} = \frac{\rho_\parallel B^2}{D}(q + \rho_\parallel[I'_{\psi_p} - \delta'_\theta]) - \frac{I}{D}\left[(\mu + \rho_\parallel^2 B)\frac{\partial B}{\partial \psi_p} + \frac{\partial \Phi}{\partial \psi_p}\right] \\ + \frac{\delta}{D}\left[(\mu + \rho_\parallel^2 B)\frac{\partial B}{\partial \theta} + \frac{\partial \Phi}{\partial \theta}\right] \end{aligned} \tag{20}$$

Notice that the function δ modifies ψ_p, θ, and $\rho_\parallel$ only through a renormalization of the time interval, through the denominator D. Thus the projection of the orbit onto the poloidal plane is independent of this function. We will show that also the toroidal precession is independent of δ, so its omission produces only a nonsecular redefinition of the guiding center.

To calculate the second order drift of a trapped particle integrate ζ over one transit of the orbit in θ. Note that for this calculation it does not matter whether the orbit is closed in the ψ_p, θ plane.

$$\Delta\zeta = \int \dot{\zeta}dt = \oint \frac{\dot{\zeta}}{\dot{\theta}}d\theta \tag{21}$$

Then from Eqs. 19, 20, to first order in $\rho_\parallel$

$$\Delta\zeta = \oint d\theta(q + \rho_\parallel[I'_{\psi_p} - \delta'_\theta + g'q]) - \oint d\theta\frac{(\mu + \rho_\parallel^2 B)}{\rho_\parallel B^2}\left[(I+gq)\frac{\partial B}{\partial \psi_p} - \delta\frac{\partial B}{\partial \theta}\right]. \tag{22}$$

Using energy conservation with $E = \rho_\parallel^2 B^2/2 + \mu B$ to find $(\mu + \rho_\parallel^2 B)\partial_\theta B = -\rho_\parallel B^2 \partial_\theta \rho_\parallel$ the terms in δ can be written as

$$-\oint d\theta \partial_\theta[\delta\rho_\parallel] = 0. \tag{23}$$

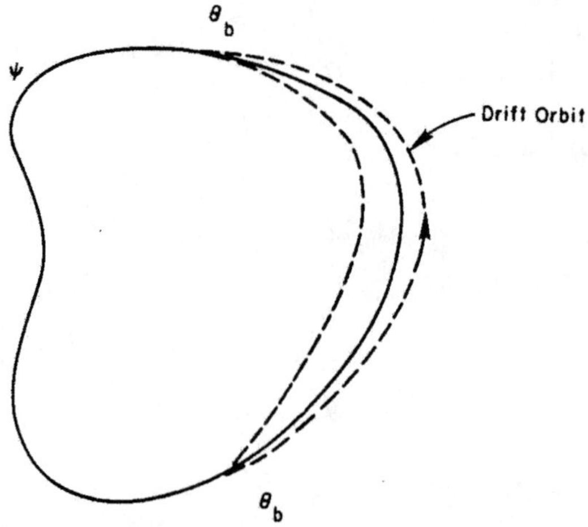

FIGURE 1. A trapped particle drift orbit.

Thus neglecting δ produces only a nonsecular change in the guiding center motion. Dropping the term in δ by a redefinition of the guiding center we have, $C = -1 + \rho_\| g'_{\psi_p}$, $F = q + \rho_\| I'_{\psi_p}$ and $D = gq + I + \rho_\| (gI'_{\psi_p} - Ig'_{\psi_p})$, the Lagrangian takes the form

$$L = (\psi + \rho_\| I)\dot{\theta} + (\rho_\| g - \psi_p)\dot{\zeta} + \mu\dot{\xi} - H, \tag{24}$$

and the canonical coordinates and momenta are immediately identified to be θ, ζ and P_θ, P_ζ, with $P_\zeta = \rho_\| g - \psi_p$, $P_\theta = \psi + \rho_\| I$. These expressions are immediately recognizable as the covariant components of the sum of particle momentum and vector potential, just as in classical electrodynamics, Eq. 1.

The equations of motion are also given by the Hamiltonian form

$$\dot{\theta} = \frac{\partial H}{\partial P_\theta} \qquad \dot{P}_\theta = -\frac{\partial H}{\partial \theta}$$
$$\dot{\zeta} = \frac{\partial H}{\partial P_\zeta} \qquad \dot{P}_\zeta = -\frac{\partial H}{\partial \zeta}. \tag{25}$$

In the absence of an electric field the particle energy takes the form

$$W = \frac{[P_\zeta + \psi_p(P_\zeta, P_\theta)]^2 B^2}{2g^2} + \mu B. \tag{26}$$

From the equations of motion we see that P_ζ is constant if in the system ζ is an ignorable coordinate. This integral of the motion considerably simplifies the particle orbits. In this case, W is a function of $P_\theta = \psi + \rho_\| I$ and θ. But $\rho_\| = (2W - 2\mu B)^{1/2}/B$ is also a function of ψ, θ so $W = constant$ defines closed curves in the ψ, θ plane. Thus the particle motion is restricted to a two dimensional drift surface characterized by the values of W, μ, regardless of the shape of the flux surface cross section.

The motion across the flux surfaces reduces to

$$\dot{\psi}_p = -\frac{g}{D}(\mu + \rho_\|^2 B)\frac{\partial B}{\partial \theta} \tag{27}$$

which is second order in ρ and usually odd in θ since B increases in the direction of decreasing X, the distance from the axis of symmetry. To first order in ρ the particle follows the field lines and to next order it drifts across the surfaces, with $\dot{\psi}_p < 0$ for $0 < \theta < \pi$ and $\dot{\psi}_p > 0$ for $-\pi < \theta < 0$, *i.e.*, inward drift in the upper half of the torus and outward drift in the lower half. As it moves around θ towards decreasing X it will normally encounter increasing field magnitude, and may encounter a mirror point, where $\rho_\parallel = 0$ because $B(\psi_p, \theta_b) = W/\mu$, θ_b the bounce angle. If such a point exists the particle is called trapped, otherwise it is called passing. Since drifts across ψ are proportional to ρ^2 the orbit of a trapped particle has the form of a banana with width $\sim \rho$, as shown in Fig. 1. To calculate the drift from the flux surface ψ_p use conservation of P_ζ and neglect variation of g along the orbit, which gives a higher order correction in ρ, to find $\Delta\psi_p = g\Delta\rho_\parallel$. For a trapped particle $\rho_\parallel$ is zero near the banana tip. The banana width along the orbit is then by conservation of P_ζ given by $\Delta\psi_p = g\rho_\parallel$ where $\rho_\parallel = (2W - 2\mu B)^{1/2}/B$. For a passing particle $\rho_\parallel$ does not vanish so the excursion from the flux surface has the form $\Delta\psi_p = g[\rho_\parallel - \rho_\parallel(0)]$ and is normally much smaller than for a trapped particle.

Although the orbit closes in the ψ_p, θ plane it does not return to the same value of ζ. The toroidal drift for a trapped particle becomes $\Delta\zeta/T$ where $T = \oint d\theta/\dot{\theta}$ is the time for one bounce period, giving

$$\frac{d\zeta}{dt} \simeq -\frac{1}{T}\oint d\theta \frac{(\mu + \rho_\parallel^2 B)}{\rho_\parallel B^2}(I + gq)\frac{\partial B}{\partial \psi_p}. \tag{28}$$

and $T = \oint d\theta D/(\rho_\parallel B^2)$. The precession is proportional to the particle energy, not the velocity, so is fairly slow except for highly energetic particles. Deeply trapped particles spend their time near the outer midplane where for a low beta equilibrium $\partial B/\partial\psi_p$ is negative, so the precession is in the direction of the toroidal current. In a high beta equilibrium the precession can be reversed. Also barely trapped particles spend most of their time near the bounce points, and if these are located on the high field side of the tokamak the precession can be reversed.

MAGNETIC PERTURBATIONS

All perturbations of the field leading to destruction of toroidal symmetry produce classes of trapped particles, and new loss mechanisms. Important interaction between plasma waves and particles involves resonances. This follows since the guiding center motion is Hamiltonian, and the orbits are closed in a poloidal cross section in the limit of axisymmetry. Without resonances there are only small perturbations of the existing KAM surfaces which produce no topological changes in them, and thus there is no large effect on particle orbits. The resonance effects are most important for high energy particles, including fusion alpha particles, injected beam particles, and particles heated through resonance with waves launched from antennae.

Ripple

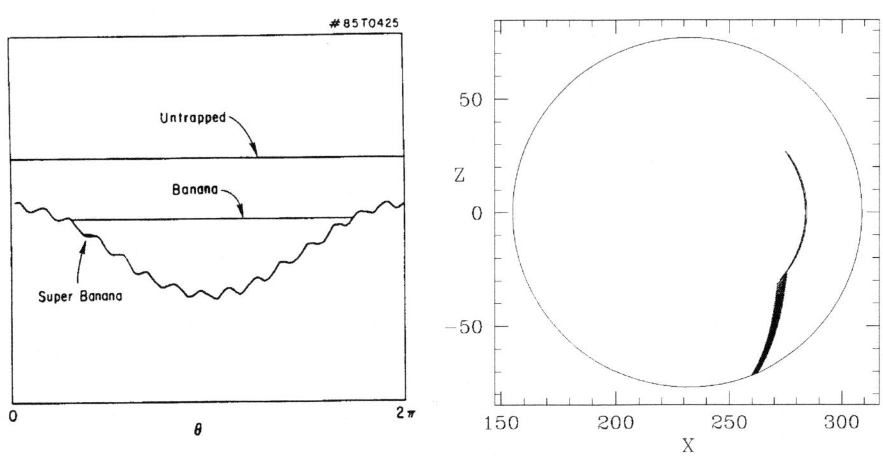

FIGURE 2. Magnetic ripple along a particle trajectory, and a loss orbit for a particle trapped in a ripple well

The equilibrium can be perturbed by allowing for a ζ dependence of B, to account for toroidal field ripple, due to the finite number of toroidal field coils, and also a toroidal dependence of the electric potential Φ. A simple model for the field magnitude including ripple in a large aspect ratio circular device is given by

$$B = B_0(1 - r cos\theta + \delta cos N\phi). \tag{29}$$

Along a field line $\phi = q\theta, r = constant$ so ripple wells exist if $r|sin\theta| < Nq\delta$. The magnitude of the field along a field line is shown in Fig. 2. If the perturbation is strong enough, magnetic wells are formed. To be trapped in one of these wells (a super banana) a particle must have $v_{\|}/v < \sqrt{\delta}$. If trapped the loss is very rapid, since the particle simply drifts down directly to the wall, as shown also in Fig. 2.

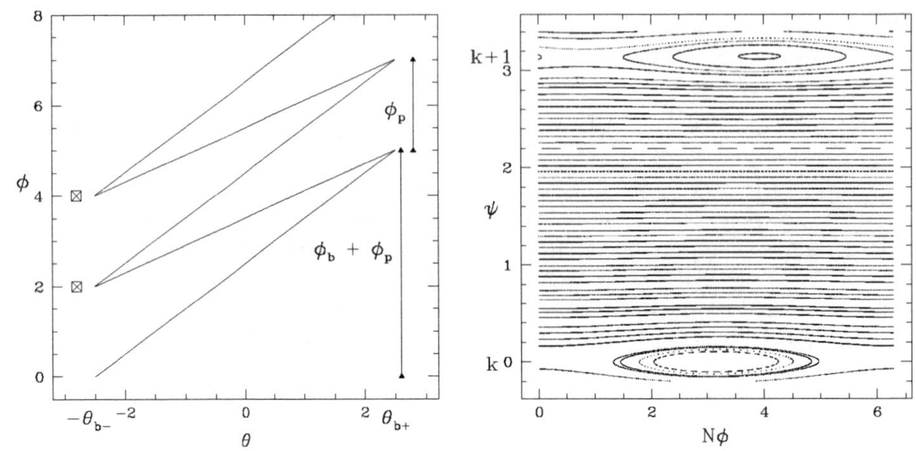

FIGURE 3. Trapped particle orbit showing resonance with field coils, and a Poincare plot showing ripple resonances

Even if ripple wells do not exist, the phase space of banana tip trajectories can become stochastic due to ripple. Because of particle drift, when the ripple causes a banana orbit to bounce either early or late it produces a small change in the radial position of the banana tip. Shown in Fig. 3 is the trajectory of a trapped particle in resonance with the field coils, shown at the left with an $\times$. Here $\phi_b = 2q\theta_b$ is the toroidal motion along the field and ϕ_p is the precessional drift motion in one bounce. For small ripple the plot of radial banana tip position versus toroidal angle is periodic, but there are resonances producing islands at radii where the precessional motion of the tip moves one coil spacing in one bounce, as shown in Fig. 3. Typically there can be one hundred such resonances across the minor radius. If the ripple is sufficiently large the islands due to these resonances overlap and the banana tip motion becomes stochastic. Because of conservation of μ and energy the tip is confined to a surface of constant B, but this means it can move vertically until the particle is lost.

The change of the bounce point due to the radial drift can be calculated using $\phi_b = 2q\theta_b$ and B at the bounce points, $B(\psi_b, \theta_b) = E/\mu$.

$$\frac{d\phi_b}{d\psi_b} = 2\theta_b \frac{dq}{d\psi_p} - 2q \frac{\partial_{\psi_p} B}{\partial_{\theta_b} B}. \tag{30}$$

The motion becomes stochastic approximately[3] when the modification of the toroidal bounce point Δ is as large as a ripple phase

$$N \frac{d\phi_b}{dr} \Delta > 1. \tag{31}$$

A more detailed examination of stochastic threshold is given in another publication[4].

This loss produces an important restriction on ripple in a reactor

$$\delta \le [r/(\pi R N q)]^{3/2} / (\rho q'). \tag{32}$$

Normally this is important only for energies above 1 MeV, so it is relevant for trapped fusion alpha-particles, high energy beam particles, and ion cyclotron resonance (ICRF) heated particles.

In Fig. 4 is shown the trajectory of a particle lost by stochastic ripple diffusion. The banana tips remain at a surface of constant field magnitude, allowing the banana tip to move vertically until the orbit contacts the wall near the outer midplane.

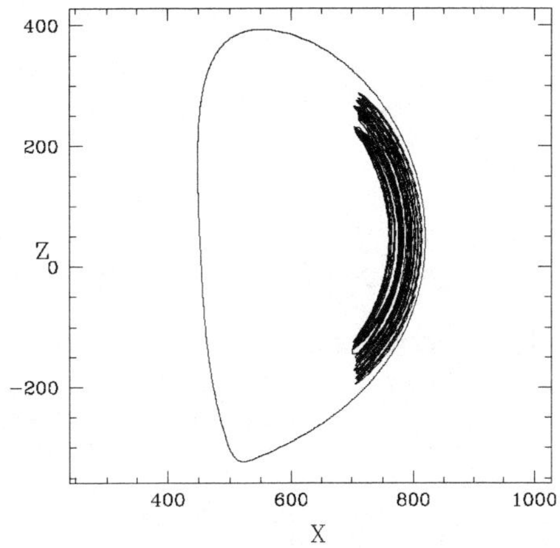

FIGURE 4. Trajectory of stochastic ripple loss banana orbit

Field modification by Magnetohydrodynamic Modes

A very small resonant perturbation of an equilibrium field can produce large islands and stochastic domains in the magnetic field. All tearing and shear Alfvèn perturbations are perturbations of $\vec{B}$ primarily orthogonal to the original $\vec{B}$. These latter can be described through

$$\delta\vec{B} = \nabla \times \alpha\vec{B} \tag{33}$$

with α an arbitrary function of ψ_p, θ, ζ. This form represents exactly the $\nabla\psi$ component of any perturbation, which is the component responsible for perturbations of the flux surfaces. If α has the form $\alpha = \alpha_{mn} sin(m\theta - n\zeta)$ it produces a magnetic island at ψ_p with $q(\psi_p) = m/n$, of width (see **Tearing modes in tokamaks** by R. White, this lecture series)

$$\delta\psi_p = 4\left(\frac{\alpha_{mn}}{s}\right)^{1/2} \tag{34}$$

with the local shear $s = q'/q$ and primes denote derivatives with respect to ψ_p.

The equations of motion are changed only by the substitution of $\vec{A}$ by $\vec{A} + \alpha\vec{B}$, giving

$$L = (\psi + \rho_{\|}I + \alpha I)\dot{\theta} + (\rho_{\|}g + \alpha g - \psi_p)\dot{\zeta} + \mu\dot{\xi} - H. \tag{35}$$

and the modification of the equations of motion are easily calculated.

HIGH ENERGY PARTICLES AND MAGNETOHYDRODYNAMIC MODES

High energy particles injected into a confined plasma have been considered for some time as a stabilizing influence. When the hot particles precess much faster than the frequency of the mode in question, they average the potentials associated with the perturbation, and also form a diamagnetic well, giving rigidity to the plasma, and stabilizing the equilibrium.

Consider an electromagnetic wave with a frequency much smaller than the gyro frequency and a wave length long compared to the gyro radius, so that the guiding center equations are a good approximation to the particle motion. Perturb the equilibrium field with a perturbation of the form $\delta\vec{B} = \nabla \times \alpha\vec{B}$ and also introduce an electric perturbation Φ. The particle motion is then described by the Lagrangian Eq. 35. We are primarily interested in motion across the flux surface, which can lead to particle loss, and changes in the particle energy. Consider a particular harmonic of the wave, of the form

$$\alpha = \alpha_{mn}e^{i\chi} \qquad \Phi = \Phi_{mn}e^{i\chi}. \tag{36}$$

with $\chi = n\zeta - m\theta - \omega t$. We then find

$$\frac{dE}{dt} = i[\rho_\| B^2 \alpha_{mn} - \Phi_{mn}]\omega e^{i\chi}, \tag{37}$$

$$\frac{d\psi_p}{dt} = -i\frac{mg+nI}{D}[\rho_\| B^2 \alpha_{mn} - \Phi_{mn}]e^{i\chi} - \frac{g(\rho_\|^2 B + \mu)}{D}\partial_\theta B. \tag{38}$$

Now carry out an average over one orbit transit (passing) or one bounce (trapped). To zero order in α and Φ the orbit closes and the $\partial_\theta B$ term integrates to zero. With orbit modifications due to alpha this term produces a correction of order $\rho^2 \alpha_{mn}$ which we neglect. We thus find that radial motion is directly related to energy change,

$$\frac{d\psi_p}{dt} = -\frac{(mg+nI)}{\omega D}\frac{dE}{dt}, \tag{39}$$

and thus for positive frequency outward motion is coupled to energy loss.

Now consider net energy transfer between a particle distribution and a wave. If the particles merely oscillate with their mean position equal to the initial one, there is no net transfer. However, at a resonance the particle motion forms an island (both in energy and in radius). In this case particles trapped in the island have a mean value of energy equal to the resonance value, not their initial value. If the particle distribution possesses a gradient (either in energy or in radius) the flattening of the distribution function by the island produces a net energy transfer to the wave

$$\Delta E \sim \omega \partial_E F - (m/r)\partial_r F. \tag{40}$$

Taking $\partial_r F \sim F/r$ and $\partial_E F \sim F/\rho^2$, the ratio of the first to the second term is $\omega r^2/(m\omega_0 \rho^2)$, typically small for high energy particles of interest, and in this case only the radial gradient of the distribution function is relevant.

For the net energy transfer between mode and ions we have Eq. 37. But the rapid mobility of the electrons shorts out the parallel electric field, $E_\| = 0$ giving $\omega B \alpha_{mn} e^{i(n\zeta - m\theta - \omega t)} - \vec{B}\cdot\nabla\Phi/B = 0$ which is

$$(gq+I)\omega\alpha_{mn} = (nq-m)\Phi_{mn}, \tag{41}$$

or

$$\frac{dE}{dt} = i\left[\rho_\| B^2 \frac{nq-m}{gq+I} - \omega\right]\Phi_{mn}(\psi_p)e^{i(n\zeta - m\theta - \omega t)}, \tag{42}$$

which becomes

$$\frac{dE}{dt} = i\left[n\dot{\zeta} - n\dot{\zeta}_d - m\dot{\theta} + m\dot{\theta}_d - \omega\right]\Phi_{mn}(\psi_p)e^{i(n\zeta - m\theta - \omega t)} \tag{43}$$

with $\dot{\zeta}_d$, $\dot{\theta}_d$ the drift terms using Eq. 35. Note that if α_{mn} represents an ideal Magnetohydrodynamic (MHD) mode it must vanish at the rational surface, so Φ is not singular. If α_{mn} describes a resistive mode, not vanishing at the rational surface, then the topology is changed and this expression is not valid for Φ_{mn} in the interior of the island.

But $i[n\dot{\zeta} - m\dot{\theta} - \omega]\Phi_{mn}(\psi_p)e^{i(n\zeta - m\theta - \omega t)} = (d/dt)\Phi_{mn}e^{i(m\theta - n\zeta - \omega t)} - \Phi'_{mn}\dot{\psi}_p e^{i(m\theta - n\zeta - \omega t)}$. So subtracting an exact time derivative $(d/dt)\Phi_{mn}e^{i(m\theta - n\zeta - \omega t)}$ we have

$$\frac{d\mathscr{E}}{dt} = i\left[-n\dot{\zeta}_d + m\dot{\theta}_d\right]\Phi_{mn}(\psi_p)e^{i(n\zeta - m\theta - \omega t)} - \Phi'_{mn}\dot{\psi}_p e^{i(n\zeta - m\theta - \omega t)} \tag{44}$$

with $\mathscr{E} = E - \Phi_{mn}e^{i(n\zeta - m\theta - \omega t)}$. The nonresonant exact derivative is simply the periodic energy change of the particle following the locally changing potential, whereas $\mathscr{E}$, entirely due to the drift motion, can be changed resonantly. Thus the transverse Alfvèn wave can resonantly exchange energy with a passing particle through the cross field drift motion only.

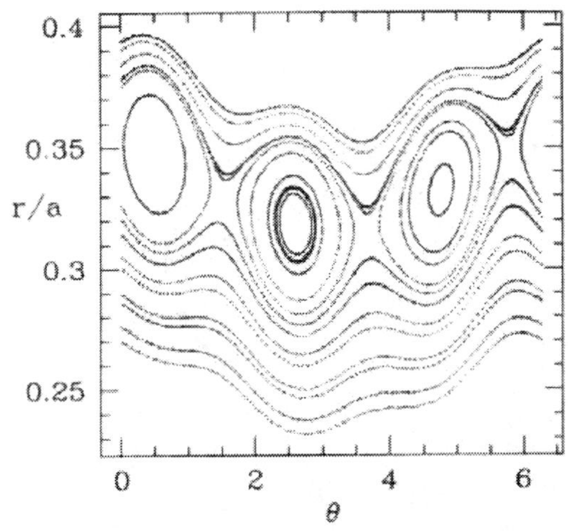

FIGURE 5. TAE induced drift island.

PASSING PARTICLE RESONANCE

There are many possibilities of exciting MHD modes which are normally stable by means of a high energy particle population. For passing particles the most important are toroidal Alfvèn modes, and there are several different branches, the most relevant being the shear Alfvèn waves. In slab geometry the waves have a continuous spectrum, but due to toroidal bending there is also a spectrum of discrete modes existing inside the gap in the continuum. In the continuum the modes are damped by dispersion, but the gap modes may be undamped except for Landau damping with the plasma, and hence are easily excited by particles with a velocity resonant with the wave[5].

Consider passing particle resonance. Use Eqs. 20, 19 giving

$$\dot{\zeta} = \omega_t + \dot{\zeta}_d \qquad \dot{\theta} = \frac{\omega_t}{q} + \dot{\theta}_d \tag{45}$$

where $\omega_t = \rho_\parallel B^2 = v_\parallel / R$ is the transit frequency, and $\dot{\zeta}_d$, $\dot{\theta}_d$ are the second order drift terms.

Using $B = 1 - rcos\theta$ gives the large aspect ratio expressions $\dot{\zeta}_d \simeq (\rho_\parallel B^2 + \mu) rcos\theta$, $\dot{\theta}_d \simeq -(\rho_\parallel B^2 + \mu) cos\theta / r$, the $\dot{\theta}_d$ term dominates and upon dropping the term in Φ'_{mn} we find for the energy evolution

$$\frac{d\mathscr{E}}{dt} = -\frac{m(\rho_\parallel B^2 + \mu)}{2r} \Phi_{mn} [cos(Q_{m+1}) + cos(Q_{m-1})]. \tag{46}$$

with $Q_m = n\zeta - m\theta - \omega t$. Thus the mode harmonic m produces two resonant surfaces at the points where $\dot{Q}_{m-1} = 0$ and $\dot{Q}_{m+1} = 0$. In general there may be many harmonics present, each producing two resonances at different surfaces. In particular a harmonic $\Phi_{m+1,n}$ produces a resonance at the surface $\dot{Q}_m = 0$ which, neglecting the drift motion, gives $[n - m/q]\omega_t = \omega$. Thus for $q > m/n$ resonance occurs with a co-moving passing particle and for $q < m/n$ it occurs for a counter-moving passing particle.

Consider the resonance $\dot{Q}_m = 0$, and note that $\dot{Q}_m$ is a function of energy. Expand it around the resonant energy $\mathscr{E}_0$ where $\dot{Q}_m = 0$. Assuming that the resonances are well separated in the vicinity of $\mathscr{E}_0$ we can drop the nonresonant $cos(Q_{m'})$ terms for $m' \neq m$ and taking for simplicity a far passing particle, *i.e.* $\rho_\parallel B^2 \gg \mu$ find

$$dQ_m = \partial_E \dot{Q} (\mathscr{E} - \mathscr{E}_0) dt, \tag{47}$$

$$d\mathscr{E} = -\frac{m\omega_t^2 \Phi_{mn}}{2r} cos(Q_m) dt. \tag{48}$$

with $\partial_E \dot{Q}$ evaluated at $\mathscr{E}_0$.

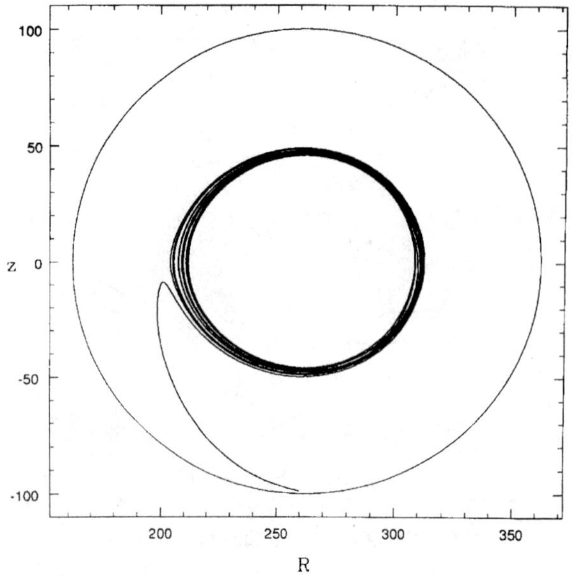

FIGURE 6. Particle loss by TAE resonance.

These equations have the same form as the equations for a magnetic island. Far from the resonance they integrate to give $\mathscr{E} \sim sin(Q_m)/(\mathscr{E} - \mathscr{E}_0)$. Near resonance we find an island given by

$$(\mathscr{E} - \mathscr{E}_0)^2 = ksin(Q_m) + c \tag{49}$$

with $k = -m\omega_t^2 \Phi_{mn}/(r\partial_E \dot{Q})$ and c an integration constant. In Fig. 5 is shown a phase space plot of a single resonance produced by a saturated TAE mode[6]. The drift is clearly seen as the $cos(\theta)$ distortion of the trajectories away from the $r = constant$ field line surfaces.

Other symmetry breaking produces further gaps in the Alfvèn continuum, for example an elliptically shaped plasma cross section [7] produces a gap and the modes are called the Elliptic Alfvèn Eigenmodes (EAE).

TAE MODE DRIVE AND SATURATION

For the most part wave particle interaction is adiabatic, with energy passing back and forth between the particles and the wave. For a single TAE mode, resonant particles exchange energy with the wave. If they lose sufficient parallel energy they can become trapped poloidally. If the resulting trapped orbit has a sufficiently large excursion this process can lead to particle loss, as shown in Fig. 6. However some particles through the formation of phase space islands as shown in Fig. 5 become trapped in the wave. This modification of the particle distribution drives the wave if the distribution has a density gradient, since as we have seen outward motion implies energy loss. In Fig. 7 is shown the average particle distribution modification due to the presence of the phase space island produced by a TAE mode. The final distribution (solid line) shows a net outward motion of part of the distribution, and hence a positive energy transfer from the distribution to the mode. This process produces the driving mechanism for the TAE mode, which in the absense of the particle drive has only a real frequency. The calculation of the growth rate depends on the details of the equilibrium and the particle distribution, and must be carried out numerically.

Due to the large alpha particle pressure in an ignited reactor size tokamak the TAE could grow to a sufficient level to cause energetic particles loss. Thus it is of great importance to know the TAE saturation level. The flattening leads to mode saturation through the classical bump on tail mechanism[8]. If the width of the phase space island is less than the spacing between nearby resonances produced by other poloidal harmonics, as is normal for typical reactor parameters, then there is very limited loss, but if the mode saturation amplitude is sufficiently large the overlap of nearby islands can lead to catastrophic avalanche loss.

For simplicity we consider a single toroidal eigenvalue, n. Due to the toroidal bending in tokamaks, all poloidal harmonics m are coupled. Due to this coupling there is a gap in the spectrum due to the cylindrical harmonics m and $m+1$. In this gap, located at the point r_0 with $q(r_0)$ between m/n and $(m+1)/n$, there exists a discrete TAE mode with a frequency given by the mode frequency at $q \simeq (m+.5)/n$, or $\omega \simeq V_A/(2qR)$, consisting primarily of the two poloidal harmonics $m, m+1$. The resonance condition between the wave and the alpha particles is $(n-m/q)\omega_t = \omega$ and thus $\omega_t = 2q\omega$.

When the trapping frequency exceeds the growth rate, trapped particles remove energy too rapidly to permit growth. The particle rotation in the island flattens the distribution function so quickly that there is no drive for the instability. To find the rotation frequency in the island, differentiate the equation determining the island to give

$$\frac{d^2Q_m}{dt^2} = (\partial_E \dot{Q})\frac{d\mathscr{E}}{dt}, \tag{50}$$

Substituting $d\mathscr{E}/dt$ we find the pendulum equation

$$\frac{d^2Q_m}{dt^2} = \omega_T^2 cos Q_m. \tag{51}$$

Using $\delta\vec{B} = \nabla \times \alpha\vec{B}$ we have $\delta B_r/B = m\alpha_{mn}/r$, and reinserting the major radius R and the gyro frequency ω_0 to restore dimensions, the trapping frequency is given by

$$\omega_T^2 \simeq \frac{\omega_t \omega}{2}\left[\frac{m^2 sR^2\omega_t^3}{qr^2\omega^2\omega_0} - 1\right]\frac{\delta B_r}{B}. \tag{52}$$

The TAE saturation level can be estimated by setting $\omega_T = C\gamma_L$, where C is a numerical factor and γ_L is the linear growth rate[6]. Since there are many resonant surfaces with somewhat different saturation characteristics, the mode saturation involves an averaging process, and will depend somewhat on the form of the distribution and profiles. However, $\gamma_L/\omega \simeq 10^{-2}$ and a single n TAE mode is very far from producing stochastic loss. If there are many such modes, there is the possibility of overlap of nearby islands, producing an avalanche loss of confinement.

FISHBONE

Now consider the effect of a rotating mode on the trapped particle population. Consider δB to be rotating in the laboratory frame. In the frame moving with the perturbation there is an electric field $\delta\vec{E} = \vec{v} \times \delta\vec{B}$ which gives rise

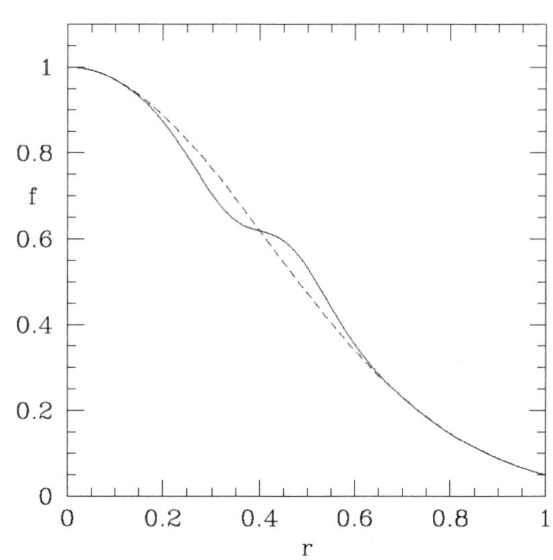

FIGURE 7. Particle distribution modification by TAE mode.

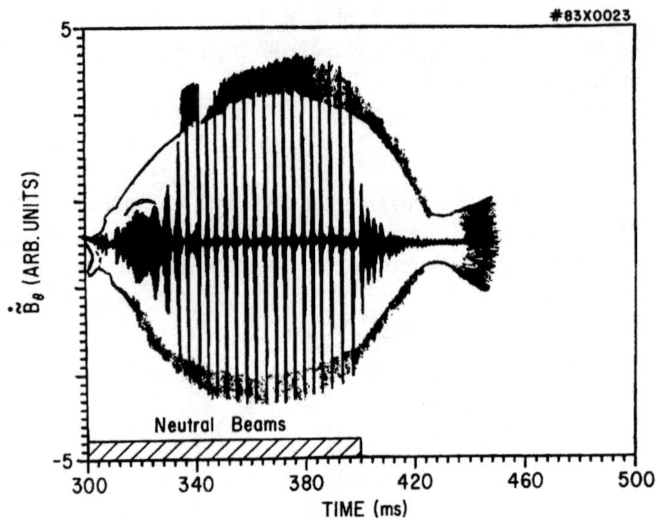

FIGURE 8. Mirnov signal during fishbone mode, outline added by artist.

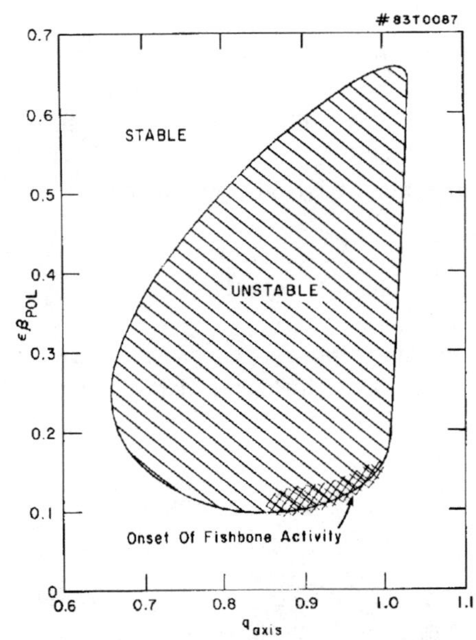

FIGURE 9. Internal kink mode stability diagram.

to a δE_θ. This in turn produces an $\vec{E} \times \vec{B}$ drift of the particles which is radial. Particles moving in phase with the perturbation experience a secular radial drift. This results in an island in energy and flux.

The first experiments involving large scale near perpendicular neutral beam injection into a high β plasma on PDX provided a surprise[9]. A very large amplitude $n = 1$ mode, with a real frequency near 20 KHz, and a few low m values, appeared in bursts with a repetition rate of a few milliseconds. The mode was called the fishbone because of the structure of the magnetic signals observed, as shown in Fig. 8. During the bursts of mode activity a sharp drop in neutron emission indicated a loss of beam particles from the plasma, and charge exchange measurements on PDX subsequently confirmed a loss of as much as one third of the beam particles in each burst.

An analytical calculation and Monte Carlo simulation[10] showed that a rotating mode with $n = 1$ and $m = 1,2,3$

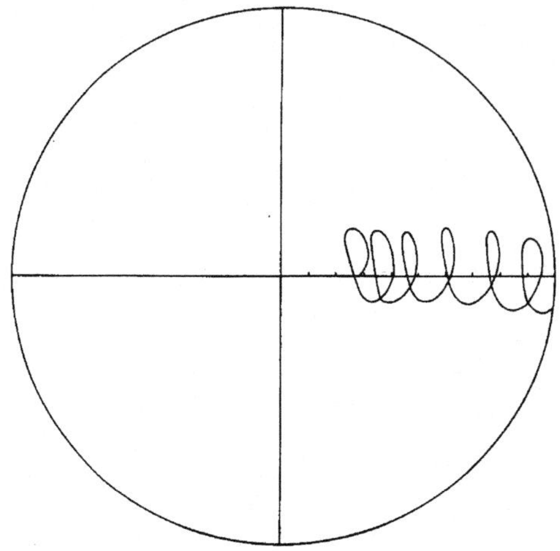

FIGURE 10. Trajectory of Fishbone ejected beam ion.

was capable of resonant ejection of trapped particles, provided that the mode frequency matched their toroidal precession rate. As seen in Fig. 9, it was noted that the plasma β and q profile were such that the plasma was marginally stable to the internal kink mode, although this mode, being purely growing, could offer no explanation for the real frequency, which appeared to be associated with the trapped particles. The mode structure obtained by resistive MHD simulations of an unstable internal kink mode agreed well with the Mirnov and soft X-ray signals associated with the mode, provided the experimental frequency was used in the simulations. Finally it was shown [11] that a trapped particle population could destabilize the internal kink mode provided it was near its ideal MHD threshold, and that the excited mode would have a real frequency and growth rate determined by the trapped particle precession rate. Also the modification of the MHD branch of the dispersion relation due to high energy particles was calculated[12].

The bounce averaged radial motion is given by Eqs. 39 and 44,

$$\frac{dr}{dt} = \frac{\omega}{(n-m/q)}\frac{\delta B_r}{B} cos(Q) J_0(m\theta_b). \tag{53}$$

For typical PDX parameters this loss mechanism is very effective, capable of ejecting a particle from deep within the plasma in 100 microseconds, as shown in Fig. 10. The sign of dr/dt depends on the particle location with respect to the mode phase $Q = n\zeta - \omega t$, and thus the ejected particles form a rotating beacon, rotating toroidally with the mode frequency.

A variational dispersion function $D(\xi)$ was found[11] of the form

$$D(\xi) = \delta W_{MHD} + \delta W_k + \delta I \tag{54}$$

where δW_{MHD} is the usual MHD expression, δW_k the energy change in the hot trapped particles due to the mode, and δI the inertia term, due to damping by Alfven resonance.

The ideal MHD energy was that giving the ideal internal kink mode

$$\delta W_{MHD} = \frac{1}{2}\int d\tau \left[\begin{array}{c} \vec{Q}_\perp^2 - \frac{j_\parallel}{B}(\vec{\xi}_\perp \times \vec{B})\cdot \vec{Q}_\perp \\ -2(\vec{\xi}_\perp \cdot \nabla p)(\vec{\xi}_\perp \cdot \vec{\kappa}) + B^2(\nabla\cdot\vec{\xi}_\perp + 2\vec{\xi}_\perp\cdot\vec{\kappa})^2 \\ +\gamma p(\nabla\cdot\vec{\xi})^2 \end{array} \right] \tag{55}$$

The kinetic energy contribution has a non self adjoint **resonance structure** of the form

$$\delta W_k = 2^{9/2}\pi^3 \int r dr d\mu dE E^{3/2} \frac{K_b \bar{J}^* Q \bar{J}}{\omega_d - \omega} \tag{56}$$

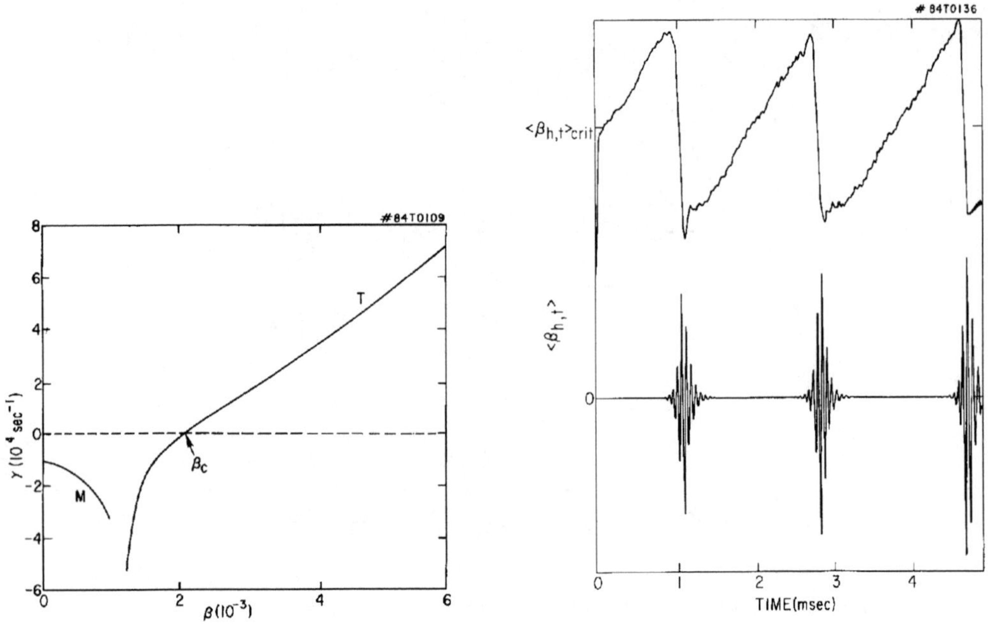

FIGURE 11. Growth rate for trapped particle and MHD branch of the dispersion relation Eq. 60, and the Fishbone cycle.

where

$$Q = \omega\frac{\partial F}{\partial E} - \frac{1}{\omega_0 r}\frac{\partial F}{\partial r}, \qquad J = \frac{\alpha B}{2}\nabla\cdot\vec{\xi}_\perp - \left(1 - \frac{3\alpha B}{2}\right)\vec{\xi}_\perp\cdot\vec{\kappa}, \tag{57}$$

The contribution to D from the inertia takes the form

$$\delta I = -\frac{\pi^2}{2}\frac{r_s^2\xi_0^2 i\omega}{\omega_A} \tag{58}$$

with $\omega_A = v_A/(\sqrt{3}Rrq')$ the effective Alfvèn frequency.

The dispersion relation then reduces to

$$0 = \frac{-i\omega}{\omega_A} + \delta\hat{W}_c + C_\beta\omega\int drdEd\alpha E^{1/2}\frac{F\omega_d\partial_r(r^2K_b)}{\omega_d - \omega} \tag{59}$$

Choice of a model slowing down particle distribution gives analytically tractable problem, $F = n(r)E^{-3/2}\delta(\alpha - \alpha_0)$ for $E < E_m$. Write $\omega_d = \omega_{dm}E/E_m$ and $\Omega = \omega/\omega_{dm}$. We then find $\beta_h = C_\beta E_m \int rdrK_b(\alpha_0)n(r)$, and the dispersion relation reduces to

$$0 = \frac{-i\Omega\omega_{dm}}{\omega_A} + \delta\hat{W}_c + C\beta_h\Omega ln\left(1 - \frac{1}{\Omega}\right) \tag{60}$$

with $C = \int drn(r)d/dr(r^2K_b)/\int drn(r)rK_b$ of order unity.

This dispersion relation has a new branch, the addition of the hot particles gives a singular perturbation theory. The singular branch goes off to infinity as the hot particle beta goes to zero, it is not a perturbation of a mode existing without the particles.

The resonant beam loss process, along with the mode destabilization mechanism, allow a description of the full fishbone cycle. If we assume that the core plasma component has been heated until it is near marginal stability for the kink mode, then the mode cycle is governed by the neutral beam deposition. In Fig. 11 is shown the growth rate of the trapped particle and MHD branches of the dispersion relation versus β_h with threshold $\beta_{h,crit}$. Hot trapped particles

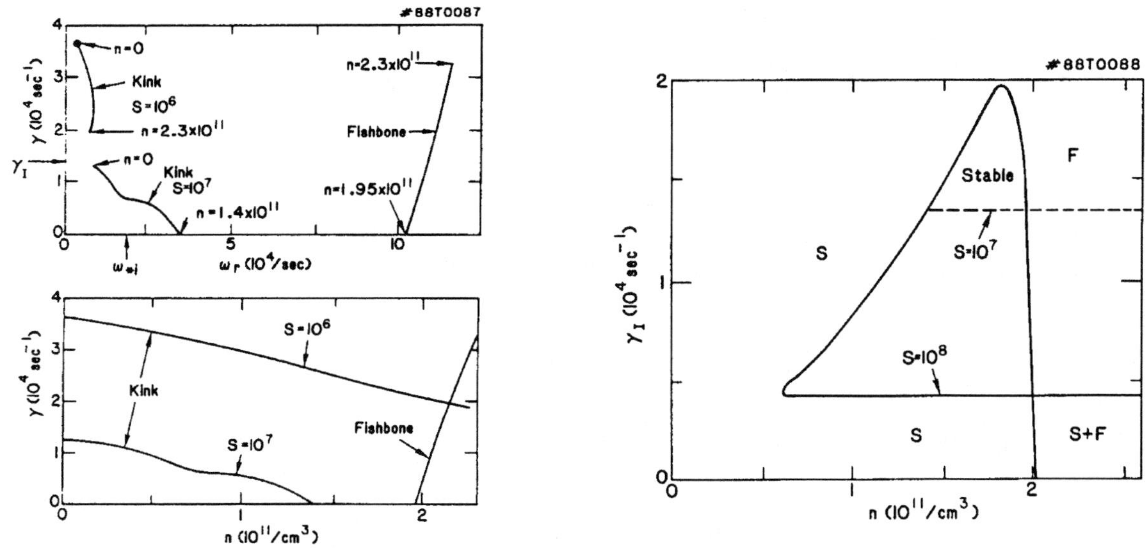

FIGURE 12. Complex frequencies and growth rate, as a function of trapped particle density, for minority heating in JET and the Stability domain for sawtooth and fishbone modes.

are added until $\beta_{h,crit}$ is exceeded, at which point the mode rapidly grows, with a real frequency and growth rate given by a weighted average trapped particle precession frequency. When the amplitude is sufficiently large rapid resonant particle loss depletes the trapped particle population. The mode stops growing when β_h is again reduced to $\beta_{h,crit}$, but resonant particle loss continues as the mode decays until β_h has been driven well below its critical value. The neutral beam injection rate is generally much slower than the loss rate at peak amplitude, so the MHD activity and the losses occur in sharp bursts, with the separation determined by the beam injection rate.

A Monte Carlo guiding center simulation, using PDX parameters with the mode growth rate and frequency calculated using the instantaneous hot particle distribution is shown in Fig. 11. There is a steady state introduction of high energy beam particles. When the beam beta reaches the threshold for the fishbone, the mode begins to grow, and rapidly ejects beam particles from the plasma when it becomes large enough. After driving the beam beta below the fishbone threshold the mode decays away, but before it becomes negligible it drives the beam beta well below threshold. The period between bursts is determined by the rate of beam replenishment.

STABILIZATION OF THE SAWTOOTH

High energy ions in banana orbits exert a stabilizing influence by adding their inertia to the field line to which they are attached. If they precess toroidally at a rate which is greater than the mode growth rate, they effectively inhibit helical displacement of the field lines. The possibility of complete stabilization of the resistive internal kink was first shown in 1989[13].

Results of a numerical solution are shown in Fig. 12 for a hydrogen minority species in the joint european totomak[14] (JET). Shown are the complex frequencies, and the growth rates, as a function of trapped particle density for an approximate JET equilibrium with $R = 296cm$, for two different values of the magnetic Reynolds number, $S = 10^6$ and 10^7. Both the resistive internal kink branch and the fishbone branch are shown. The trapped particle density n ranges from zero to $2.3 \times 10^{11}/cm^3$. The particles were taken to be a slowing down distribution with an average energy of 700keV. The toroidal field was $B = 24kG$ and the averaged trapped particle precession rate was $< \omega_d >= 1.3 \times 10^5/sec$. In Fig 12 is shown the stability domain for these same parameters. Outside of the stable triangle one would expect sawteeth (S), fishbones (F) or both (S+F). Only the lower limit on γ_I depends on the Reynolds number, and two values are shown. Parameter changes, such as a modification of the q profile or a change in the hot particle beta, can cause an abrupt loss of stabilization, thus allowing the plasma to suddenly resume sawtoothing, with what is referred to as a giant sawtooth.

ACKNOWLEDGMENTS

This work was partially supported by the U.S. Department of Energy Grant DE-FG03-94ER54271. The author is indebted to Leonid Zakharov and Allen Boozer for many useful discussions.

REFERENCES

1. White, R. B., *The Theory of Toroidally Confined Plasmas*, Imperial College Press (2006).
2. Littlejohn, R. G., Phys. Fluids 24, 1730 (1981)
3. Goldston, R.J, R. B. White, and A. H. Boozer, Phys. Rev. Lett. [47], 647 (1981)
4. White, R.B., Phys. Rev. E[58] 1774 (1998)
5. Cheng, C. Z., Physics Reports, 211 (1992)
6. Wu, Y, and R.B. White, Phys. Plasmas [1], 2733, (1994)
7. Betti, R., and J. P. Freidberg, Phys. Fluids B 3, 1865, (1991)
8. O'Neil, T., Phys. Fluids 12, 2255, (1965)
9. PDX Group, Princeton Plasma Physics Laboratory, Phys. Rev. Lett. 50, 891 (1983).
10. White, R. B., R. J. Goldston, K. McGuire, A. H. Boozer, D. A. Monticello, and W. Park, Phys. Fluids 26, 2958 (1983)
11. Chen, L., R. B. White, and M. N. Rosenbluth, Phys. Rev. Lett. 52, 1122 (1984)
12. Coppi, B., and F. Porcelli, Phys. Rev. Lett. 57, 2272 (1986).
13. White, R. B. , F. Romanelli, M. N. Bussac, Phys Fluids B 4, 745 (1990)
14. Wesson J., Tokamaks (Clarendon, Oxford 1997, pp 581-603

PHYSICS OF AND ACHIEVEMENTS WITH THE H-MODE

F. Wagner

Max-Planck-Institut für Plasmaphysik
EURATOM Association
Wendelsteinstr. 1, 17491 Greifswald, Germany

Abstract. The H-mode is a confinement mode of toroidal plasmas which may make the goals of fusion possible – the development of a clean energy source at competitive electricity costs. The most challenging aspect in the understanding of the H-mode physics is the sudden disappearance of the edge turbulence whereas its potential driving forces increase. As the physics behind the H-mode is subtle many features are not yet clarified. There is, however, substantial experimental and theoretical evidence that turbulent flows, which normally limit the confinement, are diminished by sheared poloidal flow residing at the plasma edge. There are many potential mechanisms giving rise to sheared flow. The most intriguing of these is that fluctuations themselves induce the flow, which acts back to its generating origin and annihilates the turbulence. This review summarizes some of the studies on and achievements with the H-mode both for tokamaks and stellarators.

Keywords: confinement, H-mode, tokamaks, stellarators
PACS: 52.25.Fi, 52.55.Fa, 52.55.Hc

1. INTRODUCTION

The goals of magnetic confinement is to produce plasmas which fulfil the triple product $nT\tau_E > 5\ 10^{21}$keV m^{-3} sec with density $n \sim 2\ 10^{20}$ m^{-3}, temperature T~15-20 keV and energy confinement time $\tau_E \sim$ 1-2 sec. In separate experiments, $n \sim 10^{21}$ m^{-3} [1] and $T \sim 40$ keV [2] have been achieved. The confinement time has not yet been reached under relevant plasma conditions. The necessary triple product for ignition is missed by about a factor of 6. It is one of the goals of ITER to meet this target [3].

The energy confinement time is defined by the 0-dimensional energy balance equation: $\dot{W} = P - W/\tau_E$. W and P are energy content ($W = 3/2\ (<n_e k_B T_e> + <n_i k_B T_i>)Vol$) and heating power. Under steady-state conditions, the relation $W = P\tau_E$ holds; when the heating power is turned off, W decreases as $\dot{W} = -W/\tau_E$.

The most simple heating for tokamaks is ohmic heating from the applied voltage to drive the toroidal current. Because of the temperature dependence of the electrical conductivity – increasing with electron temperature – ohmic heating alone would not provide plasmas fulfilling the triple product. Therefore, ohmic heating has to be supplemented by auxiliary heating. Figure 1, left, shows schematically a case with a pulse of auxiliary heating on top of the background ohmic heating. The energy content rises during the pulse; when the additional power is turned off, W decays with the characteristic time τ_E.

With auxiliary heating, the plasma temperatures could be increased to the above mentioned values but it was observed that τ_E strongly degrades with heating power. The plasma energy content does not increase proportional to the heating power. Such regimes were dubbed L-mode plasmas with "L" denoting low confinement.

CP1013, *Turbulent Transport in Fusion Plasmas, First ITER International Summer School*
edited by S. Benkadda

It has been observed first at the ASDEX tokamak, that τ_E could spontaneously improve yielding higher energy content at constant heating power [4]. Such a case is again schematically shown in Fig. 1 right. The confinement improves at t_{tr} (tr stands for transition). This phase has been termed H-mode, with "H" denoting high confinement.

When the heating power is turned off, W decreases with the prevailing long confinement time. It is an empirical fact that amidst the decay phase the plasma can suddenly adopt L-mode confinement. Such a transition occurs at t_{btr} (btr stands for back-transition); henceforth, W decays at a faster rate.

The total power through a flux surface is given by: $P = -2\, S\, n\chi\nabla k_B T$ with S the surface under consideration and χ being the heat diffusivity. The relation between τ_E and $<\chi>$ shows that τ_E is a measure of the thermal insulation of the plasma and τ_E is finite because of the plasma losses from the source to the sink (plasma edge). A simple practical example is: $\chi \sim 1\text{m}^2/\text{sec}$, $a \sim 1\text{m}$, $\tau_E \sim 1$ sec. At constant power flux density, the improvement of confinement (reduction of χ) causes the temperature gradient to increase. In case of sudden and distinct gradient

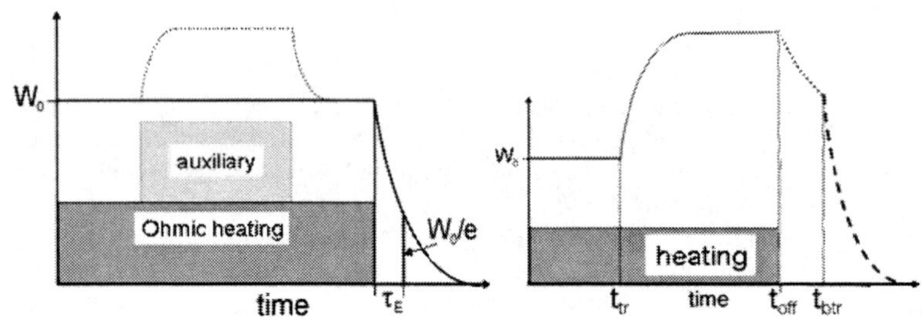

FIGURE 1. Schematic depiction of the variation of the energy content W from the ohmic level W_0 with an auxiliary heating pulse and when heating is turned off with the definition of τ_E (left); right: W in case of a confinement transition during the heating pulse (at t_{tr}) and a back-transition (at t_{btr}) after the pulsed has been switched off (at t_{off}).

steepening, a transport barrier (TB) develops. This can happen at the edge (external TB) or in the plasma core, respectively (internal TB, ITB [5]).

In this paper we will present experimental confinement and specifically H-mode results from both toroidal confinement systems, tokamaks and stellarators. These two lines share many characteristics, have the same target in terms of the triple product and require nearly identical technology. The properties of the magnetic geometry are, however, different in many aspects, which affect potentially confinement and the H-mode transition. Therefore, the comparison of tokamak and stellarator confinement serves to gain deeper understanding of the relation between confinement and magnetic characteristics specifically with respect to 2D- versus 3D- magnetic effects.

The tokamak produces the necessary poloidal field component with a toroidal current which flows inside the flux surfaces. The equilibrium geometry can be two dimensional with the toroidal angular coordinate being an ignorable one. The stellarator produces the poloidal field component by helical coils which embrace the toroidal flux surfaces. The helical coils can be physically realised like in the more traditional helical systems or the helical component can be embedded in the 3-D shapes of a set of modular coils (like in W7-AS [6], [7] or W7-X [8]). No equilibrium solution is possible with external confinement (B_θ produced by effective helical coils) in reduced dimensionality. Therefore, stellarator equilibria are three-dimensional.

The geometrical complexity of 3-D systems is nevertheless accepted because the absence of a large toroidal plasma current ensures better stability (no current driven instabilities, specifically no disruptions and neo-classical tearing modes) and allows – from the technical point of view – steady state operation. These are crucial requirements which demand for the tokamak an operational safety margin against disruptions and a conflict-free concept with integrated technology for non-inductive current drive. This task is not yet satisfactorily solved [9]. On the other hand, the classical stellarator does not fulfil the most essential requirement – sufficient confinement under the long-mean-free path conditions of a reactor. Therefore, stellarators need optimisation to remove this deficiency [10], [11].

The aspect ratio A of helical systems is generally larger than that of tokamaks reducing toroidal curvature ($A_{W7-AS} \sim 10$). In helical systems rotational transform is provided by the field component produced by the helical coils. It increases toward the plasma edge for Heliotrons/Torsatrons (CHS, LHD) and it is fairly constant for the Wendelstein stellarators (W7-AS, W7-X). Heliotrons have a strong but negative global shear and a magnetic hill; low-shear stellarators are MHD stabilised by a magnetic well. The magnetic hill is expected to allow resistive interchange activities. The flux surfaces are strongly shaped with elongation, triangularity and indentation but with large differences between helical confinement concepts causing rather individual neo-classical transport properties. Mod-B on flux surfaces can have a complex structure with local field maxima and minima giving rise to different classes of trapped particles affecting e.g. bootstrap current or trapped electron mode (TEM). The helical field ripple along the field line impairs toroidal flows – and such marginalises the $v_\phi B_\theta$-term in the radial force balance (see chapt. 3.9.).

Close to rational surfaces (5/m in W7-AS with 5 describing the toroidal and m describing the poloidal periodicity) large islands reside at the plasma edge and their inner separatrices define the plasma surface. The separatrix of W7-AS has a finite value of the rotational transform iota whereas in divertor tokamaks, $\iota \Rightarrow 0$ at the separatrix; the global magnetic shear at the edge is distinctively lower than for a typical tokamak by a factor of about 10. The field structure which causes the islands has a sensitive dependence on edge-iota rendering an edge poloidal viscous coefficient with strong variation on rotational transform. The magnetic field ripple in the classical designs is large in heliotrons and small in stellarators. This feature could contribute to the question whether ion orbit losses affect the ambipolarity conditions at the plasma edge.

As stellarators and tokamaks show the same confinement pattern – power degradation in the L-mode (see chapt. 2.3.) and H-mode transitions if specific prerequisite are fulfilled – it can be concluded that the toroidal current, the large edge-q and strong global magnetic shear of divertor tokamaks, magnetic hill or well may play, to first order, a minor role in the transition physics.

2. SOME COMMENTS TO TRANSPORT AND CONFINEMENT

2.1. Collisional and turbulent transport

L- and H-modes differ primarily in confinement which is limited by neo-classical and turbulent processes. The electric field plays a crucial role for both loss processes. The neo-classical losses can be critical for stellarators and the electric field can reduce the transport rates. Neo-classical losses can determine the ambi-polar electric field and via its gradient reduce turbulent fluxes (see chapt. 3.8.). The E-field gradient - the shear in perpendicular plasma rotation - seem to be crucial for the H-mode transition. In this way a property of collisional transport – the electric field – restores the transport – caused by turbulence - closer to the neo-classical limit.

In order to prepare the reader for the main topic of this report, I would like to first discuss the diffusivities for the two concepts which play a role here – tokamak and stellarator – and for various regimes along the simple random walk picture: The diffusivity is determined by Δx an effective step size of a scattering process and Δt, the time between two scattering processes.

Parallel to the magnetic field – the trivial case - Δx is given by the Coulomb mean free path of the particles and Δt is given by the Coulomb collision time τ_C. For fluxes perpendicular to the magnetic field, the step size is the particle Larmor radius ρ_L and again the Coulomb collision time for classical diffusion, e.g in a cylindrical magnetized plasma. In toroidal systems, the resulting field inhomogeneity affects the particle orbits in a way different for two and three dimensions, respectively. The resulting neo-classical transport is caused by the combined effects of collisions and orbits in toroidal geometry. In the axi-symmetric case, orbits of collisionless particles are confined both for free and trapped particles; the average radial drift velocity is zero. The maximal drift displacement is largest for those particles which have a low parallel velocity. These particles are deeply trapped banana particles which reside at the low field side of the plasma. With collisions, they contribute most to the radial fluxes.

In the non-axisymmetric case, particles trapped between the effective helical ripple have finite radial drifts caused by the toroidal field inhomogeneity. These collisionless orbit losses can dominate the radial fluxes.

In the fluid-like Pfirsch-Schlüter regime and the plateau regime, the generic behaviour of tokamaks and stellarators is similar though, in stellarators, the toroidal and the helical field variation plays a role. The differences are large, however, in the long-mean-free path (LMFP) regime – the collisionality regime of the reactor.

In the axi-symmetric case, the step-size Δx corresponds to the banana-width of deeply trapped particles ($\varepsilon\, \rho_{L,pol}$: $\varepsilon = 1/A$), the effective collision time is given by the particles crossing in phase space the boundary between free and trapped particles ($\varepsilon\tau_C$). The fraction of the transport relevant particles (trapped particles) has to be considered ($\varepsilon^{1/2}$). In 2-D systems, collisions in the LMFP regime lead to a radial flux: $\Gamma \sim \nu$ (ν is the collision frequency).

In stellarators, the relevant particles are those which are trapped in phase space by the field inhomogeneity of the helical field (in multi-helicity stellarators, the complete non-axisymmetric field distribution has to be considered and different classes of ripple trapped particles come into play). These particles can rapidly drift out of the confining volume. Therefore, the exact relation, $\Gamma = \Gamma(\nu)$ (and the transition into the long-mean-free-path regime), depends on the details of the magnetic configuration and is different for different helical concepts. The flux of these particles, which dominates over all other collisionally induced radial fluxes is $\Gamma \sim \varepsilon_t^2 \varepsilon_h^{3/2} v_D^2 \times 1/\nu$ (ε_t, ε_h are toroidal and helical ripple, v_D is the drift velocity). In the so-called $1/\nu$-regime, collisions provide the chance for a particle to leave the loss-cone phase space in favour of a pitch-angle yielding better confinement; therefore: $\Gamma \sim 1/\nu$. The $1/\nu$ regime of helical systems would lead to unacceptably high neo-classical losses for stellarator reactor parameters. Therefore, there is the need of optimisation, which is based on the reduction of either trapped particle fraction or of the radial drift velocity. The drift velocity is reduced by reducing the average toroidal curvature by elongation (the toroidal circumference remains, the field line length increases) and by reducing the field variation along the helical field minimum located between the coils (σ-optimisation [10]). These are measures in the frame of quasi-isodynamic systems. The trapped particle fraction subject to strong drifts is reduced by the introduction of quasi-symmetry [11] or by the linked-mirror concept (part of W7-X optimisation). An important aspect is that He-J, an optimised Heliotron following quasi-isodynamicity as optimisation principle, does reach the H-mode [64].

For completeness, we briefly comment on turbulent transport but we refer the reader to the contributions on turbulent processes and transport in this volume. The order of magnitude assessment of anomalous transport coefficients is more involved and depends on the model considered. In the most simple approach, the random-walk approach $D \sim \gamma / k_\perp^2$. γ is the growth-rate, which is a fraction f_1 of the diamagnetic frequency $\omega^* = m/r\ T_e/eBL_n$, and $k_\perp$ is the perpendicular wave number which is a fraction f_2 of the inverse Larmor radius. $k_\perp^{-1}$ and γ^{-1} can be interpreted as step size and step time.

The fluctuations residing at the plasma edge close to L-H transitions conditions are resistive ballooning and ion-temperature-gradient driven turbulence. This has been shown for ASDEX Upgrade in [12] by the analysis of the radial correlation length scaling – in one case with the cell size, in the other one with the ion gyro-radius.

2.2. The radial electric field

The self-consistent radial electric field plays a crucial role both to reduce the $1/\nu$ stellarator flows in the LMFP regime and in the physics of the H-mode. In tokamaks a low order drift kinetic expansion in the distribution function is sufficient to determine the neo-classical particle and heat fluxes. With "intrinsic" ambipolarity of the neo-classical fluxes ($\langle j_r \rangle = 0$ independent of E_r owing to angular momentum conservation with axi-symmetry), a higher order expansion is necessary to relate the parallel fluid flow and E_r to the plasma gradients. In 3-D systems, E_r is enforced by $\Gamma_e = \Gamma_i$. The neo-classical fluxes are connected to the driving forces in the following form:

$$\Gamma/n = -D_{11}(E_r)(n'/n - qE_r/T) + D_{12}(E_r)\ T'/T) \qquad (1)$$

Such a relation is given for each species. Γ depends explicitly on E_r and on on- and off-diagonal terms, with transport coefficients (D_{11}, D_{12}) which themselves depend on E_r. As a consequence, bifurcating solutions for E_r are a known feature of stellarator transport (an experimental overview is given in [13]).

There is vast experimental evidence in helical systems that the actual radial electric field is generally determined by the balance of neo-classical fluxes even in zones where the radial particle fluxes are caused by turbulent processes [14].

In tokamaks, the major emphasis of confinement improvement is directed to reduce turbulent fluxes utilising their dependence on the E_r-field gradient (see chapt. 3.7.). The emphasis of helical systems is first to improve the neo-classical fluxes utilising the E_r-field itself. Independent of the sign, the electric field affects strongly neo-classical particle and energy fluxes (in the LMFP regime). There is not a singular value of χ_e and χ_i. A given value of $|E_r|$ has a stronger impact on ion than on electron transport coefficients.

2.3. L-mode confinement and improved confinement regimes

Auxiliary heating, used in larger scale from the mid-70ies of the last century on, brought new plasma temperature records but also the sobering fact that confinement degrades with heating power. The consequence was that auxiliary heating did not allow an effective approach to the ignition conditions. Rather, the operational ranges of auxiliary heated plasmas mapped out an arc in $n\tau_E$ versus T space more or less in parallel to the ignition curve.

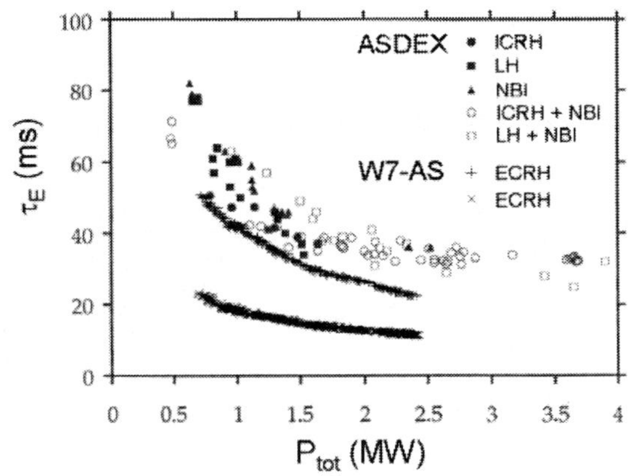

FIGURE 2. L-mode confinement for energy and particles in ASDEX at various heating scenarios and τ_E of W7-AS with ECR heating. The W7-AS τ_E data are scaled by $Vol_{ASDEX}/Vol_{W7\text{-}AS}$ (= 5.2) to allow better comparison. Results for two densities are shown: + $n_e = 2\ 10^{19}\ m^{-3}$; × $n_e = 4\ 10^{19}\ m^{-3}$.

Figure 2 shows for ASDEX the degradation of τ_E with different heating methods; for W7-AS the same behaviour is observed with ECRH. The degradation is robust, independent of concept or heating method. Also the particle confinement degrades (see Fig. 2). The degradation, the increase of the corresponding diffusivities, applied also to impurity and momentum transport. The confinement degradation with heating power is a reproducible feature, universally observed, and seemed to be an inescapable fate [15], [79]. The consequences for the future of magnetic confinement seemed to be rather bleak. The compensation of the degradation necessitates an increase in size and plasma current. Reactors operating in the L-mode needed currents around 40 MA.

Because of the importance to develop operational regimes with good confinement a large fraction of tokamak and stellarator research is devoted to the creation and understanding of improved confinement regimes [16]. Many differently dubbed improved confinement regimes exist in tokamaks: B-mode, H-mode (and variants), Z-mode, supershots, RI-mode, IOC, pellet deep-fuelled plasmas, or plasmas with internal transport barriers (ITBs)... Also in stellarators, different confinement regimes exist – the H-mode and its variants, discharges with ITBs owing to the core electron root, plasmas with peaked density profiles, improvements when the gas-flux has been reduced (so called re-heat phase), plasmas where the confinement improvement develops over many confinement times [7]. Some of these regimes are characterized by peaked density profiles requiring subtle optimisation of the edge conditions, others by the development of edge or internal transport barriers, in case of helical system as a consequence of the electron root.

3. H-MODE CONFINEMENT

Figure 3 compares global parameters of an L-mode discharge with a subsequent H-mode plasmas of identical external settings of ASDEX [24]. The figure shows from top-left to bottom-right the line average density along two chords, plasma current and loop voltage, the horizontal plasma position, H_α-radiation in the divertor, global impurity radiation, $\beta_\theta+l_i/2$ from equilibrium, H_α in the torus midplane, iron impurity line radiation and β_θ from the diamagnetic coil.

The following comments have to be made:

(1) The identification of the critical transition parameter(s) seems to be rather difficult. No resolvable difference could be noted right at the bifurcation point (arrow) of the discharge with transition to the one, which remains at the L-mode branch.

(2) The drop in H_α-radiation in the divertor indicates the H-transition.

(3) The loop-voltage decreases because of the increase in electrical conductivity.

(4) The density first decreases in spite of the additional fuelling by neutral beams caused by the initial L-mode confinement. In the H-mode case, the improvement of particle confinement causes the density to strongly rise beyond the ohmic level.

(5) The plasma horizontal position reflects the increase of plasma beta (thermal and non-thermal; at the H-transition, the β-increase causes another expansion of the major radius R_0; ELMs cause the horizontal plasma position to fluctuate.

(6) The confinement improvement applies to impurities also; their radiation increases.

(7) The improvement in en

The following papers are H-m 9], [20].

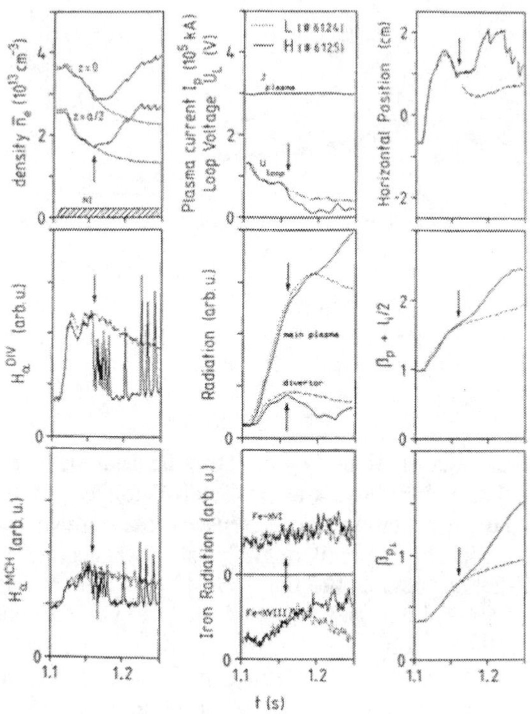

FIGURE 3. L-mode and H-mode discharges. The transition is indicated by vertical arrows. Plotted is from top left to bottom right: line-average density, current and loop voltage, plasma position, H_α in the divertor chamber, global radiation in main and divertor chamber, equilibrium signal, H_α in the main chamber, iron lines, beta from diamagnetic signal.

3.1. H-mode transition characteristics

The following is an overview over the most prominent H-mode characteristics with a brief comment on the possible implications:

(1) L- and H-modes differ in confinement time by about a factor of two: two operational branches exist for a given setting; the space in between is not accessible.

(2) Energy, particle, impurity and momentum transport improve simultaneously.

(3) The H-mode transition requires a minimal heating power: Obviously, a critical condition, the power threshold P_{th}, has to be met (see chapt. 3.3.).

(4) There is a dwell time after the heating power has been increased from the ohmic level (in case of tokamaks), before the plasma transits into the H-phase. The dwell time depends on external parameters and is shorter at higher power. It also depends strongly on the magnetic configuration and can be shorter under single- instead of double-null divertor configuration (see chapt. 3.10.).

(5) When the heating power has been switched off, the plasma remains in the H-phase for a dwell time in the order of the confinement time: After the forward L-H transition, the plasma does not hover at the transition condition but moves deeper into the H-mode domain; when the external drive is turned off, this reserve allows the plasma to initially remain in the H-phase. Of significance is that the back transition is not gradual but also occurs in a distinct step – corresponding to the gap between the H- and L-mode branches.

(6) There is also a density threshold; at low density, the critical transition condition cannot be met. Machine characteristics like radiation and mode-locking affects the low-density power threshold [21].

(7) In tokamaks, large sawteeth were found to trigger the H-mode: the critical transition parameter seems to be a local condition at the plasma edge, which can be met by a thermal wave from the core plasma.

(8) ELMs appeared in the H-mode as a new type of edge localized MHD instability [22], [30]; there are operational ways to suppress ELMs. In the quiescent H*-mode the intrinsic confinement of the H-mode can be studied. This regime is, however, normally transient because of continuous increase of impurity concentration. Because of the detrimental plasma-wall effects of ELMs, there is an intense study to avoid ELMs in H-modes and to develop operational regimes without ELMs but avoiding also impurity accumulation (see chapt. 3.5.).

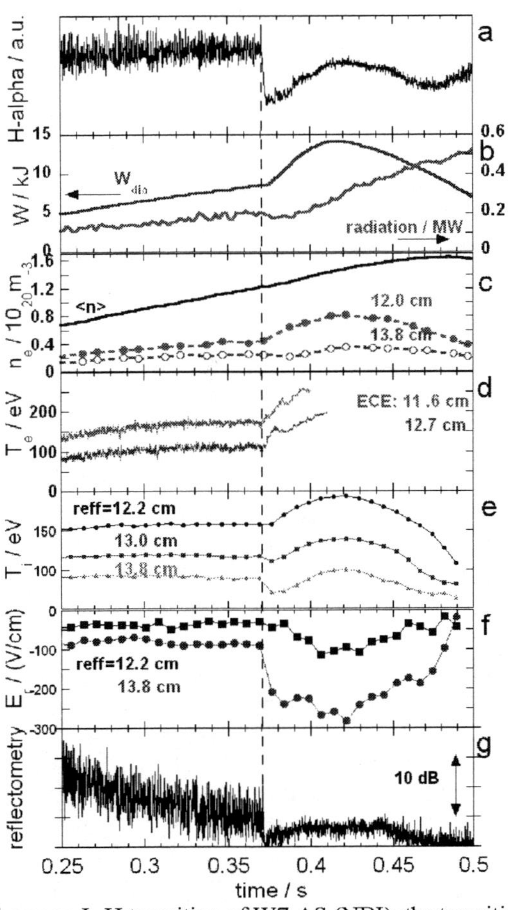

FIGURE 4. This tableau shows an L-H transition of W7-AS (NBI); the transition is indicated by the vertical line. a: H_α-radiation with the characteristic transition mark; b: Energy content W_{dia} and global radiation; c: line average density and two edge density traces; d: edge T_e traces; the lines are stopped at the cut-off; e: edge T_i traces; f: two edge traces showing the radial electric field E_r; g: scattered power from edge reflectometry.

3.2. The stellarator H-mode

The variation of global parameters in stellarator H-mode transitions is nearly identical to the one seen in tokamaks. Figure 4 includes also the variation of edge plasma parameters at the L-H transition from W7-AS stellarator [13] which again are rather similar to those observed in tokamak H-mode transitions. Figure 4a shows the H_α-trace with the distinct transition into an ELM-free phase; Figure 4b, the energy content (W_{dia}) and the global radiation; Figure 4c, the line average density and two density traces at the plasma edge (the separatrix is at $r_{eff} = 14$ cm); Figure 4d, two edge T_e-edge traces (ECE; the traces terminate at cut-off), Figure 4e, three edge T_i traces; Figure 4f, the radial electric field variation at the edge, and Fig. 4g shows logarithmically the scattered power from reflectometry P_n ($k_\theta \sim 6\text{-}10$ cm^{-1}; $n_{cutoff} = 2\ 10^{19}$ m^{-3}) indicating the level of plasma turbulence. The H-mode transition has a strong impact on all plasma parameters – on global ones like the energy and particle content and the plasma radiation, and – more distinct – on the local edge parameters. All parameters establishing the edge pressure gradient increase. Of specific importance is the sharp decrease of the fluctuation activity simultaneously with that of the edge electric field toward more negative values.

Also in W7-AS the plasma remains in the H-phase for a short period after the heating pulse has been switched off. The post-beam-pulse (PBP) dwell-time is of the order of the confinement time. With $\tau_E^H/\tau_E^L \sim 2$, $\Delta t_{H\text{-}L} \sim -\ \tau_E \ln 0.5$. Both for ASDEX and W7-AS, $\Delta t_{H\text{-}L} / \tau_E \sim 0.57$ (ASDEX: $\tau_E = 67$ ms; $\Delta t_{H\text{-}L} = 38$ ms. W7-AS: $\tau_E = 28$ ms; $\Delta t_{H\text{-}L} = 16$ ms).

3.3. The power threshold of the H-mode transition

The H-transition requires a power threshold generally larger than the available ohmic power. High-current tokamaks, however, reach the H-mode also with ohmic power. The power threshold was found to be lower for the following cases:

(1) with separatrix instead of limiter operation;

(2) in deuterium instead of hydrogen plasmas;

(3) in clean (with boronisation) instead of dirty plasmas;

(4) with gas fuelling from the divertor chamber or the high-field side instead of the low-field side;

(5) in single null (SN) plasmas with the ion-grad B-drift to (SN^+) instead of away (SN^-) from the active X-point; in double-null (DN) plasmas, P_{th} was found to lie in between [23]: Obviously, an unknown mechanism (see chapt. 3.10.) whose impact is formally expressed as ion-grad-B ($B\times\nabla B$) drift direction in relation to the X-point location, contributes with a supporting (or prohibiting, respectively) effect, which was nullified in ASDEX in the symmetric DN case. At a heating power, sufficient to get the H-mode under DN conditions, the dwell-time prior to the transition is shortened by about a factor of two if operated under SN^+ conditions (see Fig. 101 of Ref. [24]). Under SN^- conditions, the dwell-time could be longer than the NBI pulse.

(6) The power threshold depends on many external parameters (e.g. SN, DN operation) and on hidden ones difficult to quantify (e.g. wall conditions). Therefore, a precise scaling of P_{th} with plasma parameters is not possible. The following scaling relation is obtained [25]:

$P_{th} = 0.054\ n^{0.49}\ B^{0.85}\ S^{0.84}$

With P in MW, B in T, the surface S in m^2 and the density in units of 10^{20} m^{-3}. The necessary power for ITER to enter into the H-mode is expected to be about 50 MW.

(7) Stellarators: Major differences to the tokamak H-modes seem to be the low power threshold (at least in W7-AS) and the limitation of the H*-mode to specific iota windows (see chapter 3.10.). P_{th} can be as low as the technical power limits of one source for NBI or one gyrotron for ECRH, respectively – each about 0.4 MW. From the tokamak power threshold scaling [25], more than twice the power is expected.

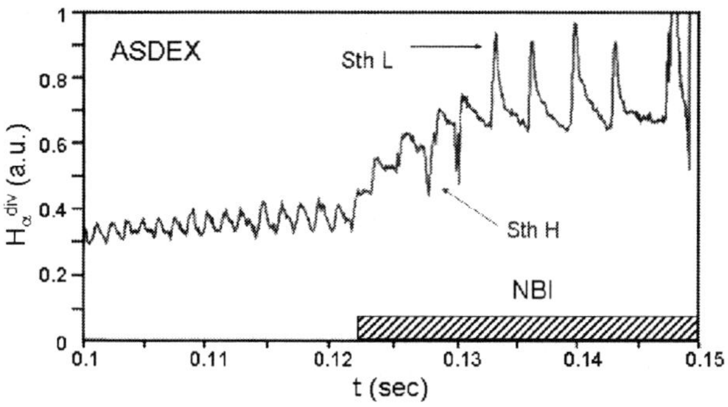

FIGURE 5. H_α of an L-mode discharges (from OH into NBI heating). The trace is modulated by sawteeth in the ohmic phase. The sawtooth amplitude increases in the auxiliary heating phase (Sth L) apart from the two pulses which induce a short H-mode (Sth H).

3.4 Development of an edge transport barrier

Direct evidence that the critical H-mode transition condition is local and resides at the plasma edge came from the observation of sawtooth-triggered H-phases. Figure 5 shows the H_α-trace of an L-mode discharges with P just below P_{th} displaying the signal variation with sawteeth in the L-phase (transient signal increase) and when short H-phases were triggered (signal decreases). The H-phases initiate an H_α-signal drop.

The sawtooth triggered H-mode helped to understand one important aspect - that the improved confinement is

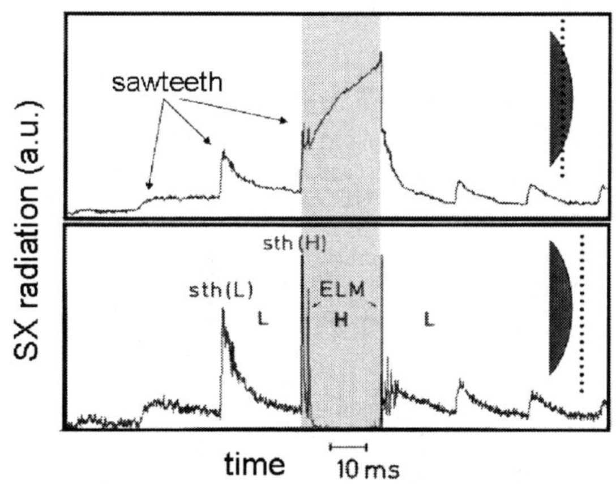

FIGURE 6. Transition from OH to NBI heating. Shown are two Soft-X-ray traces one inside the separatrix, one in the SOL (see schematics right; the dashed line represents the viewing chord).
The two signals are modulated by sawteeth with initially rising amplitude. The 4th sawtooth causes an H-transition developing a transport barrier at the edge. The short H-phase displays an ELM.

caused by the development of a transport barrier right at the edge but inside the separatrix [26]. Figure 6 shows again a sequence of sawteeth at the transition from ohmic to NBI heating. The upper trace viewed the plasma right inside the separatrix, the lower one predominantly the SOL plasma (see geometry on the right side). Auxiliary heating caused the sawtooth amplitude to initially increase. The need to overcome the power threshold translates into a critical sawtooth amplitude and when this is met (with the sawtooth labelled sth(H)), a short H-phase develops. Whereas the sawteeth in the preceding and succeeding L-phases cross the separatrix and also modulate the SX radiation emerging from the SOL, the sawtooth, which triggers a short H-phase, causes the SX signal from inside the separatrix to steadily increase at the expense of the signal from the SOL, which collapses: An edge transport barrier has been developed.

The gradual development of the edge profiles – that of the density is shown in Fig. 7 [27] - allowed to assess the reduction in heat diffusivities within the transport barrier (for the ions approximately to the level of neo-classical heat diffusivity, for the electrons still above it). With far-infrared laser scattering, it could be shown with ASDEX that the level of turbulence sharply dropped in the radial zone where the transport barrier develops [28].

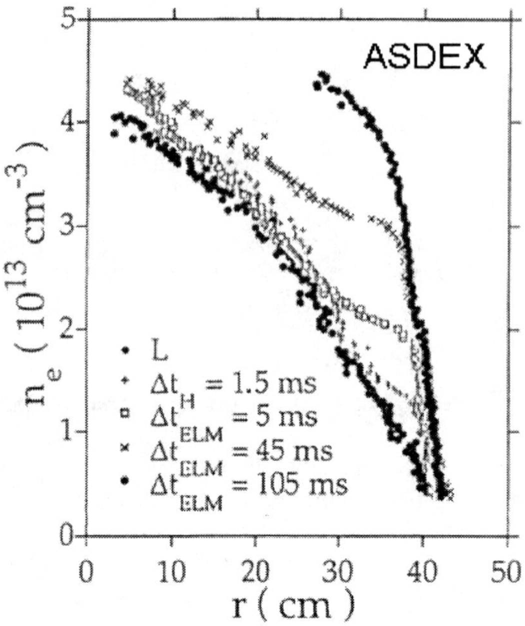

FIGURE 7. Development of the density profile after an H-transition.

Figure 8 shows the collapse of the turbulent particle flux density Γ in DIII-D as consequence of the H-transition [29]. Γ was deduced from edge density and potential fluctuations using Langmuir probes. To safely employ this tool, the study had been carried out in H-transitions of low-power ohmic discharges. The results demonstrate that the confinement improvement of the H-mode is indeed caused by a reduction of turbulent fluxes.

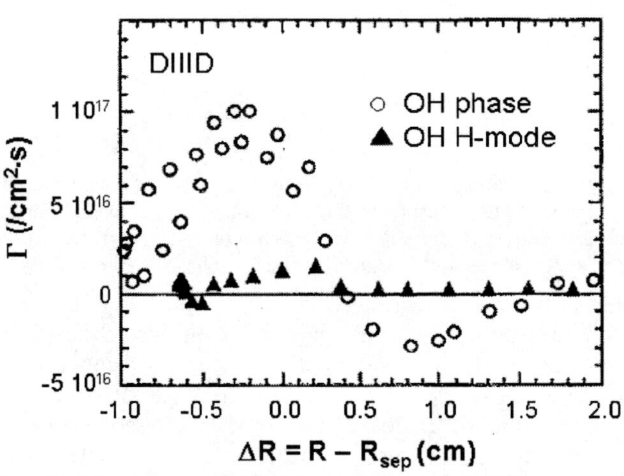

FIGURE 8. Turbulent particle flux Γ at the plasma edge for ohmically heated discharges – one with an H-mode. Γ is measured around the separatrix. The data are from DIIID [29]. Copyright: AIP.

3.5. ELMs

The evolution of the edge pressure profile in the H-mode is limited by striking an MHD instability condition. As a consequence, so-called edge localized modes (ELMs) [22], [30] appear. All confinement properties (energy, particle, impurities, and momentum) are affected. ELMs give rise to a repetitive ejection of energy pulses with $\Delta E/E \sim \Delta N/N \sim 5\%$. (N is the particle content). Under controlled conditions the relaxation process can be fairly constant. Periodically, the plasma hits the stability boundary. At low power, the process seems to be resistive, at high power more ideal. Resistive ELMs often show a growing precursor oscillation. The ELM-phase is of 200μs duration. Edge and scrape-off-layer plasma are highly turbulent in this phase.

W7-AS does show ELMs whose activity increases with heating power. There is no net induced toroidal current of substantial value to drive ELMs. The edge bootstrap current – also a potential drive for ELMs - is low owing to the high edge collisionality as the H-mode of W7-AS is generally observed at elevated densities.

The technical hazards of ELMs are the loading of the divertor plates. In JET, 1 MJ, 10 GW pulses strike the plates. In ITER, up to 20 MJ pulses are expected. The resulting wall erosion poses a major problem.

A quiescent H-mode without ELMs is possible. It has superior confinement, is hampered, however, by rising impurity radiation also from the plasma core. Nevertheless, plasmas with the highest confinement and best performance can be achieved transiently. ELMs contribute, however, also with a positive feature because they purify the plasma, providing constant impurity concentrations and allowing for steady-state operation. There seems to be a milder instability process than ELMs possible at the edge, which also limits the impurity concentration to a tolerable level but avoids the detrimental wall interaction of ELMs. These are quasi-coherent modes, which have been observed in Alcator C-mode [31] (in the EDA phase), or as edge-harmonic oscillations in DIII-D [32] (in the QH-mode), respectively, and recently also in the HDH-mode [33] of W7-AS [34] (see chapt. 3.12.). Such edge conditions may combine important features – good confinement, steady-state capability and benign plasma-wall interaction.

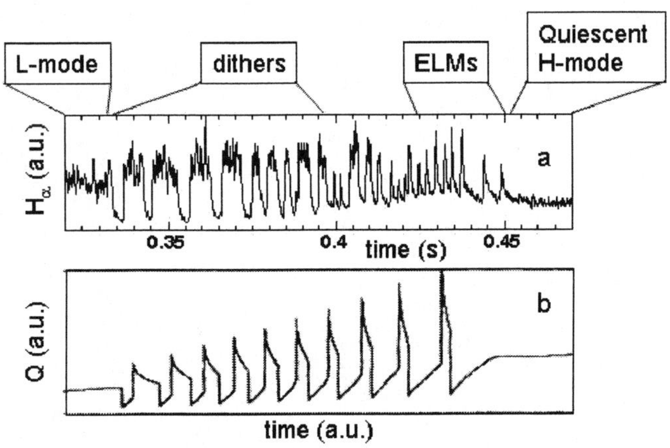

FIGURE 9. a: dithering cycle of a slow L-H transition shown by H_α radiation;
b: Model of a dithering transition taken from [35]. Copyright: APS.

3.6. The H-mode transition as a bifurcation

The question of the true nature of the transitions (L-H and H-L) actually arises because normally, the transitions are enforced by large jumps in auxiliary heating power. In W7-AS, the H-mode is generally initiated at constant heating power by slowly increasing the density (see Fig. 4). In this case the evolution remains close to transport equilibria. All these H-modes evolve more out of internal adjustment processes. Nevertheless, the formation or destruction of the transport barrier occurs in a distinct step (see Fig. 14 in case of a back-transition).

The two density traces in Fig. 3 (top, left) already indicate the nature of the H-mode transition as a bifurcation. In the H-mode case the density raises though the external gas flow was terminated; in the L-mode case the density decreased though the external gas flow was strongly increased by the gas feed-back system. The space between these density branches is not accessible at the given external parameter setting. Also, there is a hysteresis in P_{th}. The H-mode, when established, allows a reduction of the heating power by a factor of two below P_{th} [25]. One asymmetry between the L-H and the H-L back transition is that they occur from the L-mode or the H-mode confinement level, respectively. The hysteresis can therefore be explained by the confinement differences and would indicate again that the actual critical transition parameter is local and depends on $P\tau_E$.

Often, the H-transition evolves in a sequence of limit-cycle oscillations. This specific phenomenon of bifurcating states is well documented in tokamaks [35], [36]; therefore, Fig. 9a shows a case from W7-AS. The transition is characterized by dithers, which gradually transmute to ELMs. Also in this case, the plasma jumps between the two states – the L- and the H-mode branch. It is an interesting question, which of the transitions is the last dither (a bifurcation phenomenon), which the first ELM (MHD instability). The interspersed L-phases of the dithering cycle are characterised by strong, intermittent, ELM-like turbulence.

The H_α trace of Fig. 9a is reminiscent of the modelled temporal evolution of the power flux across the separatrix [35], shown in Fig. 9b. The model is based on the conception, that the transition initiates two competing processes – one which would stabilise the H-mode, the other, which would lead back to the L-mode. As the H-mode improves energy and particle confinement, one can construct a corresponding model based on τ_E, stabilizing the H-mode and on the consequences of an increase of τ_p, leading toward the L-mode.

3.7. The development of an edge E_r-field well

The potential role of the edge radial electric field E_r in the evolution of the transport barrier was first introduced by S.-I. and K. Itoh [37] actually before the first experimental observations were available. They recognized the H-transition as a bifurcation and constructed E_r-based bifurcation models to describe L- and H-mode branches. They conjectured that the losses of those ions at the edge which satisfy the toroidal resonance are not completely balanced by turbulent electron fluxes because of wave momentum losses across the plasma boundary.

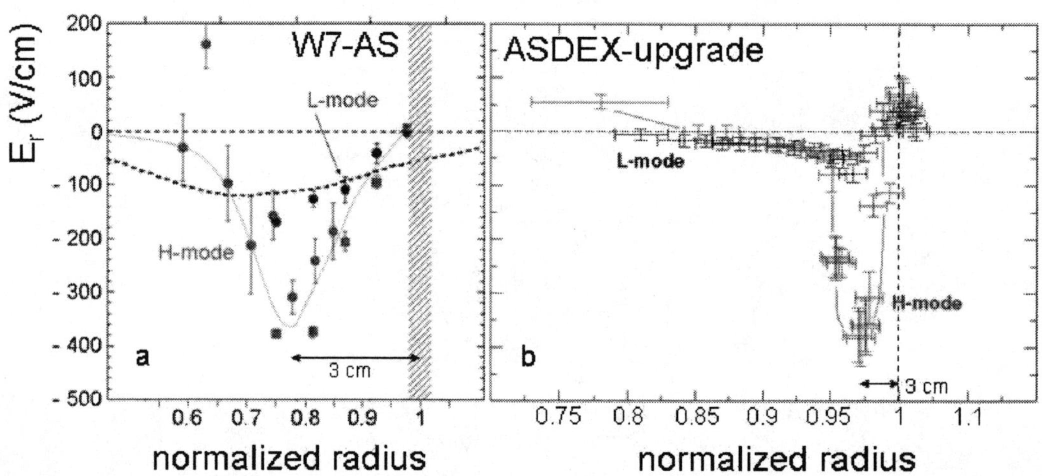

FIGURE 10. Radial electric field at the plasma edge for W7-AS [42] and ASDEX Upgrade [43] for L- and H-modes. The radial extent of E_r to its minimum is indicated. For W7-AS the ion pressure gradient is shown as dashed curve.

E_r at the edge was first measured by DIII-D [38], [39]. Figure 10 shows the development of an E_r-field well right at the edge and right after the transition both for W7-AS (Fig. 10a) [40], [41], [42] and ASDEX Upgrade (Fig. 10b) [43]. In the radial range of the well the fluctuations are quenched, the turbulent transport decreases, and steep gradients develop. Three aspects are of importance: (1) The formation of an E_r-field well right inside the separatrix, associated with strong flow shear, is a generic feature of the H-mode. (2) The spatial range of the E_r-field is in both cases about 3 cm and seems to be governed by local conditions. The overall constancy of the layer width affects models, which strongly involve B_θ. Also the radial extent of the magnetic shear does not seem to play a role. (3) E_r in the L-phase of tokamaks is low and rather flat [43]; in W7-AS there is already a substantial negative E_r field in the pre-phase, which is mostly caused by the diamagnetic contribution (dashed curve) and can be well explained by the balance of neo-classical fluxes (ion-root branch) including the ripple-trapped losses and the off-diagonal terms [42]. In addition, there is no toroidal rotation at the edge of W7-AS owing to the finite toroidal viscosity. This neo-classical pre-conditioning might be the reason for the low power threshold in W7-AS.

3.8. Turbulence decorrelation by sheared E×B flow.

The potential mechanism for the quench of this type of turbulence was developed by H. Biglari, P. Diamond and P. Terry [44] - the decorrelation of turbulent eddies by radially sheared E×B flow at the plasma edge. Such a mechanism was already proposed years ago by B. Lehnert [45]. Substantial evidence has been accumulated that turbulence decorrelation – the reduction of the radial correlation length and the correlation time and the decorrelation of density and potential fluctuations by the sheared flow of the turbulence - is indeed the decisive mechanism, which leads to a strongly reduced turbulence level and the formation of a transport barrier.

At the TEXT tokamak, the BDT-mechanism was confirmed experimentally, the radial and poloidal correlation times were reduced, the turbulence level decreased and the density gradient increased [46]. Figure 11 shows results from DIII-D.

Though it is obvious that the E_r-field is strongly affected by the H-transition we have not yet discussed whether the E_r -field is cause for or consequence of the improved confinement. The use of polarisation probes, immersed into the plasma periphery, allowed to make the radial electric field profile at the edge to a control parameter and to study the causality [47]. H-mode transitions could be triggered; the most relevant results were contributed by TEXTOR [48] and the causal relation of sheared flow and turbulence level could be demonstrated in this device [49]. The polarisation probe could initiate a drop in E_r, the development of a particle transport barrier and a modest increase in τ_E. These studies clarified the causality: E_r (or E_r') leads the chain of action.

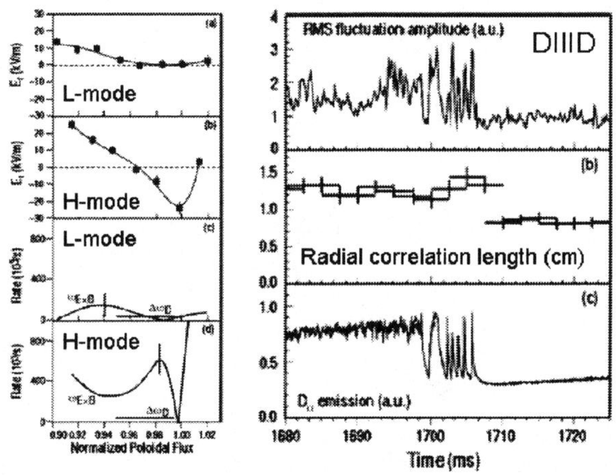

FIGURE 11. left: E_r-profiles and decorrelation rate ω_{ExB} and compared with the growthrate $\Delta\omega_D$; right: fluctuation level, radial correlation lenght and D_α emission. Data are taken from DIII-D [51] and [51, 52]. Copyright: APS and AIP.

The critical condition, $\nabla|E_r|$, was found in TEXTOR in the range 50-80 V/cm^2 [50]; on DIII-D ~ 50-100 V/cm^2 [51] and on W7-AS ~ 90 V/cm^2 [42] were observed for spontaneous transitions. $\nabla|E_r|$ can be translated into the critical transition condition – the shearing rate $\omega_{E\times B} > \gamma_{lin}$ (equation 1), γ_{lin} being the linear rise time of the turbulence. Figure 11 left shows E_r-profiles and the two critical parameters, $\omega_{E\times B}$ and $\Delta\omega_D$, the de-correlation frequency, for the L- and the H-phases of DIII-D [51]; $\Delta\omega_D$ corresponds to γ_{lin}. $\omega_{E\times B} >> \gamma_{lin}$ is observed in the H-phase. Also, as predicted by theory, the reduction of the radial correlation length of the edge turbulence within the zone of sheared flow has been demonstrated by DIII-D [51], [52] (see Fig. 11, right).

3.9. The origin of the radial electric field

- The ∇p_i - term

The steady-state radial electric field has in its most simple formulation two constituents - the diamagnetic contribution and the **v** × **B** components:

$$E_r = 1/Zen_e \nabla p_i - v_\theta B_\phi + v_\phi B_\theta \qquad \text{equ. (2).}$$

∇p_i is determined by the power balance involving heat and particle sources and sinks and – prior to the transition - L-mode transport. There is overall agreement that this component plays indeed a significant role after the H-transition: The ∇p_i– term develops within a confinement time, deepens the E_r-well at the edge and stabilises the H-mode turbulence decorrelation condition [43], [51], [53]. In case of ASDEX Upgrade, the H-mode E-field in a developed phase can be explained exclusively by the pressure gradient [43].

Right at the H-mode transition, the situation is different. The involved time scales are much shorter than those of transport. Polarisation experiments have shown that E_r is the lead term and ∇p_i is found to lag behind. The conclusion from this chain of action is that E_r (or E_r', respectively) is obviously the cause of the transition, which consecutively leads to the increase in ∇p_i.

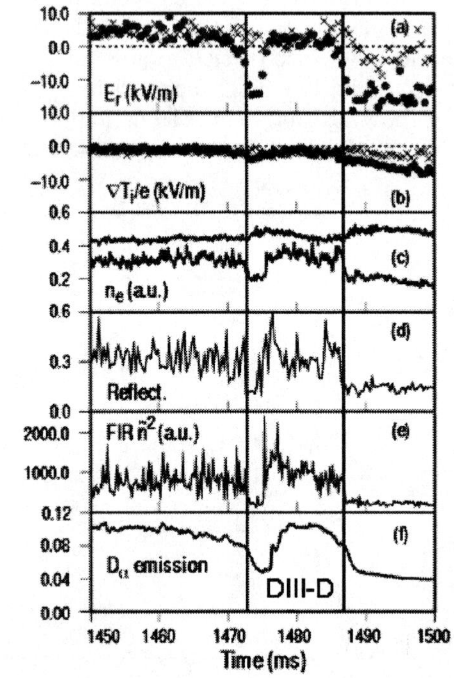

FIGURE 12. Tableau of a DIIID H-mode transition with good time resolution. The H-transition has a preceding dither. The first three traces are measured at different edge radii. a: edge radial electric field; b: edge T_i gradient; c: edge n_e; d and e: scattering signals to denote the turbulence level; f: D_α. Of importance is the slow pre-phase in E_r prior to and the rapid decrease of E_r at the transitions (in comparison of e.g. ∇T_i) [29]. Copyright: AIP.

- Time scales of transition and transport barrier development

The H-transition, induced by auxiliary heating, has a pre-phase which starts as soon as the heating power has been increased. This phase of formation is shorter in case of higher heating power, in deuterium and with proper ion-grad-B drift. In W7-AS, the pre-phase starts at constant heating power when the density rise has been completed, the plateau is reached and the plasma re-adjusts after the gas-input has been trimmed to the proper level (see Fig. 4). All these times scales allow the plasma gradients to change. Also within the H-mode, when the energy content gradually increases and the edge gradients along with it, the development is governed by the confinement time and E_r and ∇p_i develop rather jointly. The actual transition into the H-mode is governed by a much shorter time scale. Values as short as 12 μs have been measured in JFT-2M for sawtooth initiated transitions [54].

DIII-D closely monitored the H-mode transition with diagnostics with good time resolution. Such a transition is shown in Fig. 12 [29]. The data show first an L-mode, which transits into a short H-mode, continues with a short L-phase and makes a final H-mode transition. The significance of the plot is the rapid and substantial variation of the electric field in comparison to the slow variation of the ion temperature gradient and the little variation of the density (representing together the ion pressure gradient). The consequence of these studies is that the ion pressure gradient, which deepens the electric field after the transition and stabilises the H-mode and which is the reservoir for the delayed back-transition, does not primarily trigger the H-mode transition. Another mechanism, operating along a short time scale must become effective.

Also the back-transition is governed by a short time scale though it occurs without external drive in a sequence of transport equilibrium phases. The magnetic fluctuations increase at the transition in a single step within ~ 10 μs (Fig. 15). The increase of the turbulence level (induced from the scattered μ-wave power P_n) can be clearly identified within 200 μs (at a diagnostic time resolution of 4 μs). A better identification of the transition time point is not possible because P_n is modulated as is the perpendicular flow $v_\perp$ (Fig. 14).

At the back transition, E_r jumps from -240 V/cm to -160 V/cm in case of a W7-AS back-transition. The jump occurs within < 1 ms. The short transition times of O(10 μs) are not the time scales for changes of gradients or related quantities like neo-classical fluxes. It could, however, be the time scale of edge ion collisions, the adjustment of a (non-neo-classical) ambipolar electric field or that of turbulent eddies convected by a few poloidal correlation lengths.

- The $v\times B$ term

The observations reported above show that the **v** × **B** components contributes to the radial electric field and may even be the decisive transition trigger [29]. The **v** × **B** term introduces poloidal and toroidal momentum balances into the transition physics including neo-classical equilibrium characteristics and anomalous transport features. The $v_\phi B_\theta$-term is positive with co-injection and obstructs the development of a negative electric field well. With counter-NBI, the term is negative. Indeed, DIII-D observed that P_{th} is lower in ctr-NBI in comparison to co-NBI [55]. In helical systems v_ϕ is close to zero at the edge owing to the toroidal field ripple; this eases the development of a negative E_r-well with co-NBI.

The inclusion of v_θ into the transition physics opens a rich field of potential transition mechanisms [56]. v_θ is given by the poloidal momentum balance of torques from radial electric currents and poloidal damping (magnetic pumping). The steady-state momentum balance comprises of the Lorentz force term, the poloidal damping term, the friction term with the ambient neutrals and the turbulent Reynolds stress term and can be written as:

$$0 = j_r B/n_i - m_i \mu_\theta v_{\theta i} - m_i \nu_n v_{\theta i} + m_i \vartheta/\vartheta r\,(<\tilde{v}_{ri}\,\tilde{v}_{\theta i}>) \qquad \text{(equ. 3).}$$

j_r is the radial current, μ_θ is the poloidal viscosity, ν_n is the collision frequency with neutrals; $\tilde{v}_{ri}$, $\tilde{v}_{\theta i}$ are the radial and poloidal fluctuations of the velocity field.

3.10. Drives for and damping of poloidal flow.

- Ion orbit losses.

The ion loss-orbit concept [57], [58] for the H-mode is based on spontaneous losses of ions; the electron return current $j_{e,r}$ establishes the torque to spin the plasma. One of the attractions of this concept is that it would explain the short transition times. The ion loss model has to be further developed, however, to clarify whether the strong machine-to-machine variation of the edge collisionality ν_i* at the transition is in agreement with it [21], [24]. Another, possibly conflicting observation is the rather constant width of the edge transport barrier (see Fig. 10).

- The $B\times\nabla B$ drift effect on P_{th}

The flows in the scrape-off layer seem to spread momentum and to viscously drag along the plasma inside the separatrix and affect the power threshold possibly contributing to the B×∇B drift effect on P_{th}. This effect is substantial considering the ratio of $P_{th}^{SN-}/P_{th}^{SN+}$ ~ 2,6 for ASDEX (circular cross-section) [24], ~ 3 for DIII-D (elongated cross-section) [21], ~ 3.2 for Alcator C-Mod (high-field) [59], and ~ 3 for MAST (spherical tokamak) [60]. The variation of P_{th} introduced by the active X-point location is so massive that a few mm in vertical shift replace MW of heating power.

The static SOL parameters in Alcator C-Mod do not vary with the $B\times\nabla B$ direction with respect to the X-point apart from the parallel fluid flow which shows a strong dependence on configuration [59]. The SOL flow is mostly driven by an asymmetric power exhaust, which is predominantly concentrated at the bad curvature side close to the plasma midplane. A net momentum is imparted to the SOL, which couples across the separatrix and even can lead to core toroidal rotation [61]. It has to be noted that these studies are carried out with ICRH without external momentum input into C-Mod. The SOL flow emerges from different path length from the stagnation point to the outer or the inner sink (target plate), respectively. This weighting leads in the SN^+ case to a net co- and in the SN^- case to a net ctr-flow. In this way, different electric fields are set up in the SOL, which serve as boundary conditions for the E_r-field evolution inside the separatrix. The net-co flow gives rise to a more positive E-field in the SOL, which might lead to a steeper E-field gradient inside the separatrix. Ctr-flow in the SN^- case reduces the SOL E-field with the possible consequence of a higher power threshold for the H-mode transition.

- The role of the poloidal viscosity

The study of the H-mode boundaries in W7-AS gave strong evidence of the role of neo-classical poloidal damping for the H-mode transition conditions. The H-mode can be realised in W7-AS at a low power threshold but only in isolated iota windows [62]. These windows are periodically linked to the island chains 5/11, 5/10 and 5/9 who form inner separatrix. This close link of periodic magnetic structure and H-mode operational range indicates even without deeper analysis that the edge magnetic field structure is part of the H-mode transition conditions. It points to the role of a neo-classical mechanism.

The neo-classical poloidal damping [63] resulting from the specific field Fourier coefficients, which compose the field structure at the edge of W7-AS has a strong dependence on iota. Two of the three H-mode windows are found to be located at minima of poloidal damping. The actual value of poloidal damping depends, however, on collisionality and radial electric field.

Also He-J, an optimised low shear heliotron, observes the H-mode in restricted iota-windows [64] but also no perfect match with the minima in poloidal flow damping is achieved. Deviations may be caused by the reference to the vacuum field structure; the actual equilibrium with the corrugated island edge has not been reconstructed.

- The role of Reynolds Stress.

The term of the poloidal momentum balance which would be most intricately involved in the H-mode transition scenario with the sudden quench of large-amplitude L-mode turbulence is the dynamic Reynolds stress term (4th term on the RHS of equ. 3). The turbulent Reynolds stress plays an important role in normal fluid dynamics [18]. For 2-dimensional flows, the Reynolds stress couples wave energy from smaller scales to larger ones and a mean fluid flow, so called zonal flows (ZFs) [65], are self generated out of the turbulence field. The Reynolds stress and its potential to drive flow has been measured at the linear plasma device CSDX [66]. Figure 13 left top shows the radial profile of the measured Reynolds stress emerging from collisional drift waves and Fig. 13 left bottom documents the good agreement of the measured mean azimuthal velocity profile generated from the fluctuation field with the expected one. These results of a linear machine show nicely the Reynolds stress generated flows. The same effectiveness cannot be expected in toroidal systems because of damping of the associated parallel flows [67]. Therefore, separate measurements in toroidal systems are necessary, in case of tokamaks specifically at the edge with strong damping due to high q.

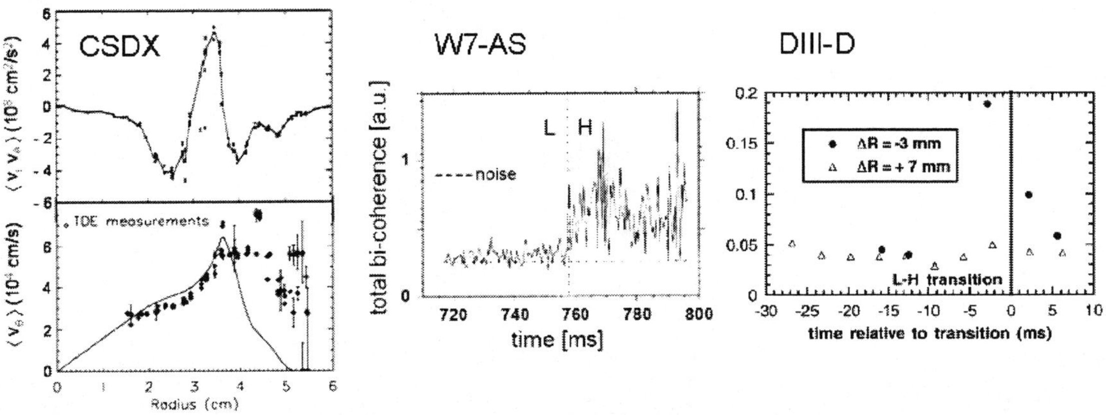

FIGURE 13. Left column top: Reynolds stress against plasma radius of the CSDX device; bottom: flow induced by the Reynolds stress; the data are compared with theory [66]. Copyright: APS. Middle and right: total bi-coherence (three-wave coupling) versus time for L- and H-phases. Middle: W7-AS [69]; Copyright: APS. Right: ibid DIIID [71]. Copyright: AIP.

The action of Reynolds stress can also be demonstrated by a totally different technique - the analysis of density and velocity fluctuations via bi-coherence spectral analysis [68] as the production of zonal flows via the Reynolds stress mechanism is equivalent to three-wave coupling [69], [70]. Figure 13 middle top and bottom show results from W7-AS [69] and DIII-D [71] which yield a distinct finite bi-coherence in the H-phase, in W7-AS throughout the H-phase, in DIII-D transitorily facilitating the bifurcation before the ∇p_i –term develops and stabilises the new mode. Also in H-modes induced by biasing, large and steady-state bi-coherence values are observed [70].

In summary, there is good experimental evidence of the action of the Reynolds stress in plasmas, the direct measurements of zonal flows [72], the formation of ZFs from the turbulence field via three-wave coupling at the plasma edge with relevance to the H-transition.

Global gyro-kinetic numerical models, which simulate the full plasma and allow the self-consistent variation of the pressure profile, can model the turbulence de-correlation by sheared flow [73]. Basis for these simulations is the zonal-flow generation via Reynolds-stress from the ambient turbulence field [65] and the turbulence de-correlation via critical shear flow (equ. 1) [44]. The non-linear evolution of the turbulence-zonal-flow system is done on the basis of ITG instabilities in toroidal geometry. With flow, the radial correlation length is strongly reduced and the turbulent eddies are nearly isotropic in r and θ.

3.11. A possible paradigm of the H-mode transition

The verification of the H-mode transition on the basis of self-induced flows, which annihilate its initially driving mechanism, requires rather detailed knowledge of spatial and temporal structures of the edge turbulence. The Langmuir probe is the only diagnostics, which can provide this information. Its safe and correct application requires, however, low power fluxes at the edge. This condition may not necessarily be provided in high-power L-mode discharges, which show the H-mode transition in the most relevant environment. Helical systems can produce plasmas with low heating power. Therefore, they have contributed to the understanding of the H-mode in a rather unique way [13].

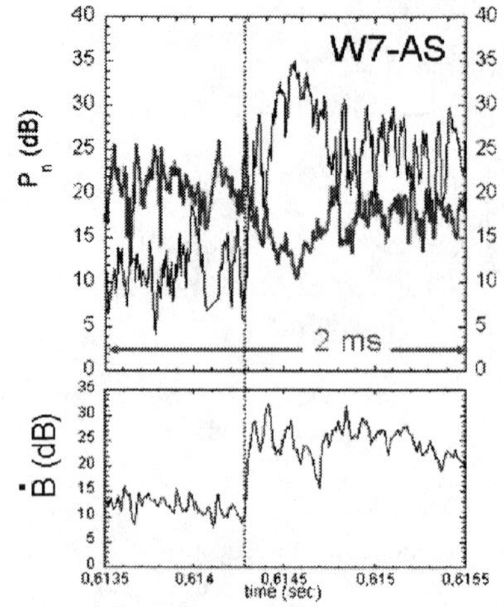

FIGURE 14. Top: Reflectometry scattering signal P_n and perpendicular flow at the edge for an H-L back transition (dashed line) of W7-AS [40].
Bottom: Mirnov signal which denotes the transition. P_n and $v_\perp$ are modulated at 11 kHz presumably by a GAM fluctuation.

In this chapter, we summarize some of the findings from H-1 heliac [74]. H-1 operates at low field, low density, with Argon plasmas and with a heating power of ~ 60 kW. Spontaneous H-mode transitions are observed which double the particle and energy content. The secondary characteristics of the H-mode are provided: sudden, sharp transition, steep edge gradients, a more negative E_r at the edge, and the disappearance of turbulence (but no ELMs).

The following picture emerges from the studies in H-1 summarized in [75]: drift wave turbulence is caused by the pressure gradient of the L-phase. A linear drift wave instability is observed at $k_i \sim 200\ m^{-1}$; this mode establishes the injection energy channel for spectral transfer of energy to other k-ranges. Energy cascades from the injection-k value, k_i, to lower and higher k-values by three-wave interaction yielding broad k-spectra in the L-phase. The modes "condense" via a non-local transfer process, connecting different size scales, into coherent structures and the spectral energy accumulates at the largest possible scale: Energy couples from k_i, energized by the pressure gradient, to the m=0 mode. Oscillating zonal flows appear, with a structure corresponding to the largest possible system size, the poloidal circumference. The oscillating zonal flows, the GAMs, represent a seed for the H-transition.

At the L-H-transition, the turbulence level on both sides of k_i strongly decreases. The drift wave instability remains, however, even intensifies because the pressure gradient increases. At the transition, the m=0 mode prevails – the remaining sheared flow. The non-local energy transfer connecting different spectral ranges avoids the development of broad-band turbulence and the plasma confinement improves. It has been confirmed that the kinetic mean flow energy in the H-mode corresponds to the electrostatic energy of the turbulence in the L-mode (within 20%).

The interaction between the turbulence, perpendicular flow and the possible interference with GAMs can also be seen in W7-AS. Figure 14 and 15 show a back-transition on a ms and sub-ms time scale [40]. The transition is marked by the sudden rise of the Mirnov signal. At the transition, the perpendicular flow decreases and the fluctuation power increases (log. scale). Unlike the Mirnov trace, the transition in P_n and $v_\perp$ occurs in steps. Both signals, P_n and $v_\perp$, are modulated by 11 kHz. This is the frequency of the GAM at the edge of W7-AS. Structural information of this mode is, however, not available. Correlation analysis shows that P_n and $v_\perp$ are anti-correlated prior to the transition: $v_\perp$ increases when P_n decreases. P_n leads this action by 10 μs [76]. This phase relation can be interpreted as evidence for the transfer of spectral power from the turbulence to the flow. More information is necessary, however, to finally conclude this. After the transition, P_n and $v_\perp$ are correlated now and increase or decrease simultaneously. The phase relation changes at the transition. This new relation could be interpreted in the sense that e.g. the pressure gradient is modulated and this causes P_n and $v_\perp$ to change simultaneously.

Figure 15 shows the back transition on a 100 μs time scale. The diagnostics, Doppler reflectometry, allows this resolution. The transition is marked and occurs within the two vertical lines. It happens after a strong positive increase of $v_\perp$ when the flow sharply swings back. In this phase the turbulence level is anti-correlated and increases. As the transition happens, the average fluctuation level does not reverse but continues to grow, the average value of $v_\perp$ further decreases, and L-mode conditions establish.

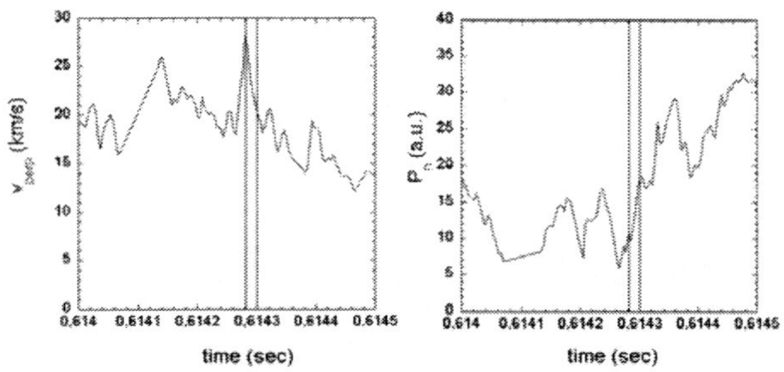

FIGURE 15. As in Fig. 14 but for a 500 ms time window. The H-L back transition occurs between the two vertical lines after $v_\perp$ has reached a maximum and P_n starts to rise.

3.12. Achievements with H-mode plasmas

- JET has achieved 16.1 MW fusion power in the quiescent, ELM-free H-mode for about 0.5 τ_E [77]. The DT-mixture was close to 1; $Q = P_{fusion}/P_{input} \sim 0.62$; $n_D T_i(0)\tau_E \sim 8.7\ 10^{20}\ m^{-3}$ keVs. The major traces (including the DT fusion power) of this world-record discharge are shown in Fig. 16.

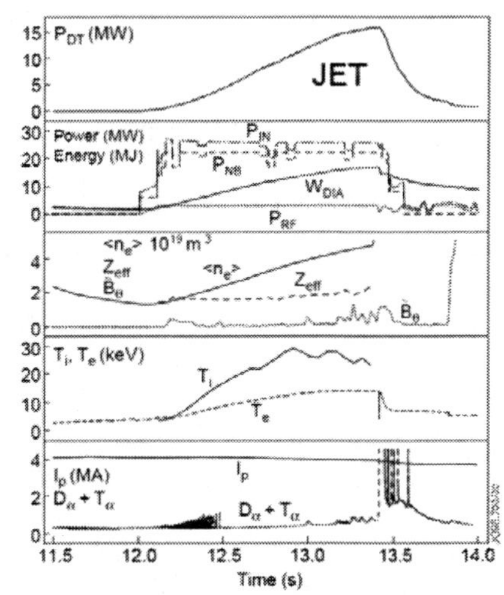

FIGURE 16. JET H-mode discharge which yielded the highest fusion power (16.1 MW) up to date [76]. Copyright: Nucl. Fusion, IOPP

- The quality of the H-mode has impressively improved since its discovery. At the 2006 IAEA conference, the paper by T. Luce et al. [78] has received a prize because it presented and discussed a nearly steady-state discharge, which would yield Q=10 when scaled to ITER. The discharge is an ELMy H-mode plasma which is shown in Fig. 17 left. The plasma current is 1.2 MA, the field is 1.7 T; $n_e = 0.4\ 10^{20}\ m^{-3}$; the heating power is 4.8 MW. The dimensionless parameters which characterize this discharge are: $H_{89} = 2.6$ (confinement enhancement factor related to 89-L-mode scaling [79]), $\beta \sim 3\%$; $\beta_N = 2.7$; $n_e/n_{eGW} \sim 0{,}4$. This state could be maintained for 36 confinement times. Scaled to ITER, this discharge type would allow it to meet its goals.

- The best performance in W7-AS was achieved in the high-density H-mode (HDH) [7], [33]. This mode develops at high density at a distinct transition, which increases τ_E (if started from the level of normal confinement) and sharply reduces τ_I, the impurity confinement time. In this mode, the confinement does not degrade approaching the n_e- and beta-limits. Figure 17 right depicts an HDH discharge with B = 0.9 T, $n_e = 2.5.\ 10^{20}\ m^{-3}$, $P_{NBI,abs} = 2.5$ MW. The parameters of this discharge are: $\beta = 3,4\%$; $\beta_N \sim 9.3$; H = 1.4 (related to ISS95-scaling [80]), $n_e/n_{eGW} = 2.5$, $\tau_I/\tau_E \sim 2$). This quality could be maintained for 85 confinement times. For details see Ref. [33].

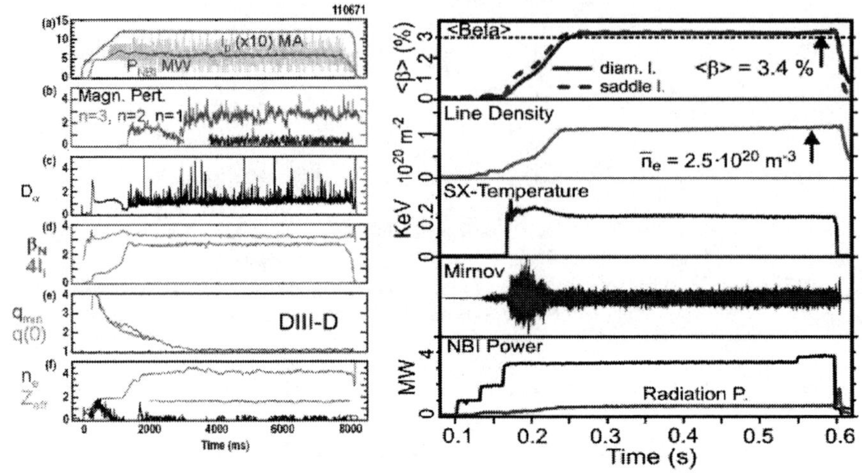

FIGURE 17. Left: High-performance, long-pulse discharge from DIIID. This discharge, when scaled to ITER, would qualify for Q=10 operation [77].
Right: High-performance, long-pulse discharge from W7-AS in the HDH regime [7, 33].

- K. Itoh et al. have analysed the costs of an ITER-like experimental reactor when the targets are reached by a design parameter setting and a plasma size, which are in accordance with L-mode operation or, alternatively, with the improved confinement of the H-mode at a smaller reactor size, respectively. The fusion reactor costs scale like $\$ \sim H^{-1.3}$. For H = 2, the ITER capital costs of 5000 Mill € would increase to 12300 Mill € for a similarly targeted device based in the L-mode.

- Several DEMO designs are available [81]. They all show that τ_E expected from H-mode confinement is sufficient for the reactor power of about 3GW. The major difference to ITER is a substantially larger β_N and a higher P/R_o ratio (R_o is the major radius). Higher β_N will ensure a larger bootstrap current. As the H-mode is characterised by broad pressure profiles, it provides better conditions for more stability at broader current profiles. The higher P/R_o ratio may demand, however, low-fluctuation edge-conditions and may necessitate quiescent, non-ELMy H-modes.

4. FINAL COMMENTS

Confinement in fusion devices obviously develops in two branches. In the L-mode branch, the free energy of the high pressure plasma leads to an increase of the overall turbulence level. A distinct bifurcation is possible, which could emerge in a self-organized fashion: the energy in the 2D-turbulence field spectrally cascades to macroscopic scales giving rise to a mean flow which acts back and its radial shear de-correlates the prevailing turbulence. This leads to the H-mode branch characterized by a very low level of edge turbulence in the medium k-range. The energy content of the plasma rises as a consequence and – what might be rather surprising for a thermodynamic system – the self-organising process brings the plasma closer to ignition conditions. The pressure profile, which is the corollary of the improved confinement, maintains the sheared flow conditions and stabilises the H-mode.

In stellarators, the H-mode develops basically with the same characteristics, which demonstrates the ubiquitous character of this regime. The neo-classical ambipolar electric field provides good starting conditions for the self-generated flow to ultimately reduce the turbulent fluxes and restore plasma transport closer to the lower-dissipation neo-classical flux level.

The H-mode provides the confinement basis for reasonably sized fusion reactors. ELMs, however, and the divertor wall erosion they cause, remain as unresolved problem. The physics of the H-mode is not yet understood in all details. To work on it is scientifically rewarding and – from the programmatic point of view of fusion – meaningful.

ACKNOWLEDGMENTS

I want to thank the members of the ASDEX and the W7-AS teams and all those who have allowed me to use their figures.

REFERENCES

1. M. Greenwald et al., *Phys. Rev. Lett.* **53**, 352 (1984).
2. T. Nishitani, et al., *Nucl. Fusion* **34,** 1069 (1994).
   C. Gormezano, et al., *Phys. Rev. Lett.* **80**, 5544 (1998).
3. D. Campbell, *this volume*.
4. F. Wagner et al., *Phys. Rev. Letters* **49**, 1408 (1982).
5. R.C. Wolf, *Plasma Phys. Contr. Fusion* **45**, R1 (2002).
6. F. Wagner et al., *Phys. Plasmas* **12**, 072509 (2005).
7. M. Hirsch et al., *Plasma Phys. Contr. Fusion* (to be published in 2008).
8. M. Wanner et al., *Nucl. Fusion* **43,** 416 (2003).
9. R. Wilhelm, *Plasma Phys. Control. Fusion* **40**, A1 (1998).
10. H. E. Mynick, T. K. Chu, and A. H. Boozer, *Phys. Rev. Lett.* **48,** 322(1982).
11. J. Nührenberg and R. Zille, *Phys. Lett.* **A 129**, 113 (1988).
12. B. Kurzan et al., *Plasma Phys. Contr. Fusion* **42**, 237 (2000).
13. F. Wagner et al., *Plasma Phys. Contr. Fusion* **48**, A217 (2006).
14. H. Maassberg, et al., *Plasma Phys. Control. Fusion* **40**, 1429 (1998).
15. R.Goldston, *Plasma Phys. Control. Fusion* **26**, 87 (1984)
16. F. Wagner and U. Stroth, *Plasma Phys. Control. Fusion* 35, 1321 (1993).
17. T.N. Carlstrom, *Fusion Sci. Technol.* **48**, 997 (2005).
18. P.W. Terry, *Rev. Mod. Phys.* **72**, 109 (2000).
19. J. Hugill, *Plasma Phys. Control. Fusion* **42**, R75 (2000).
20. F. Wagner, *Plasma Phys. Control. Fusion* **49**, B1 (2007).
21. T.N. Carlstrom and R.J. Groebner *Phys. Plasmas* **3**, 1867 (1996).
22. M. Keilhacker et al., *Plasma Phys. Contr. Fusion* **26**, 49 (1984).
23. F. Wagner et al., *Nucl. Fusion* **25**, 1490 (1985).
24. ASDEX team, *Nucl. Fusion* **29**, 1959 (1989).
25. J.A. Snipes et al., *Plasma Phys. Contr. Fusion* **42**, A249 (2000).
26. F. Wagner et al., *Phys. Review Letters* **53**, 1453 (1984).
27. M.E. Manso *Plasma Phys. Control. Fusion* **35,** B141(1993).
28. E. Holzhauer and G. Dodel, *Rev. Sci. Instruments* **61**, 2187 (1990).
29. R.A. Moyer et al., *Phys. Plasmas* **2**, 2397 (1995).
30. J. Connor, *this volume.*
31. A. Mazurenko et al., *Phys. Rev. Lett.* **89**, 225004 (2002).
32. K.H. Burrell et al, *Phys. Plasmas* **8**, 2152 (2001).
33. K. McCormick et al., *Phys. Rev. Letters* **89**, 015001 (2001).
34. E. Belonohy et al., EPS-34, *conference proceedings*.
35. H. Zohm et al., *Phys. Rev. Letters* **72**, 222 (1994).
36. S.- I. Itoh, et al., *Phys. Rev. Letters* **67**, 2485 (1991).
37. S.-I. Itoh and K. Itoh, *Phys. Rev. Lett.* **60**, 2276 (1988).
38. K.H. Burrell, et al., *Plasma Phys. Contr. Fusion* **34**, 1859 (1992).
39. R. J. Groebner, K. H. Burrell, and R.P. Seraydarian, *Phys. Rev. Letters* **64**, 3015 (1990).
40. M. Hirsch et al., *Plasma Phys. Contr. Fusion* **48**, S155 (2006).
41. J. Baldzuhn et al., *Plasma Phys. Contr. Fusion* **40**, 967 (1988).
42. H. Ehmler et al., *Nucl. Fusion* **43**, L11 (2003).
43. J. Schirmer et al., *Nucl. Fusion* **46**, S780 (2006).
44. H. Biglary, P.H. Diamond, and P. W. Terry, *Phys. Fluids* **B 4**, 1385 (1990).
45. B. Lehnert *Phys. Fluids* **9**, 1367 (1966).
46. Ch.P. Ritz et al., *Phys. Rev. Letters* **65**, 2543 (1990).
47. R.J. Taylor et al., *Phys. Rev. Letters* **63**, 2365 (1989).
48. R.R. Weynants et al., *Nucl. Fusion* **32**, 837 (1992).
49. J. Boedo et al., *Nucl. Fusion* **40**, 1397, 2000.
50. R.R. Weynants et al., *Nucl. Fusion* **32**, 837 (1992).
51. K.H. Burrell, *Phys. Plasmas* **6**, 4418 (1999).
52. M. Jakubowski, R.J. Fonck and G.R. McKee, *Phys. Rev. Lett.* **89** 265003-1 (2002).
53. F. Wagner et al., In: *Plasma Physics and Controlled Nuclear Fusion Research* 1990, **Vol. 1**, 277- 290, Nuclear Fusion Supplement, 1991, IAEA-CN-53-A-IV-2, (1991)
54. T. Ido et al., *Phys. Rev. Letters* **88**, 055066-1 (2002).

55. P. Gohil et al., *Plasma Phys. Control. Fusion* submitted.
56. J.W. Connor and H.R. Wilson, *Plasma Phys. Contr. Fusion* **42**, R1 (2000).
57. K. Itoh and S.-I. Itoh, *Plasma Phys. Contr. Fusion* **38**, 1 (1996).
58. K.C. Shaing, *Phys. Rev. Letters* **76**, 4364 (1996).
59. B. LaBombard et al., *Nucl. Fusion* **44**, 1047 (2004).
60. H. Meyer et al., *Nucl. Fusion* **46,** 64 (2006).
    T. Biewer et al., *33rd EPS Conf. on Plasma Physics* (Rome, Italy) vol 30I (ECA) P-5.112 (2006).
61. J. E. Rice et al., *Nucl. Fusion* **45**, 251 (2005).
62. M. Hirsch et al., *Plasma Phys. Contr. Fusion* **42**, A231(2000).
63. H. Wobig and J. Kisslinger *Plasma Phys. Contr. Fusion* **37**, 893 (1995).
64. F. Sano et al., *Fusion Science and Technology* **46**, 288 (2004).
65. P.H. Diamond,S.I. Itoh, K. Itoh, T.S. Hahm *Plasma Phys. Contr. Fusion* **47** R35 (2005).
    see also: K. Itoh, *this volume*
66. C. Holland et al., *Phys. Rev. Letters* **96**, 195002 (2006).
67. K. Hallatscheck, *Plasma Phys. Control. Fusion* **49** (2007); to be published.
68. G. Bonhomme, *this volume*
69. P.H. Diamond et al., *Phys. Rev. Letters* **84**, 4842 (2000).
70. G.R. Tynan et al., *Phys. Plasmas* **8**, 2691 (2001).
71. R.A. Moyer et al., *Phys. Rev. Letters* **87**, 135001-1(2001).
72. A. Fujisawa et al., *Phys. Rev. Letters* **93** 165002 (2004).
73. Z. Lin at al., *Science* **281**, 1835 (1998).
74. S. M. Hamberger et al., *Fusion Technol*ogy **17**, 123 (1990).
75. M. G. Shats et al., *Phys. Rev.* ***E*** **71**, 046409 (2005).
76. M. Endler (*private communication).*
77. M. Keilhacker et al., *Nucl. Fusion* **39**, 209 (1999).
78. T. Luce, et al., *Nucl. Fusion* **43**, 321 (2003).
79. P. N. Yuschmanov et al., *Nucl. Fusion* **30,** 1999 (1990).
80. U. Stroth, et al., *Nucl. Fusion* **36**, 1063 (1996).
81. K. Lackner, *Plasma Phys. Control. Fusion* **40**, 557 (1998).

Introduction to Turbulent Transport in Tokamak Core

T.S. Hahm

Princeton Plasma Physics Laboratory
P.O.Box 451, Princeton, NJ 08543, USA

Abstract

In the past decade, considerable progress in understanding tokamak core turbulence and transport has been made thanks to advances in experimental measurements, gyrokinetic simulations, and nonlinear theory. In this lecture, a few examples for which effective interactions among theorists and experimentalists led to significant achievements will be illustrated.

1. Introduction

One of the major goals in tokamak fusion research is to achieve high ion temperature. Experimentalists usually plot the central ion temperature as a function of the heating power. On the other hand, theoreticians prefer to think about the plasma confinement problem in terms of the flux-gradient relation which is familiar from Fick's law in thermodynamics. For instance, the ion heat flux is plotted as a function of temperature gradient. In confined plasmas, there exist many sources of free energy for instabilities and transport. These include the radial gradients of ion temperature, electron temperature, density, and toroidal rotation. One can formally write a generalized flux-gradient-relation in matrix form. We note that the radial flux of each quantity can not only be driven by the radial gradient of that quantity itself, but also by the gradients of other quantities. The contribution of the diagonal term in the transport matrix is called the diffusive term. The contributions from the off-diagonal terms are called the non-diffusive flux, or more casually, the pinch term or the convective flux.

With this elementary introduction, some confinement problems of current interest are discussed in this lecture. The ion thermal transport is discussed in Sec. II. Sec. III. and Sec. IV. cover electron thermal transport and momentum transport respectively. Conclusions are drawn briefly in Sec. V.

CP1013, *Turbulent Transport in Fusion Plasmas, First ITER International Summer School*
edited by S. Benkadda

2. Ion Thermal Transport

Ion thermal transport is better understood compared to other channels of transport. In auxiliary heated plasmas, an anomalous level of ion heat transport has been observed which is much greater than the prediction from neoclassical collisional transport theory. Ion temperature gradient driven turbulence is the best candidate for the anomalous ion thermal transport. Thanks to recent advances in gyrokinetic simulations, theoretical predictions from different codes begin to converge for simple cases, not only in numbers, but also in underlying nonlinear physics mechanisms. In the last decade, the recognition of the importance of zonal flows has caused a paradigm shift in the tokamak turbulence theory community, from the drift wave turbulence problem to the drift wave-zonal flow system problem[1]. What is a zonal flow? Zonal flows exist in nature, for instance on Jupiter and the Sun, and have been observed in many gyrokinetic simulations including Ref.[2]. Many theoreticians believe they should exist in ITER plasmas as well. Zonal flow is a toroidally symmetric, poloidally symmetric, and radially varying $\mathbf{E} \times \mathbf{B}$ flow. Zonal flow is has no radial motion and cannot contribute to the radial transport directly. Since it cannot tap the free energy in the pressure gradient, it's not a linear instability, and can only be driven by turbulence nonlinearly[3, 4]. The interaction of zonal flows with drift wave turbulence is schematically illustrated in Fig. 33 of Ref.[1]. Turbulence driven by radial gradients of pressure generates zonal flows[5, 6], and zonal flows can be damped by the collsional friction between trapped and passing ions[7]. Also, zonal flows regulate turbulence via $\mathbf{E} \times \mathbf{B}$ shearing[8, 9, 10], and so on.

Next, some examples are presented in which the zonal flows play crucial roles in determining the transport scalings with respect to key dimensionless parameters. Transport scaling with respect to the machine size, or $\rho^* = \rho_i/a$ is very important, since we want to know by how much the future larger machines will perform better than the present ones. As our understading of ITG turbulence got deeper, our predictions were modified. A recent gyrokinetic particle-in-cell simulation of a shaped tokamak plasma[11] during less than a second, seems to summarize the theoretical progress in the past 25 years. Theoretical predictions of toroidal ITG instabilities in the early 80's[12] and of their global radial structures in the early 90's[13, 14] seemed to indicate the possibility of Bohm scaling. Since the mid 90's, extensive studies of zonal flows indicated that the radially elongated global ITG eigenmodes get broken up by zonal flows into smaller eddys[15, 16, 2]. Naturally, most people expected gyroBohm scaling from the eddy size scaling. More recently, it has been indicated that, from global gyrokinetic simulations, for a moderate system size, non-negligible deviations from gyroBohm scaling can exist[17]. This can occur due to turbulence spreading into the linearly stable zone[18]. While the studies on turbulence spreading go back at least to the mid-90's,[19] it was the importance of its relation to the ρ^*-scaling which revived wide interest in the theory community[18, 20, 21, 22, 23, 24, 25, 26].

Plasma current scaling of confinement is one of the remaining puzzles of ion thermal transport. It's possible to obtain the q-dependence of the transport from zonal flow characteristics which depend on the q value. Due to their Landau damping, Geodesic Acoustic Modes can exist only in a high-q region. Therefore, in a low-q region, only stationary zonal flows will persist. Since Geodesic Acoustic Modes (GAMs) oscillate with a relatively high frequency, they are not very effective in reducing turbulence[27]. So we expect a lower level of turbulence in the low q region where static zonal flow dominates. As shown from a recent nonlinear gyrofluid simulation[28], the ion heat flux from ITG turbulence is lower for the case with lower q, since the stationary zonal flows dominate. It's confirmed by an independent gyrokinetic simulation[29]. However, the relevance of this mechanism to experiment still remains to be demonstrated.

The collisionality dependence of ion heat transport observed in tokamak experiments is a puzzle too, since the transport level was much higher than the predictions from collisional transport theory, while ITG turbulence in the core of a tokamak is essentially collisionless. Recognizing that the zonal flows regulate the turbulence level, near ITG marginality, collisional damping of zonal flows controls the transport, and the transport level can increase with an increasing ion collisionality as shown from gyrokinetic simulations[30] and confirmed by an independent gyrofluid simulation[31]. From a theoretical point of view, it's important to note that this trend can be reversed if one includes the trapped electron response, which decreases with electron collisionality[32]. While some experiments report transport increases with collisionality, further parameter scans are required to identify the underlying physics mechanisms.

3. Electron Thermal Transport

It has been well known for many years that all operational modes of tokamaks and spherical tori have exhibited anomalous electron thermal transport. There is evidence that electron temperature profiles are stiff in tokamaks. Perturbative transport experiments, including heat pulse propagagtion, and the need for an inward heat pinch term in transport analysis of auxiliarily heated plasmas, all indicate this possibility. Possible theoretical candidates for electron thermal transport include the trapped electron mode (TEM) and the ITG mode with trapped electrons, the electron temperature gradient (ETG) mode, and magnetic flutter driven transport.

Starting from the TEM, it can produce an electron thermal diffusivity of experimental relevance when strongly turbulent. One supporting result for TEM comes from the ASDEX-Upgrade ECH experiment. The transport shows a threshold behavior in agreement with the TEM theory prediction. Furthermore, the experimentally measured heat flux and the calculated TEM linear growth rate show a similar trend as functions of the electron temperature gradient. Another supporting piece of evidence for TEMs is that sometimes low-k fluctuation reduction occurs during electron ITB formation. On JET, the radial profile of the fluctuation amplitude has been mapped out using X-mode reflectometry. There is broad agreement between the radial profile of fluctuation amplitude and that of electron thermal diffusivity. The fluctuation propagates in the electron diamagnetic direction after the Doppler-shift correction. The estimated perpendicular wavelength, which was about 11 cm, was much longer than the ion gyroradius. Therefore, this fluctuation looks more like a TEM rather than anything else.

Another candidate is the ETG turbulence. To produce electron heat transport at a level of experimental relevance in tokamaks, the radial correlation length of ETG turbulence has to be significantly enhanced over the electron gyroradius. Radially elongated streamers have been observed from many nonlinear gyrokinetic simulations[35, 36]. However, the relation between the electron heat flux and the radial size of streamers, if any, still needs to be demonstrated[37]. As of now, there's no consensus in the theory community regarding whether the ETG-driven electron heat transport is high enough to explain experimental observations.

In some spherical torus plasmas, electron transport is highly anomalous even when low-**k** fluctuations are linearly stable. In NSTX neutral beam heated plasmas[38], most of the heating power from the energetic particles goes to electrons, and this is similar to a situation expected for ITER. The linear stability analysis shows that the low-k fluctuations are expected to be stable due to the high $\mathbf{E} \times \mathbf{B}$ shearing rate, leaving only high-k ETG modes unstable. Note that the ion thermal diffusivity is down to a neoclassical level, qualitatively consistent with the stabilization of low-k ITG/TEM modes. Due to its low magnetic field and consequent large electron gyroradius, NSTX is an ideal place to look for ETG-like fluctuations with good spatial resolution. Experimental measurements using high-k tangential scattering have begun to be reported. It's also noteworthy that high-k fluctuations were measured before in TFTR[39], but their frequency spectrum characteristics were inconsistent with those of ETG turbulence.

4. Momentum Transport

Now, it's well appreciated that plasma rotation plays a central role in magnetic confinement. $\mathbf{E} \times \mathbf{B}$ shear can reduce turbulence and transport, and in turn rotation can be generated by microturbulence via Reynolds stress[40]. It can stabilize MHD instabilities, such as resistive wall modes, but can be damped via nonaxisymmetric magnetic fields produced by MHD instabilties. In addition, the radial electric field interacts with energetic particles. Therefore, understanding momentum transport is important. In particular, for successful operation of ITER, we want to know whether ITER will have enough wave and neutral beam power to generate sufficient rotation for turbulence and transport reduction as well as stabilization of resistive wall modes.

Momentum transport is also usually anomalous, and roughly close to ion thermal transport as shown in TFTR in the early 90's[41]. ITG instabilties were calculated to be unstable based on a simple threshold formula[42] in these plasmas and the estimated toroidal momentum diffusivity was close to the ion thermal diffusivity, not only in its value, but also in its radial dependence. These are in broad agreement with nonlinear theory predictions[43]. These are also generic properties of electrostatic long wavelength microturbulence such as ITG and TEM in which $\mathbf{E} \times \mathbf{B}$ nonlinearity plays a dominant role. The Prandtl number is defined as the ratio between momentum diffusivity and ion thermal diffusivity. The range of Prandtl number from experiments is very wide. It was around 1 in TFTR, and also in Asdex-Upgrade more recently. In JET, it was around

0.2. In NSTX, it was smaller than one by an order of magnitude. However, in JT60-U, it was typically greater than 1. We note that, in some cases, the total momentum flux was characterized by the effective diffusivity, without separarting the momentum pinch contribution. From theory, the range of Prandtl number was somewhat narrower, although different plasmas were considered from different approaches and codes. From the gyrokinetic theory point of view, propagators which appear in the calculations of momentum and ion thermal diffusivities respectively, involve different velocity moments. Therefore, the Prandtl number can deviate from 1. From various gyrokinetic simulations treating typical tokamak core plasmas, values ranging from 0.45 to 1.6 have been reported.

A current topic of high importance is spontaneous rotation. There have been many cases in which experimentalists observed plasma rotation when it was not expected. However, systematic studies of spontaneous rotation began around 1997 when strong toroidal plasma rotation was observed in ICRH plasmas without an apparent external torque in JET[47] and Alcator C-Mod[48]. Roughly speaking, plasma rotation increases with plasma pressure or stored energy. Later, it was also found that OH plasmas also rotate spontaneously in C-Mod. Motivated by experimental observations of spontaneous rotation in ICRH plasmas, theories of ICRH spontaneous rotation have been developed. Some of these theories are based on an idea that ICRH resonance is radially inhomogeneous in space. Thus, resonant ions are transported radially. The plasma responds by rotating, either due to an effective radial current, or via an rf modified stress, or some other mechanisms. But these theories can't explain Ohmic plasma rotation. Therefore, one might pursue a theory which is universally applicable across heating schemes.

The radial electric field shear is an important player in off-diagonal, non-diffusive momentum flux. It's important to recognize that the E_r shear creates an imbalance in wave population density which yields a residual stress rather than a pinch, which spins the plasma[49]. At the same time, it reduces turbulence, which causes the anomalous flux in the first place. If the direction of rotation is right, the rotation may in turn increase E_r shear, and create a second feedback loop.

It's also worthwhile to look for a new pinch mechanism when $\mathbf{E} \times \mathbf{B}$ shear effects are weak, since spontaneous rotation has been observed in OH and electron heated plasmas. Starting from the nonlinear gyrokinetic equation with proper conservation laws[50], one can calculate the radial flux of parallel flow with toroidal effects. In particular, we keep the symplectic structure in the parallel acceleration equation contained in $\mathbf{B}*$ correction which conventional gyrokinetic equations ignore. We can demonstrate that magnetic curvature produces a momentum pinch. This effect is transparent in quasilinear calculations for fluctuations with ballooning mode structure. Therefore, it's a purely toroidal effect. The predicted direction of the turbulent convective (TurCo) pinch is likely to be inward for TEM-dominated plasmas, and for the hot electron regime. If we take moments of the gyrokinetic equations, we can show in the limit of high aspect ratio and hot electron regime, that $nU_{\|}/B^3$ or "magnetically weighted momentum density" is approximately conserved. This result strongly suggests that the pinch can be understood in terms of homogenization, a la turbulent equipartition (TEP) theory.

Table III of Ref.[51] summarizes the key expressions for the pinch velocity of the angular momentum density from this theory. There are two different contributions to the turbulent convective (TurCo) pinch velocity. The first part can be understood in terms of the TEP theory which has this flux-surface-averaged expression with a curvature drift weighted by the fluctuation amplitude. A typical magnitude of the pinch goes like one over major radius, times the momentum diffusivity. This is always inward for outward ballooning turbulence, and insensitive to the details such as the mode propagation direction. On the other hand, the curvature driven thermo-electric part arises from the ion temperature fluctuations. The ratio between the thermo-electric pinch and TEP pinch goes like a temperature ratio. Unlike the TEP pinch, the sign of the thermoelectric pinch depends on the mode propagation direction. It's inward for TEM, but for ITG it could be either inward or outward. Both TEP and thermoelectric pinches are independent of the sign of the magnetic field or the plasma current.

5. Conclusions

Finally, I would like to mention some remaining issues which need to be addressed in the near term. On the H-mode edge pedestal physics, even after many years' work there's no clear trend from empirical scalings. Furthermore, there's no credible theoretical prediction on the L-H transition power threshold either. On ELMs, beyond the usual identification of Type-I ELMs, in terms of medium mode number ideal MHD instabilities, more work is needed for the nonlinear regime and for other benign ELMs. In particular, resolving a possible turbulent cascade down to the ion gyroradius range in the ELM crash phase might be needed[52]. For particle transport, it seems that an inward pinch is needed almost ubiquitously to match experimental data, while the quasi-linear flux based on linear instabilities predicts an inward particle pinch only for particular values of collisionality and η_e. On the other hand, TEP theory predictions seem to fit the experimental observations better[53, 54]. Therefore, we need a better understanding and firmer theoretical justification for the equipartition theory. More interactions among different topical areas are badly needed for new problems such as turbulent transport of energetic ions in the presence of high mode number Alfvenic instabilities and their interactions with $\mathbf{E} \times \mathbf{B}$ flows.

Acknowledgments

The author would like to thank P.H. Diamond, K. Itoh, S.-I. Itoh, X. Garbet, Z. Lin, W. Wang, G. Rewoldt and O. Gurcan for useful interactions regarding the issues presented in this lecture. He would also like to express his appreciation of the hospitality by S. Benkadda.

This work was supported by the U.S. Department of Energy Contract No. DE–AC02–76–CHO-3073 (PPPL).

[1] P.H. Diamond, S.-I. Itoh, K. Itoh, and T.S. Hahm, Plasma Phys. Control. Fusion, **47**, R35 (2005).

[2] Z. Lin, T.S. Hahm, W.W. Lee *et al.,* Science **281**, 1835 (1998).

[3] A. Hasegawa, C.G. Maclennan, and Y. Kodama, Phys. Fluids, **22**, 2122 (1979).

[4] R.Z. Sagdeev *et al.,* Sov. J. Plasma Phys. **4**, 306 (1978).

[5] P.H. Diamond *et al.,* IAEA-CN-69/TH3/1 (IAEA, Vienna, 1998).

[6] L. Chen *et al.,* Phys. Plasmas **7**, 3129 (2000).

[7] F.L. Hinton and M.N. Rosenbluth, Plasma Phys. Control. Fusion, **41**, A653 (1999).

[8] H. Biglari, P.H. Diamond, and P.W. Terry, Phys. Fluids B **2**, 1 (1990).

[9] S.-I. Itoh and K. Itoh, J. Phys. Soc. Japan, **59**, 3815 (1990).

[10] T.S. Hahm and K.H. Burrell, Phys. Plasmas **2**, 1648 (1995).

[11] W.X. Wang, T.S. Hahm, W.W. Lee *et al.,* Phys. Plasmas **14**, 072306 (2007).

[12] W. Horton, D.-I. Choi, and W.M. Tang, Phys. Fluids **24**, 1077 (1981).

[13] S.C. Cowley, R.M. Kulsrud, and R.N. Sudan, Phys. Fluids B **3**, 1803 (1991).

[14] F. Romanelli and F. Zonca, Phys. Fluids B **5**, 4081 (1993).

[15] G.W. Hammett, M. Beer, W. Dorland *et al.,* Plasma Phys. Control. Fusion **35**, 973 (1993).

[16] R. Waltz *et al.,* Phys. Plasmas **1**, 2229 (1994).

[17] Z. Lin, S. Ethier, T.S. Hahm, and W.M. Tang, Phys. Rev. Lett. **88**, 195004 (2002).

[18] T.S. Hahm, P.H. Diamond, Z. Lin, *et al.*, Plasma Phys. Control. Fusion **46**, A323 (2004).

[19] X. Garbet, L. Laurent, A. Samain *et al.,* Nuclear Fusion **34**, 963 (1994).

[20] Z. Lin and T.S. Hahm, Phys. Plasmas **11**, 1099 (2004).

[21] L. Villard, S.J. Allfrey, A. Bottino *et al.,* Nucl. Fusion **44**, 172 (2004).

[22] R.E. Waltz and J.M. Candy, Phys Plasmas **12**, 072303 (2005).

[23] V. Naulin, A.H. Neilsen, and J. Juul Rasmussen, Phys. Plasmas **12**, 122306 (2005).

[24] O.D. Gurcan, P.H. Diamond, and T.S. Hahm, Phys. Rev. Lett. **97**, 024502 (2006).

[25] M. Yagi, T. Ueda, S.-I. Itoh, M. Azumi, K. Itoh, P.H. Diamond, and T.S. Hahm, Plasma Phys. Control. Fusion **48**, A409 (2006).

[26] F. Zonca, S. Briguglio, L. Chen *et al.,* Plasma Phys. Control. Fusion **48**, B15 (2006).

[27] T.S. Hahm, M.A. Beer, Z. Lin *et al.,* Phys. Plasmas **6**, 922 (1999).

[28] N. Miyato *et al.,* Phys. Plasmas **11**, 5557 (2004).

[29] P. Angelino et al., Plasma Phys. Control. Fusion **48**, 557 (2006)

[30] Z. Lin, T.S. Hahm, W.W. Lee *et al.,* Phys. Rev. Lett. **83**, 3645 (1999).

[31] G.L. Falchetto and M. Ottaviani, Phys. Rev. Lett. **92**, 25002 (2004).

[32] D. Mikkelsen *et al.,* IAEA-CN-71/EX-P5-03 (IAEA, Vienna, 2002).

[33] F. Ryter *et al.,* Phys. Rev. Lett. (2005).

[34] G. Conway, *Private Communication* (2002).

[35] F. Jenko *et al.,* Phys. Plasmas, **7**, 1904 (2000).

[36] W. Dorland *et al.,* Phys. Rev. Lett. **85**, 5579 (2000).

[37] Z. Lin *et al.,* Phys. Rev. Lett. **99**, 265003 (2007).

[38] S. Kaye *et al.,* Nucl. Fusion (2007).

[39] K.L. Wong *et al.,* Phys. Letters A **236**, 339 (1997).

[40] P.H. Diamond and Y.B. Kim, Phys. Fluids B **3**, 1626 (1991).

[41] S. Scott, P. Diamond, R. Fonck, *et al.,* Phys. Rev. Lett. **64**, 531 (1990).

[42] T.S. Hahm and W.M. Tang, Phys. Fluids B **1**, 1185 (1989).

[43] N. Mattor and P.H. Diamond, Phys. Fluids **31**, 1180 (1988).

[44] T. Tala, K. Crombe, P.C. de Vries *et al.,* Plasma Phys. Control. Fusion **49**, B291 (2007).

[45] M. Yoshida, Y. Koide, H. Takenaga *et al.,* Plasma Phys. Control. Fusion **48**, 1673 (2006).

[46] A. Peeters, *Private Communications* (2006).

[47] L.-G. Eriksson, E. Righi, and K.D. Zastrow, Plasma Phys. Control. Fusion **39**, 27 (1997).

[48] J.E. Rice, E.S. Marmar, F. Bombarda, and L. Qu, Nucl. Fusion **37**, 421 (1997).

[49] O.D. Gurcan, P.H. Diamond, T.S. Hahm, and R. Singh, Phys. Plasmas **14**, 042306 (2007).

[50] T.S. Hahm, Phys. Fluids **31**, 2670 (1988).

[51] T.S. Hahm, P.H. Diamond, O.D. Gurcan, and G. Rewoldt, Phys. Plasmas **14**, 072302 (2007).

[52] B.D. Scott, *Private Communication* (2006).

[53] X. Garbet, N. Dubuit, E. Asp, Y. Sarazin, C. Bourdelle, P. Ghendrih, and G.T. Hoang, Phys. Plasmas **12**, 082511 (2005).

[54] X. Garbet, *Private Communication* (2007).

Edge Localised Modes (ELMs): Experiments and Theory

J W Connor[1], A Kirk[1] and H R Wilson[2]

[1]*EURATOM/UKAEA Fusion Association, Culham Science Centre, Abingdon, Oxon, UK, OX14 3DB*
[2]*University of York, Heslington, York, UK, YO10 5DD*

Abstract. Edge Localised Modes (ELMs) are periodic disturbances of the plasma periphery occurring in tokamaks with an H-mode edge transport barrier. As a result, a fraction of the plasma energy present in the confined hot edge plasma is transferred to the open field lines in the divertor region, ultimately appearing at the divertor target plates. These events can result in high transient heat loads being deposited on the divertor target plates in large tokamaks, potentially causing damage in devices such as ITER. Consequently it is important to find means to mitigate their effects, either avoiding them or, at least, controlling them. This in turn means it is essential to understand the physics causing ELMs so that appropriate steps can be taken. It is generally agreed that ELMs originate as MHD instability caused by the steep plasma pressure gradients or edge plasma current present in H-mode, the so-called 'peeling-ballooning' model. Normally this is considered to be an ideal MHD instability but resistivity may be involved. Much less clear is the non-linear evolution of these instabilities and the mechanisms by which the confined edge plasma is transferred to the divertor plasma. There is evidence for the non-linear development of 'filamentary' structures predicted by theory, but the reconnection processes by which these are detached from the plasma core remain uncertain. In this paper the experimental and theoretical evidence for the peeling-ballooning model is presented, drawing data from a number of tokamaks, e.g. JET, DIII-D, ASDEX-Upgrade, MAST etc. Some theoretical models for the non-linear evolution of ELMs are discussed; as well as ones related to the 'peeling-ballooning' model, other candidate models for the ELM cycle are mentioned. The consequential heat loads on divertor target plates are discussed. Based on our current understanding of the physics of ELMs, means to avoid them, or mitigate their consequences, are described, e.g. the use of plasma shaping or introducing resonant magnetic perturbation coils to reduce plasma gradients at the plasma edge.

Keywords: ELMs, peeling-ballooning modes, filaments, divertor heat loads, ELM mitigation
PACS: 52.55 Fa 52.55 Rk 52.55 Tn

1. INTRODUCTION

Edge Localised Modes (ELMs) [1, 2, 3, 4] are a ubiquitous feature of tokamak H-mode discharges, the baseline operational scenario planned for ITER [3, 4]. The application of additional heating power to ohmic discharges in the 1970's led to degraded, L(low)-mode confinement, but in the early 1980's Wagner et al [5], carrying out experiments on ASDEX, discovered that an improved confinement regime, termed 'H(high)-mode', appeared above a threshold in the additional heating power. These discharges were characterised by steep edge gradients in electron density and electron and ion temperatures (i.e. steep pressure gradients), resulting in the improved confinement, edge instabilities and periodic bursts of D_α emission from enhanced plasma-wall interactions. It transpired that such behaviour is generic to divertor tokamaks: these are magnetic configurations in which divertor coils produce an X-point in the poloidal field structure so that there is a separatrix, a surface that separates closed magnetic surfaces in the interior of the plasma column from open field lines leading to divertor plates along which the plasma exhaust takes place. Such a magnetic configuration is shown in Fig. 1, which also maps the set of nested magnetic surfaces in the confined plasma region onto a radial plasma pressure profile. This profile exhibits the steep gradients in the H-mode transport barrier at the plasma edge that lead to a so-called edge pedestal.

The instabilities occurring in the plasma edge region of H-mode discharges were termed ELMs. In this paper we shall address four questions concerning them:

- What are they?

CP1013, *Turbulent Transport in Fusion Plasmas, First ITER International Summer School*
edited by S. Benkadda
2008 American Institute of Physics 0-7354-0534-9/08/$23.00

- Why do they matter?
- What causes them?
- Can we control them?

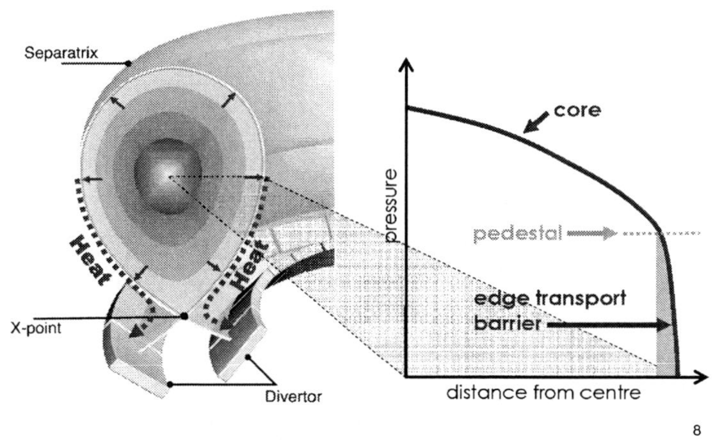

FIGURE 1. Left: The magnetic geometry of a divertor tokamak with an X-point in the poloidal field, illustrating how the heat flux emerging from the plasma core at the separatrix flows along the open fields lines to the divertor target plates; Right: A mapping of the plasma pressure on the two-dimensional magnetic surfaces onto a one-dimensional radial pressure profile, showing the steep gradients in the H-mode edge transport barrier leading to an edge pressure pedestal supporting the core profile (reproduced courtesy of Dr R. Buttery).

We shall consider these questions in detail in subsequent sections, but here we shall provide brief answers to them. ELMs are periodic disturbances of the plasma edge region, accompanied by expulsion of edge plasma and MHD activity, leading to confinement degradation. Since their main effect is to reduce the edge pedestal height this effectively changes the edge boundary condition for the core transport. In particular, for 'stiff' transport models like the turbulent transport due to the ion temperature gradient (ITG) mode, which tend to maintain the temperature profile at some marginally stable value [3, 4], this clearly lowers the core energy, exacerbating the energy confinement degradation, as shown in Fig. 2.

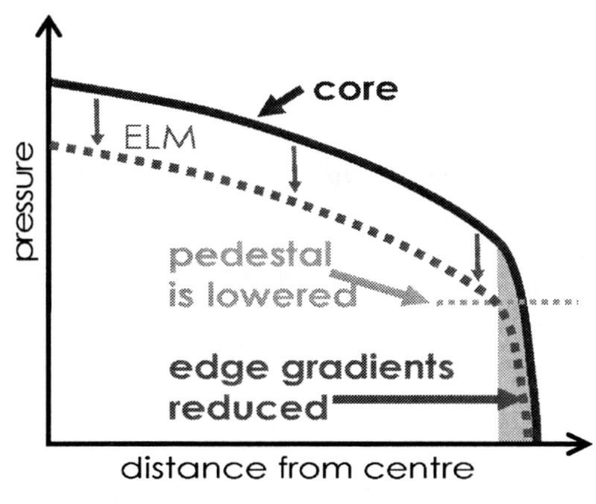

FIGURE 2. The effect of losing plasma energy from the edge plasma as a result of an ELM; not only is there a reduction in the edge gradients and a lowering of the edge pedestal, but this leads to a reduction in core confinement, particularly for stiff transport models (reproduced courtesy of Dr R. Buttery).

The deleterious effect arising from the degradation of energy confinement due to ELMs is more than offset by their beneficial effects in removing impurities and controlling plasma density, allowing one to maintain a time-averaged steady state H-mode. However there are serious consequences from ELMs: the most serious one results from the potentially high, transient heat loads they can deposit on the divertor target plates in an ITER-scale device, causing unacceptable levels of erosion [6]. This increases rapidly with the energy deposited on the target plates, as shown in Fig, 3. Another down-side of large ELMs is the danger that they might trigger core MHD modes leading to further confinement degradation or collapse of internal transport barriers (ITBs) [7].

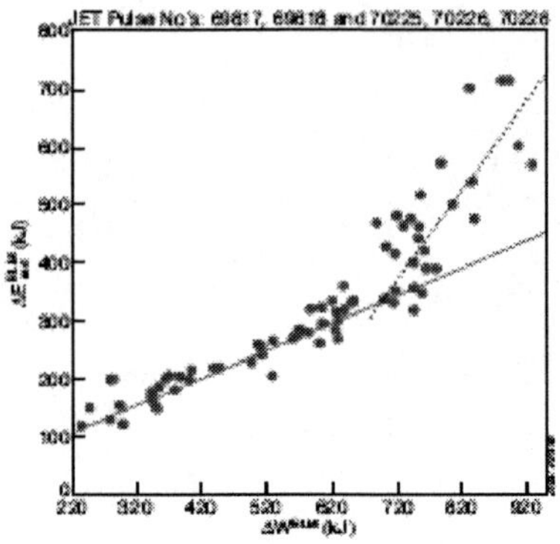

FIGURE 3. Radiation ΔE^{ELM}(kJ) from the divertor target plates due to Type I ELMs on JET as a function of ELM energy, ΔW^{ELM}(kJ); it increases sharply above a critical ELM energy loss, ~ 700kJ, implying greatly increased erosion of their surfaces [6] (reproduced courtesy of Dr A Huber and EFDA-JET).

ELMs are believed to be triggered by some form of ideal or resistive MHD instability caused by the steep edge gradients. Elucidating their nature and cause is the key to controlling them: our present understanding is suggesting control methods and operating scenarios where the benefits of H-mode confinement are retained but the more undesirable features of ELMs are absent, or at least mitigated.

2. ELM Types and Characteristics

ELMs were first characterised by the periodic enhanced plasma-wall interaction, manifested in the increased D_α (or H_α) emission observed [5]. This is a result of the rapid expulsion of the edge plasma, leading to a drop in the edge pedestals in both n_e and T_e, which then build up again on a longer timescale until the cycle repeats. In the early days only two types of ELM were identified: large amplitude, but less frequent Type I ELMs with a crash occurring on a fast timescale (~ 100 -300 μsecs) and low amplitude, more frequent Type III ELMs [1-4].

ELMs are associated with loss of edge plasma energy, δW_{ELM}, and particles, δN_{ELM}: Type I ELMs typically expel 10-15% of the pedestal plasma energy, δW_{ped}, and 4% of the particle inventory, Type III, say, 1-5%, of the plasma energy or particles [1-4]. The impact on the time-averaged global energy confinement depends inversely on the frequency, f_{ELM}, of the ELMs, so Type III (with f_{ELM} ~ 10Hz) are more serious than Type I (with f_{ELM} ~ 100 Hz) in this regard, as shown in JET for example [8]. Zohm has proposed a formula to represent the impact of ELMs on energy confinement [2]. The reduction in confinement from that in ELM-free H-mode is represented by a factor η, where

$$\eta = 1 - \left[1 - (r_{ELM}/a)^2\right](f_{ELM}\,\delta W_{ELM}/P) \qquad (1)$$

Here r_{ELM} is the inner radius of the ELM-affected region, a is the plasma minor radius and P the heating power. Typically one finds η ~ 0.85, in accord with the results arising from analysis of ELMy H-mode global energy confinement databases [3, 4].

However the ELM energy ultimately appears on the vessel wall or divertor target plates as it is transported along the open field lines. The thermal energy pulse propagates to the divertor at the sound speed [9 -11], although a burst of high energy electrons precedes this [12]. There is some correlation between the energy incident on the divertor and the loss of pedestal energy, but this is not always so [11]. Furthermore the energy appearing at the divertor increases as the collisionality, ν_*, reduces towards ITER-like values, as shown in Fig. 4 [13]. There it can cause ablation of the divertor tiles and, as the ablation process rises sharply with incident energy (say above 1 MJ, as shown in Fig.3), lead to serious erosion [6, 14]. Thus, since for fixed plasma parameters the ELM energy scales as $(\text{size})^3$, while the tile area scales as $(\text{size})^2$, Type I ELMs pose a serious threat to ITER operation, though they are not important regarding confinement degradation.

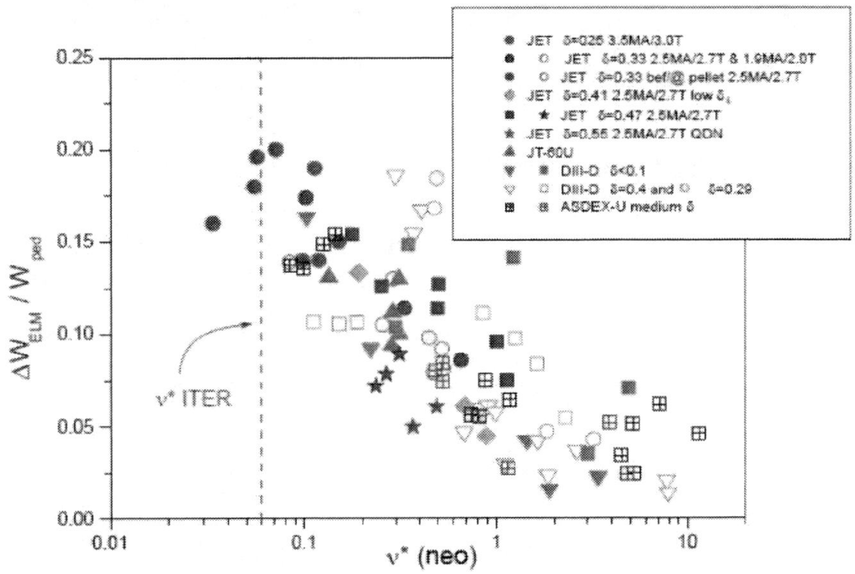

FIGURE 4. The loss of plasma energy, ΔW_{ELM}, due to an ELM event relative to the pedestal value, W_{ped}, as a function of plasma collisionality, ν_*, based on data from a wide range of tokamaks [13]; it increases as ν_* decreases towards ITER-like values, reaching (15 - 20)% (reproduced courtesy of Dr A Loarte)

Magnetic signals help to shed light on the nature and causes of ELMs. Type III ELMs are often preceded by magnetic precursors (say with poloidal mode number $m \geq 10$ and frequency $f_{ELM} \sim 100$ kHz) appearing on Mirnov coils, whereas Type I can appear as precursor-less, rapid bursts of broadband activity lasting ~ 100 μsec [1, 15]. Since the thermal energy transients appearing at the divertor plates can be preceded by fast electrons moving along open field lines [12], this suggests magnetic reconnection processes are involved in the ELM evolution. The collapse of the density associated with a Type I ELM occurs mainly on the low-field side [16, 17], suggesting a role for ballooning instability; the density perturbation then propagates to the high-field side at the sound speed. Recent developments using visible camera image, pioneered on MAST [18-20], have revealed the filamentary nature of ELMs as they emerge from the plasma column.

These filaments, which are aligned along the magnetic field lines have a radial extent ~ 5 – 10 cm and typically have toroidal mode numbers, n ~ 10 – 15; subsequently they move radially at up to a km/s, accelerating in time (> 106 m/s^2), and toroidally, decelerating in time. Such filaments are routinely observed in ELMing tokamak discharges around the world [21-24]: Fig. 5 shows examples from MAST and ASDEX Upgrade. Interestingly there were very early observations on COMPASS-D [25].

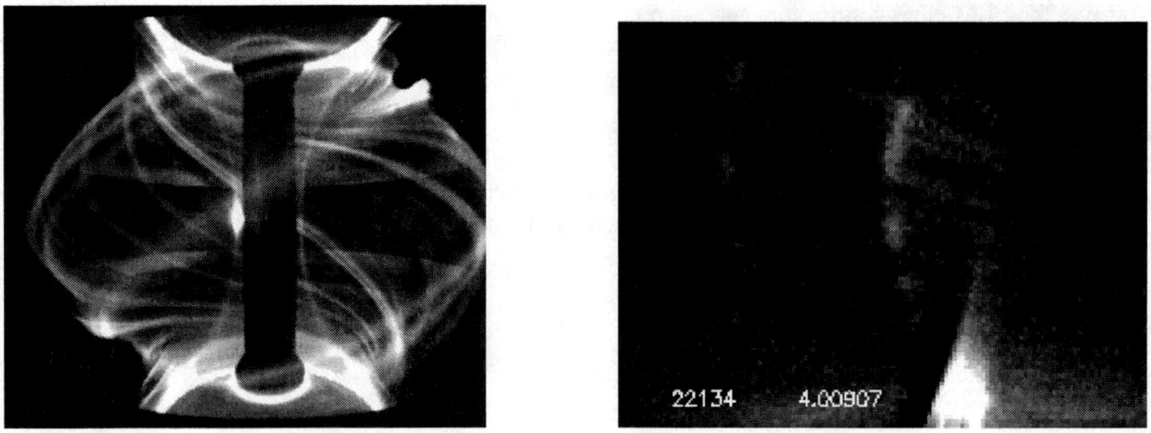

FIGURE 5. Examples of filamentary structures associated with ELMs from MAST (left) and ASDEX Upgrade (right).

Over the years a wider range of ELM types has emerged. Indeed some six types have now been identified: Types I – V and 'Grassy' ELMs. With increasing injected power, P_{inj} , one sees a transition from Type III to Type I on most machines, as shown in Fig. 6 from MAST [17]. During such a power scan the ELM frequency first drops during the Type III phase, before starting to increase with the onset of Type I ELMs. Typically the transition occurs when $P_{inj} \sim (1.5 - 2)xP_{L\text{-}H}$, where $P_{L\text{-}H}$ is the threshold power for the L-H transition [26].

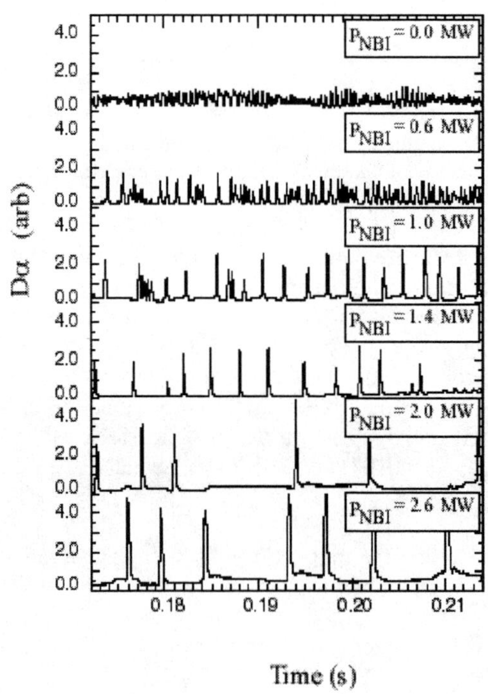

FIGURE 6. The frequency of ELMs, f_{ELM}, as manifested in D_α emission as the injection power from the neutral beams is increased on MAST; above a threshold value, $P_{inj} \sim (1.5 - 2)xP_{L\text{-}H}$, where $P_{L\text{-}H}$ is the threshold power for the L-H transition, there is a transition from Type III ELMs, where f_{ELM} decreases with P_{inj}, to Type I where it starts to increase [17].

The energy loss from the higher frequency Type III ELMs reduces confinement so that the energy confinement time is (10 - 30) % below that in Type I (or ELM-free) H-mode. An insight into the physics behind Type III ELMs might be obtained from the fact that they are observed below a critical pedestal temperature, T_{crit}, which tends to increase with the toroidal magnetic field (in ASDEX Upgrade, for example, $T_{ecrit} \sim 350eV$ [27]); this suggests they require higher electron collisionality, ν_{*e}. More revealing is the $n_{edge} - T_{edge}$ diagram pioneered by ASDEX Upgrade [28], where Type I ELMs appear above a hyperbola $n_{edge}T_{edge} \sim$ constant, the constant corresponding to the theoretically predicted onset of pressure driven, ideal MHD ballooning mode instability, while Type III occupy a region limited above by T_{ecrit} and this hyperbola. At lower density and higher T_e, i.e. lower collisionality, there is a branch of Type III ELMs which has been termed Type IV [29] - it is an open question whether this low collisionality branch is due to the same physics as the higher collisionality Type III regime. Figure 7 from DIII-D [29] illustrates the operational space for various ELM types in terms of the edge pedestal values of T_e and n_e.

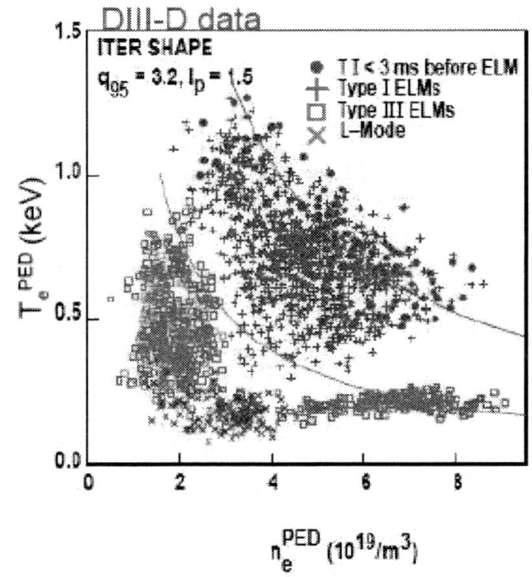

FIGURE 7. The operational diagram in the $T_e^{ped} - n_e^{ped}$ space for different ELM types in DIII-D; Type III tend to appear below a critical temperature although there is a low density branch, sometimes termed Type IV; Type I ELMs cluster around the stability boundary for pressure driven ideal MHD ballooning modes shown by the hyperbola given by $T_e^{ped} n_e^{ped}$ = constant (from Ref. 29, reproduced courtesy of IAEA and Dr T Osborne).

Yet another type of ELMs is Type II, which has even smaller energy loss per ELM than Type III [30, 31]. These require proximity to double null and highly shaped plasmas. Like Type III they require a low T_{ped}, comparable to that for Type III. They have a narrow operational window at high density, $n/n_G \sim 0.5 - 1.0$ (where n_G is the Greenwald density [3, 4]), which means the pedestal pressure, p_{ped}, is high, comparable with that for Type I, and yields Type I – like confinement; indeed Type I and Type II often co-exist. Again, like Type I, Type II ELMs are associated with broadband MHD activity. The combination of good confinement and small energy loss makes them an attractive scenario, but there is a limited operational regime for accessing them and the extrapolation to ITER is uncertain.

Another class of small ELMs with good confinement is Grassy ELMs. These have mainly been seen on JT-60U [32]; high β_p is the critical parameter (β_p is the poloidal β), but high triangularity, δ, and edge q, q_{edge}, are also required – again a narrow operational region, as shown in Fig. 8 [33]. Rotation shear is seen to be conducive to their appearance [34]. Finally there are Type V ELMs, which have been observed on NSTX at high pedestal collisionality [35]. This ELM regime has some similarities with the EDA/HRS (Enhanced D Alpha/High Recycling Steady) ELM-free H-mode regimes discussed later, but the coherent oscillation characteristic of that regime is absent.

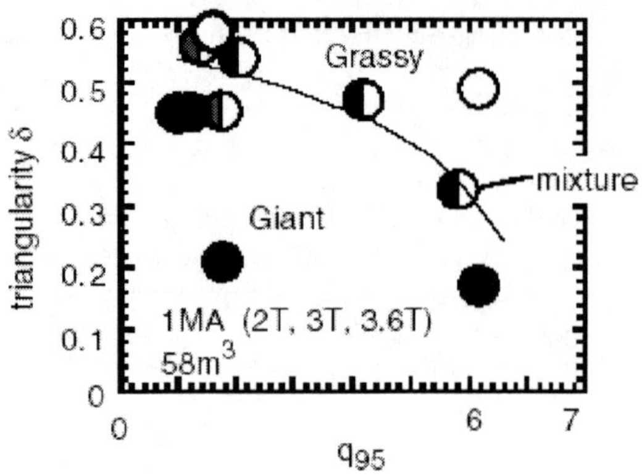

FIGURE 8. The operational diagram for small amplitude ('Grassy') ELMs in JT-60U [33]; these replace large, Type I, ELMs at higher triangularity, δ, and edge q_{95}.

3. Stationary ELM-free Regimes

While one benefits from the confinement improvement associated with H-mode operation, the ELM-free H-mode provides an unsuitable operational scenario due to the uncontrolled build-up of plasma density and impurities. Whereas one can maintain a time-averaged steady state in ELMy H-mode, we have seen ELMs bring their own problems and a stationary ELM-free steady state with H-mode confinement would be very attractive. Three such scenarios have been identified, but nevertheless have their limitations, such as limited operational windows. As a by-product, understanding the required conditions for them may provide insights into the nature and causes of ELMs.

The first is the so-called EDA (Enhanced D-Alpha) mode discovered on C-Mod [36]. It is observed at high density and low-to-modest heating power; high edge collisionality with $n/n_G \geq 1.5$ appears necessary. Its global energy confinement competes with the Type I ELMy regime. An important feature is the existence of a quasi-coherent mode (with f ~ 50-120 kHz and high m, n) which serves to maintain a constant density. An interpretation for this oscillation in terms of micro-tearing modes [37] has been proposed. In dimensionally similar experiments, ASDEX Upgrade and DIII-D have observed this mode, but have not established the steady state regime. A second candidate is the Q(quiescent)-H-mode, first seen on DIII-D [38-40] and has been robustly reproduced on ASDEX Upgrade [41]. The QH mode is only observed in counter NBI with a large plasma wall distance, suggesting a role for fast ions. The operating space corresponds to a low density regime ($n/n_G \sim 0.04$) with pedestal pressures comparable to Type I ELMy H-mode. This type of discharge also involves a steady oscillation, the Edge Harmonic Oscillation (EHO), a non-sinusoidal mode with a mix of n (n = 1, 2, 3, 4…) and f ~ 5-11 kHz, possibly accompanied by a higher frequency oscillation (350-500 kHz) [39]. Finally we mention the HRS (High Recycling Steady) H-mode, first seen on JFT-2M [42]. It is very similar to the EDA Mode and access is favoured by high density and neutral pressure, corresponding to high edge collisionality, ν_*. Again there is high frequency (> 100 kHz) edge mode activity.

4. Theoretical Models

The reproducible periodic nature of the ELM cycle and the fast crash of the ELM event itself provide challenges for theory. A successful model needs to address a number of key features:

- the trigger mechanism
- the fast crash
- the energy loss
- the periodic cycle
- the heat loads appearing on the divertor plates
- the observed operating space for various ELM types
- the observed plasma perturbations: magnetic signals, filaments etc

In addition it must be capable of predictions for the effect of ELMs on ITER and suggest how they can be avoided, or at least mitigated; this may become apparent when their nature has been elucidated.

Clues to the physics behind ELMs lie in the operating diagram of Refs. 28 and 29. Thus Type I ELMs appear to occur at the ideal MHD ballooning boundary, as found in a number of early stability calculations for various devices. (see, for example, the review [1] and also [43-45]), while resistive modes seem to be involved in Type III ELMs [46]. However there also appeared to be a role for ideal kink modes in Type I ELMs [47, 48]. Nevertheless, much earlier ELM modeling activity involved constructing 'toy' dynamical models with sufficient ingredients to capture various elements of ELM phenomenology. A number of these exploited 'limit cycle' solutions to models previously developed to explain the L-H transition [49]; such models involve coupled non-linear equations for plasma profiles and stabilizing radial electric field shear [50-53]. More sophisticated models introduced an equation for the turbulent fluctuations causing transport [54-58]. A further development in this direction, particularly relevant for Type I ELMs, were models including a role for MHD fluctuations. Thus Ref. 59 involved the interaction of ambient drift wave and MHD turbulence, mediated by the stabilising effects of plasma flow. The impact of the resulting turbulent transport on the gradients that drive instability, themselves driven by external heating, enabled the model to qualitatively reproduce much ELM phenomenology for both Types I and III ELMs. Closer to first-principles calculations, were dynamical models based on a truncation, at a low number of harmonics, of the full equations describing fluid plasma instabilities, such as resistive interchange modes in the plasma edge or scrape-off layer (SOL), relevant to Type III ELMs [60, 61]. Interestingly, one of the first models involved a numerical solution of the full non-linear equations describing resistive, current driven edge instabilities (so-called resistive 'peeling modes') [62].

In addition a number of essentially analytic models have been developed, based on some particular instability in the plasma edge or SOL, such as ideal or resistive ballooning or interchange modes, complemented by qualitative statements about their non-linear development. Thus Ref. 63 involved a two-stage process: in the first stage ideal interchange instability is excited by steep plasma gradients near the X-point, expelling a small amount of plasma that leads to an influx of impurities. In the second stage, this low conductivity region acts as an effective limiter resulting in a fast growing interchange instability, representing a Type I ELM. The model provides a reasonable estimate for δW_{ELM} and a scaling $f_{ELM} \sim PB/I_P^3$, which is consistent with JET data [64]; here P is the heating power, I_P is the plasma current and B the toroidal field. An analytic model providing a quantitative description of Type I ELMs that is based on the Current Diffusive Ballooning Mode (CDBM), a pressure driven instability, has been proposed [65-67]. It provides an explanation of the ELM trigger, the fast crash and the ELM cycle, and presents expressions for δW_{ELM} and the dependence of f_{ELM} on P that accord with observation.

However over the last decade there has been considerable support growing for the so-called 'peeling-ballooning mode' model [68-73] as the trigger for the ELM. This model builds on the two instability sources near the plasma edge: current and pressure gradients and maps out a 'triangular' operating diagram for ELMs in the space of normalized edge current, $\left(Rqsj_{\parallel}/B\right)$, and ballooning pressure gradient parameter, $\alpha = -\left(2Rq^2\mu_0/B^2\right)\left(dp/dr\right)$, see Fig. 9. For a large aspect ratio tokamak, the peeling stability criterion is given by:

$$\alpha\left[D - s\Delta'_{Sh}\right] > \left(Rqs/B\right)j_{\parallel},$$
$$j_{\parallel} = j_{ohmic} + j_{Bootstrap}, \qquad j_{Bootstrap} \sim \alpha f(\nu_{*e});$$
$$D = \varepsilon\left(1 - q_a^{-2}\right) \qquad (2)$$

and Δ'_{Sh} is the Shafranov shift., corresponding to the shifted centres of the circular flux surfaces of the s-α equilibrium model. The term D, representing favourable 'average curvature' for $q_a>1$, where q_a is the edge value of q, can be enhanced by triangular plasma shaping, δ. The ideal MHD, pressure gradient driven ballooning stability boundary is given by $\alpha = \alpha_{crit}$. Thus instability can occur when $\alpha > \alpha_{crit}$ due to ballooning modes, or when the edge current drive (responsible for the peeling mode) exceeds the stabilising effect of average curvature, itself proportional to α, as in eqn. (2).

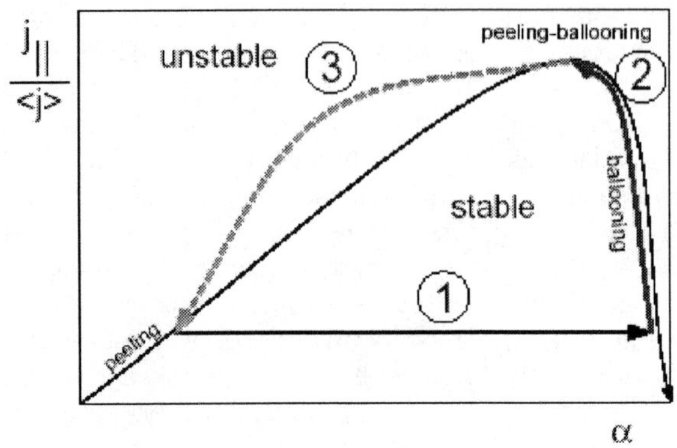

FIGURE 9. The stability diagram for peeling-ballooning modes in terms of normalized edge current and pressure gradient parameter, α. This provides a basis for a conceptual model of the ELM cycle. During phase 1 the plasma heats up until it reaches the ballooning stability boundary. This is a 'soft' limit so that during phase 2 it hovers there as the Ohmic and bootstrap currents increase with increasing temperature until the discharge reaches the peeling-ballooning boundary. At this point it plunges into the unstable region 3, rapidly losing pressure and current until it crosses the peeling stability boundary to reach a stable point at which the cycle begins to repeat.

A cartoon picture of the Type I ELM cycle, shown in Fig. 9, proposes that plasma heating drives α to α_{crit}, where it hovers; meanwhile the rising temperature drives more Ohmic current until it reaches the peeling limit, at which point pressure is expelled. Since the current diffusion time is much longer than that for energy diffusion, it takes longer for the current to reach a steady state value than does the pressure gradient. The trajectory in the stability diagram would then lead to a rapid growth of the ELM instability and further energy loss, namely the crash, until it relaxes to a stable state when the cycle begins to repeat. The effect of including the pressure driven bootstrap current would exacerbate the peeling drive at low collisionality. A less well-developed picture of Type III ELMs proposes that they follow a modified trajectory in the stability diagram. Thus they meet instability by passing through the peeling boundary at lower α, the increase in α being prevented by the onset of resistive ballooning modes at the lower temperatures relevant to Type III ELMs [74].

There have been numerous successful comparisons of the peeling-ballooning stability diagram for realistic tokamak geometry with experimental data on the appearance and type of ELMs [75-77], as shown in Fig. 10 for example [78]. In addition, the increased stabilising average curvature of highly triangular tokamak plasmas is found to correlate with the occurrence of smaller Type II ELMs. This is consistent with linear stability calculations showing that the radial mode width is smaller, implying less plasma volume would be expelled by the ELM [79-81], although this is not always so [82]. Studies on COMPASS-D showed the stabilising effect of a current ramp-down [83], as expected for a model involving peeling modes. The impact of rotation shear is only found to be significant for the modes responsible for ELMs in tight aspect ratio tokamaks [84], e.g. MAST, where flows are sonic and approach the Alfvén speed. Experimentally an effect of rotation on Grassy ELMs [34] and ELM filaments [85] is observed. The simple MHD model needs to be modified to account for diamagnetic effects in the steep edge gradients of the pedestal. Usually this is a minor effect, but can be significant at high β_p or high q_{edge} for narrow pedestals [74, 86]. The most unstable modes tend to have modest values of n, say n ~ 10, and occur where the mode structure has a mixed peeling-ballooning nature [74]. Lower n modes occur for lower values of α, when the mode is predominately peeling. Low edge current, but high α regimes, produce high-n ballooning modes which may be more benign, merely limiting the plasma pressure gradient.

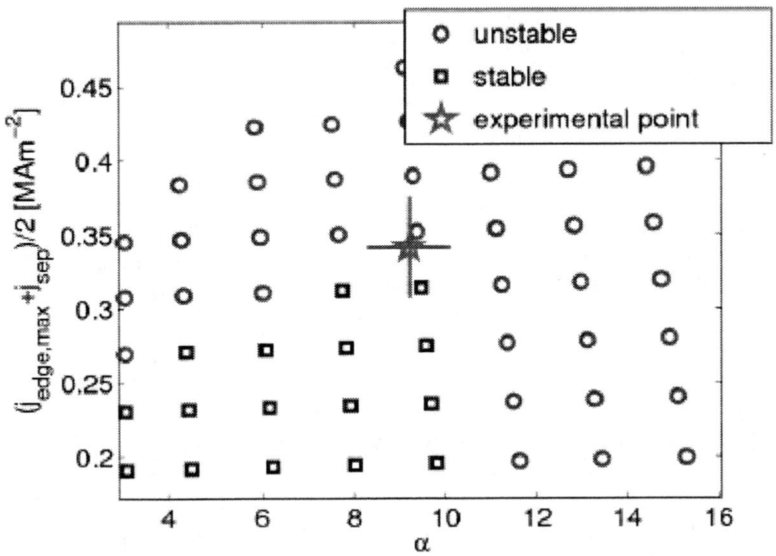

FIGURE 10. Stability diagram for a specific MAST discharge in terms of edge current and pressure gradient parameter, α, showing the experimental point lies on the boundary of the stable region [78] (provided by Dr S. Saarelma)

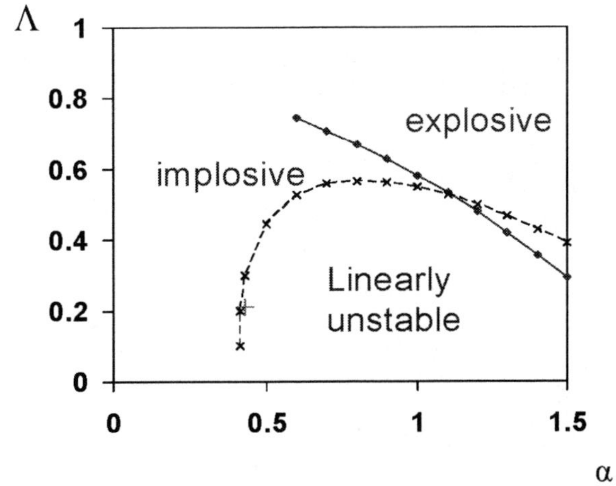

FIGURE 11. Non-linear stability diagram for ballooning modes in the space of normalized edge current, Λ, and pressure gradient parameter, α, for a set of equilibria based on a JET example, marked by a cross. The red curve marks the boundary between explosive and implosive growth of filaments, while the blue line denotes the linear stability boundary [89].

The above picture of the ELM cycle is very qualitative; what are needed are calculations of the non-linear evolution of the peeling-ballooning modes. Such a calculation has been carried out for the pure high-n ballooning modes [87, 88]. The non-linear terms produce explosive growth, although this calculation cannot be carried through to the actual crash. It is found that one can characterise the conditions for explosive growth in terms of an operational diagram in the space of a normalized edge current, Λ, and the parameter α, as shown in Fig. 11 [89]. The predicted mode structure has a filamentary structure, aligned along the magnetic field and these filaments progressively narrow, approximately in the poloidal direction, while expanding radially. The model thus predicts the expulsion of a number (~ 10 - 15) of hot filaments (flux tubes) of plasmas from the pedestal region. Two-fluid, non-linear simulations for intermediate values of n, using BOUT [90], also provide evidence for an explosive behaviour and filamentary structures, although the JOREK code finds that low-n modes appear to saturate, rather than explode [91]. The predictions of filaments were confirmed on the MAST experiment, and on a number of other tokamaks since, adding credence to the model [18-24].

However a complete ELM model needs to address how the plasma filaments emerging from the plasma column become detached and plasma energy is transported to the vessel wall or divertor plates. Clearly this requires some mechanism involving magnetic reconnection, which could be facilitated near the divertor X-point. Again some qualitative pictures have been proposed to describe this, but a quantitative model is lacking. Three suggestions for a qualitative picture are [92]:

- The filaments act as 'leaky hosepipes': within the ideal MHD model with no reconnection, the hot flux tube of plasma twists and pushes out between field lines on neighbouring surfaces; energy and particles can only be lost by diffusion from the filament into the cold scrape-off layer (SOL), this process being enhanced as the filament narrows.
- The filaments act as a 'conduit', directly connecting the hot core plasma to the divertor target plates as a result of reconnection, which preferentially occurs as the filaments cross the divertor X-point where magnetic shear is large, ripping them apart. Different mechanisms could occur in DND (Double Null Divertor) and SND (Single Null Divertor) geometries: in SND the filaments can connect directly to the divertor plates; in DND they break-off and this could occur many times during an ELM so that it would be a more benign process.
- The filaments can only be ejected if flow shear is suppressed, leading to enhanced transport and barrier collapse. However it is unclear whether the ballooning mode suppresses the flow or another mechanism suppresses flow shear, allowing filament growth. Again no reconnection is required. Flow shear suppression is observed in MAST [83] and DIII-D [93].

Filamentary structures also arise in so-called 'blob' theories [94]: if a filament of plasma breaks off from the core then it will propagate outwards due to **E×B d**rifts and this mechanism may play a role in the later phase of an ELM [95].

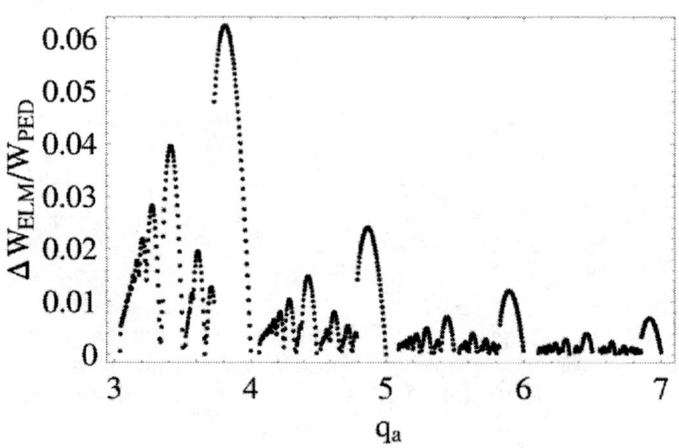

FIGURE 12. Normalized energy loss due to ELMs according to a Taylor Relaxation model, as a function of q_a the edge q, value. The sensitivity of the energy loss to the position of an external resonant surface to the plasma edge introduces a 'deterministic scatter' into the results [96] (reproduced courtesy of Dr C G Gimblett).

An alternative non-linear ELM model, involving 'Taylor Relaxation' of the peeling modes has been proposed [96]. This approach considers an initial state that is marginally peeling unstable and minimises the plasma energy, subject to the constraints of conservation of helicity and the poloidal flux in an outer annulus in the vicinity of the ELM. As a result the current flattens, producing a current pedestal, destabilising the peeling mode; stabilising skin currents form at the boundaries of the relaxed region. Peeling stability is regained when the relaxed width exceeds a critical value, this defining the extent of the ELM-affected region. This allows one to deduce the ELM energy loss. For MAST one finds an ELM width of a few % and 1-5 % of the pedestal energy is lost, but these values have a large spread depending on q_{edge}, the edge value of q – a 'deterministic scatter', see Fig. 12. This calculation ignores the effect of a separatrix in divertor geometry, which is believed to have a stabilising effect on peeling modes, as has been seen in numerical calculations [97, 98]. These calculations, in which the separatrix geometry is progressively approached, indicate that a new resistive peeling-tearing mode, which tears near the X-point, dominates; its growth rate is insensitive to q_{edge} [99]. An analytic approach based on Ref. 100 has been pursued to elucidate this point [101]; it suggests that as one approaches the separatrix, an unstable peeling mode persists but the most unstable mode number increases to higher values, as has been observed in some stability calculations for strongly shaped plasma [84].

As noted above, Type III ELMs are believed to be resistive in nature. Increases in gas-puffing are found to cause a transition to Type III ELMs in JET, consistent with this idea. A resistive interchange model driven by magnetic flutter for triggering ELMs compares well with the n_{edge} – T_{edge} operating diagram in JET [102]. A related idea involving resistive ballooning modes predicts a critical pedestal density, scaling as $B/q^{5/4}$ for the Type I-III transition [103] and also shows agreement with JET data [104]. Resistive ballooning modes are also found to be unstable in JET discharges with Type III ELMs [47, 105].

While there are first-principles models for the trigger and non-linear evolution of ELMs, the modelling of the complete ELM cycle remains primitive. Transport codes are used and involve a number of elements to represent the physics of the ELM cycle:

- A trigger given by some linear stability threshold, say for peeling-ballooning modes.
- A crash time, τ_{ELM}, either taken on a phenomenological basis or estimated on qualitative grounds [106]; e.g. $\tau_{ELM} \sim \tau_{Alfven}\left(\tau_{Res}/\tau_{Alfven}\right)^p$ with $p \sim 1/3 - 1$.
- An estimate of the energy loss δW_{ELM}:

$$\left(\delta W_{ELM}/W\right)_0 \sim \Delta_{ELM}/a \sim 1/nqs$$

  where Δ_{ELM} is the radial width affected by the ELM [106]. This estimate of Δ_{ELM} is more appropriate to kink modes: for ballooning modes one might use $\Delta_{ELM} \sim a/n^{1/2}$.
- A reduction factor on the fraction of energy lost to account for the connection of an ideal MHD mode to the divertor plates

$$\delta W_{ELM}/W \sim \left(\delta W_{ELM}/W\right)_0 \left(1+\tau_{\|}/\tau_{ELM}\right)^{-1}$$

  where $\tau_{\|} = \pi Rq(1+\nu_{*e})/c_s$ with c_s the sound speed [106], is the transit time along the open field lines to the divertor plates

Such modelling has been performed, for example with the ASTRA [106] and JETTO transport, [107] codes. The code evolves j, T_e, n_e, say using a background transport model, monitoring the trajectory in the j – α peeling-ballooning stability diagram until instability is triggered. The ELM loss, δW_{ELM}, is represented by introducing a large cross-field thermal diffusivity, $\chi_\perp$ [108], to represent turbulent transport over the ELM affected region, Δ_{ELM}, for a time τ_{ELM}, say $\chi_\perp \sim 10$ m^{2}s^{-1}. Such models produce cyclical behaviour. Naturally the ELM frequency, f_{ELM}, is given by the inverse of the reheating time, so that $f_{ELM} \propto 1/P$. However it is clear one really needs first-principles, non-linear models to provide expressions for $\chi_{\perp ELM}$, Δ_{ELM} and τ_{ELM}. In order to complete the calculation of the impact of ELMs on the divertor plates, kinetic particle-in-cell codes have been employed to model losses along the open field lines [109].

5. Active ELM Control Techniques

Since Type I ELMs pose a serious threat to ITER operation it is crucial to identify regimes with small ELMs or seek control techniques to avoid them or, at least, mitigate their consequences for divertor erosion. Three ideas are currently being explored. First is ELM control using an external magnetic coil array, the idea being to create a stochastic layer in the pedestal region so that the resulting enhanced transport limits plasma gradients in the pedestal region to below the limits for pressure gradient driven MHD instability. This was demonstrated on DIII-D by applying an n = 3 magnetic perturbation coil, which strongly impacted on the character of the edge recycling: the ELM size and energy loss were dramatically reduced without a significant loss of energy confinement, as shown in Fig. 13 [110-114]. As a consequence the energy losses due to ELMs shown in Fig. 4 are dramatically reduced. Similar effects were explored earlier on COMPASS-D [115, 116].

A second technique is ELM control using 'Pellet Pace Making', pioneered on ASDEX Upgrade [117, 118]: each injected pellet triggers an ELM at a stage where their size remains modest, thus preventing them reaching such an amplitude as to cause a large transient heat load on the divertor plates. Thirdly ELMs can be triggered by raising the edge current, a control method explored on COMPASS-D [116], TCV [119], ASDEX Upgrade [120] and JET [121]. This can be achieved, for example, by up-down movements of the plasma produced by applying voltage perturbations to the poloidal field coils for single null Type III ELMy H-modes with frequency $f \sim 2f_{ELM}$ [119, 120]. Another control mechanism may be afforded by rotation [122]. Finally the effects of toroidal magnetic field ripple are being explored on JT-60U and JET [123]. Thus on JT-60U the energy loss from most Type I ELMs is less than 6% of the pedestal energy; this is thought to be due to its large toroidal ripple. However the downside is the poorer energy confinement time.

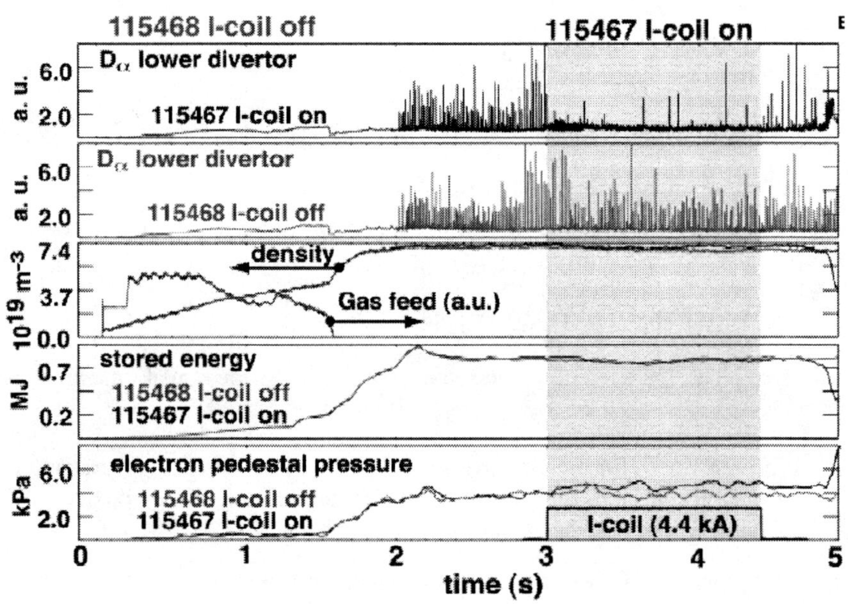

FIGURE 13. Effect of applying resonant magnetic perturbation (RMP) coils on DIII-D, showing the effect on the ELM signatures: D_α , plasma density and energy and electron pressure pedestal [111] (reproduced courtesy of Dr T E Evans).

6. Conclusions and Summary

The phenomenology and characteristics of ELMs have been discussed. There are a number of different ELM types, but they can be broadly separated into large Type I ELMs and a range of smaller ELM types, Type III being most common. While ELMs play an important role in controlling particle and impurity content and allowing a 'time-averaged' steady state, this is offset by various deleterious effects. The most serious is the danger of unacceptable erosion of divertor target plates in an ITER-scale device due to the high transient heat loads appearing there; in addition there is the potential to trigger core MHD activity with consequent confinement degradation. This means Type III, or other smaller ELM types, are to be preferred. These have an impact on global confinement but at an acceptable level. The conditions for most other small ELM types tend to correspond to limited regions of operating space: high density, high q, high-triangularity, δ; it remains unclear whether or not they are ITER-relevant. However, studying these various types can help to shed light on the physics of ELMs.

Since it is essential to avoid Type I ELMs in ITER there is current progress on devising ways to control them, normally by triggering smaller ELMs (pellet pacing; current drive) or by increasing edge transport to maintain pedestal gradients below the critical pressure gradient for instability (applying external coils to produce a stochastic region). While some ELM-free steady state regimes have been identified (EDA, HRS, QH-mode), their operational regions are limited.

Progress on ELM control or mitigation is facilitated by developing an adequate understanding and a predictive model. The peeling–ballooning model has been successful in capturing aspects of ELM occurrence and the non-linear ballooning theory predicting explosive growth of ELM filaments has been, at least qualitatively, widely vindicated by experiment. However translating this into a prediction of ELM energy losses to the divertor target plates remains a challenge, requiring an understanding of the non-linear saturation of the filaments and the process by which the energy in these flux tubes is transferred to the divertor, with only qualitative models presently available. Likewise, modelling the ELM cycle based on the peeling-ballooning model requires phenomenological (maybe with very qualitative guidance from theory) input of parameters such as the ELM crash time and the energy loss during an ELM.

ACKNOWLEDGEMENTS

Contributions from Drs Richard Buttery and Samuli Saarelma are gratefully acknowledged. This work was jointly funded by the United Kingdom Physical Sciences and Engineering Research Council and EURATOM. The views and opinions expressed herein do not necessarily reflect those of the European Commission.

REFERENCES

1. J. W. Connor, *Plasma Phys. Control. Fusion* **40** 531 (1998)
2. H. Zohm, *Plasma, Phys. Control. Fusion* **38** 105 (1996)
3. ITER Physics Basis, Chapter 2, *Nucl. Fusion* **39** 2175 (1999)
4. Progress in the ITER Physics Basis, Chapter 2, *Nucl. Fusion* **47** S1 (2007)
5. F. Wagner et al., *Phys. Rev. Lett.* **49** 1408 (1982)
6. A. Huber et al., in *Proc. 34th EPS Conf. on Plasma Physics, Warsaw 2007,* vol. **31F** , ed. P. Gasior and J. Wolowski, (Geneva: European Physical Society), ECA, paper P1.062 (2007)
7. J. W. Connor et al., *Nucl. Fusion* **44** R1 (2004) ; e.g. Appendix A.5 on JET
8. G. Fishpool et al., *Nucl. Fusion* **38** 1373 (1998)
9. A. Loarte et al., *J. Nucl. Mat.* **313-316** 692 (2003)
10. A. Kirk et al., *J. Nucl. Mat.* **313-316** 1081 (2003)
11. T. Eich et al., *J. Nucl. Mat.* **337-339** 669 (2005)
12. M. Fenstermacher et al., *Plasma Phys. Control. Fusion* **45** 1597 (2003)
13. A. Loarte et al., *'Characteristics of type I ELM energy and particle losses in existing devices and their extrapolation to ITER', Plasma Phys. Control. Fusion* **45** 1549 (2003)
14. A. Loarte et al., *Phys. Scri* **T128** 222 (2007)

15. A.W. Leonard et al, *J. Nucl. Mat.* **290-293** 1097 (2001)
16. N. Oyama et al., *Nucl. Fusion* **43** 1250 (2003)
17. A. Kirk et al., *Plasma Phys. Control. Fusion* **46** 551 (2003)
18. A. Kirk et al., *Phys. Rev. Lett.* **92** 245002 (2004)
19. R. Akers et al., *Plasma Phys. Control. Fusion* **45** A175 (2003)
20. A. Kirk et al., *Plasma Phys. Control. Fusion* **45** 1571 (2003)
21. T. Eich et al., *Phys. Rev. Lett.* **91** 195003 (2003)
22. T. Eich et al., *Plasma Phys. Control. Fusion* **47** 815 (2005)
23. M. E. Fenstermacher et al., *Nucl. Fusion* **45** 1493 (2005)
24. B. Koch et al., *J. Nucl. Mat.* **363–365** 1056 (2007)
25. M. Valovic et al., in *Controlled Fusion and Plasma Physics (Proc. 21st EPS Conf., Montpelier, 1994)*, vol. **18B**, ed. E. Joffrin et al., (Geneva: European Physical Society), Part I, 318 (1994)
26. R. Sartori et al., *Plasma Phys. Control. Fusion* **40** 757 (1998)
27. W. Suttrop et al., *Plasma Phys. Control. Fusion* **39** 2051 (1997)
28. W. Suttrop et al. in *Controlled Fusion and Plasma Physics (Proc. 23rd EPS Conf., Kiev, 1996)* vol. **20C,** ed. D. Gresillon et al. (Geneva: European Physical Society), part 1 p.47 (1996)
29. T.H. Osborne et al., in *Controlled Fusion and Plasma Physics (Proc. 24th EPS Conf., Berchtesgaden, 1997)* vol. **21A**, ed. M. Schittenhelm et al. (Geneva: European Physical Society), Part III, p.1101 (1997) and ITER Physics Expert Groups on Confinement and Transport and Confinement Modelling and Database, *Nucl. Fusion* **39** 2175 (1999)
30. T. Ozeki et al.., *Nucl. Fusion* **30** 1425 (1990)
31. J. Stober et al., *Nucl. Fusion* **41** 1123 (2001)
32. Y. Kamada et al., *Plasma Phys. Control. Fusion* **42** A247 (2000)
33. Y. Kamada et al., '*Pedestal Characteristics and extended high-β_p ELMy H-mode regime in JT-60U', Plasma Phys. Control. Fusion* **44** A279 (2002)
34. N. Oyama et al., *Nucl. Fusion* **45** 871 (2005)
35. R. Maingi et al., *J. Nucl. Mat.* **337-339** 727 (2005)
36. D. Mossessian et al., *Phys. Plasmas* **10** 1720 (2003)
37. Y-T. Lau, *Plasma Phys. Control. Fusion* **38** 1393 (1996)
38. K. H. Burrell et al., *Bull. Am. Phys, Soc.* **44** 127 (1999)
39. K. H. Burrell et al., *Phys Plasmas* **8** 2153 (2001)
40. K. H. Burrell et al., *Plasma Phys. Control. Fusion* **44** A253 (2002)
41. W. Suttrop et al., *Plasma Phys. Control. Fusion* **46** A151 (2004)
42. K. Kamiya et al., *Nucl. Fusion* **43** 1214 (2003)
43. J. Lingertat et al., in *Controlled Fusion and Plasma Physics (Proc. 24th EPS Conf., Berchtesgaden, 1997),* vol. **19C**, ed. M. Schittenhelm et al. (Geneva: European Physical Society), Part 1 p.24 (1997)
44. Y. Kamada et al., *Plasma Phys. Control. Fusion* **36** A123 (1994)
45. H. Zohm et al., *Nucl. Fusion* **35** 543 (1995)
46. G.T.A. Huysmans C.D. Challis, M. Erba, W. Kerner and V.V. Parail, in *Controlled Fusion and Plasma Physics (Proc. 22nd EPS Conf., Bournemouth, 1995)* vol **19C,** ed. B. Keen et al. (Geneva: European Physical Society) part 1 p. 201 (1995)
47. G.T.A. Huysmans, H.J. de Blanck, W. Kerner, J.P. Goedbloed and M.F.F. Nave, in *Controlled Fusion and Plasma Physics (Proc. 19th EPS Conf., Innsbruck, 1992*), vol **16C**, ed. W. Feysinger et al. (Geneva: European Physical Society), Part 1 p. 247 (1992)
48. J. Manickam , *Phys Fluids* B **4** 1901 (1992)
49. F. L. Hinton, *Phys Fluids B* **3** 696 (1991)
50. H. Zohm, *Phys Rev Lett* **72** 222 (1994)
51. S-I. Itoh and K. Itoh., *Phys. Rev. Lett.* **60** 2276 (1988)
52. S-I. Itoh , K. Itoh, A. Fukuyama and Y. Miura, *Phys. Rev. Lett.* **67** 2485 (1991)
53. S. Toda , S-I. Itoh, M. Yagi, A. Fukuyama and K. Itoh, *Plasma Phys. Control. Fusion* **38** 1337 (1996)
54. I. Voitsekhovich, A. Yu Dnestrovskij, V.V. Parail, *Nucl Fusion* **35** 631 (1995)
55. P. H. Diamond et al., *Phys. Rev. Lett.* **72** 2565 (1994)
56. H. Sugama and W. Horton, *Plasma Phys. Control. Fusion* **37** 345 (1995)
57. J-N. Leboeuf, L. A. Charlton and B. Carreras, *Phys Fluids* B **5** 2959 (1993)
58. W. Horton, G. Hu and G. Laval, *Phys. Plasmas* **3** 2912 (1996)
59. V. B. Lebedev et al., *Phys. Plasmas* **2** 3345 (1995)

60. O. Pogutse, W. Kerner, V. Gribkov, S. Bazdenkov and M. Ossipenko, *Plasma Phys. Control. Fusion* **36** 1963 (1994)
61. A. Takayama, T. Unemura and M. Wakatani, *Plasma Phys. Control. Fusion* **38** 1411 (1996)
62. The ASDEX Team, *Nucl. Fusion* **29** 1959 (1989)
63. O. Pogutse, J. G. Cordey, W. Kerner, B. Schunke, in *Controlled Fusion and Plasma Physics (Proc. 22nd EPS Conf., Bournemouth, 1995)* vol **19C,** ed. B. Keen et al. (Geneva: European Physical Society), part III, p. 277 (1995)
64. W. Kerner, O. Pogutse and R. Van der Linden, *Plasma Phys. Control. Fusion* **39** 757 (1997)
65. S-I. Itoh et al., *Phys. Rev. Lett.* **76** 920 (1996)
66. S-I. Itoh, K. Itoh, A. Fukuyama and M. Yagi, *Plasma Phys. Control. Fusion* **38** 527 (1996)
67. S-I. Itoh, K. Itoh, and A. Fukuyama, *Plasma Phys. Control. Fusion* **38** 1367 (1996)
68. C. Hegna, J. W. Connor, R. .J. Hastie, and H. R. Wilson, *Phys Plasmas* **3** 584 (1996)
69. J W. Connor, R. .J. Hastie, H. R. Wilson and R. L. Miller, *Phys. Plasmas* **5** 2687 (1998)
70. H. R. Wilson et al., *Phys. Plasmas* **6** 1925 (1999)
71. P. B. Snyder et al., *Nucl. Fusion* **44** 320 (2004)
72. H. R. Wilson et al., *Phys. Plasmas* **9** 1277 (2002)
73. H. R. Wilson et al., *Nucl. Fusion* **40** 713 (2000)
74. P. B. Snyder et al., *Phys. Plasmas* **9** 2037 (2002)
75. S. Saarelma et al., *Plasma Phys. Control. Fusion* **47** 713 (2005)
76. S. Saarelma et al., *Plasma Phys. Control. Fusion* **49** 31 (2007)
77. A. B. Mikhailovskii et al., *Plasma Phys. Reports* **23** 844 91997)
78. S. Saarelma, private communication
79. S. Saarelma et al., *Nucl. Fusion* **43** 262 (2003)
80. P. B. Snyder and H.R. Wilson, *Plasma Phys. Control. Fusion* **45** 1671 (2003)
81. L. L. Lao et al., *Nucl. Fusion* **41** 295 (2005)
82. A. Loarte et al., *Plasma Phys. Control. Fusion* **44** 1815 (2002)
83. S.J. Fielding, *Plasma Phys. Control. Fusion* **38** 1091 (1996)
84. S. Saarelma et al., *Plasma Phys. Control. Fusion* **49** 31 (2007)
85. A. Kirk et al., *Plasma Phys. Control. Fusion* **47** 315 (2005)
86. G. T. .A. Huysmans et al., *Phys. Plasmas* **8** 4292 (2001)
87. S. C. Cowley et al., *Plasma Phys. Control. Fusion* **45** A31 (2003)
88. H. R. Wilson and S. C. Cowley, *Phys. Rev. Lett.* **92** 175006 (2004)
89. H. R. Wilson et al., *Proceedings of the 21st IAEA Fusion Energy Conference, Chengdu*, on CD-ROM: IAEA-CN-I49, paper TH4-IRb (2006)
90. P. B. Snyder et al., *Plasma Phys. Control. Fusion* **46** A131 (2004)
91. G. T. .A. Huysmans, *Plasma Phys. Control. Fusion* **47** 2107 (2005)
92. H. R Wilson et al., *Plasma Phys. Control. Fusion* **48** A71 (2006)
93. J. A. Boedo et al., *Phys. Plasmas* **12** 072516 (2005)
94. S. I. Krasheninnikov et al., *Phys. Lett.* A **283** 368 (2001)
95. W. Fundamenski et al., *Plasma Phys. Control. Fusion* **48** 109 (2006)
96. C. G. Gimblett, R. J. Hastie and P. Helander, *Phys. Rev. Lett.* **96** 035006 (2006)
97. L. Degtyarev et al., *Comput. Phys. Comm.* **103** 10 (1997)
98. S. Y. Medvedev et al., in *Controlled Fusion and Plasma Physics (Proc. 30th EPS Conf., St Petersburg, 2003),* vol **27A**, ed. R. Koch and S. Lebedev (Geneva: European Physical Society), ECA, paper P-3.129 (2003)
99. G. T .A. Huysmans, *Plasma Phys. Control. Fusion* **47** 2107 (2005)
100. G. Laval, R. Pellat and J. Soule, *Phys. Fluids* **17** 835 (1974)
101. A. J. Webster et al., in *Controlled Fusion and Plasma Physics (Proc. 34th EPS Conf., Warsaw, 2007),* vol **31A** ECA, paper P4.088 (2007)
102. Yu. Igitkhanov et al., *Contrib. Plasma Phys.* **40** 368 (2000)
103. A. V. Chankin and G. Saibene, *Plasma Phys. Control. Fusion* **41** 913 (1999)
104. R. Sartori et al., *Plasma Phys. Control. Fusion* **46** 723 (2004)
105. Y. Igitkhanov et al., *Plasma Phys. Control. Fusion* **40** 837 (1998)
106. Y. Igitkhanov et al., in *Controlled Fusion and Plasma Physics (Proc. 28th EPS Conf., Madeira, 2001),* vol **25A**, ed. C. Silva et al. (Geneva: European Physical Society), ECA, 1693 (2001)
107. J-S. Lőnnroth et al., *Plasma Phys. Control. Fusion* **45** 1689 (2003)
108. R .J. Buttery et al., in *Controlled Fusion and Plasma Physics (Proc. 22nd EPS Conf., Bournemouth, 1995),* vol **19C**, ed. B. Keen et al. (Geneva: European Physical Society), part III p. 273 (1995)

109. D. Tskhsakaya et al., in *Theory of Fusion Plasmas (Proc. Joint Varenna-Lausanne Int. Workshop, Varennna, 2004),* ed. J. W. Connor et al., (Soc. Italiana di Fisica, Bologna) p. 97 (2004)
110. T. E. Evans et al., *Phys. Rev. Lett.* **92** 235003 (2004)
111. T. E. Evans et al., in *Proc.20th Int. Conf. on Fusion Energy, 2004* (Vilamoura: 2004) (Vienna: IAEA), CD – ROM file EX/2-5Ra (2004); also reproduced *in 'Progress in the ITER Physics Basis, Chapter 2: Confinement and Transport',* E. J. Doyle et al., Nucl. Fusion **47** S18 (2007)
112. T .E. Evans et al., in *Controlled Fusion and Plasma Physics (Proc. 32nd EPS Conf. on Controlled Fusion, Tarragona, 2005),* ed. C. Hidalgo and B. Ph. Van Milligen, vol **29C**, paper I5-005 (2005)
113. R. A. Moyer et al., *Phys. Plasmas* **12** 056119 (2005)
114. T. E Evans et al., *Nature Phys.* **2** 419 (2006)
115. S. J. Fielding et al., in *Controlled Fusion and Plasma Physics (Proc. 28th EPS Conf., Madeira, 2001),* vol **25A**, ed. C. Silva et al. (Geneva: European Physical Society), ECA, 1825 (2001)
116. S. J. Fielding et al., in *Controlled Fusion and Plasma Physics (Proc. 30th EPS Conf., St Petersburg, 2003),* vol **27A** , ed. R. Koch and S. Lebedev (Geneva: European Physical Society), ECA, paper P-3.129 (2003)
117. P. T. Lang et al., in *Controlled Fusion and Plasma Physics (Proc. 30th EPS Conf., St Petersburg, 2003),* vol **27A**, ed. R. Koch and S. Lebedev (Geneva: European Physical Society), ECA, paper P-1.129 (2003)
118. P. T. Lang et al., *Nucl. Fusion* **44** 665 (2004)
119. A. W. Degeling et al., *Plasma Phys. Control. Fusion* **45** 1657 (2003)
120. L. Horton et al., *Plasma Phys. Control. Fusion* **46** B511 (2004)
121. M. Becoulet et al., *Plasma Phys. Control. Fusion* **45** A93 (2003)
122. N. Oyama et al., in *Proc. 20th Int. Conf. on Fusion Energy, 2004* (Vilamoura, 2004) (Vienna : IAEA), CD-ROM file EX/2-1 (2004)
123 J-S. Lőnnroth et al., *Plasma Phys. Control. Fusion* **49** 273 (2007)

Turbulent Impurity Transport

Volker Naulin*, Jens Juul Rasmussen*, Clemente Angioni†, Carine Giroud**, Marco Valisa‡, Maria Ester Puiatti‡ and L. Carraro‡

*Association EURATOM-Risø- DTU, OPL-128, DK-4000 Roskilde, Denmark
†EURATOM-Association, MPI für Plasmaphysik, Garching D-85748, Germany
**EURATOM/UKAEA Fusion Association, Culham Science Centre, Abingdon, Oxfordshire OX14 3DB, United Kingdom
‡Consorzio RFX - Associazione Euratom-Enea sulla Fusione, I-35127 Padova,Italy

Abstract. Impurity transport is an important topic for ITER as impurity density in the core has to stay below a relative concentration of 5×10^{-5} [1] to reach ignition conditions. The reasons for that is mainly impurity radiation leading to cooling. In the edge region on the other hand site, higher impurity densities, going along with radiation energy losses are beneficial as the radiative cooling arranges for good power distribution onto the first wall.

Keywords: Impurities, turbulence, tokamaks, transport
PACS: 52.25.Vy,52.35.Ra,51.20.+d

INTRODUCTION

Thermonuclear fusion is easiest obtained using a mixture of Deuterium and Tritium as fuel. Any other material in the plasma diluting the fuel mixture is considered an impurity, this includes the Helium ash resulting from the fusion process. The Helium ash is, however, the only impurity species that has its source in the confined plasma region. All other impurities have their origin at the walls, from where they enter the plasma either as fast neutrals, originating from processes that release them from the walls, or they are already ionised in the scrape off layer, the SOL and transported across the last closed flux surface (LCFS) into the confined plasma region. Some impurities are deliberately injected into the plasma to influence the plasma properties in the SOL, for example to enhance the coupling of radiofrequency heating schemes, or are puffed or ablated, gaseous and solid impurities respectively, for transport experiments. In these cases the impurity sources are more or less known and we speak of extrinsic impurities. Materials that have their origin in the tokamak walls, either as material being deposited in layers onto the wall or wall materials themselves, are called intrinsic impurities. The sources of intrinsic impurities are badly known or not known at all.

Impurities influence the plasma performance through a number of processes. In the core the impurities dilute the plasma fuel, thus diminishing the efficiency of the fusion process. They can also influence through dilution the instabilities which drive transport and thus reduce core confinement. As impurities are usually materials with a rather high mass they efficiently radiate energy from the plasma and thus cool the plasma, an effect that is detrimental in the core, where high temperatures are needed to obtain fusion, but that is beneficial in the edge, where radiative cooling of plasma that leaves the confinement region. This distributes its energy onto a larger surface, compared to the rather small area that is "‘wetted"’ by the plasma flowing toward the divertor plates, which have to withstand tremendous heat fluxes. At the same time the presence of impurities can change the ratio between plasma density and temperature to obtain a certain critical plasma pressure at the edge, thus contributing to the H-mode physics as for example influencing the nature and frequency of edge localised modes (ELM) bursts, which pose one of the most difficult problems to obtain a technically viable operating regime for a fusion reactor.

This paper will discuss methods to investigate turbulent impurity transport only, classical and neoclassical impurity transport has been reviewed in detail in [2].

We will start by considering test particle transport in section two with a focus on simple and paradigmatic turbulence equations. A second part of these considerations will focus on the consistency of the test particle transport with bulk transport and the use of test particle densities instead of the particles themselves. The third paragraph we will describe different pinch effects, as they are of utmost importance in understanding peaked impurity profiles. Most of these pinches arrive due to finite compressibility of the advecting velocity field, with the best known being the curvature pinch originating in the compressibility of the velocity field perpendicular to the magnetic field direction, others

CP1013, *Turbulent Transport in Fusion Plasmas, First ITER International Summer School*
edited by S. Benkadda

pinch effects arising due to parallel velocity compressibilties and finally pinch effects from off -diagonal elements of the transport matrix, for example thermodiffusion, where the upgradient transport of particles is a consequence of a potentially large down gradient energy flux.

TEST PARTICLE TRANSPORT

Impurities in a plasma are never passive in the sense that they would not alter the stability of the plasma or details of the turbulence, they always participate in transport of heat, particles and momentum. However, it is a useful and instructive simplification to neglect theirinfluence on the bulk plasma and consider the impurities as species that are moved and changes by the plasma conditions around them, but exert no backreaction. Specifically, and besides energy losses through impurity radiation, this means that the impurity charge density plays only a negligible role in the quasineutrality condition expressed as $\nabla \cdot \vec{J} = 0$. The impurity contribution to quasineutrality must everywhere be smaller than each of the terms contributing to the quasi-neutrality condition from other species. Contributions to the quasi-neutrality are related to the fluctuating currents, which for an anomalous transport situation are brought about by the turbulence. Thus, a much more strict relationship has to be fulfilled, namely, the passive density must everywhere be smaller then the bulk current divergences building up in a characteristic time. For drift-wave like turbulence with a characteristic time given by ion larmor radius at electron temperature over ion sound speed ρ_s/c_s :

$$Z_{imp} n_{imp} \ll \rho_s/c_s \sum_{i,e} \nabla \cdot (n_s V_s) . \tag{1}$$

This is a strict condition, which cannot be guaranteed to be fulfilled in a real plasma at all times, as this would demand a correlation of the impurity densities with the bulk fluctuating quantities.

Noting that this condition will strictly not be fulfilled triggers an additional word of caution, if we consider self consistency. Let us make a "Gedankenexperiment" and assume we would be able to mark a certain number of bulk plasma ions, so that these marked ions fulfill the condition above. Obviously a short time after these particles have lost the memory of their starting positions, i.e., after a Lagrangian auto-correlation time, we expect and know that these particles are transported like the bulk plasma. They will be exhibiting the same gradient and the same relative flux. In fluid turbulence this allows to asses general transport properties of the turbulence from passive tracer dynamics [].

If passive tracers do not follow the bulk and do deviate from the bulk, then because the transport of passive test particles cannot tell us about f.x. the equilibrium pressure gradient. We would need the back reaction of the plasma density onto the fields governing its motion. Having stated this, we notice that the usual evaluation of anomalous transport in fluctuation based models does, however, not address the profile problem self-consistently as well. Consequently this approach will be able to to reveal pinch effects, but not self-consistently. On the positive side it has for a number of practical situations been shown, that the bulk transport and the one derived from the relative diffusion of passive test particles are indeed in agreement with each other [15, 16]. Deviations between passive tracer transport and bulk transport, as they appear, will only be resolved if we are able to solve the plasma profile problem for a multi-species plasma under turbulent action in a self-consistent manner.

Fundamentals of particle dispersion in turbulent flows

The classical results for particle dispersion in homogeneous turbulent flows dates back to the works of G.I. Taylor more than eighty years ago (see, e.g., [3, 4]). Here we present a short account of the basic features. Consider a particle released at $t = 0$ in an isotropic and homogeneous turbulent flow field characterized by the velocity fluctuations: $\vec{u}(\vec{x},t)$. The particle is assumed to ideally follow the flow as a passive tracer. The displacement of the particle from its initial position, which we may take to be the origin of the coordinate system without loss of generality, is determined as:

$$\vec{r}(t) = \int_0^t \vec{u}(\vec{r}(t'),t')dt',$$

where $\vec{v}(t) \equiv u(\vec{r}(t),t)$ designates the so-called Lagrangian velocity of the particle measured along its trajectory. The mean value of the displacement will be zero, since $\langle \vec{u}(\vec{r}(t),t) \rangle = 0$ (here $\langle \cdot \rangle$ denote average value). Thus, we shall consider the mean square displacement (the variance), which will be positive. The time derivative of the squared displacement of the particle is calculated as follows:

$$\vec{r}(t)\frac{d}{dt}\vec{r}(t) = \frac{1}{2}\frac{d}{dt}(\vec{r}(t))^2 = \vec{v}(t)\cdot\int_0^t \vec{v}(t')dt' = \int_0^t \vec{v}(t)\cdot\vec{v}(t')dt'. \tag{2}$$

We now average this expression for the single particle over an ensemble of particles that are initialized all over the flow, and obtain the mean square displacement over the ensemble of particles: $R^2(t) \equiv \langle \vec{r}^2(t) \rangle$,:

$$\frac{dR^2(t)}{dt} = 2\int_0^t \langle \vec{v}(t)\cdot\vec{v}(t')\rangle dt'. \tag{3}$$

The integrand is expressed as

$$\langle \vec{v}(t)\cdot\vec{v}(t')\rangle = C_L(t-t')\langle v^2\rangle$$

where we have introduced the normalized Lagrangian autocorrelation function, C_L, for a time stationary turbulent flows (C_L depends only on the time differences) with $C_L(0) = 1$. $\langle v^2(t)\rangle$ is the mean square velocity fluctuations. (Note: for homogeneous turbulence the mean square of the Eulerian and Langragian velocity fluctuations are equal [3]). We define the Lagrangian integral time scale:

$$\tau_L \equiv \int_0^\infty R_L(\tau)d\tau,$$

which is a characterizing the memory of the dispersing particle. From Eq. 3 we obtain for the mean squared displacement:

$$R^2(t) = 2\langle v^2\rangle \int_0^t dt' \int_0^{t'} C_L(s)ds = 2\langle v^2\rangle \int_0^t (t-s)C_L(s)ds. \tag{4}$$

We may estimate $R^2(t)$ for two limiting cases:
i) short times $t \ll \tau_L$ where Eq. 4 reduce to

$$R^2(t) = 2\langle v^2\rangle t^2,$$

ii) long times $t \gg \tau_L$ where Eq. 4 reduce to

$$R^2(t) = 2Dt.$$

Thus, for short times the particle motion is ballistic, while for large times the dispersion is characterized by a diffusive process like Brownian motion with the variance growing linearly in time, and the diffusion coefficient expressed as

$$D = \langle v^2\rangle\tau_L.$$

Thus, if the full Lagrangian statistics of the flow were known then the particle dispersion could in principle be calculated. Furthermore, if the integral of the Lagrangian autocorrelation function is finite, i.e., τ_L is finite, the dispersion process will asymptotically approach a normal diffusion process.

In addition, to the two asymptotic regimes discussed above there is a regime for intermediate times of anomalous dispersion characterized by

$$R^2(t) \propto t^\sigma$$

where $\sigma > 1$ is referred to as *superdiffusion*, and $\sigma < 1$ is referred to as *subdiffusion*. The time window of this intermidiate regime may in principle be arbitrary wide when τ_L tend to infinity. Typical values of σ found in numerical simulations of particle dispersion in two-dimensional turbulence lays in the region $1.3 < \sigma < 1.7$ [5, 6, 7]. Although it is believed that it is the intermittent trapping and de-trapping of particles in coherent vortical structures that is responsible for the anomalous dispersion, it has not been possible to find/predict a unified value of σ. Various attempts have been proposed to describe the anomalous dispersion, including stochastic models [7] and fractal kinetics [8, 9].

The Lagrangian statistics is in general not obtainable. From standard single point measurements one obtains the Eulerian statistics and the relation between Eulerian and Lagrangian statistics is not at all trivial. Numerous works have been devoted basically to derive the Lagrangian correlation from the Eulerian statistics, see e.g. [4]. For two dimensional homogeneous turbulence a particularly simple result was suggested [10] for the scaling between the Lagrangian and Eulerian correlation functions: $C_L(\beta t) \approx C_E(t)$ where $\beta < 1$ is a parameter and thus $\tau_L < \tau_E$. This result was tested numerically with good agreement. Other studies also aim at providing the scaling of the diffusion coefficient with the rms-value of the fluctuations [11], and the analytical predictions have been compared with diffusion of particles in a synthesized wave field modelling the turbulence [12]. These investigations were generalized further [13] to cases of strong turbulence where the assumptions behind the Taylor result are breaking down. This regime may be characterized by the so-called Kubo number $K > 1$, where $K = \langle v^2 \rangle^{1/2} T_c/\lambda$, T_c is a typical correlation time and λ is the averaged spatial scale of the fluctuations. The Kubo number measures the averaged distance covered by a particle during the correlation time relative to the typical spatial scale of the fluctuations. The results can be summarized by the dependence of the diffusion coefficient on the Kubo-number: $D \propto K^{\alpha}$, with $\alpha = 2$ for $K \ll 1$ and $\alpha \approx 0.7$ for $K \gg 1$. The particle trapping is found to partly hinder the diffusion, introducing a regime of subdiffusion.

Particle dispersion in plasma turbulence

The results discussed above also apply to to cross field particle dispersion in low frequency turbulence in magnetized plasmas. Here the fluctuating electric field give rise a fluctuating velocity field with the $E \times B$ velocity being the main contribution. For an overview of turbulent diffusion in magnetized plasmas see the Monograph by Balescu [14].

We shall shall here discuss a specific an example of test particle dispersion in electrostatic plasma turbulence. We base our discussions on the well-known Hasegawa-Wakatani equations (HWE) [15] for plasma edge turbulence driven by the resistive drift wave instability:

$$\partial_t n + \partial_y \varphi + \{\varphi, n\} = -C(n - \varphi) + \mu_n \nabla^2 n \,, \quad (5)$$

$$\partial_t \nabla^2 \varphi + \{\varphi, \nabla^2 \varphi\} = -C(n - \varphi) + \mu_\varphi \nabla^4 \varphi \,, \quad (6)$$

where n and φ are the fluctuating part of the density and potential, respectively. $1/C = 1/k_{\parallel}^2 L_{\parallel}^2$ is the adiabaticity parameter and contains via the parallel scale length $L_{\parallel} = (L_n T_e / m_e c_s \nu_{ei})^{1/2}$ the parallel resistivity.

It is inherently assumed that only one mode, parallel to the homogeneous magnetic field $\vec{B} = B_0 \hat{z}$, is excited, and all differential operators work in the (x,y)-plane, $\{\varphi, \psi\} \equiv \hat{z} \times \nabla\varphi \cdot \nabla\psi = \frac{\partial \varphi}{\partial x}\frac{\partial \psi}{\partial y} - \frac{\partial \psi}{\partial x}\frac{\partial \varphi}{\partial y}$. The x-coordinate is identified as the radial direction and the y-coordinate the poloidal direction and the model covers a slab region of the plasma-edge in a toroidal device. The fluctuating $E \times B$ velocity is given as $\vec{u} = (u, v) = (-\frac{\partial \varphi}{\partial y}, \frac{\partial \varphi}{\partial x})$. The normalization scales are $\rho_s = c_s/\Omega_i$ for lengths perpendicular to the magnetic field and L_n/c_s for the times, where $c_s = \sqrt{T_e/m_i}$ is the sound-speed and $L_n = (\nabla n_0(x)/n_0(x))^{-1}$ is the length scale of the radial back-ground density gradient (along x). Potential- and density-fluctuations are normalized by $(T_e/e)\ (\rho_s/L_n)$ and $n_0 \rho_s/L_n$, respectively. The electron-temperature is constant, and we assume cold ions. μ_n is designating a diffusivity and μ_φ is the viscosity. For convenience and without loss of generality we take $\mu_n = \mu_\varphi = \mu$.

The HWE Eqs. (5,6) are solved numerically on a double periodic domain by means of a de-aliased pseudo-spectral code for details see [16, 17, 18]. After the initial linear growth the system enters a saturated turbulent state, where the global statistical properties show little or no time dependence, nor any dependence on the initialization method.

To gain statistically relevant results a large number of particles, typically more then 10000, that are passively advected by the flow have to be considered. The particles are initialized at random positions uniformly over the domain, when the turbulence has reached the quasi-steady regime. The trajectories of the particle are found from:

$$\vec{r}(t) = \vec{r}_0 + \int_0^t \vec{u}(\vec{r}(t'), t')dt' \quad (7)$$

where $\vec{r}_0$ is the initial position of the particle and $\mathbf{u}$ the velocity. The principal component of $\vec{u} = (u, v)$ is the $E \times B$ velocity evaluated at the position of the particle from the electrostatic potential φ: $\mathbf{u}_E = (-\partial_y \varphi, \partial_x \varphi)$. This velocity is employed when tracking ideal "massless" particles for which inertia effects can be neglected. Inertial effects can be incorporated by adding the polarization drift, $\mathbf{u}_p$ to $\mathbf{u}_E$, which may be written in normalized form:

$$\mathbf{u}_p = -\zeta \left(\frac{\partial}{\partial t} + (\mathbf{v}_E \cdot \nabla) \right) \nabla\varphi, \quad (8)$$

where $\zeta = \frac{eM}{qm_i}\frac{\rho_s}{L_n}$, M and q are the mass and charge of the particle. For the background, singly ionized ions, with charge e and mass m_i, $\mathbf{u}_p$ is a very small correction since $\frac{\rho_s}{L_n}$ is on the order of 0.01 or less ($\rho_s = c_s/\Omega_{ci}$ and L_n is the density scale length). But for heavy ions (e.g., impurity ions, or alpha particles) it might become important. In the next section we briefly investigate the influence of the polarization drift correction on the dispersion properties and in particular the diffusion of heavy ions. The specific feature of this velocity is that it is compressible in contrast to the $E \times B$ velocity which in a homogeneous magnetic field is incompressible $\nabla \cdot \vec{u}_{E\times B} = 0$. For compressible convection velocities we have the possibility of clustering of particles and particularly of pinching, as will be discussed in the next section.

For the cases discussed above we inherently assumed that finite Larmor radius effects (FLR) could be neglected. That is, the typical scale lengths of the fluctuations are much larger than the Larmor radius of the particles. This holds well for "cold" impurity particles in the edge region. However, for high energetic particles like fusion alphas, having a Larmor radius of few cm the finite Larmor radius effect is certainly important. Then in principle the particle trajectory have to be calculated applying the full Lorenz force:

$$\frac{du_i}{dt} = -\frac{q}{m}\frac{\partial \phi'}{\partial x_i} + \Omega_c \varepsilon_{ij} u_j; \quad \frac{dx_i}{dt} = u_i; \;\; \varepsilon_{12} = -\varepsilon_{21} = 1, \varepsilon_{ii} = 0.$$

However, provide that the time scale of the turbulence is much slower than the gyro-motion one apply the standard gyrokinetic approach. That is averaging the potential over the gyro orbit of the particle, which results in the effective - gyro averaged potential - given by

$$\phi^G(\vec{r}) = \frac{1}{(2\pi)^2}\int_{-\infty}^{\infty} d\vec{k}\, \exp(i\vec{k}\cdot\vec{r})\phi(\vec{k})J_0(k\rho) \tag{9}$$

where $\phi(\vec{k})$ designates the Fourrier transform of the potential, J_0 is the Bessel function of zero order, and ρ is the Larmor radius of the impurity ion. The particles are advected by the gyro averaged $E \times B$ velocity obtained from $\phi^G(\vec{r})$.

In the case of weak turbulence one will intuitively expect that the FLR will limit the dispersion, since the particle is sampling over a varying E-field and therefore experiences and effectively smaller $E \times B$ drift. This is in agreement with results of [19] who investigated the influence of FLR on particle dispersion in drift wave turbulence, and the results of Krane et al [20]. However, recently it has been shown that for the limit of large Kubo numbers the effect of the FLR becomes less dramatic [21], and for Kubo numbers of around 10 FLR only a weak reduction of the dispersion results. Similar results are also found by [22, 23], who applied these results to explain the anomalous diffusion of fast ions originating from off-axis neutral beam injection in Asdex Upgrade [24].

Particle dispersion

First we consider a dense collection of particles initially located in a small domain around the center of the box, Fig. 1. Initially we observe a large dispersion in the radial direction (along x). After around 16 time units the poloidal diffusion starts to dominate, and at $T = 80$, the initially dense particle collection confined to an (x, y) symmetric part of the flow has been dispersed to an area in the flow that is roughly two times as long in the y direction as in the x direction.

To describe the diffusion of a large ensemble of particles we introduce the running diffusion coefficient. We split it into a radial and poloidal (x and y) component defined as

$$D_x(t) = \frac{X^2(t)}{2t}, \quad D_y(t) = \frac{Y^2(t)}{2t}, \tag{10}$$

where $X^2(t) = \langle (x(t) - x(t=0))^2 \rangle$, $Y^2(t) = \langle (y(t) - y(t=0))^2 \rangle$ are the mean square displacements of particles in radial and poloidal direction. The brackets denote an average over all particles. For a diffusive process the running diffusion coefficient $D(t)$ reaches a time independent value D, as $X^2(t) \sim t$ holds. We should note that the running diffusion coefficients ia often defined as the differential $D(t) = \frac{1}{2}\frac{dR^2}{dt}$, which also approach a constant in the diffusive limit. Applying the classical Taylor result for particle dispersion in turbulent flows (see the Introduction) for the two components of D the asymptotic diffusion coefficients can be expressed as:

$$D_x = \langle u^2 \rangle \tau_L, \;\; D_y = \langle v^2 \rangle \tau_L, \tag{11}$$

Employing the estimated value of $\tau_L \approx 2-3$ and the values of the rms velocity fluctuations $\langle u^2 \rangle \approx 1.4$, $\langle v^2 \rangle \approx 1.1$, which are only weakly dependent of C we would expect: $D_x = 4-5$ and $D_y = 2-3$ almost independent of C.

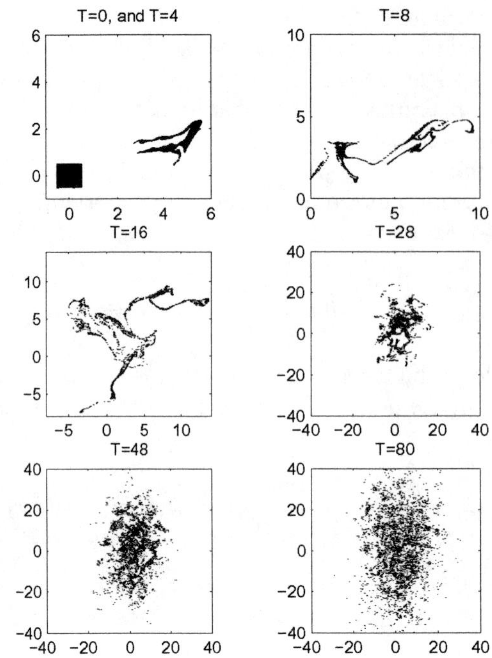

FIGURE 1. Dispersion of a dense sample of particles initially released in a square box at the center of the domain.

From the Lagrangian autocorrelation function we can estimate the Lagrangian integral time scale: $\tau_L = 2-3$, which is weakly dependent on C. Thus, the Kubo number, $K = \langle v^2 \rangle^{1/2} T_c/\lambda$, takes typical values of $K = 2-3$ for both radial and poloidal fluctuations, where the T_c is taken as τ_L and the averaged spatial scale length is estimated as $\lambda \sim 2$. Thus, the Kubo number is roughly constant for the parameters we have considered, and it is not possible to perform a Kubo number scaling of the dispersion properties.

In Fig. 2 we have shown the evolution of the radial and poloidal component of the running diffusion coefficient defined in (10) for different values of the coupling coefficient, C. We observe that D_x is strongly dependent on C, although it is found that $\langle u^2 \rangle$ only weakly decreased with C. After an initial ballistic phase, $D_x \propto t$, the coefficient decreases and finally tends to flatten out at least for $C \leq 2$. Thus, it appears that the evolution after a period of subdiffusive behaviour characterized by $X^2(t) \sim t^\sigma$ with $\sigma < 1$ approach an asymptotic diffusive regime, with a diffusion coefficient that is strongly decreasing with C. Indeed the diffusion is significantly smaller than the the "Taylor diffusion", Eq. (11). For the value $C = 5$ the subdiffusive behavior is more pronounced than at smaller C values. The poloidal diffusion is clearly larger than the radial one. For moderate values of C it is found to be in agreement with the value predicted from the Taylor-formula (11), however, for larger $C > 1$ it tends to increases with C with the tendency for superdiffusive behavior for large C-values ($C = 5$). This behavior and in particular the strong anisotropic effect for large C have also been observed for the dispersion of passive particles in the frame-work of the Hasegawa-Mima equation (see, e.g. [19, 25, 26].

The intermediate subdiffusive regime for the radial dispersion and the superdiffusive behavior for the poloidal dispersion was found to be related to trapping of particles in vortical structures [16]. A significant amount of particles are trapped in the vortical structures and convected with them for some time (typically t = 10 - 15 for moderate values of C). The vortical structures propagate predominantly in the poloidal direction with a poloidal velocity about 0.45, which is 3 times higher than their typical radial velocity 0.15. Thus, it may be estimated that the convection by the vortices will enhance the poloidal spreading and limit the radial spreading of the particles.

In order to investigate the radial subdiffusive and poloidal superdiffusive behavior in more detail for the case $C = 5$ we have plotted the mean square displacement both in radial and poloidal direction in Fig. 3a. For Y we observe a clear superdiffusive behavior in an intermediate time span ($5 < t < 100$), $Y^2 \sim t^\sigma$ with an exponent $\sigma \simeq 1.4$, while it is approaching a normal diffusion for $t > 100$ ($\sigma = 1$). For X it is more subtle, a naive fit would indicate a sub-

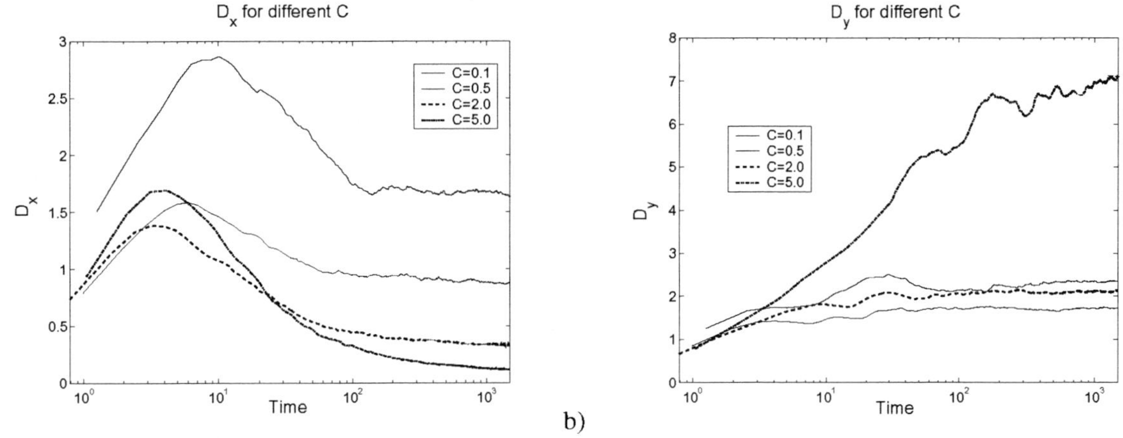

a) b)

FIGURE 2. a) Running diffusion coefficient in the radial direction, D_x, and b) in the poloidal direction, D_y, versus time for various values of C.

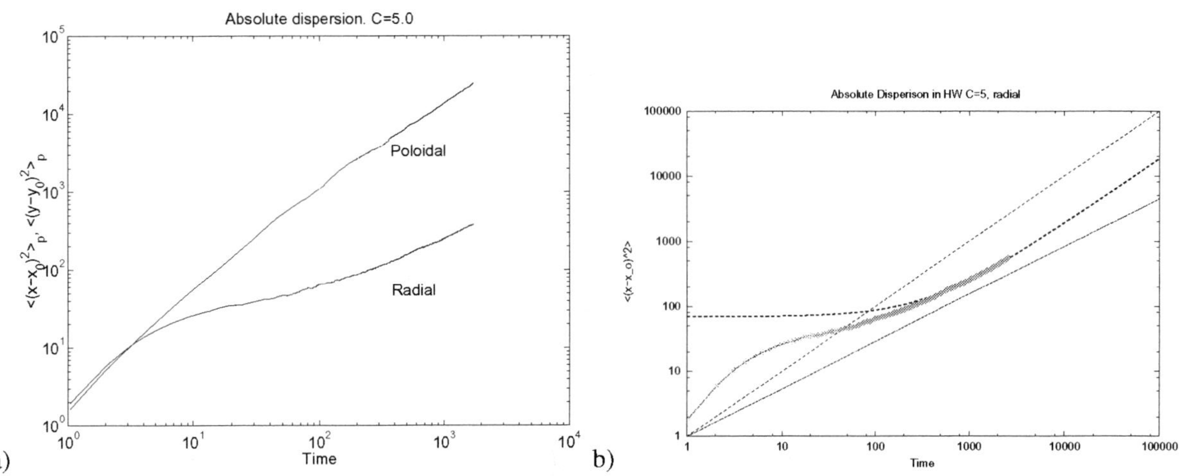

a) b)

FIGURE 3. a) The mean square radial and poloidal particle displacement $X^2(t)$ and $Y^2(t)$ versus time for $C = 5$. For the radial displacement a subdiffusive regime is clearly visible for $t > 5$. For the poloidal/meridional displacement a superdiffusive regime is visible for $5 < t < 500$. b) A detailed fit to $X^2(t)$. The dashed curve is a least mean square fit of the form $At^{\sigma} + B$ to $X^2(t)$ for $t > 400$, where $\sigma = 1, A = 0.184, B = 68.75$. The dashed line has the slope 1 and the dot-dashed line has the slope 0.73.

diffusive behavior continuing for $t > 100$ with $X^2 \sim t^{\sigma}$, $\sigma \simeq 0.73$ following a stronger subdiffusive regime, $\sigma \simeq 0.5$, for $10 < t < 100$ However, a detailed fit, see Fig. 3b, shows that after the subdiffusive behavior for $10 < t < 500$ there is a a slow transition to the asymptotic diffusive behavior. The dashed curve is the best fit of a function of the form $At^{\sigma} + B$ for $t > 400$ with $\sigma = 1$. The "naive" fit in the regime $200 < t < 2000$, $X^2(t) \sim t^{0.73}$, is shown by the dot-dashes line.

Consistency with background transport

We would like to point out, that there are two ways to determine the plasma particle transport. One way is to determine the particle transport is to derive a diffusion coefficient from following passive tracer particles convected by the turbulent flow, as described above. The other way is to look directly at the convected density, leading to the familiar expression $\vec{\Gamma} = n\, \vec{u}_{E\times B}$ for the local turbulent particle flux. The radial component of $\vec{\Gamma}$ is the cross-field transport as usually measured in experiments and simulations of electrostatic turbulence, see e.g., [27, 28, 29, 30]. Note that an effective diffusion coefficient related to the flux is defined through Fick's law, $D^{eff} = \Gamma_0/\nabla n_0$, with our normalizations

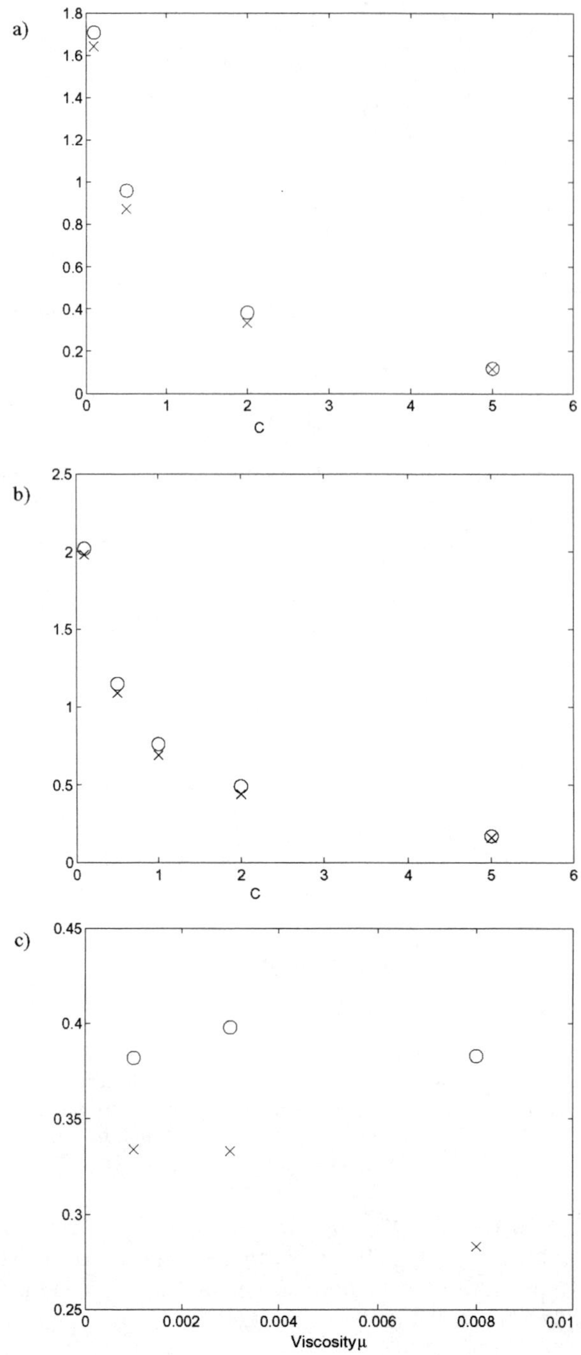

FIGURE 4. Comparison of radial diffusion coefficient D_x denoted by $\times$ and the diffusion coefficient obtained from the particle flux D_Γ (◯), for various values of C. a) Newtonian viscosity $\mu = 0.001$. b) Hyper-viscosity with the operator: ∇^6. c) For different values of the Newtonian viscosity and $C = 2$ (from [17]).

reduce to $D^{eff} = \Gamma_0$. Here Γ_0 is the flux averaged over the whole domain. A more elaborated discussion of the relation between the $E \times B$-flux and the particle diffusion is presented by [17]. Here we just show in Fig. 4 that the effective diffusion coefficient obtained from calculating the the mean flux and the particle diffusion coefficient are consistent. Thus for the present model with fixed and given density gradient the asymptotic diffusion coefficient for ideal tracers indeed is consistent with the transport of the bulk plasmas.

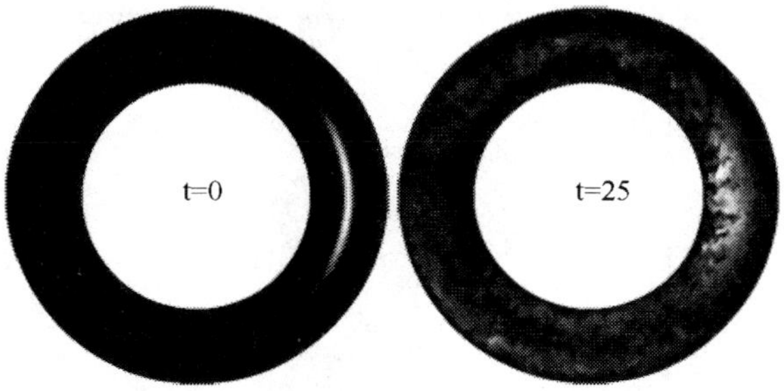

FIGURE 5. Poloidal crosssection showing impurity concentration initially (left) and after 25 time units (right) (from [35]).

DIFFERENT PINCH EFFECTS

Curvature pinch

The pinch effects may be generally ascribed to the compressibility of the advecting velocity field. As a simple example, we first consider a nonuniform magnetic field mimicking the toroidal field in toroidal devices, then the spatial dependence of the magnetic field will result in a compressibility of the $E \times B$ velocity $\nabla \cdot \vec{u}_{E\times B} = \hat{z} \times \varphi \nabla (1/B)$ for a simple slab geometry. The the pinch effect may simply be understood by considering the continuity equation for the impurity particle density N [31],

$$\left(\frac{\partial}{\partial t} + \frac{1}{B} \hat{z} \times \nabla \phi \cdot \nabla \right) \frac{N}{B} = 0, \tag{12}$$

which implies that N/B is a Lagrangian invariant advected by the compressible electric drift $\hat{z} \times \nabla \phi / B$. Now, assuming that the impurities are effectively mixed by the turbulent velocity fluctuations, this invariant will be uniformly distributed in space and the poloidally averaged impurity density $N_0(x)$ varies like $B(x)$. Thus, impurities released in the outboard midplane region will on average experience an inward transport, which corresponds to the the so-called inward pinch. This effect is usually referred to the curvature pinch, since the magnetic field inhomogeneity is originating from the curved toroidal geometry. The effect was clearly demonstrated in two-dimensional interchange instability simulations by employing the so-called ESEL code [32], and it was observed that impurities originating far out in the scrape-off-layer effectively was transported across the last closed flux surface and into the main plasma [33]

In a three dimensional flux tube geometry the curvature pinch was also clearly demonstrated for the transport of impurities in drift-Alfvén turbulence by employing the TYR code [34, 35]. Fig. 5 shows the poloidal projection of the impurity density development in that case. The analysis is based on expressing the flux Γ of the impurity species by the standard parameters used in modeling and in evaluation of transport experiments: a diffusion coefficient D and a convection velocity V, which is associated to a pinch effect:

$$\Gamma(s) = -D(s) \partial_x < n >_y + V(s) < n >_y \,. \tag{13}$$

The turbulence is radially homogeneous, so there is no radial dependence of D and V. Profiles $< \cdot >_y$ are obtained by averaging over y, corresponding to averaging toroidally over the domain extend in the drift plane, with s representing the coordinate along the magnetic field projected onto the poloidal plane. It is readily observed that by rewriting Eq. 13 in the form:

$$\Gamma(s) / < n >_y = -D(s) \partial_x < \ln n >_y + V(s) \,, \tag{14}$$

scatter plots of $\Gamma(s)/ < n >_y$ versus $\partial_x < \ln n >_y$ will provide values for $D(s)$ and $V(s)$. Figure 6 shows a scatter plot with a fit, and from the slope we obtain the value of D, while V is given by the offset at zero gradient. A comparison of the evolution of the impurity density profile with its presumed evolution using the values for D and V obtained from the above analysis is shown in Figure 7. This shows that while the scatter plot has a significant amount of noise, the fitted values for D and V are reasonably represents the overall transport process. Directly using such a kind of fit, would be an alternative method to determine the D and V values.

The dependence of D and V on s) is significant and as shown in Fig. 8. The effective convection velocity $V(s)$ changes sign and is at the high field side directed out of the plasma ($V(s)$ is directed towards the symmetry axis of the torus for all s). This pinching velocity is due to curvature and can be consistently explained within the framework of

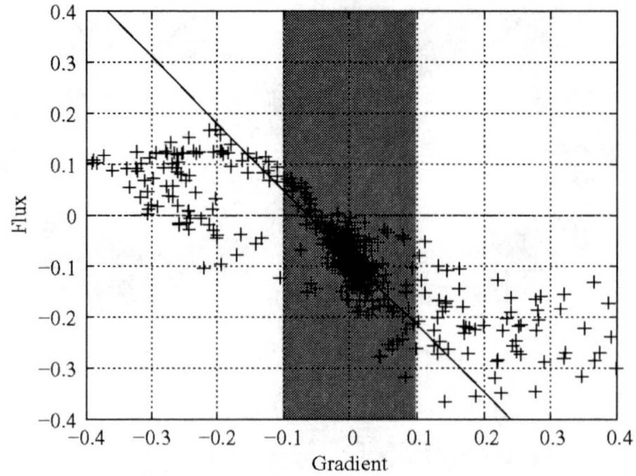

FIGURE 6. Scatter plot (at high field side) of flux versus gradient with linear fit, gray area indicates gradients admissed for the fitting procedure(from [35]).

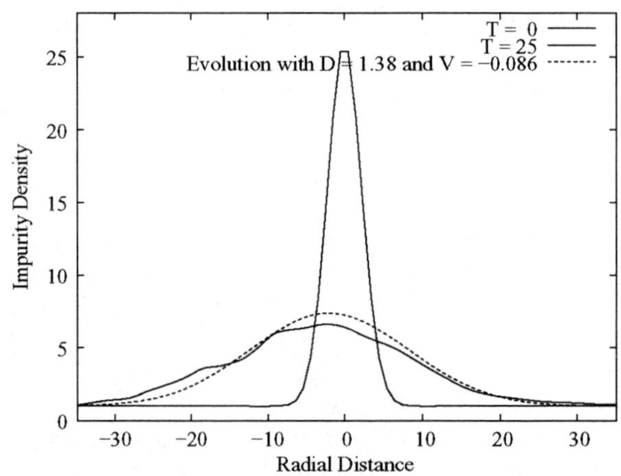

FIGURE 7. Impurity density profile at the outboard midplane at $t = 0$ and after 25 time units compared to the expected evolution of a Gaussian due to coefficients D and V obtained by the fitting procedure (from [35]).

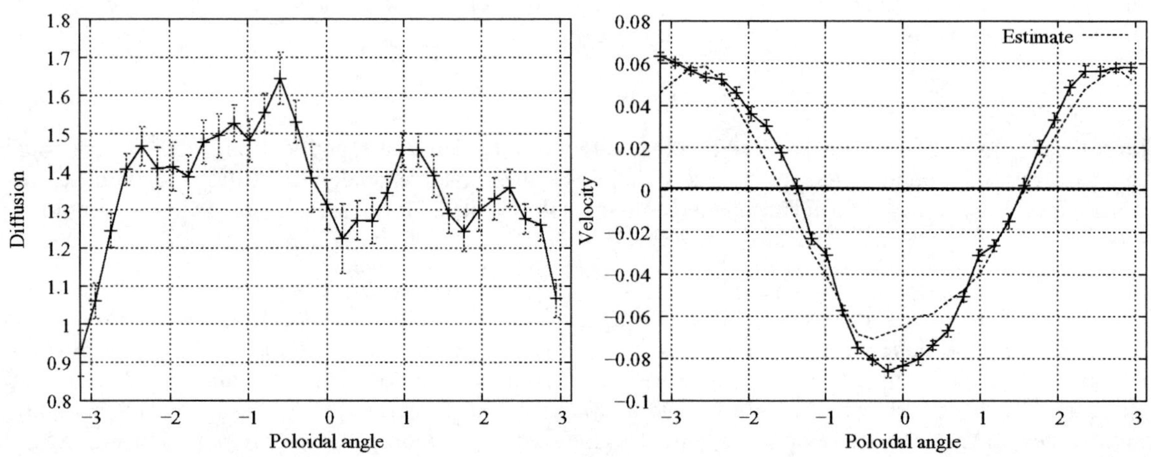

FIGURE 8. Impurity diffusion D (top) and pinch velocity V (bottom) over poloidal position (s) with error-bars. The pinch velocity is compared to $\omega_b * \cos(s) * D(s)$ (from [35]).

Turbulent EquiPartition (TEP) [36, 37], similar to the arguments in connection with Eq. 12 In the absence of parallel convection, finite mass effects and diffusion, Eq. (**??**) has the following approximate Lagrangian invariant

$$L(s) = \ln n_{imp} + \omega_B cos(s)x - \omega_B sin(s)y \,, \tag{15}$$

where $\omega_B = 2\frac{L_\perp}{R}$ governing the toroidal magnetic field inhomogeneity, R is the tokamak major radius and $L_\perp$ the mean gradient length of the pressure profile. TEP assumes the spatial homogenization of L by the turbulence. Since the parallel transport is weak, each drift plane $s = const.$ homogenizes independently. This leads to profiles $< L(s) >_y = const(s)$. At the outboard midplane ($s = 0$) the impurities are effectively convected radially inward leading to an impurity profile ($< \ln n_{imp} >_y \propto const. - \omega_B x$), while at the high field side they are effectively convected outward ($< \ln n_{imp} >_y \propto const. + \omega_B x$). One should note that this effective inward or outward convection is not found as an average $E \times B$ velocity, but is mitigated by the effect of spatial homogenization of L under the action of the turbulence. The strength of the "pinch" effect is consequently proportional to the mixing properties of the turbulence and scales with the measured turbulent diffusion. We arrive at the following expression for the connection between pinch and diffusion:

$$V(s) = -\alpha \omega_B \cos(s) D(s) \,. \tag{16}$$

Considering a stationary case with zero flux and combining Eq. (15) and Eq. (13) we obtain $\alpha = 1$. The ballooning in the turbulence level causes the inward flow on the outboard midplane to be stronger than the effective outflow on the high-field side. Therefor, averaged over a flux surface and assuming poloidally constant impurity density, a net impurity inflow results. This net pinch is proportional to the diffusion coefficient D in agreement with experimental observations [38]. The proportionality constant will, however, depend on the amount of ballooning of the transport, which has been investigated f.x. in [39, 40, 34]. The level of ballooning of the transport increases with the transition from the drift regime to the MHD regime, as $\hat{\beta}$ increases to large values. Here, for small $\hat{\beta}$ the ballooning of the transport is weak at about 70%.

Inertial pinch

The equation for the impurities can be rewritten in the form [41]:

$$d_t(\ln \theta - \zeta \omega) = \zeta \nabla_\perp \ln \theta \cdot d_t \nabla_\perp \varphi + \frac{\mu_\theta}{\theta} \nabla_\perp^2 \theta \,. \tag{17}$$

If the diffusivity μ_θ is of order $\zeta \ll 1$ and fluctuations θ_1 of the impurity density measured relative to a constant impurity background θ_0 do not exceed a corresponding level, the quantity $\ln \theta - \zeta \omega$ is approximately a Lagrangian invariant. Turbulent mixing will homogenise Lagrangian invariants in TEP states [42, 37], leading to

$$\ln \theta - \zeta \omega \approx \text{const}, \tag{18}$$

which constitutes a prediction about the effect of compressibility on the initially homogeneous impurity density field. The conservation of impurity density yields

$$\frac{\theta}{\theta_0} \approx 1 + \zeta \omega, \tag{19}$$

which conforms with the assumed ordering. We thus predict a linear relation between impurity density θ and vorticity ω, the proportionality constant being the mass–charge ratio ζ. This is related to, but not the same as, the aggregation of dense particles in vortices in fluids due to the Coriolis force [43]. The prediction is verified by numerical simulations of inertial impurities in saturated HW-turbulence for $\mathscr{C} = 1$. The simulations are performed on a $[-20, 20]^2$ domain, using 512×512 gridpoints, and impurity diffusivity 0.02. The impurity density field is initially set to unity. The impurity density field for $\zeta = 0.01$ is presented together with vorticity in Figure 9. Figure 10 shows a scatter plot of the point values of impurity density and vorticity at time 150 for three different values of ζ. The proportionality factor $\theta = 1 + K\omega$ is determined to be slightly below one: $K \simeq 0.82\,\zeta$.

The role of inertia for a radially inward pinch is investigated by considering the collective drift of impurities.

$$\vec{\Gamma}_\theta \equiv \vec{\Gamma}_\theta^E + \vec{\Gamma}_\theta^p \equiv \int \vec{x} \theta \vec{\hat{z}} \times \nabla_\perp \varphi + \int \vec{x} \theta \, \zeta \, d_t \nabla_\perp \varphi, \tag{20}$$

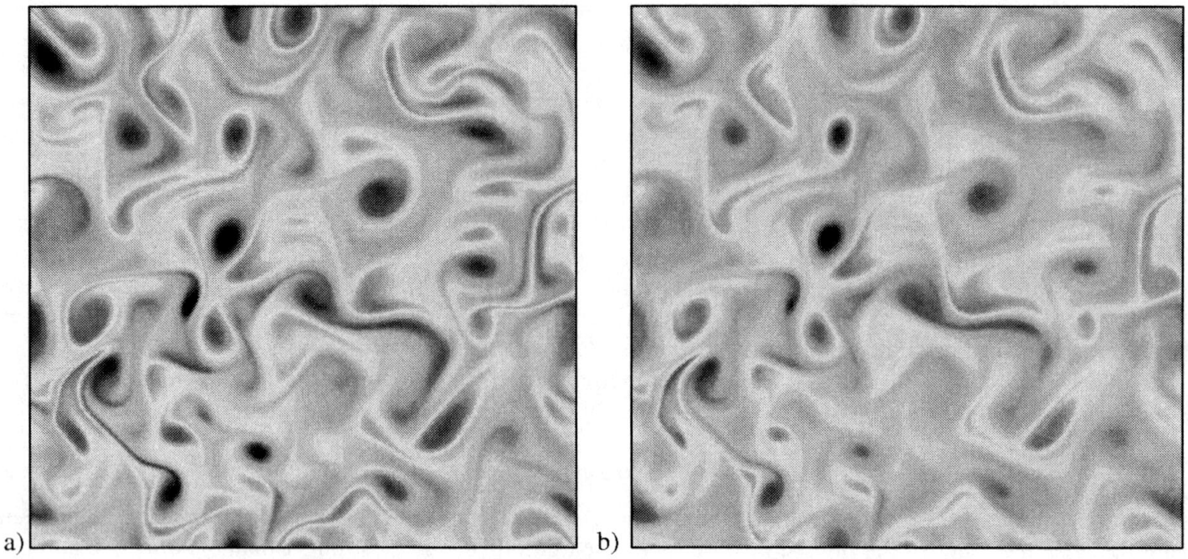

FIGURE 9. (a) Vorticity and (b) density of inertial impurities in the saturated state with $\mathscr{C} = 1$ and $\zeta = 0.01$, L = 40. Other parameters: $\mu_n = \mu_\omega = \mu_\theta = 0.02$.

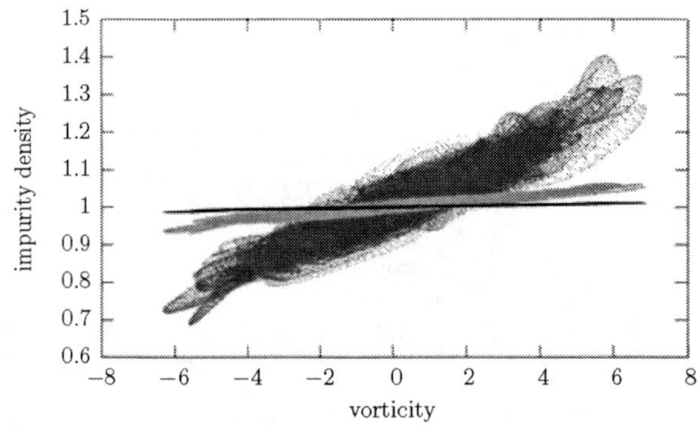

FIGURE 10. Scatter plot of impurity density and the vorticity field at $t = 100$ for different values of the mass–charge ratio ζ in the saturated state in HW with $\mathscr{C} = 1$: $\zeta = 0.05$ (red), $\zeta = 0.01$ (green), and $\zeta = 0.002$ (blue).

TABLE 1. Radial drift velocity of impurities for different values of the mass–charge ratio ζ in the saturated state in HW with $\mathscr{C} = 1$. Calculated as the average value between $t = 25$ and $t = 150$. Parameters: $\mu_n = \mu_\omega = \mu_\theta = 0.02$.

ζ	radial drift
−0.010	8.67×10^{-4}
0.001	-8.66×10^{-5}
0.002	-1.73×10^{-4}
0.005	-4.35×10^{-4}
0.010	-8.69×10^{-4}
0.020	-1.75×10^{-3}
0.050	-4.55×10^{-3}

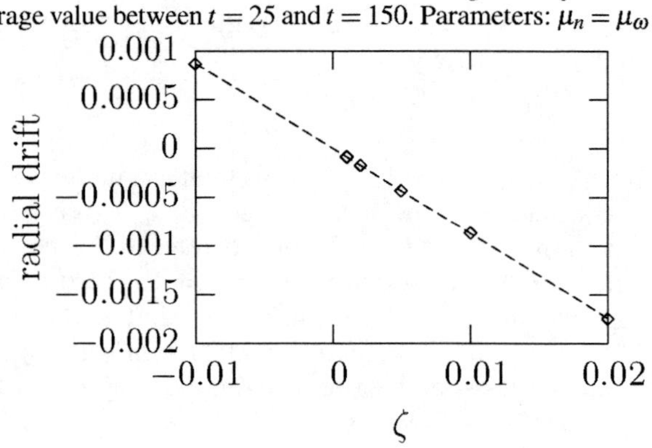

In terms of the integral impurity flux, we define an overall impurity drift velocity $\vec{V}_\theta$ by

$$\vec{V}_\theta \equiv \vec{V}_\theta^E + \vec{V}_\theta^p \equiv \frac{\vec{\Gamma}_\theta^E}{\int \vec{x}\theta} + \frac{\vec{\Gamma}_\theta^p}{\int \vec{x}\theta} = \frac{\vec{\Gamma}_\theta}{\int \vec{x}\theta}, \tag{21}$$

Ideal impurities do on average not experience a drift, but this is not the case for inertial impurities, since compressibility effects arrange for a correlation between θ_1 and ω. Note that only the deviations from the above discussed linear relationship $\theta = 1 + K\omega$ result in a net flow, as $\int K\omega v_r dx = 0$ for periodic boundary conditions. The radial drift velocity has a definite sign that depends on the sign of ζ, which was first discussed in [44]. There is a continuous flow of impurities in a definite direction (inward for positively charged impurities). This resembles the anomalous pinch observed in magnetic confinement experiments [45]. Average radial drift velocities computed using the values of the drift from $t = 25$ to $t = 150$ are presented in Table 1. The scaling of the average radial drift with ζ is seen to be remarkably linear.

Parallel compressibility pinch

A similar pinch effect results from compressibility along magnetic field lines, that is when parallel advection is taken into consideration, details can be found in [46].

Thermodiffusion

Additional pinch effects appear from off-diagonal elements of the transport matrix. Obviously this needs the introduction of a finite impurity temperature and thus a finite impurity pressure. To elucidate this effect we consider, for illustration only, the system of impurity pressure and density in a system with an effective curvature given via ω_B [47] :

$$d_t n_Z = -\omega_B \left(n_Z \nabla\phi + \frac{1}{Z}\nabla p_Z \right) \tag{22}$$

$$d_t p_Z = -\Gamma\omega_B \left(p_Z \nabla\phi + \frac{1}{Z}\nabla p_Z^2/n_Z \right) \tag{23}$$

with $\Gamma = 5/3$. We evaluate the quasilinear fluxes, assuming that the impurity density response is due to an external potential perturbation caused by some instability and arrive at the following expression for the particle ($Gamma_Z$) and heat (Q_Z) fluxes:

$$\begin{pmatrix} \frac{\Gamma_Z}{n_Z} \\ \frac{Q_Z}{n_Z T_Z} \end{pmatrix} = -\begin{bmatrix} D_n n & D_n T \\ D_n t & D_T T \end{bmatrix} \begin{pmatrix} \partial_r n_Z/n_Z + \omega_B \\ \partial_r T_Z/T_Z + \omega_B(\Gamma - 1) \end{pmatrix} \tag{24}$$

in which case we see that the quasilinear particle flux recovers the curvature pinch (occurence of $omega_B$), but through the off diagonal element picks up a contribution proportional to the gradient of the impurity temperature ($D_n T$). Expressing the flux now in terms of a diffusive contribution, namely everything that is proportional to ∇n_z and a pinch (the rest), this term leads to a pinch effect. The physics behind this easy, namely the plasma can increase entropy more efficiently by exchanging a larger number of colder impurity ions with a smaller number of hotter ions. The effective energy flux is down gradient in this case.

It should be noted, however, that in the quasilinear estimate the actual direction of the pinch effects (including thermodiffusion) depends on the propagation velocity of the underlying unstable modes.

QUASI–LINEAR GYROKINETIC CALCULATIONS OF IMPURITY TRANSPORT (AND COMPARISON WITH EXPERIMENTS)

Recently, the need of explaining and predicting impurity transport observations in fusion experiments which cannot be accounted for by neoclassical transport, lead to calculations of turbulent impurity transport adopting a physical description as realistic as possible, by the application of gyrokinetic models, and related codes.

In order to show how these calculations can be performed, first we recall that the δf gyrokinetic equation for the perturbed distribution function of particles of a given species is linear in the radial logarithmic gradients of the unperturbed (equilibrium) density and temperature of particles of the same species. Therefore, for small charge concentration of the impurity, which is usually the case for large part of the impurities in the experimental conditions, impurities should behave close to the passive (tracer) limit. In this case, a linear relationship holds between the particle flux of the impurity, and its density and temperature logarithmic gradients, which we express in terms of the dimensionless ratio of the major radius R to the gradient length L, namely

$$\frac{R\Gamma}{n_s} = D\frac{R}{L_{ns}} + D_T\frac{R}{L_T} + RV_P. \tag{25}$$

Here the diagonal diffusive coefficient D, the off-diagonal thermodiffusive coefficient D_T and the pure convection term V_p can be identified directly.

This property can be used in the framework of quasilinear gyrokinetic calculations in order to compute the previous contributions separately and measure the relative weight of each one and perform comparisons with the experiments. In experimental stationary conditions, in the absence of sources, the flux of the impurity is close to zero, which implies that for practical purposes it is useful to introduce the coefficients $C_T = D_T/D$ and $C_P = RV_P/D$. Then, the parameter $V/D \approx \frac{R}{L_{ns}}$ usually directly accessible in the experiments can be obtained by the sum of the thermodiffusive and pure convection contributions, in the form

$$\frac{RV}{D} = C_T\frac{R}{L_T} + C_P. \tag{26}$$

This quantity can be computed directly with a gyrokinetic code and compared to the experimental observations. Such investigations have been performed using the gyrokinetic code GS2 [48, 49]. As already presented in the previous section, it has been noted that the signs of the coefficients C_T and C_P may change depending on the direction of propagation of the background plasma microinstability [50, 46, 47, 51, 52]. In particular the thermodiffusive term provides an outward/inward contribution for modes propagating in the ion/electron diamagnetic direction [53, 50, 46], and decreases with increasing charge of the impurity, while the pure convection coefficient is usually directed inwards by E×B compression. Parallel dynamics adds a contribution to this term [46] which can lead to a total particle flux which is directed outwards, in the case of modes propagating in the electron diamagnetic direction. Conditions of outward impurity convection are extremely interesting from the experimental side, particularly in the development of ITER relevant scenarii. Therefore, some recent work has focused on the conditions for which an outward impurity convection is measured experimentally (e.g. [54, 55, 56]). Here we show specific gyrokinetic simulations of the previously introduced pure convection coefficient for the experimental conditions presented in Ref. [55]. In these experiments in ASDEX Upgrade (AUG), with Si laser ablation, a central outward total impurity convection was measured in the presence of strongly localized central electron cyclotron heating, while in the absence of this, the impurity convection was measured directed inwards. Such a transition in the behaviour cannot be explained by neoclassical transport. For this reason, corresponding investigations with a gyrokinetic model have been performed [52]. Input parameters of the gyrokinetic calculations were taken using experimental measurements. It has been found that in the plasma centre, in the presence of strong electron heating, producing locally strong electron temperature gradients, modes propagating in the electron diamagnetic direction can be destabilized. These modes feature an extremely elongated eigenfunction along the field line, shown in Fig. 11(a), and are generated by the non-adiabatic passing electron response of passing electrons. Passing electrons are transported inward in these conditions. Such an electrostatic potential develops an eleongated structure of parallel velocity fluctuations by parallel force balance, shown in Fig. 11(b). The latter, by parallel compression, leads to the development of a radial impurity flux which is directed outwards, as shown in Fig 11(c), for these modes, which have phase velocity in the electron diamagnetic direction. The mechanism of outward impurity convection introduced in Ref. [46] is found to occur in the specific simulation of these AUG experimental conditions. We mention also that such an outward convection mechanism is not specific exclusively of very long wave numbers, as erroneously pointed out in [51], but covers the full usual long wave length spectrum, as specifically demonstrated in [52]. We have to mention as well that the destabilization of these extremely elongated eigenfunctions is possible only in the very central region of the plasma, and that this kind of modes cannot be thought to be responsible of the outward convection observed at mid-radius in JET in Ni laser ablation experiments [56]. In that case, usual ballooning ITG modes were found as dominant modes, producing an inward convection of the impurity, in disagreement with the experimental observations. An outward impurity flux of the impurity was found in the gyrokinetic simulations of these experiments only by switching off the ITG instability and singling out the effect of the otherwise subdominant trapped electron mode.

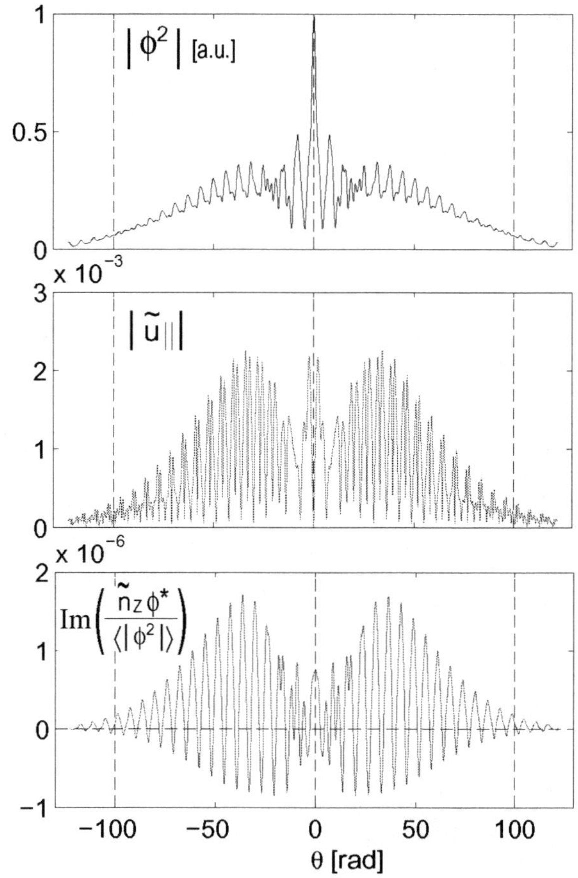

FIGURE 11. Linear mode structure along the field line of the electrostatic potential fluctuations $|\phi^2|$ (a), of the Si parallel velocity $|\tilde{u}|$ (b), and of the normalized cross phase relationship between Si denity fluctuations and electrostatic potential fluctuations, proportional to the Si radial flux, (c) as a function of the extended ballooning angle θ for a calculation with experimental input parameters for a AUG H–mode plasma with central electron heating [52].

In conclusion, the study of turbulent impurity transport by means of gyrokinetic models has made significant progresses and first non-trivial qualitative agreement between the experimental observations and the theoretical predictions have been found. However, the way is still long before a model for impurity transport is indeed assessed for reliable ITER prediction purposes.

ACKNOWLEDGMENTS

One of the authors (VN) would like to acknowledge discussions with X. Garbet and Y. Yankov on pinch effects and thermodiffusion.

REFERENCES

1. A. Kallenbach, R. Neu, R. Dux, H. Fahrbach, J. Fuchs, L. Giannone, O. Gruber, A. Herrmann, P. Lang, B. Lipschultz, C. Maggi, J. Neuhauser, V. Philipps, T. Putterich, V. Rohde, J. Roth, G. Sergienko, A. Sips, and ASDEX Upgrade Team, *Plasma Phys. Controlled Fusion* **47**, B207–B222 (2005).
2. S. P. Hirshman, and D. J. Sigmar, *Neoclassical transport of impurities in Tokamak plasmas*, Unknown, 1981.
3. H. Tennekes, and J. L. Lumley, *A First Course in Turbulence*, MIT Press, Cambridge, Massachusetts, 1972, ISBN 0-262-20019-8.
4. W. D. McComb, *The Physics of Fluid Turbulence*, Clarendon Press, Oxford, 1990.

5. D. Elhmaïdi, A. Provenzale, and A. Babiano, *J. Fluid Mech.* **257**, 553–558 (1993).
6. A. Provenzale, *Annu. Rev. Fluid. Mech.* **31**, 55–93 (1999).
7. C. Pasquero, A. Provenzale, and A. Babiano, *J. Fluid Mech.* **439**, 279–303 (2001).
8. A. V. Milovanov, *Phys. Rev. E* **63**, 047301 (2001).
9. G. M. Zaslavsky, *Phys. Rep.* **371**, 461 – 580 (2002).
10. J. Lynov, A. Nielsen, H. Pécseli, and J. Rasmussen, *J. Fluid Mech.* **224**, 485–505 (1991).
11. J. H. Misguich, R. Balescu, H. Pécseli, T. Mikkelsen, S. Larsen, and Q. Xiaoming, *Plasma Phys. and Contr. Fusion* **29**, 825–856. (1987).
12. M. Pettini, A. Vulpiani, J. H. Misguich, M. D. Leener, J. Orban, and R. Balescu, *Phys. Rev. A* **38**, 344–363 (1988).
13. M. Vlad, F. Spineanu, J. H. Misguich, and R. Balescu, *Phys. Rev. E* **58**, 7359–7368 (1998).
14. R. Balescu, *Aspects of Anomalous Transport in Plasmas*, Taylor and Francis Group, CRC Press, Paris, 2005.
15. A. Hasegawa, and M. Wakatani, *Phys. Rev. Lett.* **50**, 682–686 (1983).
16. V. Naulin, A. H. Nielsen, and J. Juul Rasmussen, *Phys. Plasmas* **6**, 4575 – 4585 (1999).
17. R. Basu, T. Jessen, V. Naulin, and J. Juul Rasmussen, *Phys. Plasmas* **10**, 2696–2703 (2003).
18. R. Basu, V. Naulin, and J. Juul Rasmussen, *Comm. in Nonlinear Science and Numerical Simulation* **8**, 477–492 (2003).
19. G. Manfredi, and R. O. Dendy, *Phys. Rev. Lett.* **76**, 4360 (1996).
20. B. Krane, H. L. Pécseli, and J. Trulsen, *Phys. Fluids* **15**, 211–226 (2003).
21. M. Vlad, F. Spineanu, S. I. Itoh, M. Yagi, and K. Itoh, *Plasma Phys. Controlled Fusion* **47**, 1015–1029 (2005).
22. T. Hauff, and F. Jenko, *Phys. Plasmas* **13**, 102309 (2006).
23. T. Hauff, and F. Jenko, *Phys. Plasmas* **14**, 092301 (2007).
24. S. Guenter, G. Conway, S. DaGraca, H.-U. Fahrbach, C. Forest, M. Munoz, T. Hauff, J. Hobirk, V. Igochine, F. Jenko, K. Lackner, P. Lauber, P. McCarthy, M. Maraschek, P. Martin, E. Poli, K. Sassenberg, E. Strumberger, G. Tardini, E. Wolfrum, H. Zohm, and Asdex Upgrade Team, *Nucl. Fusion* **47**, 920–928 (2007).
25. S. V.Annibaldi, G. Manfredi, R. Dendy, and L. Drury, *Plasma Phys. Controlled Fusion* **42**, L13–L22 (2000).
26. S. V.Annibaldi, G. Manfredi, and R. Dendy, *Phys. Plasmas* **9**, 791–799 (2002).
27. T. Huld, A. H. Nielsen, H. L. Pécseli, and J. J. Rasmussen, *Phys. Fluids B* **3**, 1609 – 1625 (1991).
28. M. Endler, H. Niedermeyer, L. Giannone, E. Holzauer, A. Rudyj, G. Theimer, N. Tsois, and Asdex-Team, *Nucl. Fusion* **35**, 1307–1339 (1995).
29. B. A. Carreras, C. Hidalgo, E. Sánchez, M. Pedrosa, R. Balbín, García-Cortés, B. van Milligen, D. E. Newman, and V. E. Lynch, *Phys. Plasmas* **3**, 2664–2672 (1996).
30. V. Naulin, and K. H. Spatschek, *Phys. Rev. E* **55**, 5883–5893 (1997).
31. M. B. Isichenko, and N. V. Petviashvili, *Phys. Plasmas* **2**, 3650–3654 (1995), URL `http://link.aip.org/link/?PHP/2/3650/1`.
32. O. E. Garcia, V. Naulin, A. H. Nielsen, and J. Juul Rasmussen, *Phys. Plasmas* **12**, 062309 (2005).
33. O. E. G. V. Naulin, A. H. Nielsen, and J. J. Rasmussen, *Physics AUC* **17**, 263 (2007).
34. V. Naulin, *Phys. Plasmas* **10**, 4016–4028 (2003).
35. V. Naulin, *Phys. Rev. E* **71**, 015402 R (2005).
36. J. Nycander, and V. V. Yankov, *Phys. Plasmas* **2**, 2874–2876 (1995).
37. V. Naulin, J. Nycander, and J. Juul Rasmussen, *Phys. Rev. Lett.* **81**, 4148 – 4151 (1998).
38. M. E. Perry, N. H. Brooks, D. A. Content, R. A. Hulse, M. ali Mahdavi, and H. W. Moos, *Nucl. Fusion* **31**, 1859 – 1874 (1991).
39. B. D. Scott, *Plasma Phys. Control. Fusion* **39**, 471 – 504 (1997).
40. B. D. Scott, *Plasma Phys. Control. Fusion* **39**, 1635 – 1668 (1997).
41. M. Priego, O. E. Garcia, V. Naulin, and J. Juul Rasmussen, *Phys. Plasmas* **12**, 062312.1–062312.11 (2005).
42. V. V. Yankov, *Physics-Uspekhi* **40**, 477–493 (1997).
43. A. Bracco, P. H. Chavanis, A. Provenzale, and E. A. Spiegel, *Phys. Fluids* **11**, 2280–2287 (1999).
44. V. Naulin, J. Juul Rasmussen, and M. P. Wood, "Pinch effects in turbulent impurity transport," in *Theory of Fusion Plasmas*, edited by J. W. Connor, O. Sauter, and E. Sindoni, Editrice Compositori, Bologna, 2004, pp. 365–371.
45. R. Dux, *Fusion Science and Technology* **44**, 708 – 715 (2003).
46. C. Angioni, and A. Peeters, *Phys. Rev. Lett.* **96**, 095003 (2006).
47. N. Dubuit, X. Garbet, T. Parisot, R. Guirlet, and C. Bourdelle, *Phys. Plasmas* **14**, 042301 (2007).
48. M. Kotschenreuther, G. Rewoldt, and W. Tang, *Computer Phys. Comm.* **88**, 128 (1995).
49. W. Dorland, F. Jenko, M. Kotschenreuther, and B. Rogers, *Phys. Rev. Lett.* **85**, 5579 (2000).
50. X. Garbet, N. Dubuit, E. Asp, Y. Sarazin, C. Bourdelle, P. Ghendrih, and G. Hoang, *Phys. Plasmas* **12**, 082511 (2005).
51. C. Bourdelle, X. Garbet, F. I. andA. Casati, N. Dubuit, R. Guirlet, and T. Parisot, *Phys. Plasmas* **14**, 112501 (2007).
52. C. Angioni, R. Dux, E. Fable, and A. G. Peeters, *Plasma Phys. Controlled Fusion* **49**, 2027 (2007).
53. M. Frojdh, Liljestrom, and H. Nordman, *Nucl. Fusion* **32**, 419 (1992).
54. E. Scavino, J. Bakos, and H. Weisen, *Plasma Phys. Controlled Fusion* **46**, 857 (2004).
55. R. Dux, R. Neu, A. Peeters, G. Pereverzev, F. R. A. Mück, and J. Stober, *Plasma Phys. Controlled Fusion* **45**, 1815 (2003).
56. M. Puiatti, M. Valisa, C. Angioni, L. Garzotti, P. Mantica, M. Mattioli, L. Carraro, I. Coffey, and C. Sozzi, *Phys. Plasmas* **13**, 042501 (2006).

Non-diffusive transport modeling: statistical basis and applications.

D. del-Castillo-Negrete

Oak Ridge National Laboratory
Oak Ridge TN, 37831-8071

Keywords: non-diffusive transport, fusion plasmas, fractional diffusion, Levy processes
PACS: 52.25.Fi, 05.40.Fb, 47.27.T-, 02.50.-r, 05.60.-k

INTRODUCTION

The key issue to achieve controlled fusion is to confine a very high temperature plasma for a sufficiently long time. However, plasma confinement is limited by thermal transport processes, and therefore, one of the main goals of plasma physics research is to understand transport. Coulomb collisions provide the most basic, and natural transport mechanism in a plasma. However, transport calculations based only on collisional effects (i.e., classical and neoclassical transport theories) tend to underestimate the transport levels experimentally observed in magnetically confined fusion plasmas. It is believed that plasma instabilities and turbulence play a critical role in the observed enhanced transport. The study of turbulence-driven transport, also known as anomalous transport, has therefore attracted a lot of attention. Although significant progress has been made in the classical, neoclassical, and anomalous theories of transport, there are important open problems likely to play a role in the design and operation of present and future fusion devices including ITER. A common feature of these problems is that they seem to defy our intuition rooted in the concept of diffusive transport.

Classical and neoclassical transport theories, as well as most of the anomalous transport theories are based on the standard diffusion paradigm that establishes a local relationship between the fluxes (e.g., particle and heat fluxes) and the thermodynamic forces driving them (e.g., density and temperature gradients). In the simplest case of a single field, T, (e.g., the temperature), in a one-dimensional domain, x, (e.g., the radial direction in a tokamak) the diffusion paradigm assumes the Fourier-Fick's prescription,

$$q = -\chi \partial_x T \,, \tag{1}$$

where q denotes the flux and χ the diffusivity. This allows to close the continuity equation,

$$\partial_t T = -\partial_x q + S \,, \tag{2}$$

where S is a source, to get the well-known diffusion model

$$\partial_t T = \partial\left(\chi \partial_x T\right) + S \,. \tag{3}$$

Within this approach, all the efforts are then focused on the modeling of χ, which can in principle depend on x, t, and in the case of non-linear models, on T and its gradient.

Despite the appeal and widespread use of the diffusion model, it is important to realize that its cornerstone, the Fourier-Fick's prescription in Eq. (1), is based on highly restrictive assumptions that might breakdown in problem of interest to magnetically confined fusion plasma. Numerical examples of this type of transport, known as non-diffusive transport, include Hasegawa-Mima turbulence [1], 3-dimensional, resistive, pressure gradient-driven plasma turbulence in cylindrical [2, 3, 4] and toroidal geometry [5], and gyrokinetic plasma turbulence [6]. Non-diffusive transport has also been reported in reduced Hamiltonian models of $\mathbf{E}\times\mathbf{B}$ chaotic transport in the presence of zonal flows and drift waves [7, 8]. Experimental findings exhibiting deviations from the standard diffusion paradigm include the anomalous scaling of the confinement time in low confinement mode plasmas [9], the non-Gaussianity of experimentally measured fluctuations [10, 11], and fast propagation phenomena in perturbative transport experiments

CP1013, ***Turbulent Transport in Fusion Plasmas, First ITER International Summer School***
edited by S. Benkadda

[12, 13]. Recently, particle transport in a toroidal plasma confinement device was shown to be non-diffusive in the presence of magnetic chaos [14].

This paper presents a tutorial on the modeling of non-diffusive transport using fractional diffusion. We review the statistical basis of fractional diffusion models and discuss applications to magnetically confined plasmas. Fractional diffusion models are formulated using fractional derivative operators, which are integro-differential operators that extend the notion of derivatives to fractional orders. As we will explain in detail below, in the context of transport, these operators appear in the description of stochastic processes without characteristic spatio-temporal scales in general, and in the description of intermittent (avalanche-like) turbulent transport in plasmas in the presence of coherent structures (e.g., vortices and zonal flows) in particular. In addition, fractional diffusion operators exhibit self-similar non-diffusive scaling, and are intimately related to non-Gaussian (Lévy) distributions believed to play a role in plasma turbulence. Contrary to diffusive models in which the flux at a give point depends only on the properties of the scalar field and its gradient at that point, fractional models incorporate a more general non-local, flux-gradient relation of the form

$$q = -\chi \partial_x \int K(x-x')\, T(x',t)\, dx' , \tag{4}$$

where the function $K(x-x')$ accounts for the non-local contribution of T at x' to the flux at x. The rate of decay of K determines the degree of non-locality. As expected, for $K(x-x') = \delta(x-x')$, the local Fourier-Fick's law in Eq. (1) is recovered. In the case of fractional diffusion, $K(x-x') = 1/|x-x'|^{\zeta}$, where $\zeta > 0$ is a constant determined by the statistical properties of the underlying stochastic process.

Non-local flux-gradient relations of the form in Eq. (4) have been used before in the study of parallel electron heat transport in magnetized plasmas [15, 16], and in the study of transport due to long scale-length fluctuations [17]. However, the physics behind the non-local models discussed here is different, and it is based on the theory of non-Gaussian stochastic processes. From the statistical mechanics point of view, the diffusion equation can be understood as a macroscopic transport model describing a Brownian random walk, which at a microscopic level assumes an uncorrelated, Gaussian stochastic process. Similarly, as we will discuss in detail below, the non-local models discussed here correspond to macroscopic models describing transport processes governed by a generalization of the Brownian random walk model known as the Continuous Time Random Walk (CTRW) model. Within this formalism, non-Markovian (memory) effects can also be naturally included in the dynamics, leading to flux-gradient relations of the form in Eq. (4) with an additional integro-differential operator in time. These properties make fractional diffusion operators a particularly appealing option from the physics point of view, compared to other type of integral operators used in the modeling of non-local transport. From the analytical and computational point of view fractional diffusion operators also present advantages since the mathematical theory of these operators is well-understood and there are efficient numerical methods to solve fractional equations, see e.g. Refs.[18, 19]. An alternative way to incorporate non-Gaussian stochastic processes in models of non-diffusive transport is based on the direct use of the Montroll-Weiss master equation, and nonlinear generalization to take into account critical gradient threshold conditions [20].

There are several very good reviews and books that the interested reader is encouraged to consult to complement the material presented here. When doing this, it is important to keep in mind important differences in the terminology used. In plasma physics "anomalous transport" usually refers to diffusive transport processes with anomalously large diffusivities (compared to the classical and neoclassical predictions). However, in non-equilibrium statistical mechanics, dynamical systems, and in most transport studies outside plasma physics "anomalous transport" refers to non-diffusive transport. For an introduction to non-diffusive transport in the context of disorder media see Ref. [21]. A very readable introduction to the CTRW, accompanied by interesting historical remarks can be found in Ref. [22]. For a discussion on the connection between the CTRW and fractional diffusion see Ref. [23]. The connection between chaotic transport and fractional diffusion is discussed in Ref. [24]. A recent collection of tutorial and research papers reviewing the foundations and various application of non-diffusive transport and fractional diffusion can be found in Ref. [25]. The general problem of anomalous transport in plasma physics is covered in Ref. [26] which is a sequel of the Refs. [27] where the classical and neoclassical theories of transport are explained in detail. A classic reference on Lévy distribution is Ref. [28]. For an introductory level, clear description of stochastic processes in physics see Ref. [29]. Two recommended books on fractional calculus are Ref. [30] at an introductory level, and Ref. [31] for a more advanced discussion. The latter reference is also recommended for a brief historical outline of fractional calculus.

The rest of this paper is organized as follows. The next section introduces basic concepts of non-diffusive transport in the context of a simple model of chaotic transport by drift-waves in zonal flows. Following this, we review the foundations of non-diffusive transport models on the theory of stochastic processes. It is shown that the well-established formalism of fractional calculus provides a unifying framework for the formulation of non-diffusive transport models incorporating non-Gaussian processes, non-locality, non-Markovian (memory) effects, and anomalous self-similar scaling. The section "Applications of fractional diffusion", is devoted to several applications of fractional diffusion to plasma transport. The use of fractional operators to model (in a renormalization sense) the propagator of particle transport in pressure gradient driven plasma turbulence is discussed. Also, it is shown that fractional diffusion models naturally exhibit non-diffusive transport phenomenology typical observed in fusion plasmas including the anomalous scaling of confinement time with system's size, profile peaking, up-hill transport, and fast pulse propagation phenomena. A discussion on a recent application of the fractional model to perturbative experiments conducted at the Joint European Torus (JET) tokamak is also presented. A brief summary and the conclusions are presented in the last section of the manuscript.

NON-DIFFUSIVE TRANSPORT

The goal of this section is to motivate and define basic concepts of non-diffusive transport in terms of a relatively simple transport problem of interest to magnetically confined plasmas. The problem concerns $\mathbf{E}\times\mathbf{B}$ test particle transport due to drift waves in the presence of zonal shear flows. We will show that the combination of trapping effects of $\mathbf{E}\times\mathbf{B}$ eddies and large displacements induced by the zonal flows naturally results in non-diffusive transport that cannot be described using the diffusion equation with an effective diffusivity.

The dynamics of test particles is determined by the equation

$$\frac{d\mathbf{r}}{dt} = \mathbf{V}\left[\mathbf{r}(t),t\right], \tag{5}$$

where $\mathbf{r}$ denotes the particle's position, and $\mathbf{V}$ is the velocity of the prescribed flow. In the case of $\mathbf{E}\times\mathbf{B}$ transport,

$$\mathbf{V} = -\frac{\mathbf{E}\times\mathbf{B}}{B^2}, \tag{6}$$

where $\mathbf{B}$ is the magnetic field, and $\mathbf{E}$ the electrostatic field [32]. By a test particle we mean a particle that is simply carried by the flow without modifying it. Although this approximation neglects self-consistent coupling effects, it is a good starting point to study transport in general, and non-diffusive processes in particular. In the case of a constant magnetic field, $\mathbf{B} = B_0\hat{\mathbf{z}}$, and a two-dimensional electrostatic field, $\mathbf{E} = -\nabla\phi(x,y,t)$, substituting Eq. (6) into Eq. (5), and rescaling time to eliminate the constant B_0, leads to the Hamiltonian system

$$\frac{dx}{dt} = -\frac{\partial\phi}{\partial y}, \qquad \frac{dy}{dt} = \frac{\partial\phi}{\partial x}, \tag{7}$$

with the electrostatic potential playing the role of Hamiltonian, and the spatial coordinates, (x,y), playing the role of canonical conjugate phase space variables. In the present context, x corresponds to the direction of the density gradient driving the drift-wave instability, and y corresponds to the direction of propagation of the drift-waves. In toroidal geometry, x corresponds to a normalized coordinate along the minor radius, and y to a poloidal-like angular coordinate. However, here we will assume a slab approximation and treat (x,y) as Cartesian coordinates.

In the Hasegawa-Mima model for drift waves [33], the electrostatic potential is determined from

$$\left[\partial_t + (\mathbf{z}\times\nabla\phi)\cdot\nabla\right]\left(\nabla^2\phi - \phi - \beta x\right) = 0, \tag{8}$$

where we have used dimensionless variables, and $\beta = n_0'/n_0$ is assumed constant with $n_0(x)$ the plasma density. Although the main emphasis of this paper is transport in plasmas, it is interesting to mention that there is a very close analogy between the dynamics of drift-waves in the Hasegawa-Mima approximation and the dynamics of Rossby waves in rapidly rotating neutral fluids in the quasigeostrophic approximation [34]. In this analogy, the role of the strong magnetic field is played by the rapid rotation, the plasma electrostatic potential corresponds to the fluid stream function, the plasma density corresponds to the fluid vorticity, and the plasma density gradient corresponds to the topography gradient or equivalently to a gradient in the Coriolis force in the case of a rotating sphere [35, 36]. Based on this analogy, experimental [37] and theoretical [38, 7] transport studies on rotating fluids can be applied to $\mathbf{E}\times\mathbf{B}$ transport in plasmas as discussed in Ref. [8].

In principle the Hamiltonian ϕ in Eq. (7) should be obtained by solving the nonlinear Hasegawa-Mima Eq. (8) with the appropriate boundary conditions. However, finding analytic solutions with nontrivial spatio-temporal variability of this equation is difficult and one has to resort to simplified approaches to model the Hamiltonian ϕ. Here, following Refs. [38, 7, 8], we model ϕ using the linear eigenmodes of Eq. (8),

$$\phi = \varphi_0(x) + \sum_{j=1}^{N} \varepsilon_j \, \varphi_j(x) \, \cos k_j(y - c_j t) \,, \tag{9}$$

where $\varphi_0(x)$ is the electrostatic potential (streamfunction) of the equilibrium zonal shear flow, and $\varphi_j(x)$ are the eigenfunctions of the linear problem with wave numbers, k_j, and frequencies, $c_j k_j$, for $j = 1, \ldots N$. We will consider monotonic zonal flows of the form

$$u_0(x) = \tanh(x) \,, \tag{10}$$

In this case, as discussed in Ref. [8], depending on the parameter values, there is a band of unstable modes bounded by two regular neutral modes with eigenfunctions

$$\varphi_j = [1 + \tanh x]^{\frac{1-c_j}{2}} \, [1 - \tanh x]^{\frac{1+c_j}{2}} \,, \tag{11}$$

Since these modes are neutral, c_j, for $j = 1$ and 2, are real, and the corresponding values of k_j are obtained from the linear dispersion relation. Neutral model are important because they lay at the stability boundary and describe the dynamics near marginal stability. To construct the Hamiltonian model we consider a traveling wave perturbation of the first neutral mode in the co-moving reference frame

$$\phi = \ln(\cosh x) + \varphi_1(x) \, [\varepsilon_1 \cos k_1 y + \varepsilon_2 \cos(k_1 y - \omega t)] - c_1 x, \tag{12}$$

where the first term is the potential of the shear flow in Eq. (10), and ω is the frequency of the perturbation.

When $\varepsilon_2 = 0$ the Hamiltonian is time independent, the $\mathbf{E} \times \mathbf{B}$ motion is integrable and the test particles follow the ϕ =constant contours shown in Fig. 1-(a). In this case, particles inside the separatrix remain trapped, and those outside the separatrix are always untrapped. However, when there is a time dependent perturbation, e.g. when $\varepsilon_2 \neq 0$ in Eq. (12), the $\mathbf{E} \times \mathbf{B}$ motion is in general not integrable. In this case, as shown schematically in Fig. 1-(b) the separatrix breaks and forms a *stochastic layer* where test particles alternate chaotically between being untrapped and being trapped inside the vortices. Figure 2 shows typical test particle trajectories in this case.

Because of the stochastic component of the dynamics, it is tempting to assume that the test particles in the stochastic layer follow a Brownian type random walk in the poloidal y-direction jumping between $\mathbf{E} \times \mathbf{B}$ eddies. Based on this assumption, one would conclude that the probability distribution function (pdf), P, of this ensemble of particles could be modeled by a diffusion equation of the form

$$\partial_t P = \chi \partial_y^2 P, \tag{13}$$

with an effective diffusivity, χ, defined as

$$\chi = \lim_{t \to \infty} \frac{\sigma^2(t)}{2t}, \tag{14}$$

where $\sigma^2 \langle [\delta y - \langle \delta y \rangle]^2 \rangle$ is the variance of the test particle motion, $\langle \rangle$ denotes average over the ensemble of initial conditions, and $\delta y = y(t) - y(0)$ is the particle displacement in the direction perpendicular to the density gradient. Because particle motion in the x-direction is bounded by the $\mathbf{E} \times \mathbf{B}$ zonal flows, transport in this direction is not relevant. Note that, in order for χ to have a finite, non-zero value, the second moment must exhibit the diffusive asymptotic scaling, $\sigma^2 \sim t$.

Although this random Brownian walk diffusive point of view has been used successfully to describe some problems, it fails when the underlying stochastic process involves long-range correlations, anomalously large displacements described by jump probability distribution function with divergent second moments, or when memory (non-Markovian) effects are present in the dynamics [21]. In this case, the second moment exhibits the anomalous scaling

$$\sigma^2 \sim t^\gamma, \tag{15}$$

with $\gamma \neq 1$. Depending on the values of the exponent γ several cases are distinguished. When $0 < \gamma < 1$ the variance grows slower than in the diffusive, Brownian random walk and the system exhibits sub-diffusion. In this case, the effective diffusivity in Eq. (14) vanishes, $\chi = 0$. On the other hand, when, $1 < \gamma < 2$, the system exhibits super-diffusive transport and the effective diffusivity diverges, $\chi = \infty$. The fact that in the presence of anomalous scaling the

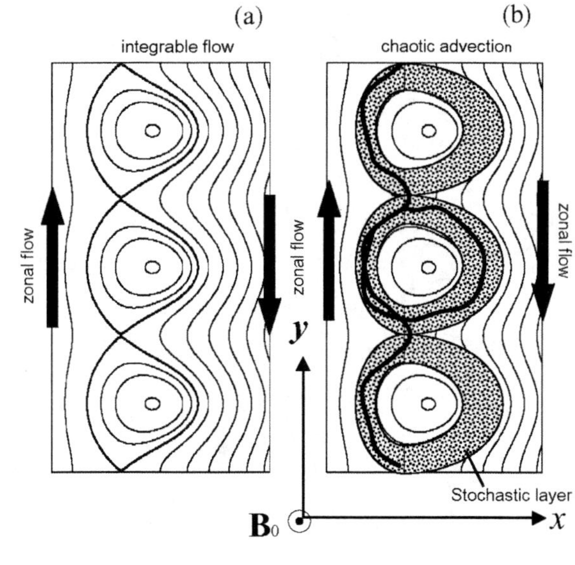

FIGURE 1. When $\varepsilon_2 = 0$, the $\mathbf{E} \times \mathbf{B}$ Hamiltonian model in Eq. (12) is integrable and test particles follow the contour of constant ϕ shown in panel (a). However, for $\varepsilon_2 \neq 0$, the Hamiltonian is not integrable, the system exhibit chaotic transport, and test particles in the stochastic layer (shaded region in panel (b)) alternate in a seemingly random way between being trapped inside the vortices and moving following the zonal shear flow whose direction is denoted by the arrows.

transport coefficient χ either vanishes or becomes infinite indicates that in this case the dynamics can not be modeled using the diffusion model. This is precisely the case here. As shown in Figure 3-(a) the second moment of the particles in the stochastic layer exhibit super-diffusive anomalous scaling with $\gamma = 1.9$.

In the case of normal diffusion, e.g. a Brownian random walk, the pdf of particle displacements, $P(\delta y, t)$, converges asymptotically to a Gaussian distribution, which, as it is well known, exhibits the diffusive scaling $P = t^{-1/2} f(Y/t^{1/2})$ where $Y = \delta y - \langle y \rangle$ and f is a Gaussian. From the statistical point of view, this result is a direct consequence of the Central Limit Theorem according to which the pdf of a sum of independent, random variables identically distributed according to a pdf p, asymptotically converge to a Gaussian distribution provided the second moment of p is finite [29]. In the case of non-diffusive transport, this Gaussian scaling breaks down and the asymptotic pdf, P^*, exhibits the more general anomalous scaling

$$P^*(Y,t) = t^{-\gamma/2} f\left(Y/t^{\gamma/2}\right), \tag{16}$$

where f is a non-Gaussian function that typically exhibits slowly decaying tails. Consistent with the definition $\sigma^2 = \int Y^2 P dY$, the scaling exponent in Eq. (16) is the same as the scaling exponent of the variance in Eq. (15). From Eq. (16) it follows that

$$P^*(Y,t) = \lambda^{\gamma/2} P^* \left(\lambda^{\gamma/2} Y, \lambda t\right), \tag{17}$$

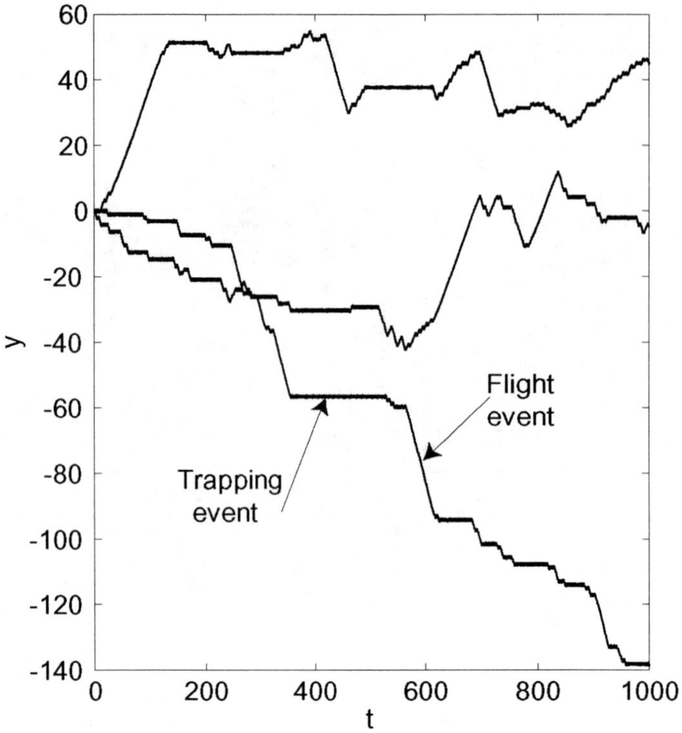

FIGURE 2. Time series of the y-coordinate of three different particles in the stochastic layer. Each time series consists of a sequence of trapping events (corresponding to horizonal sections of the curves) during which particles encircle the $\mathbf{E} \times \mathbf{B}$ vortices, and flight events (corresponding to diagonal sections of the curves) during which particles follow the zonal flow.

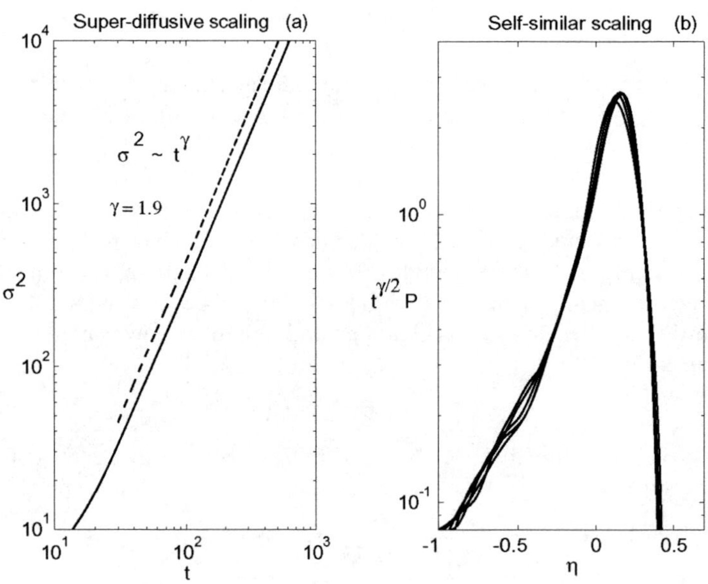

FIGURE 3. Panel (a) shows the variance, σ^2, of particle displacements in the $\mathbf{E} \times \mathbf{B}$ Hamiltonian in Eq. (12) with $\omega = 6$. Pane (b) shows the corresponding probability distribution function, $P(\delta y, t)$, as function of the self-similarity variable $\eta = (\delta y - \langle \delta y \rangle)/t^{\gamma/2}$ for different times. The collapse of the curves at successive times indicates that, at large times, P relaxes to a self-similar non-Gaussian distribution of the form in Eq. (16).

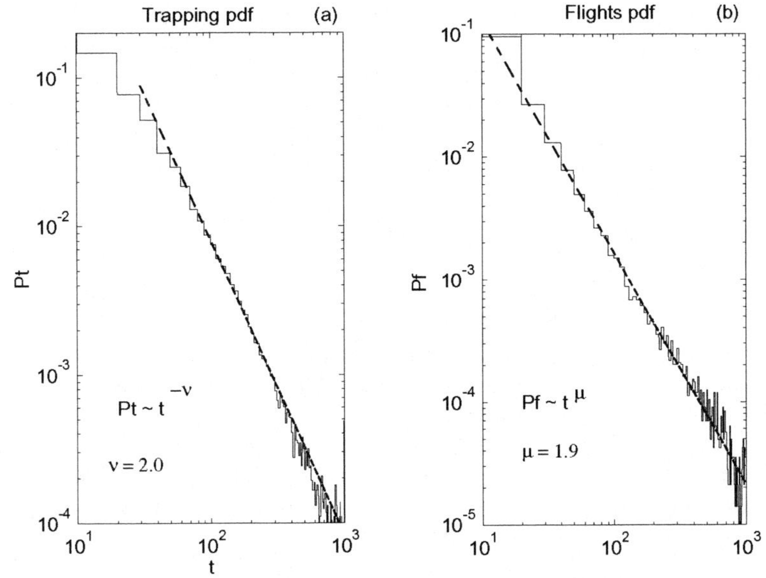

FIGURE 4. (a) Probability distribution function of trapping events, $P_t(t)$, and (b) probability distribution function of flight events, $P_f(t)$, in the $\mathbf{E} \times \mathbf{B}$ Hamiltonian in Eq. (12) for the parameter values corresponding to Fig. 3.

where λ is a real number. That is, up to a scale factor, the asymptotic pdf, P^* is invariant under the space-time renormalization $(Y,t) \rightarrow (\lambda^{\gamma/2}Y, \lambda t)$. Figure 3 shows $t^{\gamma/2}P$ at different times as function of the scaling variable $\eta = Y/t^{\gamma/2}$ for the $\mathbf{E} \times \mathbf{B}$ test particle problem. The departure from Gaussian behavior, and the presence of anomalous scaling consistent with Eq. (16) is evident from the collapse of the curves at different times. This result provides further evidence of the inadequacy of the diffusion model to describe the $\mathbf{E} \times \mathbf{B}$ transport problem considered here.

Of particular interest are the slowly decaying tails of the pdfs. These tails result from the presence of anomalously large displacements caused by the zonal flow. To quantify this effect, it is convenient to introduce the pdf of trapping events, P_t, of duration t, and the pdf of flights events, P_f, of duration t. A trapping event is defined as a portion of a trajectory that encircles an $\mathbf{E} \times \mathbf{B}$ vortex at least once without leaving it, and a flight event is defined as any portion of a trajectory that is not a trapping event. As observed in Fig. 2, a typical stochastic trajectory consists of a sequence of trapping and flight events of different duration. When accounting for the flight events it is important to realize that a particle jump can be either positive, $\delta y > 0$, or negative, $\delta y < 0$. In the general asymmetric case, the pdf of positive flights is different from the pdf of negative flights.

Figure 4 shows the trapping and negative flights pdfs corresponding to the asymmetric pdf of particle displacements in Fig. 3. It is observed that both pdfs exhibit algebraic decay of the form

$$P_f \sim t^{-\mu}, \qquad P_t \sim t^{-\nu}, \tag{18}$$

with $\nu = 2$ and $\mu = 1.9$. The algebraic decay of the trapping and flights pdfs has a number of interesting consequences. The slow decay of algebraic tails is directly linked to intermittency, i.e. , the significant presence of very large trapping or flight events. Also, the scale invariance of power laws implies that the distribution of trapping and flights events is statistically self-similar. Most importantly, when the tails of the pdf exhibits algebraic decay, high order moments diverge. In particular, if the pdfs scale as in Eq. (18), the m-th moment, $\langle t^m \rangle = \int t^m P_{flight} dt$, diverges for $m \geq \mu - 1$, and the m-th moment, of the trapping PDF diverges for $m \geq \nu - 1$. Assuming that the jump velocity of particles is constant, the case $\mu < 3$ corresponds to Lévy flights for which the second moment diverges in an unbounded domain.

STATISTICAL BASIS OF NON-DIFFUSIVE TRANSPORT

Since the work of Einstein, Smoluchowsky, and Bachelier in the early 1990, it is known that the diffusion equation is closely related to the Brownian random walk. In a similar way, the non-diffusive models of interest here are closely related to a generalization of the Brownian random walk known as the Continuum Time Random Walk (CTRW) model originally developed in a seminal paper by Montroll and Weiss [39]. The CTRW in its simplest form extends the usual Brownian model by introducing generalized waiting time and jump distribution functions. To make this notion precise consider the random motion of a particle that at times $t_1, t_2, \ldots t_i \ldots$ is located at $x_1, x_2, \ldots x_i \ldots$. Let $\Delta x_i = x_i - x_{i-1}$ denote the particle jump at the i-th iteration of the process and $\Delta t_i = t_i - t_{i-1}$ the waiting time before the i-th jump is taken. Both Δt_i and Δx_i are assumed to be independent, identically distributed, random variables. We denote by $\eta(x)dx$ the probability that a jump is in the range $x < \Delta x_i < x + dx$, and by $\psi(t)dt$ the probability that the waiting time Δt_i satisfies $t < \Delta t_i < t + dt$. In the Brownian random walk model it is assumed that η is a Gaussian processes and ψ a Poisson distribution. These assumptions guarantee the existence of well defined spatio-temporal transport scales and imply that rare events occur with very low probability. The CTRW relaxes these restrictive assumptions and allows the use of general probability distribution functions incorporating non-Gaussian processes and non-Markovian (memory) effects. The need to incorporate these generalization is motivated by the study of non-diffusive transport. For example, in the transport problem discussed in the previous section, the need to go beyond Gaussian jump distributions is motivated by the significant probability of anomalously large particle displacements caused by zonal flows and reflected in the algebraic decay of the pdf of flights in Eq. (18). On the other hand, the incorporation of the trapping effect of eddies requires the inclusion of non-Poisson waiting time distributions consistent with the algebraic decaying wainting time distribution in Eq. (18). We will limit attention to integrable CTRW models. For a discussion on non-integrable models, and their relation to fractional diffusion see Ref. [40].

The Montroll-Weiss master equation

One of the main goals of this paper is to review the statistical foundation of non-diffusive transport models in the context of the CTRW [39, 23]. The starting point is the Montroll-Weiss master equation describing the pdf $P(x,t)$ of having a particle at position x at time t in terms of the pdfs η and ψ,

$$P(x,t) = \delta(x) \int_t^\infty \psi(t')\, dt' + \int_0^t \psi(t-t') \left[\int_{-\infty}^\infty \eta(x-x')\, P(x',t')\, dx' \right] dt' . \qquad (19)$$

To understand the meaning of this equation note that since $\psi(t)$ is the waiting time pdf, the probability that a particle does not move during a time interval $(0,t)$ is $\int_t^\infty \psi(t')dt'$. Therefore, assuming that the initial condition is a Dirac delta function, $P(x, t=0) = \delta(x)$, the first term on the right hand side (rhs) of Eq.(19) represents the persistence of the initial condition. That is, the contribution to P from particles that have not moved from their initial location. On the other hand, the second term on the rhs of Eq.(19) represents the contribution from particles moving from x' to x. This term includes, in addition to the transition probability $\eta(x-x')$, the time integral of $\psi(t-t')$ to guarantee the arrival of the particles at location x at time t.

The integral operators in the Montroll-Weiss Eq.(19) take a particularly simple form in Fourier-Laplace variables. Introducing

$$\mathscr{F}[\eta] = \hat{\eta}(k) = \int_{-\infty}^\infty e^{ikx} \eta(x) dx , \qquad \mathscr{L}[\psi] = \tilde{\psi}(s) = \int_0^\infty e^{-st} \psi(t) dt , \qquad (20)$$

and using the convolution theorem of Fourier-Laplace transforms, Eq.(19) reduces to an algebraic equation that can be trivially solved to yield

$$\hat{\tilde{P}}(k,s) = \frac{1-\tilde{\psi}}{s} \frac{1}{1-\tilde{\psi}\hat{\eta}} . \qquad (21)$$

Equation (21) can be used to write Eq. (19) as an evolution equation of the form $\partial_t P = \ldots$. To do this it is convenient to introduce an auxiliary function ϕ, whose physical meaning will become apparent later, such that $\tilde{\phi} = s\tilde{\psi}/(1-\tilde{\psi})$. In terms of $\tilde{\phi}$, Eq. (21) becomes

$$s\hat{\tilde{P}} - 1 = \hat{\tilde{P}} \tilde{\phi} (\hat{\eta} - 1) . \qquad (22)$$

Using the basic property of the Laplace transform,

$$\mathscr{L}[\partial_t P] = s\tilde{P} - P(x, t=0) , \qquad (23)$$

and assuming a delta function initial conditions, $P(x,t=0) = \delta(x)$, the inversion of the Fourier-Laplace transform of Eq.(22) leads to

$$\partial_t P = \int_0^t dt'\, \phi(t-t') \int_{-\infty}^\infty dx' \left[\eta(x-x')P(x',t') - \eta(x'-x)P(x,t') \right] . \qquad (24)$$

where we have use $\mathscr{F}[\delta(x)] = 1$. The master Eq. (24) has the structure of a generalized rate equation in which ϕ plays the role of the memory function and η determines the transition probability. The first term inside the square bracket on the rhs of Eq. (24) determines the flux of particles towards x and the second term the flux of particles away from x. In the Markovian case, i.e. in the absence of memory effects,

$$\phi(t-t') = \frac{1}{\tau}\delta(t-t'), \tag{25}$$

and Eq. (24) reduces to the standard master equation

$$\partial_t P = \frac{1}{\tau}\int_{-\infty}^{\infty} dx' \left[\eta(x-x')P(x',t') - \eta(x'-x)P(x,t')\right] . \tag{26}$$

In this case, $\tilde{\phi} = 1/\tau$, and ψ corresponds to a Poisson distribution function

$$\tilde{\psi} = \frac{1}{1+\tau s}, \qquad \psi = \frac{1}{\tau}e^{-t/\tau}, \tag{27}$$

where $\tau = \langle t \rangle$ denotes the mean waiting time.

Diffusion limit

The Montroll-Weiss master equation is a kinetic type description and as such contains "too much" information that might be irrelevant for the spatio-temporal scales of interest to macroscopic transport. In particular, this description requires the specification of the η and ψ pdfs. However, it is natural to expect that the effective dynamics at the macroscopic level depends only on generic properties of these distributions and not on the specific details of their functional form. To illustrate this point, consider the application of Eq. (26) to diffusive transport

$$\partial_t P = \frac{1}{\tau}\int_{-\infty}^{\infty} dx' \eta(x') P(x-x',t) - \frac{P}{\tau}, \tag{28}$$

where we have changed the integration variable and used $\int \eta dx = 1$. Expanding P in Taylor series around $x' = 0$ gives

$$\partial_t P = \frac{1}{\tau}\int_{-\infty}^{\infty} dx' \eta(x') \left[P(x,t) - x'\partial_x P + \frac{x'^2}{2}\partial_x^2 P \ldots\right] - \frac{P}{\tau}. \tag{29}$$

Assuming that η is a sharply localized symmetric distribution, we get to leading order the diffusion equation

$$\partial_t P = \chi \partial_x^2 P, \tag{30}$$

where $\chi = \langle x^2 \rangle/(2\tau)$. That is, in this limit the macroscopic transport properties of the system are insensitive of the form of η and depend only on its second moment. The localization approximation, which is a key assumption of the diffusive model, is justified by the typically small steps taken by the particles in a Brownian random walk. However, as we will discuss at length later on, this assumption breaks down when the underlying stochastic process incorporates anomalously large displacements.

It is instructive to re-derive the diffusion equation in Fourier space. The Fourier transform of the Markovian Montroll-Weiss Eq. (28) gives

$$\partial_t \hat{P} = \frac{\hat{P}}{\tau}\left[\hat{\eta} - 1\right]. \tag{31}$$

In this case, the transition to the diffusion limit is achieved by expanding in Taylor series $\hat{\eta}$ and keeping the lowest order terms

$$\partial_t \hat{P} \approx \frac{\hat{P}}{\tau}\left[\hat{\eta}(0) + \hat{\eta}'(0)k + \frac{\hat{\eta}''(0)}{2}k^2 + \ldots - 1\right]. \tag{32}$$

The moments of η are related to the derivatives of $\hat{\eta}$ according to

$$\langle x^n \rangle = (-i)^n \hat{\eta}^{(n)}(0), \tag{33}$$

where $\hat{\eta}^{(n)}$ denotes the n-th derivative of $\hat{\eta}$. The normalization condition implies $\hat{\eta}(0) = 1$, the assumed symmetry of, $\eta(x) = \eta(-x)$, implies $\hat{\eta}'(0) = 0$, and $\hat{\eta}''(0) = -\langle x^2 \rangle$. Thus, using the identity

$$\mathscr{F}[\partial_x^n P] = (ik)^n \hat{P}, \tag{34}$$

the inversion of the Fourier transform in Eq. (31) leads to leading order in k to the diffusion Eq. (30).

Physically, the low k-limit corresponds to a coarse grained, long-wavelength description that allows us to go from the kinetic-like description based on the master equation to a fluid-like, continuum description based on the diffusion equation. This alternative derivation provides further evidence of the fact that, in the macroscopic limit, only the behavior of the lowest order moments of the jump pdf matter, and that the details of η are largely irrelevant. This situation is reminiscent, and closely related to the central limit theorem. In the study of diffusive transport, the only information kept is the second moment of the jump pdf which defines the diffusion coefficient. This explains the widespread use, and apparent generality of the diffusive model. It seems to be the simplest and most natural model one can get from a long-wavelength expansion of the master equation. However, when it comes to approximations, "the devil is in the details". A critical, and often overlooked, approximation implicit in the derivation of the diffusive model is the localization hypothesis used in Eq. (29). The validity of this assumption is based on the absence of rare events. Rare events manifests as slowly decaying tails in the pdf η that assign a significant probability to anomalously large displacements. When this is the case, the moments of the pdf might not exist. In particular, if η exhibits slow algebraic decay of the form $\eta \sim x^{-(1+\alpha)}$, the integral used in the computation of $\langle x^2 \rangle$ diverges and the diffusion model makes no sense since the diffusivity $\chi = \langle x^2 \rangle / (2\tau)$ is infinite! This is the case found in the example discussed in the Sec. "Non-diffusive Transport". The question then is, what is the correct macroscopic transport model in this case? Before addressing this question we have to take a brief detour into the study of non-Gaussian pdfs used to describe rare events.

Lévy stable distributions

Probability distribution functions are determined through their Fourier transform known as characteristic function and written as

$$\hat{\eta} = e^{\Lambda(k)}. \tag{35}$$

The exponent Λ determines the type of pdf. In the case of a Gaussian

$$\eta = \frac{1}{\sigma\sqrt{2\pi}} e^{-x^2/2\sigma^2}, \tag{36}$$

we have

$$\hat{\eta} = e^{-k^2\sigma^2/2}, \qquad \Lambda_G = -k^2\sigma^2/2. \tag{37}$$

However, there is a wider class of distributions known as Lévy stable distributions with characteristic functions [28, 29]

$$\hat{L}_{\alpha\theta} = e^{\Lambda_{LS}}, \qquad \Lambda_{LS} = -\sigma^\alpha |k|^\alpha \left[1 + i\theta\, sign(k)\, \omega\right], \tag{38}$$

where, $0 < \alpha < 2$, $sign(k) = k/|k|$, and ω given by

$$\omega = \begin{cases} \tan(\pi\alpha/2) & \text{for } \alpha \neq 1, \\ -(2/\pi)\ln|k| & \text{for } \alpha = 1. \end{cases} \tag{39}$$

The Lévy stable distributions have their origin in the generalization of the central limit theorem (CLT). According to this theorem, the pdf of a sum of independent, identically distributed random variables asymptotically converge to a Gaussian pdf provided the pdf of the random variables has a well defined second order moment. When this is not the case, it can be shown that the resulting pdf converges to a Lévy distribution. Thus, Lévy distributions play the role of the Gaussian distribution in the study of stochastic processes involving rare events. Because of this, Lévy distributions play a prominent role in the study of non-diffusive transport.

Lévy distributions are characterized by two parameters: the stability index $\alpha \in (0,2)$, and the skewness parameter $\theta \in (-1,1)$. From Eq. (38) it follows that $L_{\alpha,-\theta}(x) = L_{\alpha,\theta}(-x)$. The case $\theta = 0$ corresponds to symmetric distributions, and the cases $\theta = \pm 1$ correspond to extremal distributions. For $0 < \alpha < 1$ these distributions are one-sided, in particular, for $\theta = -1$ ($\theta = 1$) they are only defined for $x > 0$ ($x < 0$) are vanish for $x < 0$ ($x > 0$). The stability index determines the asymptotic decay of the Lévy distributions. In the case of symmetric distributions

$$L_{\alpha 0} \sim |x|^{-(1+\alpha)}, \qquad x \to \pm\infty. \tag{40}$$

for $0 < \alpha < 2$.

Fractional diffusion equation in space

Our goal in this subsection is to carry out the fluid, continuum limit of the Montroll-Weiss equation in the case when η is a Lévy stable distribution. In this case, the $k \to 0$, long-wavelength, fluid limit is based on the approximation $\exp\Lambda_{LS} \approx 1 + \Lambda_{LS}$ which according to Eq. (31) gives

$$\partial_t \hat{P} = \frac{\hat{P}}{\tau}\Lambda_{LS} = \frac{-\sigma^\alpha}{\tau}\left[1 + i\theta\, sign(k)\, \omega\right] |k|^\alpha \hat{P}. \tag{41}$$

The challenge now is to invert the Fourier transform to get an evolution equation for P in physical space. In the Gaussian case this was particularly simple due to the relation between the integer powers of k and the derivatives of P in Eq. (34). Despite the fact that the problem in Eq. (41) is more complicated because α is in general not integer, we can formally extend Eq.(34) to arbitrary powers and *define* an operator $D^\alpha_{|x|}$ such that

$$\mathscr{F}\left[D^\alpha_{|x|} P\right] = -|k|^\alpha \hat{P}. \tag{42}$$

Using Eq. (42), the inversion of the Fourier transform in Eq. (41) gives, in the symmetric, $\theta = 0$, case

$$\partial_t P = \chi_\alpha D^\alpha_{|x|} P, \qquad \chi_\alpha = \sigma^\alpha/(2\tau). \tag{43}$$

By analogy with Eq. (34), we will refer to the operator $D^\alpha_{|x|}$ as the "fractional derivative" of order α. However, it must be bear in mind that, as we will discuss in below, $D^\alpha_{|x|}$ is in fact an integro-differential operator.

To deal with the general $\theta \neq 0$ case, we introduce the operators ${}_{-\infty}D^\alpha_x$ and ${}_xD^\alpha_\infty$ such that

$$\mathscr{F}\left[{}_{-\infty}D^\alpha_x P\right] = (-ik)^\alpha \hat{P}, \qquad \mathscr{F}\left[{}_xD^\alpha_\infty P\right] = (ik)^\alpha \hat{P}. \tag{44}$$

Using the fact that, for $\alpha \neq 1$, Λ_{LS} in Eq. (41) can be equivalently written as

$$\Lambda_{LS} = \sigma^\alpha \left[l(-ik)^\alpha + r(ik)^\alpha\right], \tag{45}$$

where

$$l = -\frac{(1-\theta)}{2\cos(\alpha\pi/2)}, \qquad r = -\frac{(1+\theta)}{2\cos(\alpha\pi/2)}, \tag{46}$$

we obtain from the inverse Fourier transform of Eq. (41) for a general value of θ,

$$\partial_t P = \chi_\alpha \left[l\, {}_{-\infty}D^\alpha_x + r\, {}_xD^\alpha_\infty\right] P. \tag{47}$$

After some algebra, it can be shown that the operators in Eq.(44) are a special case of integro-differential operators

$${}_aD^\alpha_x f = \frac{1}{\Gamma(m-\alpha)} \frac{\partial^m}{\partial x^m} \int_a^x \frac{f(y)}{(x-y)^{\alpha+1-m}}\, dy, \tag{48}$$

$${}_xD^\alpha_b f = \frac{(-1)^m}{\Gamma(m-\alpha)} \frac{\partial^m}{\partial x^m} \int_x^b \frac{f(y)}{(y-x)^{\alpha+1-m}}\, dy, \tag{49}$$

where m is a positive integer such that $m - 1 \leq \alpha < m$. The operators ${}_aD^\alpha_x$ and ${}_xD^\alpha_b$ are called the left and right Riemann-Liouville fractional derivatives of order α [30, 31]. The parameters a and b determine the lower and upper limits of the integral operators, in the case of Eq.(44) $a = -\infty$ and $b = \infty$. The symmetric fractional diffusion operator formally introduced in Eq. (43) can be written as

$$D^{\alpha}_{|x|} = \frac{-1}{2\cos(\pi\alpha/2)}\left[{}_{-\infty}D^{\alpha}_{x} + {}_{x}D^{\alpha}_{\infty}\right]. \tag{50}$$

Although in the derivation of Eq. (47) we started from the characteristic function of the Lévy distribution, it is important to realize that, in the fluid limit, the only thing that matters is the asymptotic behavior of the jump pdf and not its specific functional form. Accordingly, it can be shown that a properly taken distinguish fluid limit will lead to the Eq. (47) for any jump pdf exhibiting the asymptotic scaling $\eta \sim x^{-(1+\alpha)}$ [41].

Fractional diffusion equation in time: non-Markovian transport

Up to now we have neglected memory effects and have restricted attention to Markovian transport models, see Eq. (25). However, as discussed in Ref. [42], non-Markovian effects play an important role in the study of subdiffusive transport of charged particles in fluctuating magnetic fields. Also, as we will discuss in the next section, the modeling of ballistic and superdiffusive $\mathbf{E}\times\mathbf{B}$ transport requires to include both, the non-Gaussian (Lévy) effect discussed above, and non-Markovian effects. For the sake of simplicity, in this subsection we will limit attention to non-Markovian effects and assume that the statistics of the spatial jumps is Gaussian. The following subsection will address the general case.

Given the waiting time pdf $\psi(t)$, let

$$\Psi = \int_t^{\infty} \psi(t')dt', \qquad \tilde{\Psi} = \frac{1-\tilde{\psi}}{s}, \tag{51}$$

be the survival pdf giving the probability that a particle waits a time longer than t before jumping to a new location. As expected, according to Eq. (27), in the Markovian case this function decays exponentially, $\Psi = e^{-t/\tau}$. One natural way to incorporate memory into the system is by using pdfs with slowly decaying algebraic tails to allow the possibility of arbitrary long waiting periods at all scales. There is a very interesting function that accomplishes this in a particularly elegant way. It is called the Mittag-Leffler function and it is defined as

$$E_{\beta}(z) = \sum_{n=0}^{\infty} \frac{z^n}{\Gamma(\beta n+1)}, \tag{52}$$

where $0 < \beta < 1$. A key property of E_β, which makes it a good candidate for the modeling of process with anomalously long waiting limes, is its asymptotic behavior,

$$E_{\beta}(z) \sim -\frac{z^{-1}}{\Gamma(1-\beta)} - \frac{z^{-2}}{\Gamma(1-2\beta)}\cdots, \tag{53}$$

which for $z = t^\beta$ gives $E_\beta(-ct^\beta) \sim t^{-\beta}$ [30]. Note that according to Eq. (52), for $\beta = 1$, E_β reduces to the exponential, $E_1(z) = e^z$. Based on the fact that the gamma function generalizes the factorial function, $n!$, to non-integer values, $E_\beta(z)$ can be naturally viewed as a "fractional" generalization of the exponential. This intuition is reinforced by the observation that the Laplace transform of $E_\beta(z)$,

$$\mathscr{L}\left[E_{\beta}\left(ct^{\beta}\right)\right] = \frac{s^{\beta-1}}{s^{\beta}-c}, \tag{54}$$

reduces to the Laplace transform of the exponential in Eq. (27) for $\beta = 1$.

Motivated by these appealing properties of the Mittag-Leffler function, we consider a survival pdf of the form [43, 41]

$$\Psi = E_{\beta}\left[-(t/\tau)^{\beta}\right]. \tag{55}$$

Using Eqs. (54) and (51), the Montroll-Weiss equation in Fourier-Laplace variables, Eq. (21), becomes

$$s^{\beta}\tilde{\hat{P}} - s^{\beta-1} = \frac{1}{\tau^{\beta}}(\hat{\eta}-1)\tilde{\hat{P}}. \tag{56}$$

Since for the time being we are restricting attention to Gaussian pdfs for the jumps, e.g. Eq. (37), the long wavelegth (fluid) limit, $\hat{\eta} \approx 1 - k^2\sigma^2/2$, and the inversion of the Fourier transform using Eq. (34) is straightforward and leads to

$$s^{\beta}\tilde{P} - s^{\beta-1} = \chi_{\beta}\,\partial_x^2\tilde{P}, \qquad \chi_{\beta} = \sigma^2/(2\tau^{\beta}). \tag{57}$$

The problem now is how to invert the Laplace transform to get an evolution equation for P in time. To accomplish this we follow a similar idea to that used in the study of the fractional diffusion in space. We formally introduce the operator ${}_0^cD_t^{\beta}$ such that

$$\mathscr{L}\left[{}_0^cD_t^{\beta}P\right] = s^{\beta}\tilde{P} - s^{\beta-1}P(x,t=0), \tag{58}$$

and write Eq. (57) as

$${}_0^cD_t^{\beta}P = \chi_{\beta}\partial_x^2 P. \tag{59}$$

Comparing Eqs. (23) and (58), it is natural to refer to ${}_0^cD_t^{\beta}$ as the fractional derivative in time. The explicit form of this integro-differential operator is

$${}_0^cD_t^{\beta}P = \frac{1}{\Gamma(1-\beta)}\int_0^t \frac{\partial_{t'}P}{(t-t')^{\beta}}\,dt', \tag{60}$$

for $0 < \beta < 1$. However, note that this operator, commonly called Caputo fractional derivative, is different to the Riemann-Liouville fractional derivate in Eq. (48) for $\alpha = \beta$ [30]. For $0 < \beta < 1$, the two operators are related by the identity ${}_0^cD_t^{\beta}P = {}_0D_t^{\beta}\left[P - P(t=0)\right]$ which allows to rewrite Eq.(59) in the equivalent form

$$\partial_t P = {}_0D_t^{1-\beta}\chi_{\beta}\partial_x^2 P. \tag{61}$$

where we have used the identity ${}_0D_t^{1-\beta}\left({}_0^cD_t^{\beta}\right) = \partial_t$.

Before concluding this subsection it is important to keep in mind that, although the discussion of the non-Markovian effects was based for simplicity on the use of the Mittag-Leffler function, the range of applicability of Eq. (59) is wider. In particular, as in the discussion of the spatial part of the dynamics, what is important is not the specific form of the pdfs of the underlying stochastic process, but their asymptotic properties. In the study of the temporal part of the dynamics, what matters for the time scales of interest to macroscopic transport, is the long time asymptotic behavior of $\Psi(t)$, which in Laplace variables corresponds to the $s \to 0$ limit. It can be shown that a properly taken distinguished limit of the Montroll-Weiss equation for a general survival pdf with asymptotic scaling $\Psi(t) \sim t^{-\beta}$ leads to Eq. (59) [43, 41].

Most general case

The discussion in the previous subsections has treated standard diffusion (resulting from Gaussian processes) , fractional diffusion in space (resulting from Lévy processes), and fractional diffusion in time (due to non-Markovian effects) independently. However, in many applications, in particular in non-diffusive $\mathbf{E}\times\mathbf{B}$ transport, these different transport mechanisms act simultaneously, and therefore it is of interest to study the Montroll-Weiss equation and its corresponding fluid limit for the most general class of stochastic processes. This exercise not only allows the simultaneous inclusion of Gaussian and α-stable Lévy processes, but it also allows the incorporation of a wider class of jump pdfs. To accomplish this we start form the Lévy-Khintchine formula which is the most general representation of the characteristic function of a stochastic process [44]

$$\Lambda = \ln\hat{\eta} = aik - \frac{1}{2}\sigma^2k^2 + \int_{-\infty}^{\infty}\left[e^{ikx} - 1 - iku(x)\right]w(x)dx. \tag{62}$$

According to Eq. (62), a general stochastic process consists of a combination of a drift component, aik, a Brownian motion (Gaussian) component, $-\sigma^2k^2/2$, and a jump component represented by the integral on the right hand side. These three components are completely determined by the Lévy-Khintchine triplet (a,σ^2,w). The function w is called the Lévy density and it determines the frequency and magnitude of jumps. The function $u(x)$ is a truncation function used to remove the singularity of the integrand at the origin to guarantee the convergence of the integral. It can be shown that the specific form of this function is irrelevant and that different choices only manifest as different rescaling of the constant a. As expected, in the absence of jumps, $w = 0$, and without drift, $a = 0$, Eq. (62) reduces to the characteristic function of the diffusion problem in Eq. (37). On the other hand, for a Lévy density corresponding to an α-stable process,

$$w_{LS}(x) \quad = \quad \begin{cases} \sigma^{\alpha} \frac{(1+\theta)}{2} |x|^{-(1+\alpha)} & \text{for } x < 0, \\ \sigma^{\alpha} \frac{(1-\theta)}{2} x^{-(1+\alpha)} & \text{for } x > 0, \end{cases} \tag{63}$$

the integral on the right hand side of Eq. (62) leads to the characteristic function in Eq. (38) used in the study of fractional diffusion in space.

The long wavelength fluid limit for a general Λ is base on the approximation

$$\hat{\eta} = e^{\Lambda} \approx 1 + \Lambda + \ldots . \tag{64}$$

We will assume a Mittag-Leffler survival pdf of the form in Eq. (55). However, as mentioned before, this is mostly done for simplicity, and the results apply to general survival pdfs provided they exhibit algebraic decay of the form $\Psi \sim t^{-\beta}$. Substituting the approximation in Eq. (64) in Eq. (56), and inverting the Laplace transform using Eq. (58) leads to

$${}_{0}^{c}D_{t}^{\beta} \hat{P} = \frac{1}{\tau^{\beta}} \Lambda \hat{P} . \tag{65}$$

Substituting Eq. (62) into Eq. (65) and taking the inverse Fourier transform yields [45, 46, 47]

$${}_{0}^{c}D_{t}^{\beta} P = -V \partial_x P + \chi_{\beta} \partial_x^2 P + \int_{-\infty}^{\infty} \left[P(x-y,t) - P(x,t) + u(y) \partial_x P \right] w(y) dy , \tag{66}$$

where $V = a/\tau^{\beta}$. Equation (66) is the macroscopic transport equation describing the continuum, fluid limit of a CTRW for a general stochastic process. For the special case of an α-stable process with density in Eq. (63) this equation reduces to

$${}_{0}^{c}D_{t}^{\beta} P(x,t) = -V \partial_x P + \chi_{\beta} \partial_x^2 P + \chi_{\alpha\beta} \left[l \, {}_{-\infty}D_{x}^{\alpha} + r \, {}_{x}D_{\infty}^{\alpha} \right] P , \tag{67}$$

where $\chi_{\alpha\beta} = \sigma^{\alpha}/(2\tau^{\beta})$. The model in Eq. (67) includes drift, normal diffusion, fractional diffusion space, and non-Markovian effects.

Truncated Lévy processes

The general transport model in Eq. (66) allows to go beyond the standard fractional model in Eq. (67). As an example consider a stochastic process described by the density [48]

$$w_{ET}(x) = \begin{cases} \sigma^{\alpha} \frac{(1+\theta)}{2} |x|^{-(1+\alpha)} e^{-\lambda |x|} & \text{for } x < 0, \\ \sigma^{\alpha} \frac{(1-\theta)}{2} x^{-(1+\alpha)} e^{-\lambda x} & \text{for } x > 0, \end{cases} \tag{68}$$

$0 < \alpha \leq 2$, $c > 0$, $-1 \leq \theta \leq 1$ and $\lambda \geq 0$. For $\lambda \neq 0$ this density allows the incorporation of truncation effects in the dynamics.

In the α-stable case in Eq.(63) it is assumed that the jumps can be arbitrarily large. In fact, as mentioned before, in this case $\langle x^2 \rangle = \infty$ and there is not a characteristic scale for the jumps. Although the presence of large displacements has been well document in numerical simulations of turbulent transport in plasmas, it is clear from the physical point of view that particle displacements cannot be arbitrarily large. In particular, it is expected that some sort of decorrelation in the particles' trajectories or boundary effects in finite size systems will eventually lead to the truncation of otherwise arbitrary large transport events. The need to incorporate this kind of truncation effects is the main motivation to study the CTRW with truncated Lévy densities of the form in Eq. (68). Substituting Eq. (68) into Eq. (62) and integrating gives [48]

$$\Lambda_{ET} = iak - \frac{\sigma^2}{2} k^2 - \frac{\sigma^{\alpha}}{2\cos(\alpha\pi/2)} \begin{cases} (1+\theta)(\lambda+ik)^{\alpha} + (1-\theta)(\lambda-ik)^{\alpha} - 2\lambda^{\alpha} , \\ (1+\theta)(\lambda+ik)^{\alpha} + (1-\theta)(\lambda-ik)^{\alpha} - 2\lambda^{\alpha} - 2ik\alpha\theta\lambda^{\alpha-1} , \end{cases} \tag{69}$$

for $0 < \alpha < 1$ and $1 < \alpha \leq 2$ respectively. Substituting Eq. (69) into Eq. (65) and taking the inverse Fourier transform using the identities

$$\mathscr{F}\left[e^{-\lambda x}\,{}_{-\infty}D_x^{\alpha}\,e^{\lambda x}P\right] = (\lambda - ik)^{\alpha}\,\hat{P}\,, \qquad \mathscr{F}\left[e^{\lambda x}\,{}_xD_{\infty}^{\alpha}\,e^{-\lambda x}P\right] = (\lambda + ik)^{\alpha}\,\hat{P} \tag{70}$$

leads to the following macroscopic transport model describing the interplay between memory, long jumps and truncation effects [45]

$${}_0^CD_t^{\beta}P = -U\partial_x P + \chi_{\beta}\partial_x^2 P + \chi_{\alpha\beta}\mathscr{D}_x^{\alpha,\lambda}P - \nu P\,, \tag{71}$$

where the λ-truncated fractional derivative operator of order α, $\mathscr{D}_x^{\alpha,\lambda}$, is defined as

$$\mathscr{D}_x^{\alpha,\lambda} = le^{-\lambda x}\,{}_{-\infty}D_x^{\alpha}\,e^{\lambda x} + re^{\lambda x}\,{}_xD_{\infty}^{\alpha}\,e^{-\lambda x}\,, \tag{72}$$

where $U = V$ for $0 < \alpha < 1$, and $U = V - v$ for $1 < \alpha < 2$ with

$$v = \frac{\sigma^{\alpha}\alpha\theta\lambda^{\alpha-1}}{|\cos(\alpha\pi/2)|}\,, \qquad \nu = -\frac{\sigma^{\alpha}\lambda^{\alpha}}{\cos(\alpha\pi/2)}\,. \tag{73}$$

Fundamental solutions and classification of transport regimens

Using Eqs. (65) and (58), it follows that the fundamental solution (Green's function), G, of the general macroscopic non-diffusive transport model in Eq. (66) is

$$\hat{\tilde{G}} = \frac{s^{\beta-1}}{s^{\beta} - \Lambda}\,. \tag{74}$$

Inverting the Fourier-Laplace transforms with the aid of Eq. (58) leads to

$$G(x,t) = \frac{1}{2\pi}\int_{-\infty}^{\infty} e^{-ikx}\,E_{\beta}\left[\Lambda(k)t^{\beta}\right]dk\,. \tag{75}$$

The specific form of the solution in Eq. (75) depends on Λ, and in this subsection we will limit attention to α-stable Lévy processes in Eq. (38). In this case, the scaling relation $\Lambda_{LS}(\mu k) = \mu^{\alpha}\Lambda_{LS}(k)$ implies

$$G(x,\mu t) = \mu^{-\beta/\alpha}G(\mu^{-\beta/\alpha}x,t)\,. \tag{76}$$

That is, the solution at time μt can be obtained from a rescaling of the solution at time t. In renormalization group language, G is a fixed point of the renormalization transformation $(G;x,t) \rightarrow (\mu^{-\beta/\alpha}G;\mu^{-\beta/\alpha}x,\mu^{-1}t)$. Introducing the similarity variable $\eta = x(\chi_{\alpha\beta}^{1/\beta}t)^{-\beta/\alpha}$, the Green's function in this case can be written as [49, 50]

$$G(x,t) = (\chi_{\alpha\beta}^{1/\beta}t)^{-\beta/\alpha}K(\eta)\,, \qquad K(\eta) = \frac{1}{2\pi}\int_{-\infty}^{\infty} e^{-i\eta k}\,E_{\beta}\left[\Lambda_{LS}(k)\right]dk\,. \tag{77}$$

As expected, in the absence of memory effects, $\beta = 1$, $E_{\beta}(z) = e^z$, and $\hat{K} = \exp\Lambda_{LS}$ which according to Eq. (35) is the characteristic function of an α-stable process. That is, the Lévy α-stable distributions are the Green's functions of the spatial fractional diffusion equation for $0 < \alpha < 2$.

It follows from Eq. (77) that the moments of the pdf scale as

$$\langle x^q\rangle = I(\chi_{\alpha\beta}^{1/\beta}t)^{q\beta/\alpha}\,, \qquad I = \int |\eta|^q K(\eta)d\eta\,. \tag{78}$$

This result allows the classification of different transport regimes depending on the values of α and β. In the case $\alpha = 2$ and $\beta = 1$, the fractional model reduces to the standard diffusion equation and, as expected, Eq. (78) gives the well-known diffusive scaling $\langle x^q\rangle \sim t^{q/2}$. For $(\alpha,\beta) \neq (2,1)$ three transport regimens can be distinguished. The case $2\beta > \alpha$ is called *super-diffusion* since the moments grow faster than in the diffusive case. On the other hand, when $2\beta < \alpha$ the situation is exactly the opposite and we have *sub-diffusion*. The third case corresponds to $\alpha = \beta$. We refer to this regime as *ballistic* since the moments exhibit the "free-streaming" scaling $\langle x^q\rangle \sim t^q$. In the computation of the moments it is important to bear in mind that, for $0 < \alpha < 2$, the moments of the pdf in an unbounded domain are infinite for $q > \alpha$. However, when this is the case, one can still invoke this classification by restricting attention to low order, fractional moments, $\langle |x|^q \rangle$.

Of particular interest is the asymptotic decay of the Green's function. It can be shown that for $\eta \gg 1$, K, in Eq. (77) decays as $K \sim \eta^{-(1+\alpha)}$. This implies that or a fixed $t = t_0$,

$$G(x,t_0) \sim x^{-(1+\alpha)}, \qquad x \gg \left(\chi_f^{1/\beta} t_0\right)^{\beta/\alpha}. \tag{79}$$

Introducing the time similarity variable $\xi = t\,|x|^{-\alpha/\beta}$, the Green's function can be written as

$$G(x,t) = |x|^{-1}\xi^{-\beta/\alpha} K(\xi^{-\beta/\alpha}). \tag{80}$$

Using the large η and small η behavior of $K(\eta)$ it can be show that for fixed $x = x_0$,

$$G(x_0,t) \sim \begin{cases} t^{\beta} & \text{for} \quad t \ll \left(\chi_f^{-1} x_0^{\alpha}\right)^{1/\beta} \\ t^{-\beta} & \text{for} \quad t \gg \left(\chi_f^{-1} x_0^{\alpha}\right)^{1/\beta}. \end{cases} \tag{81}$$

That is, according to Eq. (79) the order of the fractional derivative in space, α, determines the algebraic asymptotic scaling of the propagator in space for a fixed time, and according to Eq. (81) the order of the fractional derivative in time, β, determines the asymptotic algebraic scaling of the propagator in time for a fixed x. These two properties provide a useful "rule of thumb" to construct fractional models given the spatio-temporal asymptotic scaling properties of the propagator.

APPLICATIONS OF FRACTIONAL DIFFUSION MODEL

In the previous section we discussed the statistical basis of non-diffusive transport modeling based on the use of fractional diffusion operators. The goal of this section is to present applications of fractional diffusion models to non-diffusive transport problems of interest to magnetically confined plasmas.

Turbulent transport

Understanding turbulence-driven transport is one of the fundamental problems in magnetically confined plasmas. To make progress in the study of this challenging problem, it is valuable to start with simplified descriptions of the dynamics. Of particular interest is the study of passive tracers in prescribed flows. By a passive tracer we mean a scalar quantity that does not react back on the underlying flow while being transported. An example of a passive tracer is a low density impurity in a fluid or plasma. Although the temperature, density, and pressure, are strictly speaking not passive since they are self-consistently coupled to the dynamics, the study of passive tracers in plasmas is an important step in the study of the transport of these quantities.

The study of transport of passive tracers involves two steps. The first step is to construct the flow field, $\mathbf{V}$, transporting the tracer. This is usually accomplished by solving a fluid model. In the case of magnetized plasmas, typical examples of interest include Hasegawa-Mima, Hasegawa-Wakatani, or resistive pressure-gradient driven reduced models. Once the advection velocity field $\mathbf{V}$ has been computed, the next step is to compute the evolution of the tracer transported by the flow. This can be done following a continuum Eulerian approach or a particle-based Lagrangian approach. In the Eulerian approach the tracer is described by a scalar field P evolving according to the advection equation

$$(\partial_t + \mathbf{V}\cdot\nabla)P = 0. \tag{82}$$

In the Lagrangian approach, the passive tracer is viewed as an ensemble of "test particles" following the characteristics of Eq. (82),

$$\frac{d\mathbf{r}}{dt} = \mathbf{V}\left[\mathbf{r}(t),t\right]. \tag{83}$$

The complexity of the transport problem depends on the dynamics of $\mathbf{V}$. In the case of fully developed, isotropic, homogeneous turbulence, one can invoke strong, statistical simplifying assumptions and treat Eq. (83) as a Langevin (stochastic) differential equation driven by an un-correlated, zero-mean, Gaussian process. Under these conditions, the problem reduces to the Brownian random walk, and the evolution of P can be described using the diffusion model in Eq. (30) with a properly defined "effective diffusivity" χ. The passage from the advection equation in (82) to the diffusion equation (30) can be viewed as a sort of "renormalization" procedure,

$$-\mathbf{V}\cdot\nabla \to \chi\nabla^2\,, \tag{84}$$

in which all the spatio-temporal dynamics of the velocity field $\mathbf{V}$ is encapsulated into the single parameter, χ. In more precise terms, the renormalization approach to turbulent transport consists of finding an effective operator $\mathscr{D}$ such that $\mathscr{D}\langle P\rangle(x,t)$ and $\langle(\mathbf{V}\cdot\nabla)P\rangle(x,t)$ are phenomenological equivalent from the transport point of view, where x denotes the coordinate direction of interest to transport and $\langle\rangle$ denotes average over the remaining coordinate directions. By phenomenological equivalent we mean that the Green's function of the operator $(\partial_t+\mathscr{D})$ (which gives the transition probability or propagator) reproduces the average response $\langle P\rangle$ of the advection operator $(\partial_t+\mathbf{V}\cdot\nabla)$ to localized initial conditions.

Equation (84) is certainly a very convenient simplification, since solving the diffusion equation is considerably simpler than dealing with the advection equation in (82). However, as we have discussed at length before, the un-correlated, Gaussian statistics assumption is a big assumption and in many cases it breaks down. Our goal in this subsection is to discuss a specific plasma physics example when this happens, and to describe how the fractional diffusion formalism can provide an alternative quantitative description. In particular, we show that the pdf of particle displacements (propagator), P, of test particles in pressure gradient driven plasma turbulence is strongly non-Gaussian, and that its spatio-temporal evolution can be described using the fractional diffusion equation in Eq. (67) with $V=\chi_\beta=0$, which, using Eq. (61), can be write in the equivalent form

$$\partial_t P = \chi_{\alpha\beta}\,{}_0D_t^{1-\beta}\left[l\,{}_{-\infty}D_x^{\alpha}+r\,{}_xD_{\infty}^{\alpha}\right]P\,. \tag{85}$$

Comparing Eqs.(82) and (85), this amounts, in"renormalization" language, to

$$-\mathbf{V}\cdot\nabla \to {}_0D_t^{1-\beta}\left[l\,{}_{-\infty}D_x^{\alpha}+r_x D_{\infty}^{\alpha}\right]. \tag{86}$$

According to this prescription, the spatio-temporal complexity and non-Gaussian statistical properties of the advection operator $\partial_t+\mathbf{V}\cdot\nabla$ are described in terms of a non-local (in space and time) transport operator.

The turbulence model of interest here is based on an electrostatic approximation of the reduced resistive magneto-hydrodynamic equations, and it describes the evolution of the electrostatic potential, $\tilde{\Phi}$, and the pressure, $\tilde{p}$,

$$\left(\frac{\partial}{\partial\tau}+\tilde{\mathbf{V}}\cdot\nabla\right)\nabla_\perp^2\tilde{\Phi} = -\frac{1}{\eta m_i n_0 R_0}\nabla_\parallel^2\tilde{\Phi}+\frac{B_0}{m_i n_0}\frac{1}{r_c}\frac{1}{r}\frac{\partial\tilde{p}}{\partial\theta}+\mu\nabla_\perp^4\tilde{\Phi}\,, \tag{87}$$

$$\left(\frac{\partial}{\partial\tau}+\tilde{\mathbf{V}}\cdot\nabla\right)\tilde{p} = \frac{\partial\langle p\rangle}{\partial r}\frac{1}{r}\frac{\partial\tilde{\Phi}}{\partial\theta}+\chi_\perp\nabla_\perp^2\tilde{p}+\chi_\parallel\nabla_\parallel^2\tilde{p}\,. \tag{88}$$

The instability drive is the flux-surface averaged pressure gradient, $\partial\langle p\rangle/\partial r$, determined according to

$$\frac{\partial\langle p\rangle}{\partial\tau}+\frac{1}{r}\frac{\partial}{\partial r}r\langle\tilde{V}_r\tilde{p}\rangle = S_0+D\frac{1}{r}\frac{\partial}{\partial r}\left(r\frac{\partial\langle p\rangle}{\partial r}\right). \tag{89}$$

The magnetic field corresponds to an equilibrium with toroidal component B_0. The tildes indicate fluctuating quantities (in space and time), and the angular brackets, $\langle\rangle$, denote flux surface averaging over a cylinder at a fixed radius. The equilibrium density is n_0, the ion mass is m_i, the averaged radius of curvature of the magnetic field lines is r_c, and the resistivity is η. The sub indices "$\perp$" and "$\parallel$" denote the direction perpendicular and parallel to the magnetic field respectively. The function S_0 represents a source of particles and heat which we model using a parabolic profile, $S_0=\bar{S}_0\left[1-(r/a)^2\right]$. Further details on this model can be found in Ref. [3].

The numerical solution of the turbulence model gives the fluctuating electrostatic potential, $\tilde{\Phi}$, from which the $\mathbf{E}\times\mathbf{B}$ advection velocity is computed using $\mathbf{V}=-(1/B_0^2)\nabla\tilde{\Phi}\times\mathbf{B}$. Since the magnetic field is static, the tracers provide a Lagrangian description of the electrostatic turbulence in the model. The equation of motion in (83) was solved for an

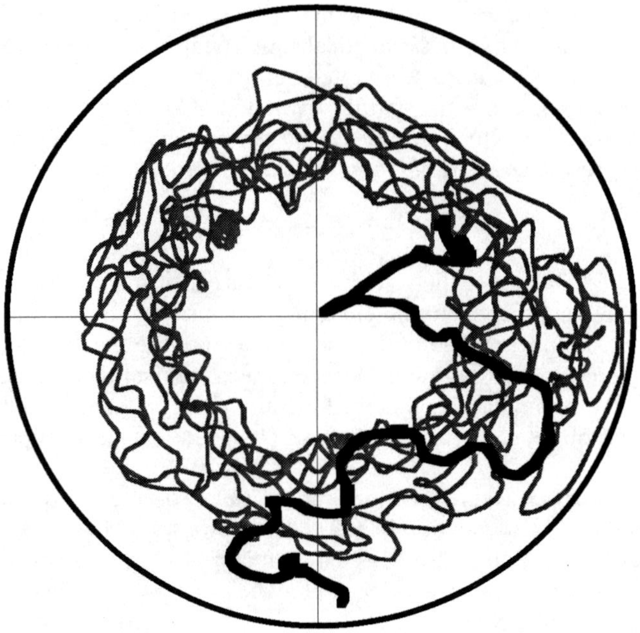

FIGURE 5. Test particles in pressure-gradient-driven 3-dimensional turbulence exhibiting trapping due to turbulent eddies (see Fig. 6) and large displacements due to avalanche-like events. The combination of these two effects is the seed of non-diffusive transport in this system.

ensemble of 25×10^3 tracer particles with initial conditions randomly distributed on θ and z on the surface a cylinder of radius $r = a/2$. As shown in Fig. 5, tracer particles either get trapped in eddies for long times, or jump over several eddies in a single "flight". As in the case of the $\mathbf{E} \times \mathbf{B}$ vortex chain discussed in the Sec. "Non-diffusive Transport", the particle trapping is caused by the electrostatic eddies shown in Fig. 6. The long flight events in this system are caused by avalanche-like pressure relaxation events. Fig. 7 shows the pdf of particle displacements, P. As expected, it is strongly non-Gaussian and exhibits slowly decaying algebraic tails. As discussed in Refs. [3, 4], and shown in Fig. 7, the shape and spatio temporal evolution of P in this case is well reproduced by the solution of the fractional diffusion equation in Eq. (85) with $\alpha = 3/4$, $\beta = 1/2$, $\theta = 0$ and $\chi_{\alpha\beta} = 0.09$. The model also reproduces the super-diffusive scaling of the moments observed in the numerical simulations. These results lead numerical support to the idea that fractional diffusion operators can indeed be used in a renormalized description of turbulent transport in the sense of Eq. (86). In Ref. [51] it was shown how this idea can be put in a more formal basis in the context of quasi-linear type renormalization calculations that do not assume the restrictive localization hypothesis upon which quasi-linear diffusion rests. On a parallel development, in Ref. [52] it was shown that when the velocity field in the Langevin Eq. (83) is stochastic but exhibits algebraic decaying correlations, the effective dynamics can be described using a fractional operator in time, i.e. with $\alpha = 2$ and $\beta \neq 1$.

In an effort to isolate the main ingredients responsible for non-diffusive transport in $\mathbf{E} \times \mathbf{B}$ flows, and in particular, to explore the role of zonal flows in a simple setting, it is instructive to return to the Hamiltonian model introduced in Eq. (12). As discussed before, this model is constructed using linear modes of the Hasegawa-Mima equation in the presence of a strong shear flow. By construction, this is not a turbulent system in the Eulerian sense since in this case $\mathbf{V}$ is a smooth, deterministic function of (x,y) and t. However the system exhibits Hamiltonian chaos, and the particle orbits are stochastic. This type of stochasticity, that manifests in the orbits of the traces but not in the underlying velocity field, is usually called chaotic advection or Lagrangian turbulence. Despite these differences, the simple Lagrangian turbulence transport problem based on Eq. (12), and the more complex Eulerian turbulence transport problem described above share the common feature that in both systems non-diffusive transport is due to the

FIGURE 6. Fluctuating electrostatic potential, $\tilde{\Phi}$, at a fixed time obtained from the numerical integration of the pressure-gradient-driven 3-dimensional turbulence model in Eqs. (87)-(88). The observed eddies are responsible for the trapping of the tracer (see Fig. 5) that leads to the observed non-diffusive transport.

presence of coherent structures, e.g. vortices and shear flows. Thus, it is of interest to compare the numerical results discussed in the Sec. "Non-diffusive Transport", with the fractional model. This exercise has been carried out in Ref. [53]. Figure 8-(a) shows that, in this case, the pdf of particle displacements is one-sided and decays algebraically according to Eq. (40) with $\alpha \approx 0.9$. This result, together with the scaling of the variance in Figure 3-(a), $\gamma \approx 1$, implies $\beta \approx \alpha$. This non-diffusive transport regime is usually called "neutral diffusion" and according to Eq. (78) it exhibits a ballistic scaling. On the other hand, contrary to the $\theta = 0$ symmetric fractional diffusion process in plasma turbulence discussed before, in this case, $\theta = 1$, and the fractional operator is strongly asymmetric. As discussed in Ref. [53], these differences can be traced back to the role played by the strong zonal flow and the asymmetry of the stochastic separatrix layer in the Hamiltonian model. Figure 8-(a) shows the comparison of the numerically determined pdf with the solution of the corresponding initial value problem of the neutral, $\alpha = \beta$, fractional diffusion model.

Variable diffusivity, and finite size domains

The fractional diffusion model used in the previous subsection assumed a constant fractional diffusivity, $\chi_{\alpha\beta}$, and an unbounded domain. These assumptions are justified by the relative simplicity of the passive scalar transport problem. However, before attempting to apply the fractional diffusion model to the study transport in more realistic settings, the model has to be generalized to incorporate spatio-temporal dependent diffusivities and finite-size domains with prescribed boundary conditions. The incorporation of variable diffusivities can be accomplished by writing the model

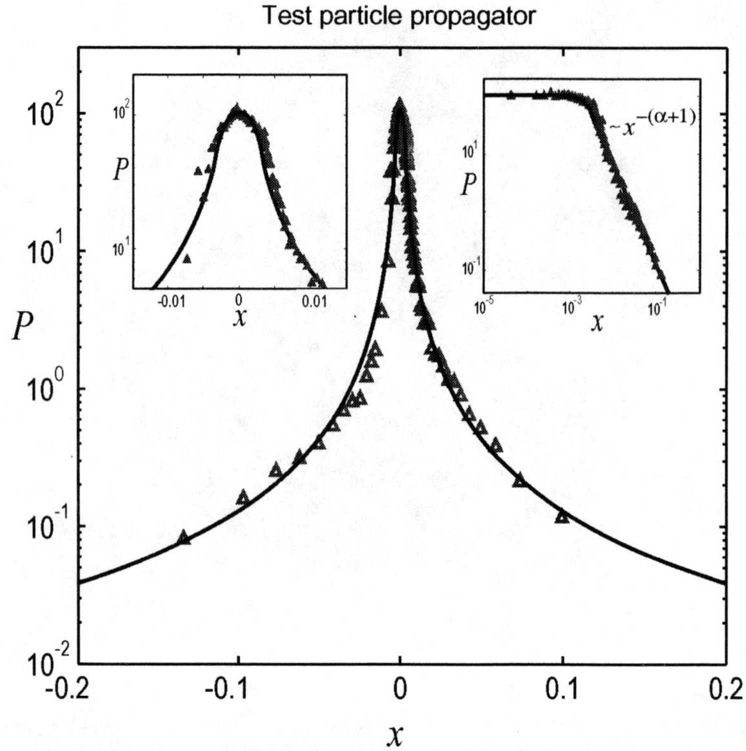

FIGURE 7. Non-Gaussian pdf of test particles in plasma turbulence. The triangles denote the results from the turbulence model in Eqs. (87)-(89). The solid line is the solution of the symmetric ($\theta = 0$) fractional diffusion model in Eq. (85) with $\alpha = 3/4$, $\beta = 1/2$ and $\chi = 0.09$. The insert on the left shows the parabolic dependence of the core of the pdf, and the log-log insert on the right shows the algebraic decay of the tails with exponent $1+\alpha = 7/4$.

in the equivalent flux conserving form

$$\partial_t T = -\partial_x [q_l + q_r + q_d] + S\,, \tag{90}$$

where

$$q_d = -\chi_d\, \partial_x T\,, \tag{91}$$

and q_l and q_r are non-local fluxes defined according to

$$q_l = -l\chi_f\, {}_0D_t^{1-\beta}\, {}_{-\infty}D_x^{\alpha-1}\, T\,, \qquad q_r = r\chi_f\, {}_0D_t^{1-\beta}\, {}_xD_\infty^{\alpha-1}\, T\,. \tag{92}$$

where χ_d and χ_f can now depend on x, and in the case of nonlinear models (e.g. critical gradient models) also on T and the gradient of T. For the diffusive transport channel, the Fourier-Fick's law in Eq. (91) imposes a local relation between the flux and the gradient of T. However, according to Eqs. (92), fractional diffusion incorporates spatial, non-local couplings between the flux and T through the integro-differential operators ${}_aD_x^{\alpha-1}$ and ${}_xD_b^{\alpha-1}$, and the possibility of non-Markovian (memory) effects through the operator ${}_0D_t^{1-\beta}$. Note that we have added the source term S on the right hand side of Eq. (90), this term will play an important role in the applications of the fractional model.

The incorporation of finite-size domains is more delicate. In principle one could do this by simply replacing the lower, $-\infty$, and upper, $+\infty$, limits in the integrals defining the fractional derivatives in space by the finite values a and b, where (a,b) is the domain of the system. However, the problem with this approach is that the left (right) Riemann-Liouville derivative, ${}_aD_x^{\alpha}$, (${}_xD_b^{\alpha}$) is in general singular at the lower, $x = a$, (upper, $x = b$) limit for finite a (b) [30]. As discussed in Ref. [19] this difficulty can be circumvent by regularizing the fractional operators according to the prescription

$$q_l = -l\chi_f\, {}_0D_t^{1-\beta}\, {}_aD_x^{\alpha-1} \left[T(x) - T(a) - T'(a)(x-a)\right]\,, \tag{93}$$

$$q_r = r\chi_f\, {}_0D_t^{1-\beta}\, {}_xD_b^{\alpha-1} \left[T(x) - T(b) + T'(b)(b-x)\right]\,. \tag{94}$$

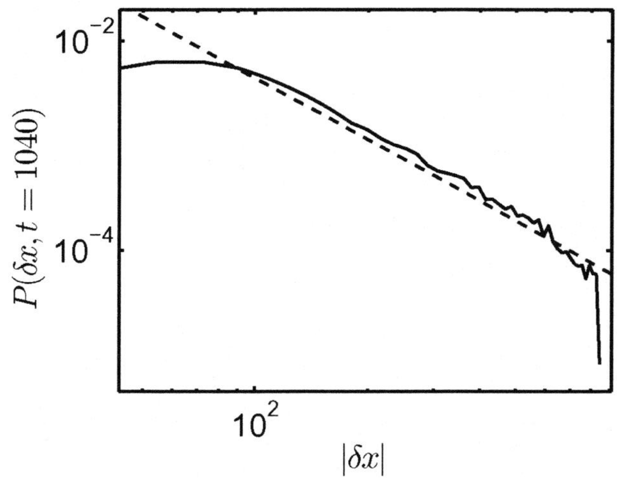

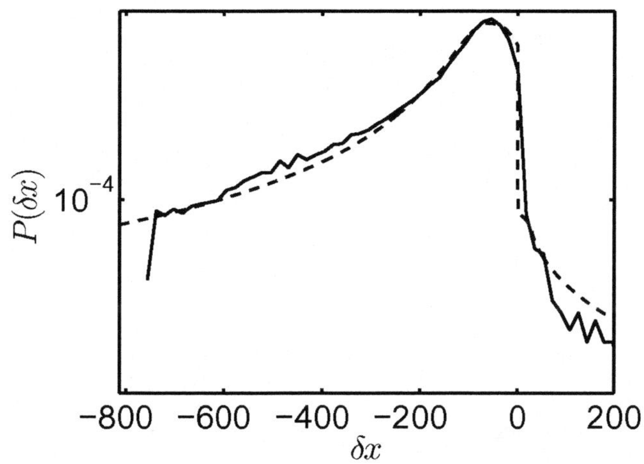

FIGURE 8. Non-Gaussian pdf of test particles in $\mathbf{E} \times \mathbf{B}$ chaotic advection by drift-waves in zonal flows obtained from the numerical integration of the Hamiltonian model in Eq. (12) for a large ensemble of initial conditions. Panel (a) shows the scaling of the tail of the pdf. The solid line is the data and the dashed line an algebraic decaying fit of the form $\sim 1/|x|^{-(1+\alpha)}$ with $\alpha \approx 0.90$. The solid line in panel (b) denotes the pdf obtained from the data and the dashed line the solution of the fractional model.

This prescription is natural in the sense that, for constant χ_f, $\partial_x q_l = -l\chi_f\, {}_0D_t^{1-\beta}\, {}_aD_x^{\alpha} f$ (an similarly $\partial_x q_r$) involves a fractional operator of order $1 < \alpha < 2$ which, consistent with the definition of the fractional derivatives in the Caputo sense, requires the subtraction of the boundary values of the field, $T(a)$, and its derivative, $T'(x)$, for the regularization.

Steady state and anomalous confinement time scaling

As a first application of the finite-size domain fractional model to a situation directly relevant to magnetic confinement, we compute steady state solutions of Eqs. (90)-(91) and (93)-(94). The numerical results presented in this and the following subsections are based on the finite-difference numerical method for integrating the fractional diffusion equation presented in Ref. [19]. For the rest of this paper we will restrict attention to Markovian models and assume $\beta = 1$ and $1 < \alpha < 2$. In the present context, T denotes the plasma temperature. The variable x represents a normalized radial coordinate of the poloidal cross-section of a toroidal confinement device with $x = a = 0$ denoting the magnetic axis, and $x = b = L$ the plasma edge, where L is the minor radius. Consistent with the observation that transport in the

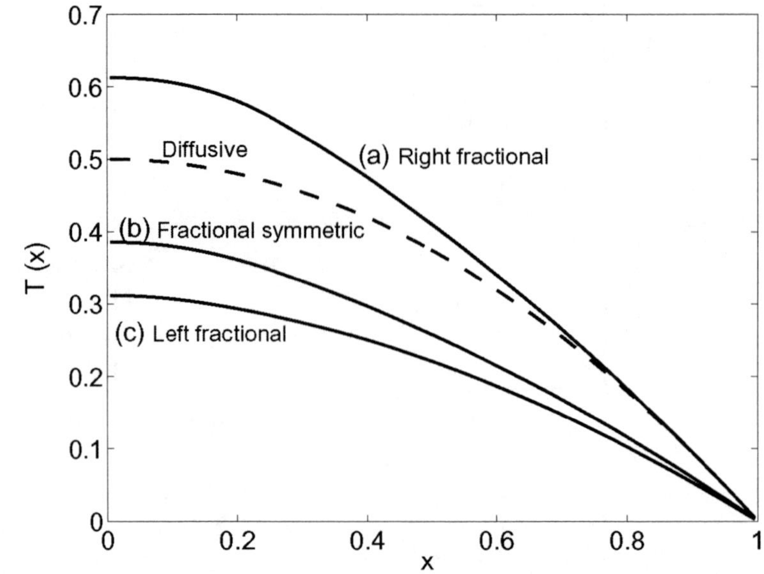

FIGURE 9. Steady state profiles according to the non-local, finite-size domain fractional transport model in Eqs. (90)-(91) and (93)-(94) with $S = S_0$ constant source, and $\alpha = 1.5$. Curve (a) corresponds to an asymmetric right-fractional operator with $l = 0$ and $r = \sqrt{2}$, curve (b) corresponds to a symmetric fractional operator with $l = r = \sqrt{2}$, and curve (c) corresponds to an asymmetric left-fractional operator with $l = \sqrt{2}$ and $r = 0$. For reference, the dashed line curve shows the profile with only standard diffusion ($\chi_{f0} = 0$).

core $x \sim 0$ is dominated by standard diffusive processes we will assume

$$\chi_d = \chi_{d0} \qquad \chi_f = \frac{\chi_{f0}}{2}\left[1 + \tanh\left(\frac{x-A}{B}\right)\right], \tag{95}$$

where χ_{d0}, χ_{f0}, A and B are constants such that $\chi_f(x = 0) = 0$. The boundary conditions are $T'(x = 0,t) = T(x = L,t) = 0$, which imply the vanishing of the total flux at the magnetic axis,

$$[q_d + q_l + q_r]_{x=0,t} = 0, \tag{96}$$

as required by the physics of the problem.

Figure 9 shows steady state profiles in the presence of a time independent, spatially uniform source, $\alpha = 1.5$, and different values of the asymmetry parameter θ. In all cases we choose $\chi_{f0}/L^{\alpha} = \chi_{d0}/L^2$ to assign equal weight to fractional and Gaussian diffusion. It is observed that fractional diffusion has a significant effect in the equilibrium profile. In particular, as evidenced in case (a) of the figure, the right fractional flux in Eq. (94) has a favorable effect on the confinement properties of the system, whereas, as shown in case (c), the left fractional flux in Eq. (93) is detrimental to confinement. The dependence of the confinement properties on the asymmetry of the fractional operator can be explained in terms of the different radial profiles of the left and the right fluxes. In particular, as explained in Ref. [19], in this case, the right (left) fractional flux is negative (positive) and contributes to the inward (outward) transport responsible for the increase (decrease) of the confinement properties of the system.

The energy confinement time, τ, defined as

$$\tau = \frac{1}{P}\int_0^L T(x)dx, \qquad P = \int_0^L S(x)dx, \tag{97}$$

is a key parameter in determining the feasibility of controlled fusion in a magnetically confinement device. It is intuitively clear that τ should increase with the size, L, of the device. No matter what transport mechanism is at play, the farther away the walls of the device are from the core of the plasma, the longer it will take heat to be lost. However,

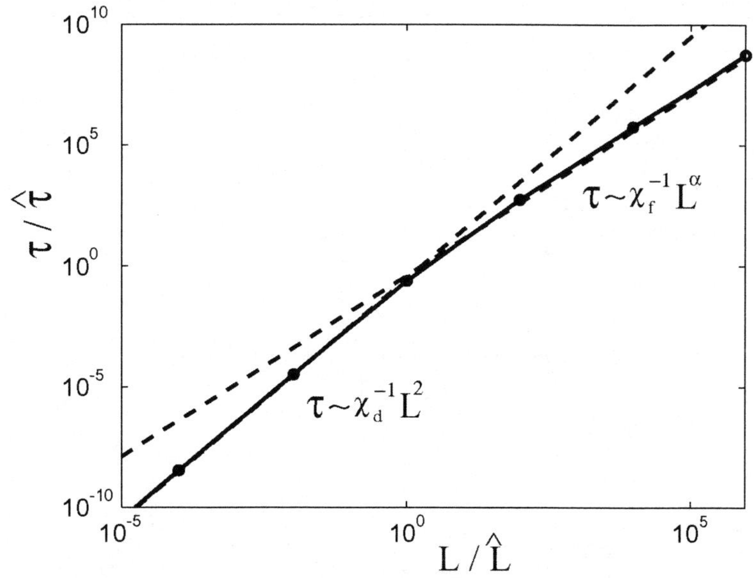

FIGURE 10. Confinement time scaling with domain size for the steady state solution in curve (b) of Fig. 9 according to the non-local transport model in Eqs. (90)-(91) and (93)-(94) with $S = 1$ constant source, and $\alpha = 1.5$. The Figure shows the normalized confinement time $\tau/\hat{\tau}$ as function of the normalized domain size $L/\hat{L}$ with $\hat{\tau} = \left(\chi_d^{\alpha}/\chi_f^2\right)^{1/(1-\alpha)}$ and $\hat{L} = (\chi_d/\chi_f)^{1/(1-\alpha)}$. The dots joined by the solid lines denote the numerical results. The dashed lines are power law fits corresponding to diffusive scaling, $\tau \sim \chi_d^{-1} L^2$, (steeper line), and non-diffusive scaling, $\tau \sim \chi_f^{-1} L^{\alpha}$ (less steep line) with $1 < \alpha < 2$.

what is critical, and highly nontrivial to determine, is the scaling of τ with L. If transport is completely dominated by pure diffusive processes, simple scaling arguments of the diffusion equation lead to the scaling $\tau \sim L^2$. This scaling is very favorable, doubling the size of the device increases by four the confinement time. However, the situation in real life is considerably worse. For example, it has been experimentally observed [54] that low-confinement mode plasmas satisfy the less favorable anomalous scaling

$$\tau \sim L^{\alpha}, \qquad 1 < \alpha < 2. \tag{98}$$

Although at present there are no first principle calculations that determine the precise dependence of the scaling of τ on the parameters of the systems, it is instructive to explore minimal models for radial transport consistent with Eq. (98). Here we show that fractional diffusion is a natural choice to model the anomalous scaling in Eq. (98).

Figure 10 shows the dependence of the confinement time on the system size obtained from the numerical integration of the fractional model with a time independent, spatially uniform source, $\alpha = 1.5$, and $\theta = 0$. This case corresponds to the equilibrium profile (b) in Fig. 9. Other numerical results for different values of θ can be found in Ref. [19]. There are two well-defined scaling regions. Defining,

$$\hat{\tau} = \left(\frac{\chi_{d0}^{\alpha}}{\chi_{f0}^{2}}\right)^{1/(1-\alpha)}, \qquad \hat{L} = \left(\frac{\chi_{d0}}{\chi_{f0}}\right)^{1/(1-\alpha)}, \tag{99}$$

ot is observed that for $L/\hat{L} \ll 1$, the system exhibits diffusive scaling $\tau \sim \chi_{d0}^{-1} L^2$. However, for $L/\hat{L} \gg 1$, there is a transition to non-diffusive scaling $\tau \sim \chi_{f0}^{-1} L^{\alpha}$. For fixed χ_{d0} and χ_{f0}, the transition to the non-diffusive scaling regime occurs for large L, and for fixed L the transition takes place for $\chi_{f0} \gg \chi_{d0}$. That is, when regular and fractional diffusion are both present, the detrimental effects of fractional diffusion play a role provided the system size is large enough to accommodate the anomalously large transport events typical of non-diffusive processes.

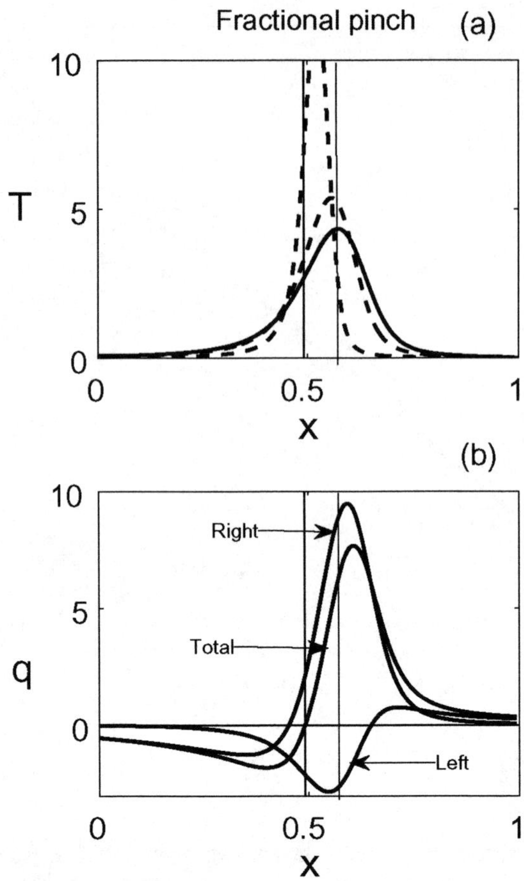

FIGURE 11. Up-hill transport in a decaying pulse according to the fractional model in Eqs. (90)-(91) and (93)-(94) with $\chi_d = 0$, $\chi_f \neq 0$. Panel (a) shows T, and panel (b) the left q_l, right q_r and total $q_l + q_r$ fluxes in the fractional asymmetric case $(\alpha, \theta) = (1.3, 0.5)$. The solid line in panel (a) shows the profile at the final time, and the dashed lines the profiles at earlier times. The vertical lines in (b) demark the region of up-hill transport.

Up-hill transport, and profile peaking

The cornerstone of the diffusion model, the Fourier-Fick's prescription $q_d = -\chi_d \partial_x T$, implies that transport is always "down-hill", i.e. in the direction opposite to the local gradient. However, due to the non-locality of the fractional derivative operators, the situation in the case of fractional diffusion can be fundamentally different. To illustrate this, Fig 11-(a) shows the time evolution of a localized initial condition in the fractional model with $\alpha = 1.3$ and $\theta = 0.5$. In this case, it is observed that the peak of the distribution shifts to the right. To understand the origin of this drift, Fig 11-(b) shows the spatial profiles of the left, q_l, the right, q_r, and the total $q_l + q_r$ fractional fluxes. A region of "up-hill" transport, in which the flux and the local gradient of the profile have the *same* sign is observed in the interval $x \approx (0.5, 0.57)$. The positive flux in this region is responsible for the observed positive drift of the profile. The relation between the flux and the gradient in fractional diffusion can in fact be highly nontrivial. Figure 12 shows the relation between q and $-\partial_x T$ corresponding to the q and T profiles in Fig 11. It is observed that, contrary to the linear Fourier-Fick's prescription, the relation between q and $-\partial_x T$ in fractional diffusion can be multivalued. The left-upper quadrant, $q > 0$ and $-\partial_x T < 0$, is the region of up-hill transport. The multivalueness of the relation between q and

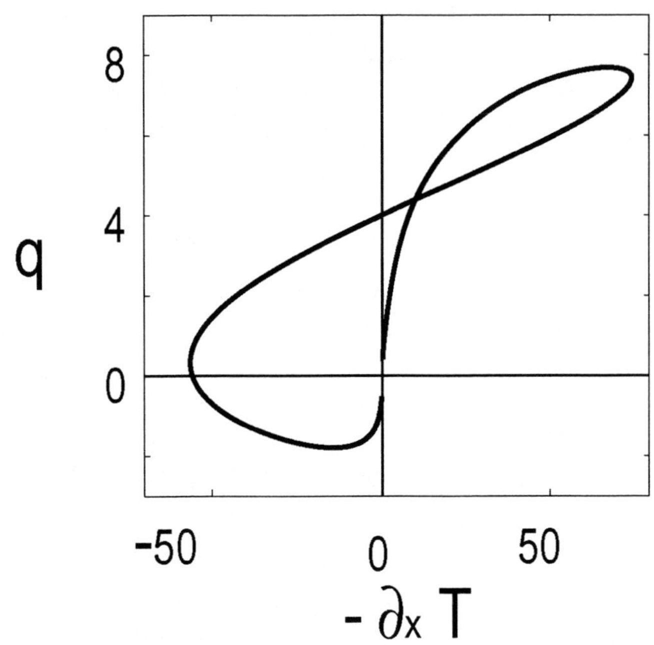

FIGURE 12. Flux-gradient relation for the decaying pulse in Fig. 11. Contrary to the Fourier-Fick's linear relation, $q = -\chi \partial_x T$, in the fractional model the relation between q and $\partial_x T$ is nonlinear and multivalued. The top, left quadrant, $-\partial_x T < 0$ and $q > 0$ corresponds to up-hill transport.

$-\partial_x T$ is a generic feature of non-local transport models. However, the presence of a drift requires and asymmetry between the left and the right fluxes, i.e., $\theta \neq 0$. It can be shown that the position of the peak of the profile, x_m, during the drift satisfies [19]

$$x_m(t) = \eta_m \chi_f^{1/\alpha} t^{\beta/\alpha} , \tag{100}$$

where

$$\eta_m = \theta \left(\frac{\alpha+1}{2\alpha} \right) \alpha^{1/\alpha} \left| \tan \left(\frac{\alpha \pi}{2} \right) \right| . \tag{101}$$

As expected, in the $\theta = 0$ symmetric case and in the $\alpha = 2$ diffusive limit, the drift vanishes.

To make evident the fundamental limitations of the standard diffusion model to describe non-local transport processes, Fig. 13 shows a plot of the "effective diffusivity" corresponding to Figs. 11-12. The use of effective diffusivities is a handy technique in transport modeling. Given a profile, $T(x,t)$, and the corresponding flux, $q(x,t)$, the basic idea is to work backwards and define a spatio-temporal dependent diffusivity according to $\chi_{eff}(x,t) = -q/\partial_x T$. By construction, the diffusion model with this diffusivity reproduces the observed transport. Although, based on this, one might claim that given a problem it is always possible to "describe" it a posteriori using a diffusive model, the effective diffusivity can in some case be physically meaningless. This is the case in Fig. 13 where it is observed that χ_{eff} can become negative, discontinuous, and even infinite! However, it should be pointed out that in the standard approach to transport modeling it is usual to invoke in addition to an effective diffusivity, and effective pinch or drift velocity V_{eff}. This additional field gives extra room to the a posteriori modeling of transport in terms of physically meaningful advection-diffusion equations. However, in the search for the simplest minimal transport model, in this paper we have limited attention to diffusive and fractional-diffusive process and have not considered the role of additional advection operators of the form $V_{eff} \partial_x T$. For a comparative discussion on the range of applicability of standard and fractional diffusion models see Ref. [55].

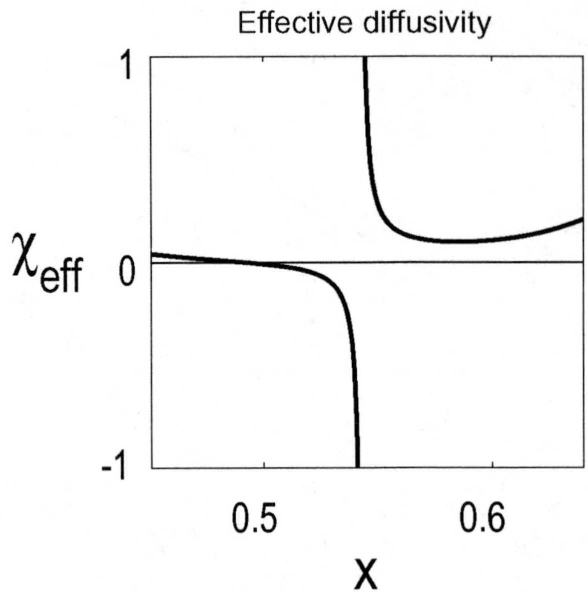

FIGURE 13. "Effective" standard diffusivity χ_{eff} for the decaying pulse in Fig. 12 evolving according to the fractional transport model in Eqs. (90)-(91) and (93)-(94) .

A problem related to up-hill transport is "profile peaking". In its simplest form, this phenomena manifests as an increase of the temperature (or density) at the core of the plasma in the absence of a direct source [56]. This effect can not be described within the diffusion paradigm in the absence of an external additional pinch velocity, V_{eff}. This is because the Fourier-Fick's prescription and the boundary condition, $\partial_x T(x=0)=0$, implies a flat profile, $\partial_x T=0$, in the regions where the source locally vanishes. However, in the context of fractional diffusion, profile peaking is a natural phenomena that results from the non-local relation between the flux and the profile. To illustrate this, Fig. 14 shows steady state solutions of the fractional model with a localized, off-axis source of the form

$$S(x) = S_0 \exp\left[-\frac{1}{2}\left(\frac{x-\mu_s}{\sigma_s}\right)^2\right], \tag{102}$$

with $S_0 = 1/(\sqrt{2\pi}\sigma_s)$, $\sigma_s = 0.025$, $\mu_s = 0.75$, and the diffusivities in Eq. (95) with $A=0.2$ and $B=0.1$. As expected, the asymmetry of the fractional diffusion operators has a critical influence on the central peaking of the profile. In particular, left-asymmetric operators ($\theta = 1$) lead to flat profiles, whereas right-asymmetric operators ($\theta = -1$), and symmetric operators ($\theta = 0$) lead to peaked profile.

Fast pulse propagation

Since the early work in Ref. [12], pulse perturbative experiments have provided a valuable tool to explore non-diffusive transport processes in magnetically confined plasmas. These experiments probe the transient response of the plasma to localized temperature perturbations. Of particular interest is the speed at which a perturbation introduced at the edge of the plasma propagates to the core. In the case of cold pulses, it is routinely observed that the perturbation propagates on a time scale of few ms, i.e. much faster than expected on the basis of diffusive time scales compatible with plasma confinement [13, 57]. A problem of current interest to plasma transport is to develop models that account for this anomalous fast propagation. In this subsection we show that the fractional diffusion model is capable of reproducing the basic phenomenology of fast pulse propagation.

The results are based on the numerical integration of Eqs. (90)-(91), and (93)-(94). A steady state was first computed with boundary conditions $q(0) = T(1) = 0$, and a localized, on-axis source of the form in Eq. (102) with $\mu_s = 0$, and

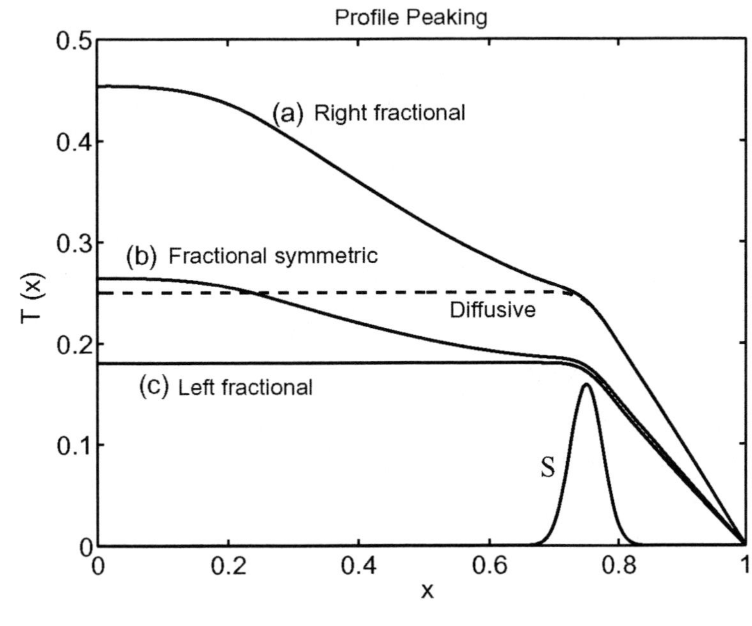

FIGURE 14. Profile peaking with off-axis fueling. Steady state profiles according to the nonlocal transport model in Eqs. (90)-(91) and (93)-(94) for the localized off-axis source in Eq. (102), $\alpha = 1.5$, spatial dependent fractional diffusivity in Eq. (95), $\chi_{d0} = 1$ and (a) $\theta = 1$, (b) $\theta = 0$ and (c) $\theta = -1$. For reference, the dashed line curve shows the profile in the diffusive case ($\chi_f = 0$) where peaking does not take place. S shows the source profile. The central peaking in cases (a) and (b) is a consequence of the strong up-hill transport caused by fractional diffusion.

$\sigma_s = 0.075$. We considered a symmetric, $\theta = 0$, fractional operator, and two values of the non-locality parameter α, $\alpha = 1.75$ and $\alpha = 1.25$. A fractional diffusivity of the form in Eq. (95) with $A = 0.1$, $B = 0.025$, and a constant Gaussian diffusivity with $\chi_{d0} = \chi_{f0}$ was assumed. The amplitude S_0 of the source was adjusted so that $T(0) = 1$ in all the cases considered. Once the equilibrium profile is established, a localized cold pulse perturbation of the form

$$\delta T(x,0) = -C \exp\left[-\frac{(x-\mu_p)^2}{2\sigma_p^2}\right], \tag{103}$$

with $C = 0.3$, $\mu_p = 0.75$ and $\sigma_p = 0.03$ was introduced, and the evolution of the temperature perturbation, $\delta T(x,t) = T(x,t) - T_0(x)$, followed in time.

The spatio-temporal evolution of the perturbed temperature, $\delta T(x,t)$ is shown in Fig. 15. In addition to the $\alpha = 1.75$ and $\alpha = 1.25$ cases, we included, for reference purposes, the evolution in the case when only standard diffusion is present. In this latter case the pulse spreads on a diffusive time scale and the effect of the perturbation at the core is negligible. However, as the second and third panels of Fig. 15 show, when the value of α is reduced, non-local transport yields a fast drop of the temperature at the core. Of particular interest is the case $\alpha = 1.25$ for which the non-local response gives rise to a temperature drop at the core larger than the drop experienced at intermediate places as evidenced by the detached "blob" observed near the core for $t \sim 0.05$. The time evolution of the normalized temperature perturbation is shown in Fig. 16 for different values of x. At the location of the pulse, $x = 0.75$, the temperature relaxation is dominated by Gaussian diffusion, and very similar behavior is observed independent of the value of α. However, at the core, $x = 0$, a significant delay of the signal is observed in the diffusive case. We define the time delay, δt, as the time required for the core temperature to exhibit a temperature drop of size δT_c, i.e., $\delta T(0, \delta t) = \delta T_c$. The value of the threshold δT_c is a matter of convention. Here we use $\delta T_c = -0.0375$. The average pulse propagation speed is then defined as $V = 1/\delta t$. For the parameter values considered, it is observed that the propagation speed in the diffusive case, $V = 11$, can increase considerably in the presence of fractional diffusion. In particular, for $\alpha = 1.75$, $V = 69$ and for $\alpha = 1.25$, $V = 105$. A detailed numerical study of the dependence of the pulse propagation speed for a wide range of parameter values can be found in Ref. [58]. Pulse propagation, and other non-diffusive transport phenomena has also been modeled using probabilistic transport models in which non-locality

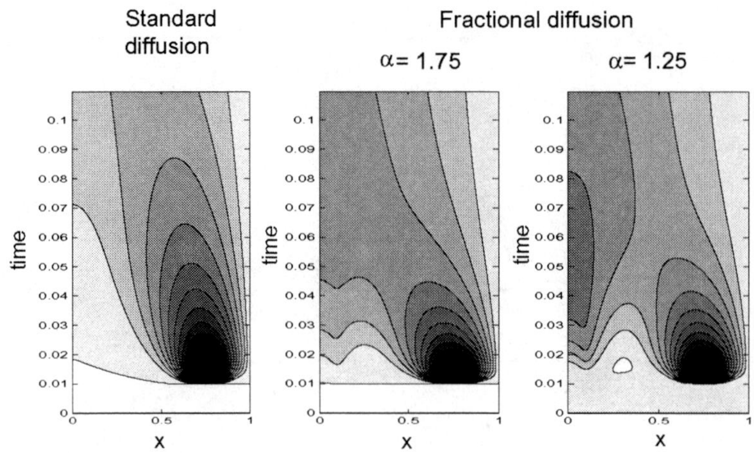

FIGURE 15. Dependence of the evolution of the cold pulse on the non-locality parameter α for $\theta = 0$ and $\chi_{d0} = \chi_{f0}$. The three contour plots show the spatio-temporal evolution of the perturbed temperature $\delta T = T(x,t) - T_0(x)$ with black corresponding to the minimum of δT, and white corresponding to $\delta T = 0$. The first column corresponds to the standard diffusive model. Figure 16 shows the corresponding temperature traces.

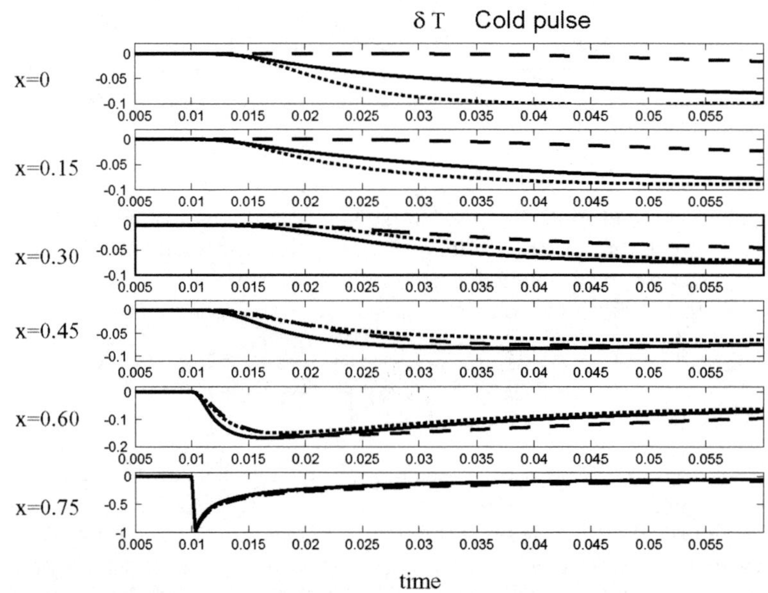

FIGURE 16. Traces of normalized temperature perturbation $\hat{\delta T} = \delta T / |\min[\delta T(x, 0.01)]|$ at various spatial locations corresponding to the simulations shown in Fig. 15 with $\theta = 0$ and $\chi_{d0} = \chi_{f0}$ and $\alpha = 1.75$ (solid curve) and $\alpha = 1.25$ (dotted curve). The dashed curve corresponds to the standard diffusive result.

is incorporated through non-Gaussian (Lévy type) probability distribution functions [20, 59]. For a discussion on the connection between probabilistic and fractional diffusion models see Ref. [19].

Application to experiments conducted at JET

The success of the fractional diffusion model in exhibiting fast pulse propagation phenomena rests on the non-locality of the integro-differential operators used to define the fractional fluxes. However, fast propagation phenomena can also be incorporated in the context of local models with the proper inclusion of non-linear terms. Thus, an open

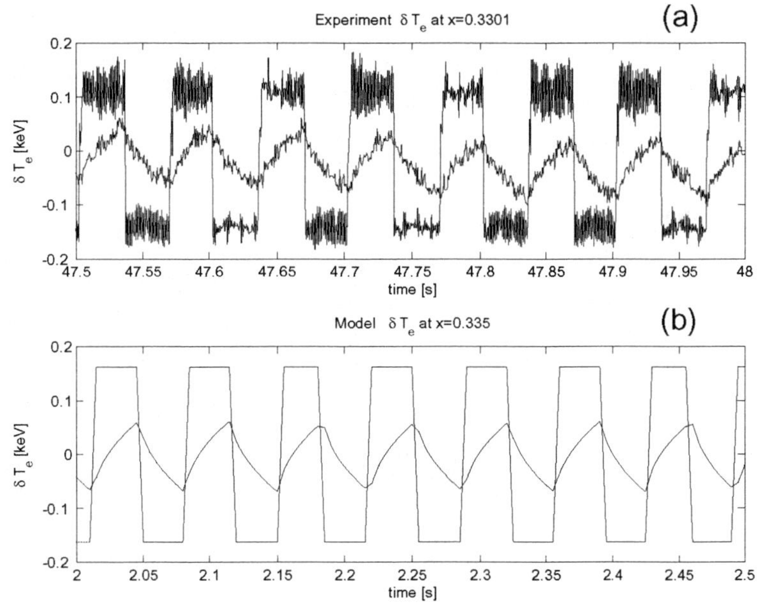

FIGURE 17. Traces of electron temperature perturbation and power modulation in JET (a) and in the fractional model (b). The square-wave periodic signal corresponds to the on-off power amplitude modulation, and the other signal corresponds to the electron temperature perturbation.

problem of interest is to decide if the fast propagation of pulses observed in experiments is due to non-local transport mechanisms, or if it could be explained on the basis of non linear but local transport models. To discriminate between these two classes of transport models conclusively, it is necessary to consider, in addition to cold pulses, other type of perturbations with different propagation properties. The use of radio frequency (RF) power modulation perturbations is a natural choice as it has been experimentally observed that the resulting heat waves exhibit clear propagation asymmetries compared to pulses. In particular, cold pulses can propagate fast even in regions where heat waves slowdown [13, 61]. A promising model validation approach is then to focus on experiments in which both type of perturbations, cold pulses and power modulation, are present for the same plasma. This kind of experiments have been carried out in JET [60, 61], and are the main object of discussion in the present subsection.

The modeling of this type of experiments involves two independent steps. The first step is to calibrate the parameters of the model by best fit of the modulation data. Once this is done, the second step is to use the model (with no free parameters left) to predict the propagation speed of the cold pulse. The attempts to follow this prescription using local models have not been satisfactory since they consistently underestimate the propagation speed of the pulses. Also, attempts to use theory based local models have so far failed to reproduce the asymmetry of behavior between heat modulation and edge cold pulses.

To apply the fractional model to these experiments, we integrated numerically Eqs. (90)-(91) and (93)-(94) with $\alpha = 1.25$, $\beta = 1$, $\theta = 0$ and a source taken directly from the experiment, including the steady state contribution from Ohmic heating, neutral beam injection, and RF heating through mode conversion and fast waves. As in the experiment, the source includes a time dependent, 14.5 Hz., square-wave periodic modulation. Details of the calculations can be found in Ref. [58]. Figure 17 compares the time evolution of the temperature perturbation in the experiment and in the fractional model at a fixed point in space. A good agreement between the experimental signal and the fractional model is observed. The response of the system to the power modulation is studied in a more quantitative basis by Fourier analysis of the temperature perturbation, $\delta T = T(x,t) - \langle T \rangle(x)$, where

$$\delta T(x,t) = \sum_{n=1}^{\infty} A_n(x) \cos\left[2\pi n \nu t + \Phi_n(x)\right] , \tag{104}$$

and $\langle T \rangle = 1/t \int_0^t T dt$ is the time averaged equilibrium profile. Figure 18 shows the spatial profiles of the amplitude, $A_n(x)$, and the phase, $\Phi_n(x)$, for the dominant $n = 1$ and $n = 3$ harmonics. By symmetry, the second harmonic is negligible in this case. This figure shows that for the chosen parameters, the fractional model reproduces the profiles

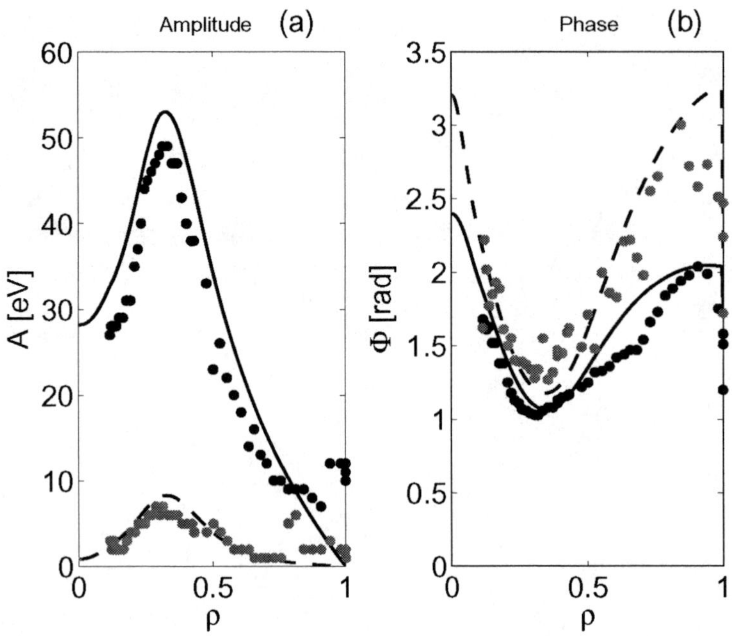

FIGURE 18. Experimental (dots) and fractional model (lines) profiles of the amplitude A and phase delay Φ corresponding to the 1st (solid curve) and the 3rd (dashed curve) Fourier harmonics of the electron temperature.

of the amplitude and the phase of the dominant harmonics of the electron temperature corresponding to the propagation of heat waves excited by the ICRH modulation in JET. However, this is just half of the story. As mentioned before, the key issue is to reproduce simultaneously the fast propagation of the cold pulse observed in the experiment for the *same* parameters of the model. This is precisely where the local, non-linear models have not been effective, since they significantly underestimate the speed of the cold pulses for the parameter values for which they are able to reproduce the modulation data. On the other hand, the fractional model can successfully accommodate the propagation of pulses with speeds comparable to those observed in the experiment while still retaining the slower propagation of the modulation heat pulses. In particular, as Fig. 19(b) shows, the fractional model exhibits a ~ 30 eV drop in the core temperature is about 4 ms.

SUMMARY AND CONCLUSIONS

In this article we have discussed an approach to modeling non-diffusive transport based on a class of integro-differential operators known as fractional derivatives. We have restricted attention to transport of a single scalar in a one-dimensional domain. In this context, diffusive transport is characterized by the Fourier-Fick's prescription $q = -\chi \partial T$, where T is the scalar being transported, e.g., the temperature, q is the flux, and χ denotes the diffusivity. From the statistical mechanics point of view, diffusive transport assumes an underlying un-correlated, Markovian, Gaussian stochastic process, e.g. a Brownian random walk. However, as we have discussed at length, these restrictive assumptions break down on some transport problems of interest to magnetically confined plasmas.

There are several signatures of non-diffusive transport. In the study of particle transport, non-diffusive processes manifest in the anomalous scaling of the moments of the particle's displacement pdf, $\langle x^q \rangle \sim t^{q\gamma/2}$, where $\gamma \neq 1$. When this is the case, the concept of effective diffusivity defined in terms of the second moment of the pdf is meaningless because it is either zero or infinite. However, in the context of fractional-diffusion, anomalous scaling is a natural concept that follows from the spatio-temporal, non-diffusive self-similar scaling of fractional derivative operators. An object of particular interest in the Lagrangian study of transport is the probability distribution function of particle displacements, also known as the propagator. From the modeling point of view, the propagator characterizes the response of the system to a localized initial condition. Mathematically, the propagator corresponds to the Green's function of the model. In a standard diffusive process the propagator is a Gaussian which, as it is well-know, is

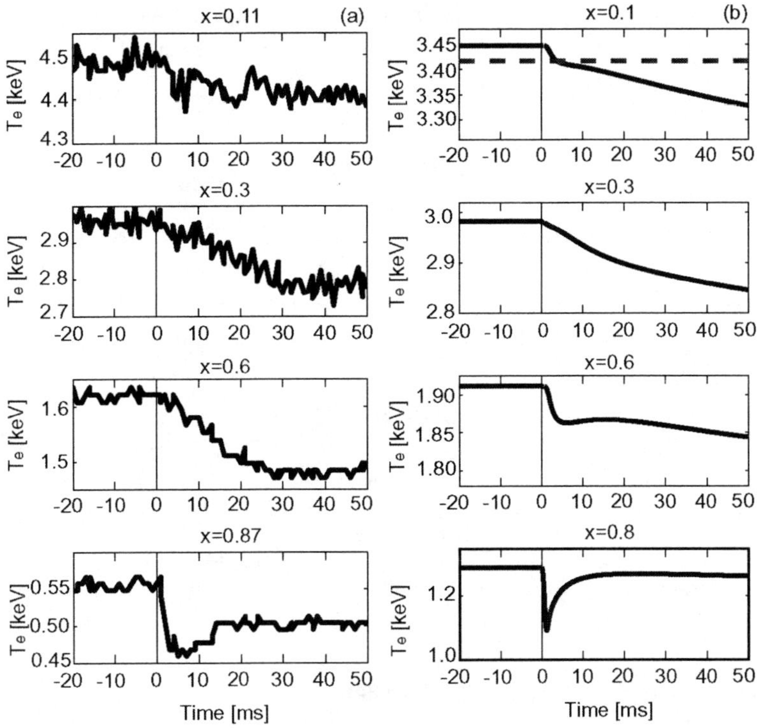

FIGURE 19. Comparison between the temperature traces in the experiment (left panel) and the model (right panel). Consistent with the experiment, the model exhibits a drop of 30 eV (corresponding to the dashed line) in about 4ms.

the Green's function of the diffusion equation with constant diffusivity. The need to go beyond the diffusion model is clearly motivated by the appearance of non-Gaussian propagators in numerical studies of no-diffusive particle transport. These non-Gaussian propagators cannot be obtained from solution of the diffusion equation. Here we have discussed two examples in the context of $\mathbf{E}\times\mathbf{B}$ test particle transport, one corresponding to chaotic transport by drift-waves in zonal flows, and the other corresponding to transport in 3-dimensional, resistive pressure-gradient driven plasma turbulence. We have shown that in both cases the fractional-diffusion model is able to reproduce the shape and spatio-temporal scaling of the propagator.

Experimental signatures of non-diffusive transport include the anomalous scaling of the confinement time with the size of the system, $\tau \sim L^{\alpha}$, profile-peaking, up-hill transport, and fast propagation phenomena. Here we have presented numerical evidence that fractional diffusion models are capable of reproducing these phenomenology. These transport processes could in principle be described using diffusion-like, local models that incorporate additional pinch terms, nonlinearities, and effective diffusivity with ad-hoc spatio-temporal dependences. However, the appeal of the fractional diffusion paradigm is that it provides a unifying framework to describe all these processes (as well as the numerical turbulent transport results discussed above) using minimal assumptions. The recent application of the fractional diffusion model to perturbative experiments conducted at JET provides further evidence of the capabilities of the model to describe non-diffusive transport at a quantitative level. These experiments are particularly valuable since they involve two different kinds of perturbations (cold pulses and heat waves) with markedly different propagation properties that have proved difficult to describe using local diffusive models, even with the incorporation of non-linear effects (e.g., turbulence spreading and critical gradient thresholds) and spatially dependent diffusivities.

Although a "first principles" derivation of fractional diffusion operators starting from "primitive" plasma transport equations is not feasible at this point, the well-established foundation of the fractional diffusion model in the context of the non-equilibrium statistical mechanics of non-Gaussian processes is an important first step. Here we have reviewed how the fractional diffusion model follows naturally from the fluid limit of the master equation describing the continuous time random walk. In principle one could model transport using directly the master equation. However, as mentioned before, the master equation is a kinetic description which in a sense contains too much information that might be irrelevant to the spatio-temporal scales of interest to transport. For example, whereas the fractional model is fully determined by only two parameters, α and β that depend on the space-time asymptotic properties of the underlying stochastic process, the master equation requires the detailed specification of the whole waiting time and jump probability distribution functions.

ACKNOWLEDGMENTS

It is a pleasure to thank Sadruddin Benkadda for the organization of the first ITER summer school on turbulence and transport, in Aix en Provence where this work was originally presented. This work was sponsored by the Oak Ridge National Laboratory, managed by UT-Battelle, LLC, for the U.S. Department of Energy under contract DE-AC05-00OR22725.

REFERENCES

1. S. V. Annibaldi, G. Manfredi, and R. O. Dendy, Phys. of Plasmas **9** (3): 791-799 (2002).
2. B. A. Carreras, V. E. Lynch VE, and G. M.. Zaslavsky, Phys. of Plasmas **8** (12) 5096-5103 (2001).
3. D. del-Castillo-Negrete, B. A. Carreras, and V. E. Lynch, Phys. Plasmas **11**, 3854 (2004).
4. D. del-Castillo-Negrete, B. A. Carreras, and V. E. Lynch, Phys. Rev. Lett., **94**, 065003 (2005)..
5. L. Garcia L, B. A. Carreras, Phys. of Plasmas, **13** (2), (2006).
6. T. Hauff, F. Jenko, and S. Eule Phys. Plasmas 14, 102316 (2007).
7. D. del-Castillo-Negrete, Phys. Fluids **10**, 576 (1998).
8. D. del-Castillo-Negrete, Phys. Plasmas **7**, 1702 (2000).
9. R. J. Goldstone. Plasma Phys. Controlled Fusion **26**, 87 (1984).
10. R. Jha, P. K. Kaw, D. R. Kulkami, J. C. Parikh, and the ADITYA Team, Phys. Plasmas **10**, 699 (2003).
11. V. Yu. Gonchar, et al., Plasma Phys. Rep. **29**, 380 2003 .
12. K. W. Gentle, G. Cima, H. Gasquet, G. A. Hallock *et al.*, Phys. Plasmas **2**, 2292 (1995).
13. P. Mantica and F. Ryter, C.R. Physique **7**, 634-649 (2007).
14. G. Spizzo, R. B. White, and S. Cappello Phys. Plasmas **14**, 102310 (2007)
15. G. W. Hammett, and F. W.Perkins. Phys. Rev. Lett., **64**, 25 (1990).
16. E. D. Held, J. D. Callen, C. C. Hegna, and C. R. Sovinec, Phys. Plasmas **8**, 1171 (2001).
17. A. Yoshizawa, S. I. Itoh, and K. Itoh, *Plasma and Fluid Turbulence*, (Institute of Physics, Series in Plasma Physics, 2003).
18. V. Lynch, B. A. Carreras, D. del-Castillo-Negrete, K. M. Ferreira-Mejias, and H. R. Hicks, J. Comput. Phys. **192**, 406-421 (2003).
19. D. del-Castillo-Negrete, Phys. Plasmas **13** 082308 (2006).
20. B. Ph. van Milligen, R. Sanchez, and B. A. Carreras, Phys. Plasmas **11** 2272 (2004).
21. J. P. Bouchaud and A. Georges, Phys. Rep. **195**, 127 (1990).
22. E. W. Montroll, and M. F. Shlesinger, in *Nonequilibrium Phenomena II.From Stochastics to Hydrodynamics*. Eds. J. L. Lebowitz and E. W. Montroll. Elsevier Science Publishers BV, 1984.
23. R. Metzler, and J. Klafter, Phys. Rep., **339**, 1, (2000).
24. G. M. Zaslavsky, Physics Reports **371**, 461-580 (2002).
25. R.Klages, G.Radons, I.M.Sokolov (Eds.), *Anomalous Transport* (Wiley-VCH, Weinheim, 2008)
26. R. Balescu, *Aspects of anomalous transport in plasmas* (Institute of Physics, Bristol, 2005)
27. R. Balescu, *Transport process in plasmas, Vols. I and II* (Elsevier Publishers, B.V., 1998)
28. G. Samorodnitsky, and M. S. Taqqu, *Stable non-Gaussian random processes* (Chapman & Hall, New York, 1994).
29. W. Paul, and J. Baschnagel, *Stochastic Processes: From Physics to Finance* (Springer-Verlag, New York, 1999).
30. I. Podlubny, *Fractional Differential Equations* (Academic Press, San Diego, 1999).
31. S. G. Samko, A. A. Kilbas, and O. I. Marichev, *Fractional Integrals and Derivatives*, (Gordon and Breach Science Publishers, Amsterdam, 1993).
32. D. R. Nicholson, *Introduction to Plasma Theory* (John Wiley & Sons Inc , NY, 1993).
33. A. Hasegawa and K. Mima, Phys. Fluids **21**, 87 (1978).
34. J. Pedlosky, *Geophysical Fluid Dynamics*. 2nd ed. (Springer-Verlag, New York, 1992).
35. V. I. Petviashvili, and O. A. Pokhotelov, *Solitary Waves in Plasmas and in the Atmosphere*. (Gordon and Breach, PA, 1992).
36. W. Horton and A. Hasegawa, Chaos, **4**, 227 (1990).
37. T. H. Solomon, E. R. Weeks, and H. L. Swinney, Phys. Rev. Lett. **71**, 3975 (1993).
38. D. del-Castillo-Negrete and P. J. Morrison, Phys. Fluids A **5**, 948 (1993).
39. E. W. Montroll, and G. H. Weiss, J. Math. Physics, **6**, 167, (1965).
40. R. Sanchez, B. A. Carreras, and B. Ph. van Milligen, Phys. Rev. E, **71**, 011111, (2005).
41. E. Scalas. R. Gorenflo, and F. Mainardi, Phys. Rev. E, **69**, 011107, (2004).
42. R. Balescu, Phys. Rev. E, **51** 4807 (1995).
43. R. Hilfer, and L. Anton, Phys. Rev. E, **51**, R848, (1995).

44. K. Sato, *Lévy Processes and Infinitely Divisible Distributions* (Cambridge University Press, UK, 1999).
45. A. Cartea and D. del-Castillo-Negrete, Phys. Rev. E **76**, 041105 (2007).
46. B. Meerschaert and B. Scheffler, Phys. Rev. E **65**, 041103 (2002).
47. A. Piryatinska, A. I. Saichev, and W. A. Woyczynski, Physica A **349**, 375 (2005).
48. I. Koponen, Phys. Rev. E., **52**, 1197 (1995).
49. A. Saichev and G. M. Zaslavsky, Chaos **7** (4) 753 (1997).
50. F. Mainardi, Y. Luchko, and G. Pagnini, Fractional Calculus and Applied Analysis, Vol. 4, 153-192 (2001).
51. R. Sanchez, B. A. Carreras, D. E. Newman, V. E. Lynch, and B. Ph. van Milligen, Phys. Rev. E **74** 016305 (2006).
52. R. Balescu, Chaos Solitons and Fractals, **34**, 62-80 (2007).
53. K. Gustafson, D. del-Castillo-Negrete, and W. Dorland, Non-diffusive transport in $\mathbf{E} \times \mathbf{B}$ zonal flows: finite Larmor radius effects and fractional diffusion modeling. Preprint (2008).
54. S. M. Kaye, C. W. Barnes, M. G. Bell, et al., Phys. Fluids B **2**, 2926 (1990).
55. D. F. Escande, and F. Sattin, Phys. Rev. Lett. **99**, 185005 (2007).
56. T. C. Luce. CC. C. Petty, and J. C. M. de Hass. Phys. Rev. Lett., **68**, 52 (1991).
57. J. D. Callen, and M. W. Kissick. Plasma Phys. Controlled Fusion **39**, B173 (1997).
58. D. del-Castillo-Negrete, P. Mantica , V. Naulin, J.J. Rassmussen and JET EFDA contributors, "Fractional diffusion models of non-local perturbative transport: numerical results and applications to JET experiments." Submitted to Nuclear Fusion (2008).
59. B. Ph. van Milligen, B. A. Carreras, V. E. Lynch, and R. Sanchez, Nucl. Fusion **47** 189-195 (2007).
60. P.Mantica et al., Proc.19th Intern. Conf. on Fusion Energy, Lyon [IAEA,Vienna,2002] EX/P1-04.
61. P.Mantica et al., "Core transport studies in JET". Accepted for publication in FS & T (2008).

On the physics of plasma rotation in fusion plasmas

Carlos Hidalgo

Laboratorio Nacional de Fusion, EURATOM-CIEMAT, Madrid, Spain

Abstract

A view of the mechanisms underlying momentum transport in fusion plasmas devices is presented. Heat and energy transport in fusion plasmas is partly due to turbulent processes having an intermittent character and presenting some universal features. The investigation of the mechanisms underlying the generation of plasma flows (via turbulence and neoclassical effects) is key area of research to control transport in magnetically confined plasmas. Projections of ITER plasma performance require better understanding of plasma rotation.

CP1013, ***Turbulent Transport in Fusion Plasmas, First ITER International Summer School***
edited by S. Benkadda

I. Introduction

A confined thermonuclear plasma is far from equilibrium, due to strong external drives that maintain steep gradients in the thermodynamic parameters. The most obvious relaxation mechanism is via Coulomb collisions; in toroidal devices (tokamaks and stellarators) the resulting particle and energy flows are described by neoclassical transport theory [[i]]. However, experimentally observed transport does not, in general, agree with the calculated (neoclassical) values. In particular, electron energy transport can be over two orders of magnitude larger than the predictions from neoclassical theory; transport is anomalous.

The heat and particle transport in fusion plasmas is generally due, in part, to turbulent process associated with small-scale instabilities driven by the inhomogeneity of density and temperature profiles in the direction normal to the magnetic surfaces. The magnitude of turbulent transport is probably the dominant parameter affecting the global confinement properties and hence the economical performance of a fusion reactor.

The mechanisms underlying the generation of plasma flows play a crucial role to control transport in magnetically confined plasmas (e.g. development of transport barriers). It is well known the importance of driving shear in plasma rotation in the development of transport barriers. Toroidal rotation can be driven by external forces such as momentum from Neutral Beam Injection (NBI). However, in large scale devices like ITER (where the available NBI power is limited and the energy of injected neutrals must be high to reach the core plasma region) the NBI driven rotation will be limited. Then, it is important to study the possible role of other mechanisms which can drive plasma rotation. However, compared to energy and particle transport, there has been much less effort to study momentum transport in fusion plasmas. Understanding the interplay between flows, transport and fluctuations remains one of the key issues in magnetic fusion research.

II. Turbulent transport control: role of sheared flows and magnetic topology

Radial electric fields play a key role in the explanation of transport in fusion plasmas. Sheared poloidal flows can influence the turbulence via shear decorrelation mechanisms and, as a consequence, modify the induced transport. A strong radial gradient in the radial electric field (E_r) produces strong shear in the particle ($E_r \times B$) drift velocity (B being the toroidal magnetic field). This tears apart the coherent cellular pattern of the unstable modes. This effect can be expressed by a decorrelation time which is proportional to the inverse of the radial gradient of the radial electric field, $\tau_c \approx B(dE_r/dr)^{-1}$. On the other hand, the decorrelation time due to background diffusion is estimated by $\tau_b \approx L_c^2/D$, where L_c is the radial scale length of fluctuations and D is the diffusion coefficient. When τ_c is smaller than τ_b, the sheared flow reduces the radial scale (correlation) length of the fluctuations, and consequently the level of anomalous transport. Historically, this mechanism for reducing turbulent transport was proposed for the explanation of the transition to the High Confinement mode (H-mode), which is now obtained routinely at the plasma edge and corresponds to the formation of a local transport barriers.

The force balance relation (for any plasma species) shows that the equilibrium radial electric field depends on the ion pressure gradient (∇P_i), equilibrium flow velocity ($U_{i\phi}, U_{i\theta}$) and the magnetic field (B_ϕ, B_ϕ) in the toroidal and poloidal direction respectively,

$$E_r = \frac{1}{eZ_i n_i}\nabla P_i + U_{i\phi} B_\theta - U_{i\theta} B_\phi$$

Then, poloidal and toroidal flows play a crucial role in the development of radial electric fields and in the control of turbulence transport via sheared flows.

Figure 1 shows how instabilities driven by different free energy sources (i.e. gradients) can produce turbulence (via non-linear saturation mechanisms). Sheared flows (electric fields) can damp the turbulence level, and the resulting transport can influence the gradients (the free energy source) themselves – giving rise to a feedback loop. The discovery of the stabilizing effect of sheared flows on plasma turbulence brought on a new era in nuclear fusion research. Recently, the best fusion performance to date has been obtained in plasma conditions where turbulent transport reduction by shear decorrelation is taking place. It should be noted that magnetic configurations with reversed shear (defined as $s = (r/q)dq/dr$), where q is the safety factor) has also strong influence (via the twisting of plasma convective cells) on plasma instabilities [ii]. For the first time, it has been possible to reduce the rate of ion thermal transport to the minimum level set by particle collisions (neoclassical transport) over the whole plasma. These results represent a revolutionary step forward in the control of plasma turbulence and transport.

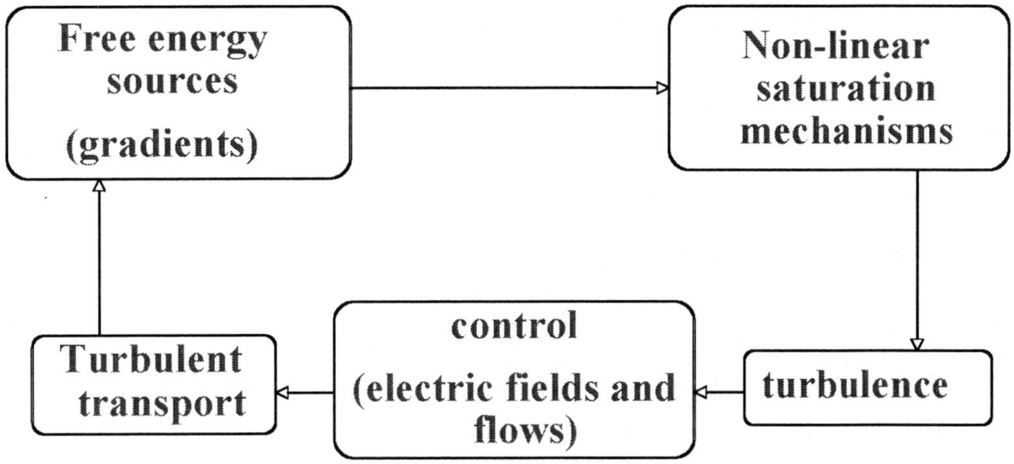

Fig. 1 Link between plasma free energy sources, turbulence, sheared flows and turbulent transport.

Although understanding the physics underlying the generation of flows is a key issue in the plasma physics fusion community, the interplay between sheared flows, intermittent transport events and turbulence still remains an active area of research. The influence of sheared flows on the statistical properties of fluctuations has been studied, showing that probability distribution functions become more Gaussian in the presence of perpendicular sheared flows. Also, fluctuation signals show a bursty character with spikes that are asymmetric in time. This time asymmetry of fluctuation events is minimum close to the shear layer [[iii]]. These characteristics have been observed both in fusion and low temperature plasmas.

However, visualizing the predicted tilt of convective cells and eventually their break-up once the shear flows reach the critical value ($\tau_b \approx \tau_c$), using e.g. 2-D imaging techniques, remains an open issue. The critical test for this basic prediction of the shear decorrelation model (put forward more than 15 years ago) is a real challenge for experimentalists. On the other hand, turbulent events (blobs) may also provide an additional mechanism for flow generation via Reynolds stress (e.g. eddy tilting) (see section III). Quantifying the importance of such mechanisms is an active area of research, while zonal flow generation has already been identified in the plasma edge region [[iv,v]] (see section IV).

III. Driving and damping mechanisms of plasma rotation

The resistance of fluids to shearing motion is a well known observation. The tendency of sheared motion to be reduced with the passage of time, if no other forces are at work to maintain it, leads to the concept of (positive) coefficient of viscosity, the constant of proportionality relating the stress to the shear. In a turbulent flow, when the momentum flux perpendicular to the mean flow direction is directed from regions of larger values toward regions of smaller values of mean flow, it is said that a turbulent (eddy) viscosity is present.

The concept of a reverse effect (e.g. negative viscosity) is something which appears to be again common sense [vi]. However, for certain kind of flows (e.g. planet´s atmosphere [vi], plasmas [v]) evidence of negative viscosity effects have been reported. In this case the mean flow can gain kinetic energy from the turbulence with direct impact in the development of sheared flows.

Some conditions must be fulfilled in the system to show negative viscosity behaviour in steady state plasmas. First eddies which transport the momentum contrary to the gradient of mean flow must have a supply of turbulent kinetic energy (otherwise they will die out). Second, the mean flow should experience some form of braking (e.g. positive viscosity) so that its value does not increase without limit. However, this braking should be low enough to allow the generation of differential rotation. Third, some kind of turbulent irregularity must be present. This ingredient is illustrated in figure 2, showing a flow with some hypothetical pattern producing a convergence of momentum into the mid-channel [vi]. The essential features are the elliptical circulation and the systematic tilts of their major axes, which can be expressed as gradients in quadratic terms of fluctuating velocities. In steady state the turbulent drive is equal to the damping (i.e. for the poloidal dynamics $d\langle \tilde{v}_r \tilde{v}_\theta \rangle / dr = \mu V_\theta$) [v]). In particular, the development of edge sheared flows in the TJ-II stellarator can be explained using a second-order phase transition model (i.e. flows driven by turbulence) [vii].

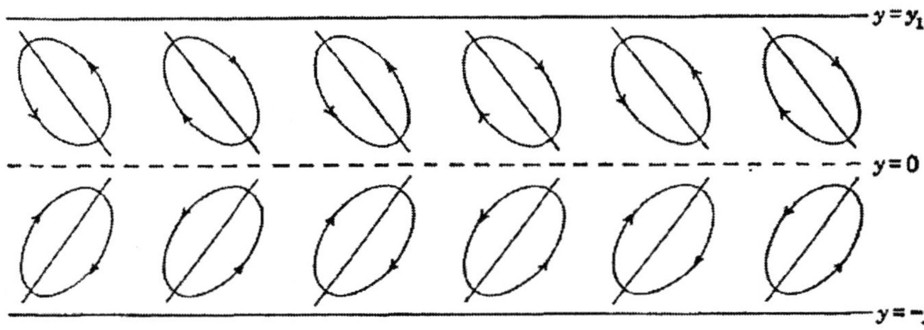

Fig. 2 Some kind of turbulent irregularity must be present to get flow driven by instabilities in plasmas. This ingredient is illustrated in the figure, showing a flow with some hypothetical pattern producing a convergence of momentum into the mid-channel [6].

The invariance of the magnetic moment (μ) implies that the (perpendicular) velocity component increases as the particle moves along the magnetic field lines into a region of higher magnetic field strength. If the magnetic field is high enough (at the high field side of the torus) the parallel velocity becomes zero and then the particle can reflected by toward the weak-field region (giving rise to particle orbits called banana orbits). The poloidal rotation can damped by the magnetic pumping (i.e. the plasma rotation in the presence of inhomogeneous magnetic field heats up the plasma and leads to an irreversible transformation of the kinetic energy of rotation into thermal energy which damps the rotation) [[viii]]. Also, a neoclassical theory of plasma rotation is available and the main result is that the poloidal rotation is proportional to the gradient of the ion temperature [[ix]]

$$U_{i\theta} = k U_{Ti} \qquad U_{Ti} = \frac{1}{M_i \omega_{ci}} \frac{\partial T_i}{\partial r}$$

where the coefficient k depends on the confinement regime.

IV. Plasma rotation in fusion plasmas

Significant changes in the poloidal velocity during the development of transport barriers have been identified in the JET tokamak [[x]]. The experimental velocities turn out to be up to an order of magnitude larger than the neoclassical values, generally used in the past in simulations, suggesting the possible role of turbulent driven rotation mechanisms. However, good agreement between experimental and neoclassical values of poloidal rotation has been reported in small TCABR tokamak [ix]. Clarifying the role of both neoclassical and turbulent mechanisms in determining (e.g. poloidal) plasma rotation is currently an active area of research.

Recent experiments have pointed out the possible influence of turbulence as a component of the anomalous parallel flows observed in the plasma boundary region [xi, xii]. Compensating flows caused by a strong turbulent transport poloidal asymmetry (interchange instability) have been considered as a flow driving mechanism in the SOL [xii]. Figure 3 shows the way in which this instability can be developed in the plasma edge region in the presence of a magnetic field which comes out of the page and decreases with the distance to the plasma boundary (Fig.3a). In Fig. 3b, the plasma has developed a sight hump; the ions and electrons drift in opposite directions under the influence of the inhomogeneous magnetic field (∇B drift), resulting in an electric field (Fig. 3c). Under the combined influence of this electric field and the magnetic field the small bump will tend to grow, creating an unstable situation in the plasma region in which the magnetic field intensity decreases as one moves radially out of the plasma boundary region. As a consequence of this mechanism, strong poloidal asymmetries are expected in the amplitude of fluctuations and turbulent transport, the maximum corresponding to those regions where the magnetic field intensity decreases as one moves away from the plasma. This simplified picture is globally consistent with experimental results. The fluctuation level has been studied on the inboard (high field) and outboard (low field) side in several devices. Recent experiments show that the fluctuation level can be more than a factor of ten higher on the low field side as compared to the high field side, showing clear evidence of strong poloidal asymmetries in the level of fluctuations [xiii]. Parallel flows in the Scrape-Off-Layer (SOL) of the TCV tokamak seem to be well explained by Pfrisch-Schlüter flows with an additional component due to turbulent driven flows [xiv].

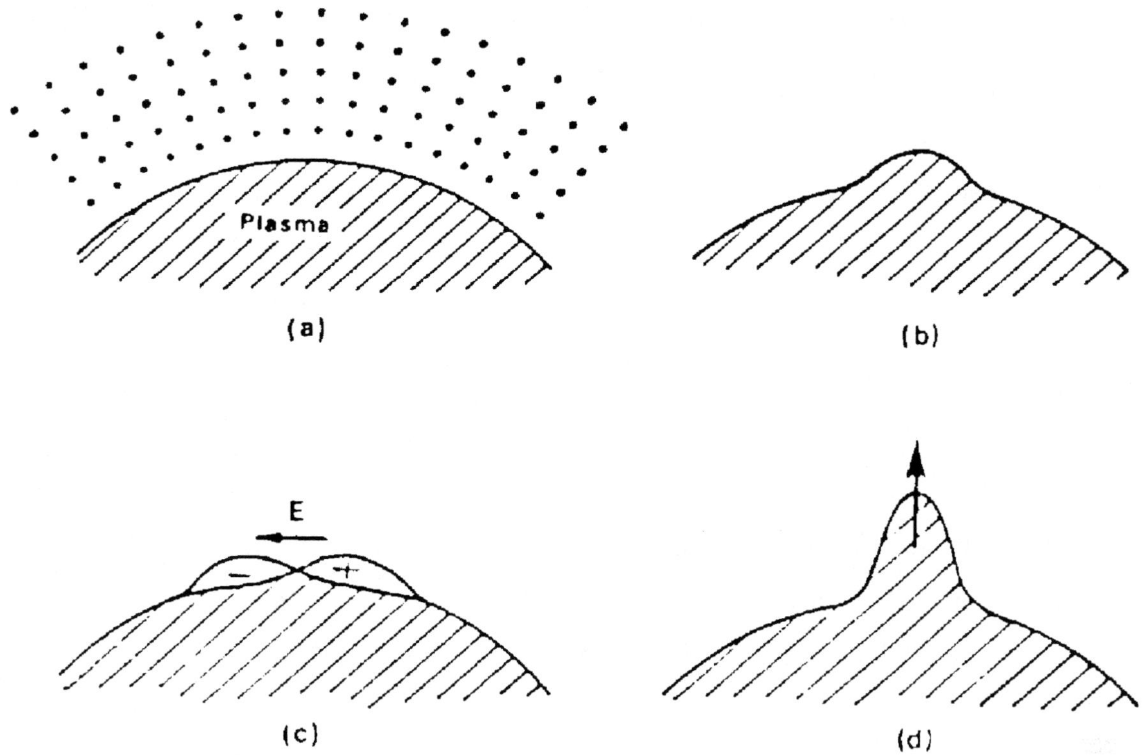

Fig. 3 Development of interchange instability in a toroidally confined plasma.

In the plasma core region, evidence of anomalous toroidal momentum transport has been reported in different tokamak devices [xv, xvi, xvii, xviii]. Effective ion momentum and thermal diffusivities are comparable in magnitude and, in some cases, have a similar parameter dependence. Perturbative experiments allows to evaluate the momentum diffusivity (χ_ϕ) and the convection velocity (v_{pinch}) and then the effective momentum diffusivity is given by $\chi_{\phi,eff} = \chi_\phi + L_u v_{pinch}$. Recent experiments in JET [xix] tokamak show $\chi_\phi / \chi_i \approx 1$ and $v_{pinch} \approx 15\, m/s$ and χ_ϕ increasing with heating power in JT-60U [xx]. These observations show that heat and momentum transport are not independent of each other but have a complex coupling.

The dominant role of external momentum (e.g. NBI) to explain plasma rotation in fusion plasmas is called into question by experiments showing evidence of significant toroidal rotation with no-momentum input. Experiments carry out in Alcator C-mod tokamak show that core toroidal rotation velocity is observed to propage in from the plasma edge plasma and the resulting core rotation was found to depend strongly on (edge) magnetic configuration. Different mechanisms have been proposed to explain these results, including neoclassical effects [xxi], turbulence driven models [xxii, xxiii] and, in the case of ICRF heating, fast particle effects. Spontaneous toroidal flow not driven by neutral beams has also been observed in stellarator devices [xxiv]. The flow reversal observed in the CHS stellarator can be explained by the spontaneous flow driven by large radial electric fields. Other experiments [xxv] also show regimes with spontaneous core rotation.

In a turbulent flow, energy can be interchanged between the mean flow (large scales) and the turbulence (small scales) [v]. Experimentally, pionner works were focused on a direct measurement of the radial-poloidal Reynolds stress component in the boundary region of fusion plasmas, in the search of poloidal turbulent flow driving mechanism. More recent works have been focused on the study of the formation of the so-called zonal flows in plasmas [xxvi].

Recently, SOL flow measurements have shown a possible link with the magnetic configuration dependence on the power threshold for the L-H transition [xxvii]. Further experiments are needed to investigate the role of boundary conditions (e.g. SOL flow) in the power threshold to trigger edge transport barriers.

Parametric scalings of the spontaneous toroidal rotation (i.e. not driven by external momentum input) observed on a large number of tokamaks have been combined with an eye toward revealing the underlying mechanisms and extrapolation to future devices. For an ITER discharge toroidal rotation may be expected (based of the empirical scaling deduced on existing tokamak devices) high enough to stabilize resistive wall modes without external momentum input [xxviii].

V Conclusions

There have been significant changes in the understanding of mechanisms underlying particle, energy and momemtum transport in fusion plasmas devices. It is now clear that intermittent transport cannot be understood without considering the coupling with plasma flows. Projections of ITER plasma performance require better understanding of plasma rotation.

Hopefully, experimental and theoretical efforts will come together into a coherent global picture, connecting transport and flows.

References

[i] J. Weson, *"Tokamaks"*, Oxford University Press, Oxford 1997
[ii] T.M. Antonsen et al., Phys. Plasmas 3 (1996) 2221.
[iii] E. Sánchez et al., Plasma Phys. 7 (2000) 1408.
[iv] G.R. McKee et al., Phys. Plasmas 10 (2003) 1712.
[v] P. Diamond et al., Plasma Phys. Control. Fusion 47 (2005) R35.
[vi] V. P. Starr, "Physics of Negative Viscosity Phenomena" (1968) McGraw-Hill.
[vii] M.A. Pedrosa et al., Plasma Phys. Control Fusion 49 (2007) B303.
[viii] A. B. Hassam and R. Kulsrud Phys. Fluids 21 (1978) 2271.
[ix J. H. F. Severo et al., Nuclear Fusion 43 (2003) 1047
[x] K. Crombé, Y. Andrew, M. Brix et al., Phys. Rev. Letters **95** (2005) 155003.
[xi] C. Hidalgo et al., Phys. Rev. Letters 91 (2003) 065001.
[xii] B. Labombard et al., Nuclear Fusion 44 (2004) 1047.
[xiii] J.L. Terry et al., Phys. of Plasmas 10 (2003) 1739
[xiv] R. Pitts et al., Journal of Nuclear Materials 363 (2007) 505.
[xv] K. H. Burrell et al., Nuclear Fusion 28 (1988) 3.
[xvi] S. D. Scott et al., Phys. Rev. Lett. 64 (1990) 531.
[xvii] W. D. Lee et al., Phys. Rev. Lett. 91 (2003) 205003.
[xviii] J.E. Rice et al., Nucl. Fusion, 44 (2004) 339.
[xix] T. Tala et al., Plasma Phys Controlled Fusion 49 (2007) B313.
[xx] M. Yoshida et al., Nucl. Fusion 47, 856 (2007).
[xxi] A. L. Rogister et al., Nucl. Fusion 42 (2002) 1144.
[xxii] B. Coppi, Nucl. Fusion 42 (2002) 1.
[xxiii] X. Garbet et al., Phys. Plasmas 9 (2002) 3893
[xxiv] K. Ida et al., Phys. Rev. Lett. 86 (2001) 3040.
[xxv] B P Duval et al., Plasma Physics Controlled Fusion 49 (2007) B183
[xxvi] K Hallatschek Plasma Phys. Control. Fusion 49 (2007)B137-B148
[xxvii] A. Hubbard et al., Phys. of Plasmas 14 (2007) 056109.
[xxviii] J.E. Rice et al 21st Fusion Energy Conference, Chengdu (2006)

Basic Principles of Stochastic Transport

K.H. Spatschek

Institut für Theoretische Physik, Heinrich-Heine-Universität Düsseldorf
D-40225 Düsseldorf, Germany

Abstract.
Transport in stochastic magnetic fields is reviewed. In the first part, the topic is motivated by commenting on the intricatenesses of known (nonlinear) transport theories. In the second part, non-integrable magnetic field line systems, their generation and Hamiltonian description are discussed. The symplectic mapping is introduced as the adequate tool for the analysis of the statistics of magnetic field lines. Transport along the unstable and stable manifolds of hyperbolic fixed points is an effective mechanism for heat transfer from the hot core to the plasma boundary. The third part deals with anomalous test particle transport theories starting from stochastic Liouville-type models. Several theories are based on the V-Langevin equation in the guiding center limit. In fusion devices, the mean magnetic fields are sufficiently strong to support the small gyro-radii assumption over a broad area, at least for the electrons. The question remains in what way finite Larmor radii influence the transport, especially in regions where the guiding center assumption fails. Indeed, in tokamaks such areas can be found, e.g. in the vicinity of hyperbolic points. Then, the more general A-Langevin equation has to be used. Based on the latter description, first for small Kubo numbers, the well known transport coefficients (formulated by Rechester and Rosenbluth, Kadomtsev and Pogutse, and others) are recovered. Second, for large Kubo numbers, new transport regimes are identified.

Keywords: Kewwords
PACS: PACS

INTRODUCTION

Nonlinear signatures of particle and heat transport are generic in, e.g., fluid dynamics, plasma physics, and astrophysics. Transport anomalies reach from Bohm-like diffusion in gas discharges up to low-energy cosmic ray penetration into the heliosphere. In the classical picture, large magnetic guiding fields reduce the perpendicular diffusion decisively. Collisions appear as an obstacle for the free motion along the field lines, on the one hand, and increase, on the other hand, the transport in perpendicular direction. The specific reason for the extraordinary interest in the mechanisms of diffusion in stochastic fields lies in the unexpected large losses caused by anomalous transport [1, 2]. Anomalous particle and heat losses have enormous consequences for practical applications. For example, in magnetic fusion devices, heat fluxes should be relatively low and well-distributed over the wall to reach tolerable local power depositions. Therefore, it is desirable to identify the basic processes which rule anomalous losses from open nonlinear (chaotic) systems. Recently, stochasticity has also been discussed in relation to ELM (edge localized mode) mitigation in tokamaks. Edge stochastization became a candidate for the plasma-wall-interaction control [3]. Auxiliary coils are being added to existing configurations to control transport in several tokamaks. These additional coils are new and dominating sources of stochasticity. Examples can be found on the tokamaks Tore-Supra, DIII-D, JET, and TEXTOR, and are being planned for ASDEX-UPGRADE and ITER.

Anomalous transport analysis belongs to the key problems in plasma research and nuclear fusion applications [4]. During the last decades, considerable progress was reported in understanding the basic features [5]. Linear transport theory has been modified to take care of the geometrical effects and large mean free paths. The neoclassical theory (see [6] and references therein) is a great intellectual and practical success. Nevertheless, it is not able to resolve all the problems. As it is well-known [7], in many cases the strong deviation of the diffusion rate from classical or neoclassical predictions is due to nonlinear effects caused by (electrostatic as well as electromagnetic) fluctuations. In the past, several attempts have been made for a self-consistent theory of nonlinear transport; see [4, 8] and references therein. Quasi-linear theory and transport estimates based on the weak-turbulence description are by far the most successful analytical approaches. However, it is well-known [7] that they are of limited applicability. Strong plasma turbulence is a very complex and complicated problem. Analytical evaluations are generally too difficult, and therefore numerical simulations become more and more important. They lead to a rich data base with many hints for fundamental

CP1013, *Turbulent Transport in Fusion Plasmas, First ITER International Summer School*
edited by S. Benkadda

transport scalings.

Anomalous charged particle transport is also a long-standing problem in astrophysical issues [9, 10, 11, 12, 13, 14, 15, 16, 17, 18, 19]. A variety of problems, such as low-energy cosmic ray penetration into the heliosphere, the transport of galactic cosmic rays in and out of the interstellar magnetic field, the Fermi acceleration mechanism, and so on, are on the top of the astrophysical agenda [20]. In most cases, astrophysical transport is collisionless. Usually, no dominant guiding fields exist.

A possible approach to the problem of anomalous plasma transport is to consider the (passive) motion of (test) particles under the influence of given perturbations; see [21] and references therein. Such a treatment is quite common in fluid turbulence [22] where passive motion of scalars, vectors, particles, etc has been investigated extensively. In magnetic fusion research, there exists an additional, qualitatively important reason to investigate particle motion in given stochastic fields. Perturbations in the magnetic field structure are more or less unavoidable because of errors in the coil arrangements of the devices. In addition, and recently that aspect became very important, additional coils are being installed in tokamaks to control the particle and heat loads on the walls via magnetic stochastization of the edge [23, 24, 25, 26]. By the title stochastic transport we mean transport in stochastic fields produced by some external means.

The strength of the magnetic field fluctuations may be quite small, e.g. less than one-tenth of a percent of the zeroth-order confining field (which defines the *parallel* direction). Nevertheless, magnetic fluctuations have a strong influence on *perpendicular* transport. Particles will propagate in the perpendicular direction significantly faster than predicted by classical theory, as has been shown in meanwhile classical papers [27, 28, 9, 29, 30]. According to these findings, the perpendicular diffusion coefficient shows a decisive dependency on the parallel diffusion coefficient. This is caused by the fact that perpendicular fluctuations in the magnetic field are small channels of parallel diffusivity in perpendicular direction.

The kinetics of test particle diffusion with special emphasis on the collisional limit was discussed within a modified perturbation theory [31]. Myra et al. [32] showed, both analytically and by Monte Carlo simulation, that the diffusion coefficient can be sensitive to the spectral width of the magnetic turbulence. A general variational principle for particle diffusion across braided magnetic surfaces was established by Laval [33]. Models for magnetic field line configurations were discussed by several authors, e.g. in Ref. [34, 35]. Statistical plasma transport theory [36, 37] reexperienced strong interest [21, 38]. The Langevin equation model [39] was used for analyzing the subdiffusive behavior of charged particles in stochastic magnetic fields [40, 41]. The relation to continuous time random walks was elucidated [42]. The reliable transition from Euler to Liouville correlation functions turned out to be a very important step [43, 44, 45, 41, 46]. Effects of time-dependence of the magnetic fluctuations [47] and plasma flows [48] on particle diffusion were investigated. Trajectory structures and transport became a main topic [49, 50, 51, 52, 53, 54, 55, 56, 57, 58, 59, 60, 61]. Larmor radius effects have been shown to be important in the percolative regime [62, 63, 64].

For the calculation of effective perpendicular transport one needs the parallel diffusion coefficient. In the collisionless case, classical theory predicts a superdiffusive transport in parallel direction. In fact that transport will not occur, as long as magnetic perturbation fields act as an additional source for collision-like interactions. Even without binary collisions, the parallel motion should show diffusive behavior. That behavior was recently discussed for plasmas in the presence of magnetic fluctuations.

The manuscript is organized as follows. The first part contains some comments on the applicability of existing transport theories. The second part mainly deals with the statistics of magnetic field lines. Anticipating preferential motion of particles along the field lines, some direct consequences can be drawn for the overall losses of particles and heat. The third part introduces stochastic transport theory. The so called A-Langevin model and the Euler statistics of the magnetic field fluctuations are discussed. The Green-Kubo-formalism is used to derive a differential equation for the diffusion coefficient and the mean-squared-displacement (MSD). The decorrelation trajectory method (DCT) is presented. The analytical calculations are complemented by numerical simulations of the A-Langevin equation.

PART I: GENERAL ASPECTS OF NONLINEAR TRANSPORT

Linear transport predictions

The simplest estimate of a diffusion coefficient D in a quiescent, unmagnetized plasma can be obtained via the random walk formula

$$D \equiv \chi_{\|} = \frac{(\Delta x)^2}{2\Delta t} \approx \nu \lambda_{\mathrm{mfp}}^2 \approx \frac{v_{th}^2}{\nu}\,, \tag{1}$$

where λ_{mfp} is the mean free path, v_{th} the thermal velocity, and ν the collision frequency. Here, we do not indicate the particle species which could be easily done by introduced appropriate indices (e for electrons and i for ions). It is evident how to generalize that procedure to the transport perpendicular to an external magnetic field. The characteristic step size is the (thermal) Larmor radius ρ_L, leading to

$$D \equiv \chi_\perp = \frac{(\Delta x)^2}{2\Delta t} \approx \nu \rho_L^2 \approx v_{th}^2 \frac{\nu}{\Omega_L^2}\,, \tag{2}$$

where $\Omega_L = \frac{eB}{mc}$ is the gyro-frequency. Obviously one predicts from here

$$\chi_\perp = \frac{\nu m c^2 T}{e^2 B^2} \sim \frac{T}{B^2}\,. \tag{3}$$

The temperature T is measured in eV. In most fusion devices, magnitude and scaling of the transport do not agree with the B^{-2}–prediction.

The more systematic way to determine (linear) transport coefficients is to use kinetic theory and subsequent transport truncations. Kinetic theory is based on a so called kinetic equation. The latter follows for classical, non-relativistic plasmas from the Liouville equation. Using the Liouville density we can derive distribution functions which obey the BBGKY hierarchy. Truncations in the kinetic regime lead to plasma kinetic equations like the Vlasov equation, the Landau-Fokker-Planck, the Balescu-Lenard equation, respectively. Averaging over gyro-motion leads to the drift-kinetic or gyro-kinetic equations. In principle, a kinetic equation is a closed equation for the one-particle distribution function f, which symbolically can be written in the form [65]

$$\frac{\partial f}{\partial t} + \mathscr{D} f = \mathscr{J}(f|f)\,. \tag{4}$$

It contains a convective ($\mathscr{D}$) and a collision ($\mathscr{J}$) term. From such an equation we can derive a hierarchy of equations for the plasmadynamical moments. Again we need truncations, now called transport truncations. Besides very systematic formalisms, shortcuts exist [4] such as

$$\text{for} \quad u_{E\times B} \sim v_{th}: \; f \approx f_M(\vec{v} - \vec{u}) + \mathscr{O}(\delta) \tag{5}$$

and

$$\text{for} \quad u_{E\times B} \sim \delta v_{th}: \; f \approx f_M(\vec{v}) \left[1 + 2\frac{u_\parallel v_\parallel}{v_{th}^2}\right] - \vec{\rho}\cdot\nabla f_M + \mathscr{O}(\delta^2)\,, \tag{6}$$

where $\delta = \rho_L/L$ is the ratio of the Larmor radius to the length L of the system; f_M designates a Maxwellian, and $u_{E\times B}$ is the $E \times B$ velocity. The mentioned truncations lead immediately to the MHD and drift models, respectively.

Classical transport theory assumes that the collisional mean free path is smaller than the characteristic dimension of the system. However, the classical theory for a homogeneous and stationary magnetic field does not apply as a local theory of transport in a tokamak. Global geometric characteristics of the confining elements have a strong influence on transport. Fortunately, another smallness parameter, namely the ratio of the gyro-radius over the dimension of the system, can be used. Three regimes of collisionality are characteristic of the *neoclassical* transport theory: the banana regime (electronic diffusion increases starting from zero), the plateau regime (diffusion almost independent of collisionality), and the Pfirsch-Schlüter regime (diffusion increases with collisionality). Within this theory, cross-coefficients lead to effects which have no classical or Pfirsch-Schlüter counterparts, e.g. a parallel electric field produces an inward radial electron flux, a parallel electric field produces outward electron and ion heat fluxes, a parallel electric current is produced by an radial ion heat flux or by a radial electron/ion temperature gradient, radial pressure gradients may drive (bootstrap) currents, and so on.

Nevertheless, open problems still exist. For example, it appears that sometimes the electron thermal conductivity deviates from theoretical predictions by a factor $10 - 10^3$, the ambipolar diffusion coefficient may have an anomalous factor of 10^2, and the ion thermal conductivity may be different by a factor of 10.

Nonlinear transport phenomenology

Nonlinear effects are responsible for the differences between linear theoretical predictions and experimental facts. A very simple argument sheds light on the appearance of nonlinearities in transport theory. The continuity equation for the particle current density $\vec{\Gamma}$ is

$$\partial_t \vec{\Gamma} = -\nabla \cdot (\vec{u}\vec{\Gamma}) - \nabla P - \nabla \cdot \Pi + \frac{e}{mc}\vec{\Gamma} \times \vec{B} + \frac{e}{m}\vec{E} + \frac{1}{m}\vec{R} . \tag{7}$$

On the right-hand-side the convective term, pressure and field contributions as well as the resistive term appear. We decompose each variable into an averaged and a fluctuating part, respectively, such as

$$X := \langle X \rangle + \delta X . \tag{8}$$

A short calculation shows the averaged $x-$component of the current density in the form

$$\langle \Gamma_x \rangle = \underbrace{\frac{1}{m\Omega}\langle R_y \rangle}_{classical} + \underbrace{\frac{1}{\Omega}\langle (\hat{b} \times \nabla \cdot \Pi) \cdot \hat{x} \rangle}_{neo-classical} + \Gamma_x^{anomalous} + \Gamma_x^{time-dependent} . \tag{9}$$

The first part, caused by the friction term, is the classical one. The second one, from the pressure tensor, is the neoclassical one. The third one on the right hand side is the term we are presently mainly interested in,

$$\Gamma_x^{anomalous} = \frac{c}{B}\langle \delta n \, \delta E_y \rangle + \frac{1}{B}\left\{ \langle \delta T_\parallel \delta B_x \rangle - \langle \delta \Gamma_x \delta B_\parallel \rangle \right\} . \tag{10}$$

The fourth, time-dependent term is usually neglected.

We should emphasize that this simple argument does not cover the import aspect of gyro-averaging. The latter is the central part of gyro-kinetic models.

Weak-turbulence theory turned out to be a very successful approach, being specifically developed for plasma fluctuations with a characteristic frequency. Essentially, in its simplest form, it results from perturbation theory. It is thus restricted to weak fluctuation levels. Introducing, e.g. for the second moment of fluctuating electrostatic potentials, the spectral density $n(\vec{k},t)$,

$$\langle \delta\varphi_{\vec{k}} \delta\varphi^*_{\vec{k}'} \rangle = n(\vec{k},t)\delta(\vec{k}-\vec{k}') , \tag{11}$$

we obtain a wave-kinetic equation in the form

$$\frac{\partial n(\vec{k},t)}{\partial t} = \mathscr{I}(\vec{k},t) + \Gamma(\vec{k})n(\vec{k},t) , \tag{12}$$

where on the right-hand-side a nonlinear term ($\mathscr{I}$) and a damping term (Γ) appears. When damping can be neglected in a certain (inertial) region of $\vec{k}-$space, we may write for the spectral energy density $E(k)$ in isotropic turbulence

$$\frac{\partial E(k)}{\partial t} + \frac{\partial P(k)}{\partial k} = 0 . \tag{13}$$

This equation can be used for analyzing wave number spectra.

Let us repeat the simple dimensional analysis known from fluid theory. In $3-$dimensional space we can write the dimensions of energy E and spectral energy density $E(k)$ as

$$[E] = \left[\frac{gL^2}{T^2}\right] = \left[\frac{\rho L^5}{T^2}\right] \quad \Rightarrow \quad [E(k)] = \left[\frac{\rho L^3}{T^2}\right] \text{ and } [P(k)] = \left[\frac{\rho L^2}{T^3}\right] , \tag{14}$$

since we define

$$\frac{E}{V} = \int E(k)dk \tag{15}$$

independent of the dimension of the system. The famous argument of Kolmogorov [66] states that (in fluids) the time T can be eliminated via the constant transfer function P leading to

$$[E(k)] = \left[\frac{\rho L^3}{T^2}\right] \sim \rho^{1/3} P^{2/3} k^{-3+4/3} \sim k^{-5/3} , \tag{16}$$

i.e. the famous Kolmogorov spectrum.

Because of the characteristic frequencies appearing in plasmas, there the situation is different. The characteristic time follows from the finite frequency $\omega = \omega(k)$, and we can formulate a non-dimensional variable ξ in the form

$$\xi = \frac{Pk^2}{\rho\omega^3} . \tag{17}$$

Now more freedom appears since we may introduce an arbitrary function f of the dimensionless quantity to obtain for the spectral density

$$E(k) \sim \rho\omega^2(k)k^{-3}f(\xi) \tag{18}$$

When $f(\xi) \sim \xi^{1/3}$ one recovers some fundamental aspects of drift-wave turbulence. However, in the strict sense, the latter functional form is not uniquely derived.

There are many attempts to generalize weak-turbulence theory to strong plasma turbulence via renormalization procedures [67, 68, 69, 70]. Although intellectually convincing, for practical applications the evaluation is in most cases technically very ambitious (and perhaps more time-consuming than solving the original gyro-kinetic problem).

Especially for drift-wave turbulence semi-empirical models exist which are published under different names like mixing length theory, the marginal stabilization scenarios, etc. Let us elucidate a typical scenario along the following lines.

For electrostatic turbulence, the nonlinear current density in radial direction r is

$$\Gamma_r \approx \langle \delta n \, \delta u_r \rangle \hat{=} -D\frac{\langle n \rangle}{L_n} \tag{19}$$

where we expressed on the right-hand-side the current density as an effective diffusion current with a diffusion coefficient D and an inhomogeneity length L_n. For drift waves, the linear growth-rate, frequency, density fluctuation, and $E \times B-$velocity are

$$\frac{\gamma_k}{\omega_k} \sim \delta_k, \; \omega_k \approx \frac{ckT}{eBL_n}, \; \delta n_k \approx \langle n \rangle \frac{e\phi_k(1-i\delta_k)}{T}, \; \delta u_r \approx ick\frac{\phi_k}{B} , \tag{20}$$

respectively. Putting these terms together, we arrive at

$$\Gamma_r \approx \langle n \rangle \frac{cT}{eB}\sum_k k\delta_k \left| \frac{\delta n_k}{\langle n \rangle} \right|^2 . \tag{21}$$

The latter expression contains the still unknown fluctuation spectrum. For the level of the latter we assume that it increases until the corresponding local density change eliminates the instability driving density gradient:

$$\nabla \delta n \hat{=} k\delta n_k \sim \frac{\langle n \rangle}{L_n} . \tag{22}$$

We finally obtain

$$D \approx \frac{\gamma_{\max}}{k_{\max}^2} \tag{23}$$

when we approximate the sum by its maximum term.

According to Prandtl the turbulent diffusion coefficient can be written as

$$D \approx \frac{L_r^2}{\tau_c} \tag{24}$$

where L_r is the characteristic length (mode radial correlation length), and τ_c is the eddy turnover time (autocorrelation time for the turbulent field).

Now we may have two extremely different scenarios. First, small-scale turbulence with $L_r \approx \rho_s \sim \rho_i$ and the diamagnetic frequency

$$\tau_c^{-1} \approx \omega_T^* \equiv \frac{k_\theta T}{BL_T} \sim \frac{v_{ti}}{a}, \quad k_\theta \approx \rho_i^{-1} . \tag{25}$$

Combining these characteristic scales, we obtain

$$D \sim \rho_i^2 \, \omega_T^* \sim D_{Bohm} \frac{\rho^*}{L_T^*} \sim T^{3/2} B^{-2} \,, \tag{26}$$

which is known as the gyro-Bohm scaling.

On the other hand, for large-scale turbulence, e.g. with $L_r \approx \sqrt{a\rho_s}$, where a is the dimension of the system (minor radius of the tokamak), we estimate

$$D \sim \rho_i v_{ti} \sim D_{Bohm} \left[\frac{m^2}{s}\right] = \frac{T[eV]}{16B[T]} \sim TB^{-1} \,. \tag{27}$$

This is the celebrated Bohm scaling.

Other scalings have been proposed in addition. For example the Goldstone regime $D \sim \sqrt{\rho_i}$ in the L-regime.

The main question remains: How can we predict the characteristic scales. Several procedures have been proposed during the last decades. A quite convincing one is the following.

Let a system be modeled by a nonlinear partial differential equation, with dissipation, for the field u. The system contains (at least) two quadratic or higher-order conserved quantities in the absence of dissipation. One of the conserved quantities, let us say $A(u)$, decays faster than the other(s), e.g. $B(u)$. The modal cascade is expected to reach a quasi-stationary state which minimizes A for constant B, i.e.

$$\delta A - \lambda \, \delta B = 0 \,. \tag{28}$$

This formulation, as well as arguments for it, can be found in the review article by Hasegawa [71].

Besides the evident progress in analytical theory one should have in mind that (really) self-consistent models are generally very difficult to solve and often show their power only qualitatively, i.e. in scaling predictions. Heuristic, quasi-self-consistent models are often ad hoc, are good for rough interpretations, but are, strictly speaking, not predictive. Non-locality, intermittency, interaction with coherent structures, etc are are open problems which could not be solved so far in full generality. Many new insights rely on numerical simulations. The progress in numerical codes is enormous; unfortunately, analytical theory is behind but has remained indispensable.

Evidence of effective parallel transport

Another approach to nonlinear plasma transport, being favored in the present presentation, is based on the following physical picture. Suppose that magnetic fluctuations are present in the plasma such that the total field consists of a (strong) zeroth order contributions and fluctuations,

$$\vec{B} = \vec{B}_0 + \delta\vec{B}, \quad \vec{b} = \frac{\delta\vec{B}}{B} \,. \tag{29}$$

We assume that locally components of the magnetic field being perpendicular to the zeroth order (main) magnetic field do exist. To zeroth order, heat transport is mainly along the (zeroth order) magnetic field, since $\kappa_\perp \ll \kappa_\parallel$ holds for the heat conductivities. We may write

$$\frac{\partial T}{\partial t} \approx \kappa_\parallel (\hat{n} \cdot \nabla)^2 T \,. \tag{30}$$

The normal vector $\hat{n}$ is along the total magnetic field. We have

$$\hat{n}_0 \equiv \frac{\vec{B}_0}{B_0}, \quad \kappa_\parallel = \frac{v_{th}^2}{v} \gg \kappa_\perp \,. \tag{31}$$

Introducing fluctuations, we obtain an additional heat transport in the perpendicular direction with respect to the zeroth order magnetic field. Assuming that on the (zeroth order) magnetic surfaces the temperature is uniform, we may write for the averaged temperature

$$\frac{\partial \langle T \rangle}{\partial t} = \kappa_\parallel \langle b^2 \rangle \frac{\partial^2 \langle T \rangle}{\partial x_\perp^2} + \kappa_\parallel \left\langle b \frac{\partial}{\partial x_\perp} (\hat{n}_0 \cdot \nabla) \delta T \right\rangle \,. \tag{32}$$

This means that due to the (perpendicular) fluctuations, a quite effective perpendicular transport appears which is caused by an effective parallel heat conduction.

PART II: DESCRIPTION OF STOCHASTIC (MAGNETIC) FIELDS

The phenomena summarized so far gave rise to the "old" paradigm: Plasma stochasticity predominantly originates from plasma instabilities in high-dimensional phase space. Then it is counter-productive to good confinement, i.e. enhanced losses appear (due to micro-instabilities, MHD instabilities, ripple loss, etc). Nonlinear dynamics taught us that already a few degrees of freedom system may become stochastic. In plasmas sources of that type of stochasticity are confinement, heating, shaping and correction coils, MHD control coils and boundary layer coils, ergodic divertor coils, and so on. The "new" paradigm arose: Stochastization may have positive aspects, since it is manageable in some respect, and may help for ELM mitigation, particle exhaust, zonal flows, and so on. Specific coil arrangements have been tested already in Tore-Supra, W7-AS (island divertor), DIII-D, Textor, and JET, and are being planned for ASDEX-U and ITER.

Clebsch representation and field line characteristics

For the magnetic field we use the Clebsch representation [72]

$$\vec{B} = \nabla\psi \times \nabla\theta - \nabla H \times \nabla\varphi \,. \tag{33}$$

This representation follows, e.g., from the vector potential in cylindrical coordinates

$$\vec{A} = A_\rho \nabla\rho + A_\theta \nabla\theta + A_\varphi \nabla\varphi \,. \tag{34}$$

Introducing

$$A_\rho = \frac{\partial G(\rho,\theta,\varphi)}{\partial\rho}\,, \quad \psi = A_\theta - \frac{\partial G(\rho,\theta,\varphi)}{\partial\theta}\,, \quad -H = A_\varphi - \frac{\partial G(\rho,\theta,\varphi)}{\partial\varphi} \tag{35}$$

we arrive at the Clebsch form. Using the Clebsch representation, it is quite obvious that the magnetic field line system is of Hamiltonian form. Starting from the differentials

$$\begin{aligned}
d\psi &= \vec{B}\cdot\nabla\psi = -(\nabla H \times \nabla\varphi)\cdot\nabla\psi && (36)\\
&= -\left(\frac{\partial H}{\partial\theta}\nabla\theta\times\nabla\varphi\right)\cdot\nabla\psi = -\frac{\partial H}{\partial\theta}(\nabla\psi\times\nabla\theta)\cdot\nabla\varphi\,, && (37)\\
d\theta &= \vec{B}\cdot\nabla\theta = -(\nabla H\times\nabla\varphi)\cdot\nabla\theta && (38)\\
&= -\left(\frac{\partial H}{\partial\psi}\nabla\psi\times\nabla\varphi\right)\cdot\nabla\theta = \frac{\partial H}{\partial\psi}(\nabla\psi\times\nabla\theta)\cdot\nabla\varphi\,, && (39)\\
d\varphi &= \vec{B}\cdot\nabla\varphi = (\nabla\psi\times\nabla\theta)\cdot\nabla\varphi\,, && (40)
\end{aligned}$$

we immediately get the symplectic structure

$$\frac{d\psi}{d\varphi} = -\frac{\partial H}{\partial\theta}\,, \qquad \frac{d\theta}{d\varphi} = \frac{\partial H}{\partial\psi}\,. \tag{41}$$

Compared to classical mechanics, φ corresponds to "time", ψ to "momentum", and θ is the "coordinate". Within the Hamiltonian formalism, one can introduce action (in the following denoted $\vec{J}$) and angle (in the following denoted $\vec{\theta}$) coordinates. Generally speaking, when the Hamiltonian consists of an integrable part H_0 (leading to closed magnetic flux surfaces) and a perturbation

$$H(\vec{J},\vec{\theta}) = H_0(\vec{J}) + \delta H(\vec{J},\vec{\theta})\,, \tag{42}$$

the canonical equations read

$$\frac{\partial H}{\partial\vec{J}} = \dot{\vec{\theta}} = \vec{\Omega}(\vec{J}) + \frac{\partial\delta H}{\partial\vec{J}}\,, \qquad \frac{\partial H}{\partial\vec{\theta}} = -\dot{\vec{J}} = \frac{\partial\delta H}{\partial\vec{\theta}}\,. \tag{43}$$

Within a Fourier decomposition

$$\delta H(\vec{J},\vec{\theta}) = \sum_{\vec{m}} e^{i\vec{m}\cdot\vec{\theta}} H_{\vec{m}} \tag{44}$$

we get the following approximate solution

$$\vec{\theta} \approx \vec{\theta}_0 + \vec{\Omega} t\,, \qquad \Delta\vec{J} = -\int dt \frac{\partial \delta H}{\partial \vec{\theta}} \approx -i \sum_{\vec{m}} \frac{\vec{m} H_{\vec{m}} e^{i\vec{m}\cdot\vec{\theta}_0}\left(e^{i\vec{m}\cdot\vec{\Omega} t} - 1\right)}{\vec{m}\cdot\vec{\Omega}} \tag{45}$$

with resonances for $\vec{m}\cdot\vec{\Omega}(\vec{J}) = 0$. As it is well-known, the resonances result in breaking of magnetic surfaces. Stochastic field line motion can occur.

For tokamak applications, the perturbation terms may be written with toroidal and poloidal mode numbers in the form $\delta H \sim \varepsilon H_{mn} \cos(m\theta + n\varphi + \chi_{m0})$. Resonances occur at the rational magnetic surface ψ_{mn} with $q(\psi_{mn}) = m/n$, creating chains of islands. The island width is [73]

$$W_{mn} = 4 \left| \frac{\varepsilon H_{mn}(\psi_{mn})}{dq^{-1}/d\psi} \right|^{1/2}, \tag{46}$$

and chaos sets in when

$$\sigma_{Chir} = \frac{W_{mn} + W_{m+1n}}{|\psi_{m+1n} - \psi_{mn}|} \geq 1\,. \tag{47}$$

This (qualitative) theoretical prediction has been successfully tested in many numerical applications and is also very helpful for interpreting experimental results.

Symplectic mapppings

The conventional method to determine the magnetic field topology is numerical field line tracing and subsequent *Poincaré* display. In the mapping approach to integrate the Hamiltonian system one wishes to avoid small time–step integration replacing it by return mappings of variables (θ, ψ) to a certain poloidal section of torus $\varphi = \varphi_k$. Below we define a mapping of field lines in a toroidal system. To be more general we consider the case when a toroidal mode number n takes values proportional to a certain number N ($N \geq 1$) which corresponds to the N–fold symmetry of magnetic perturbations along the toroidal angle φ, i.e., $n = sN$, $(s = \pm 1, \pm 2, \ldots)$. *Let* $(\theta = \vartheta_k, \psi = \psi_k)$ *be the poloidal angle and toroidal flux* at the poloidal sections $\varphi = \varphi_k = 2\pi k/N$, $(k = 0, \pm 1, \pm 2, \ldots)$. We define the map

$$(\vartheta_{k+1}, \psi_{k+1}) = \hat{M}(\vartheta_k, \psi_k), \tag{48}$$

which relates the variables (θ, ψ) at the successive sections φ_k and φ_{k+1}. Then the map $(\hat{M})^N$ defines the Poincaré return map. The flux-preserving property of the mapping (48) is expressed by

$$\left| \frac{\partial(\vartheta_{k+1}, \psi_{k+1})}{\partial(\vartheta_k, \psi_k)} \right| = 1. \tag{49}$$

Symplectic mappings for integrating Hamiltonian systems, based on the Hamilton–Jacobi method and perturbation theory, have been developed in [74, 75, 76, 77]. The general form of the mapping (48) for the Hamiltonian system is given by the following symmetric flux-preserving form (see [76]):

$$\Psi_k = \psi_k - \varepsilon \frac{\partial S^{(k)}}{\partial \vartheta_k}, \qquad \Theta_k = \vartheta_k + \varepsilon \frac{\partial S^{(k)}}{\partial \Psi_k}, \tag{50}$$

$$\Psi_{k+1} = \Psi_k, \qquad \bar{\Theta}_k = \Theta_k + w(\Psi_k, \varepsilon)(\varphi_{k+1} - \varphi_k), \tag{51}$$

$$\psi_{k+1} = \Psi_{k+1} + \varepsilon \frac{\partial S^{(k+1)}}{\partial \vartheta_{k+1}}, \quad \vartheta_{k+1} = \bar{\Theta}_k - \varepsilon \frac{\partial S^{(k+1)}}{\partial \Psi_{k+1}}, \tag{52}$$

where $w(\Psi, \varepsilon) = \partial \mathscr{H}/\partial \Psi$ is the frequency of perturbed motion, and $S^{(k)} \equiv S(\vartheta_k, \Psi_k)$ is the value of the generating function $G(\vartheta, \Psi, \varphi, \varphi_0; \varepsilon)$ taken at sections $\varphi = \varphi_k$, i.e., $S(\vartheta_k, \Psi_k) = G(\vartheta_k, \Psi_k, \varphi_k, \varphi_0; \varepsilon)$. The generating function obeys the Hamilton-Jacobi equation,

$$H\left(\Psi + \varepsilon \frac{\partial G}{\partial \vartheta}, \vartheta, \varphi\right) + \varepsilon \frac{\partial G}{\partial \varphi} = \mathscr{H}(\Psi, \varepsilon), \tag{53}$$

in the finite interval $\varphi_k < \varphi < \varphi_{k+1}$, satisfying the condition $G|_{\varphi=\varphi_0} = 0$ at the initial value φ_0 ($\varphi_k < \varphi_0 < \varphi_{k+1}$).

In the first order of ε the frequency $w(\Psi, \varepsilon)$ is determined by the inverse safety factor $q(\Psi)$:

$$w(\Psi, \varepsilon) = \frac{1}{q(\Psi)}, \tag{54}$$

and the generating function $G(\vartheta, \Psi, \varphi, \varphi_0)$ in the finite interval $\varphi_{k+1} < \varphi < \varphi_k$ is given by [76]

$$G(\vartheta, \Psi, \varphi, \varphi_0) = -(\varphi - \varphi_0) \sum_{m,n} H_{mn}(\Psi) \left[a(x_{mn}) \sin(m\vartheta - n\varphi + \chi_{mn}) + b(x_{mn}) \cos(m\vartheta - n\varphi + \chi_{mn}) \right], \tag{55}$$

where

$$a(x) = \frac{1 - \cos x}{x}, \qquad b(x) = \frac{\sin x}{x}, \qquad x_{mn} = \left(\frac{m}{q(\Psi)} - n \right) (\varphi - \varphi_0). \tag{56}$$

The free parameter φ_0 lies in the interval $\varphi_k \le \varphi_0 \le \varphi_{k+1}$.

The mapping (50) can be also considered as the alternative method of the symplectic integration of Hamiltonian systems. Having this in mind, we can use it for other Hamiltonian systems, e.g. the drift motion of particles in stochastic magnetic fields. We have analyzed the onset of chaotic drift motion due to resonant magnetic field perturbations [78]. The chaotic region at the plasma edge is formed due to interactions of several resonant drift islands. Their radial positions with respect to the divertor coils, and the distances between them, are important factors in formation of particle chaos. These factors are determined not only by the plasma parameters, as in the case of magnetic field line stochasticity, but also by the particle parameters. It was shown that due to a toroidal drift the positions of resonant drift surfaces are shifted with respect to resonant magnetic surfaces. The shift depends not only on the energy of particles but also on their directions of motion. For co-passing particles the resonant surfaces are shifted inwardly, while for counter-passing particles outwardly. Fig. 1 shows an example. An equilibrium electric field and a rotating magnetic field also significantly vary the positions of resonant drift surfaces, acting selectively on the co-passing and counter-passing particles, respectively. For instance, a rotating field of frequency $f = 10^4$ Hz shifts the resonant drift surface outward (inward) up to 2 - 3 cm for the co-passing (counter-passing) particles. The shift of resonant drift surfaces significantly affects the formation of the stochastic zone of drift motion since the magnetic field perturbation strongly decays in the inward radial direction. Particularly, it has been shown that the ergodic regions for co-passing and counter-passing particles are different. This difference becomes more pronounced with an increase in particle energy.

Looking for the deviations of two, initially adjacent field lines, the deviation

$$d\vec{l}_k = \begin{pmatrix} d\psi_k \\ d\vartheta_k \end{pmatrix}, \qquad d\vec{l}_{k+1} = J_k d\vec{l}_k, \qquad J_k = \begin{pmatrix} \frac{\partial \psi_{k+1}}{\partial \psi_k} & \frac{\partial \psi_{k+1}}{\partial \vartheta_k} \\ \frac{\partial \vartheta_{k+1}}{\partial \psi_k} & \frac{\partial \vartheta_{k+1}}{\partial \vartheta_k} \end{pmatrix} \tag{57}$$

follows via the Jacobian matrix. The eigenvalues λ of the latter

$$\begin{vmatrix} \frac{\partial \psi_{k+1}}{\partial \psi_k} - \lambda & \frac{\partial \psi_{k+1}}{\partial \vartheta_k} \\ \frac{\partial \vartheta_{k+1}}{\partial \psi_k} & \frac{\partial \vartheta_{k+1}}{\partial \vartheta_k} - \lambda \end{vmatrix} = 0, \qquad \lambda^{(k)} = \max\left(\lambda_1^{(k)}, \lambda_2^{(k)}\right) \tag{58}$$

determine the Liapunov exponent. We have a global Liapunov exponent for unstable orbits

$$\sigma = \lim_{N \to \infty} \frac{1}{N} \ln \prod_{k=1}^{N} \lambda^{(k)} > 0 \tag{59}$$

and a finite-time Liapunov exponent for finite connection lengths

$$\sigma_N = \frac{1}{N} \ln \prod_{k=1}^{N} \lambda^{(k)} > 0\,. \tag{60}$$

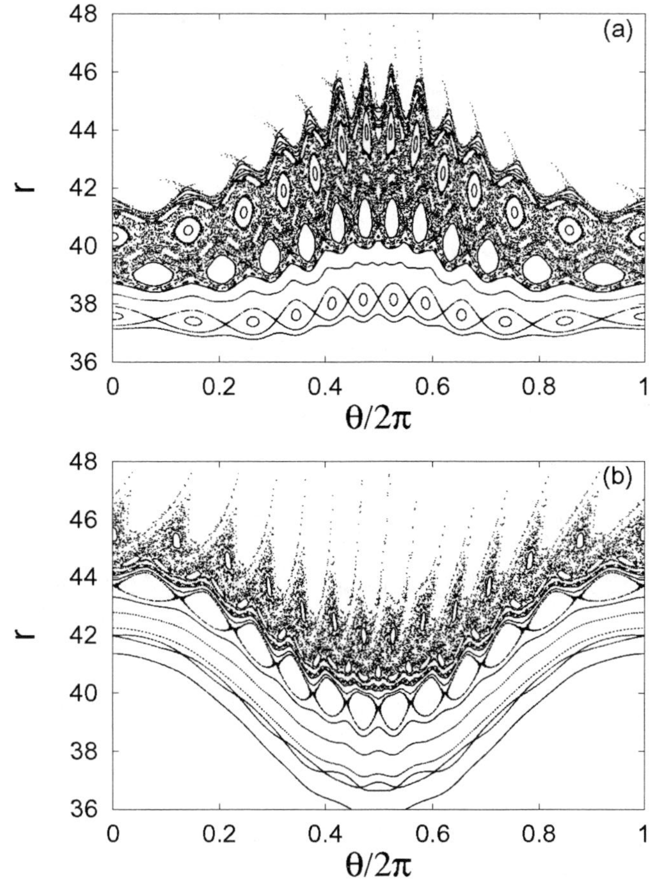

FIGURE 1. Drift surfaces and chaotic drift motion for (a) co- and (b) counter-passing particles, respectively

From the latter we can define a local e-folding length

$$L_N = \frac{l}{\sigma_N} \tag{61}$$

where l is the length of a field line per one map step. Averaged over magnetic surfaces (zeroth order) with label ρ (radius) we get the Kolmogorov length

$$L_K(\rho, N) = \bar{L}_N = \frac{l}{\bar{\sigma}_N} \approx \pi q R_0 \left(\frac{\pi \sigma_{Chir}}{2}\right)^{-4/3} \tag{62}$$

where the latter expression is an approximation in the quasi-classical limit. The deviation of field lines becomes exponential according to

$$\langle |\Delta \vec{x}_\perp(\zeta)|^2 \rangle \sim |\Delta \vec{x}_\perp(0)|^2 \exp\left\{\frac{2\zeta}{L_K}\right\} . \tag{63}$$

Here we have introduced the toroidal length $\zeta \sim \varphi$.

Quasi-linear field line diffusion

From Poincaré sections for magnetic field line motion (which qualitatively look similar to Fig. 1) we can recognize different areas, some of them are stochastic, others belong to still integrable regions. In the (completely) stochastic regions, standard theories can be applied to investigate the field line wandering. Here, we elucidate the quasi-linear description. Starting from the Hamiltonian formulation (41), i.e.

$$\frac{d\psi}{d\zeta} = -\frac{\partial \Psi_p}{\partial \theta}, \quad \frac{d\theta}{d\zeta} = \frac{\partial \Psi_p}{\partial \psi} \tag{64}$$

with the Hamiltonian

$$H \equiv \Psi_p = H_p(\psi) + \varepsilon H_1(\theta, \psi, \zeta), \quad V_1 = -\frac{\partial H_1}{\partial \theta} \tag{65}$$

we can derive a magnetic diffusion coefficient from quasi-linear theory in the following way. The kinetic equation for the field line distribution function f is

$$\frac{df}{d\zeta} = \frac{\partial f}{\partial \zeta} + \dot{\theta}\frac{\partial f}{\partial \theta} + \dot{\psi}\frac{\partial f}{\partial \psi} = \frac{\partial f}{\partial \zeta} + \frac{\partial H}{\partial \psi}\frac{\partial f}{\partial \theta} - \frac{\partial H}{\partial \theta}\frac{\partial f}{\partial \psi}. \tag{66}$$

By standard multiple-scale analysis [79] we find for the averaged zeroth order distribution function

$$\frac{\partial \bar{f}_0}{\partial \zeta_2} = \frac{\partial}{\partial \psi}\left(D\frac{\partial \bar{f}_0}{\partial \psi}\right), \tag{67}$$

where

$$D \equiv D(\psi) = \left\langle \int_0^{\zeta_0} d\zeta_0' V_1\left[\theta - \Omega\zeta_0', \psi, \zeta_0 - \zeta_0'\right] V_1\left[\theta, \psi, \zeta_0\right] \right\rangle_{\theta, \zeta_0} \tag{68}$$

is the magnetic diffusion coefficient. We can estimate the latter by the correlation length L_{corr} for field line fluctuations in the form

$$D_{m(agnetic)} = \int_0^\infty d\zeta \langle b_x[\vec{x}_\perp(\zeta), \zeta] b_x[\vec{x}_\perp(0), 0]\rangle \sim b^2 L_{corr}. \tag{69}$$

Introducing the Eulerian perpendicular ($\lambda_\perp$) and parallel ($\lambda_\|$) correlation lengths, respectively, we can distinguish the two cases:

$$\Delta x < \lambda_\perp \quad : \quad D_m\lambda_\| < \lambda_\perp^2 \rightarrow L_{corr} \approx \lambda_\| \rightarrow D_m \approx b^2\lambda_\|, \tag{70}$$

$$\Delta x > \lambda_\perp \quad : \quad D_m\lambda_\| > \lambda_\perp^2 \rightarrow D_m L_{corr} \approx \lambda_\perp^2 \rightarrow D_m \approx \frac{b^2\lambda_\perp^2}{D_m} \rightarrow D_m \approx b\lambda_\perp. \tag{71}$$

Summarizing we have

$$D_m \approx \begin{cases} b^2\lambda_\| & \text{for} \quad b < \frac{\lambda_\perp}{\lambda_\|}, \\ b\lambda_\perp & \text{for} \quad b > \frac{\lambda_\perp}{\lambda_\|}. \end{cases} \tag{72}$$

The first result $D_m \approx b^2\lambda_\|$ for $\lambda_\perp \rightarrow \infty$ is known in literature as the quasi-linear result, whereas the second one $D_m \approx b\lambda_\perp$ for $\lambda_\| \rightarrow \infty$, $\lambda_\perp$ finite is due to Kadomtsev and Pogutse [30]. As we shall see, there is a problem with the second result since it does not take into account correctly trapping.

Non-diffusive magnetic field line motion

As has been mentioned already, quasi-linear diffusion can only be expected in the fully stochastic region. In an (open) chaotic system, there exist (laminar) regions with small connection lengths of magnetic field lines. There, the motion is non-diffusive. Experimental hints exist for that non-diffusive areas exist. For example, characteristic spatial heat flux patterns have been observed in bounded stochastic plasmas [80, 81]. They show a characteristic number

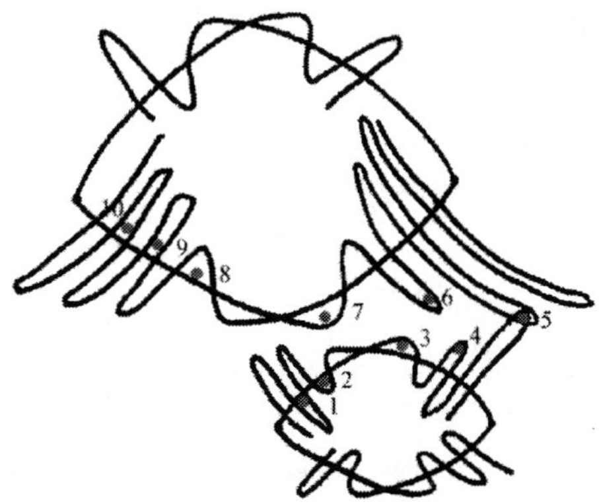

FIGURE 2. Sketch of the stable and unstable manifolds of periodic hyperbolic fixed points and their relation to field line wanderung 1, 2, 3, ... in a Poincaré plot.

of stripes on the target plates when plotted in the (φ, θ)-plane, where φ is the toroidal angle and θ is the poloidal angle of the torus. In addition, when varying the edge safety factor q_a, and projecting on the (q_a, θ)-plane, generic butterfly-like structures appear. In order to analyze and classify the spatial structures of heat flux patterns [81], one can use the concept of magnetic footprints [82, 83] together with an analysis of the stable and unstable manifolds of hyperbolic periodic points [84, 85, 86, 87] of magnetic resonances present in the stochastic plasma edge region. Field line wandering occurs along the stable and unstable manifolds of hyperbolic fixed points, as indicated in Fig. 2

This is a very efficient process near the plasma boundary. Typical structures in the laminar zones can be already seen in the upper parts of Fig. 1 a and b. A careful geometrical analysis of the “fingers” seen there explains many of the complex heat deposition patterns observed in tokamaks.

PART III: STOCHASTIC TRANSPORT THEORIES

V-Langevin equations

The starting point for a systematic stochastic transport theory for electrons and ions is a stochastic differential equation. Balescu and co-workers [8] advanced the field considerably by using the so called V-Langevin equation

$$\frac{dx_p(t)}{dt} = b_x[x_p(t), y_p(t), z_p(t)]\frac{dz_p(t)}{dt} + \eta_{\perp x}(t)\,, \tag{73}$$

$$\frac{dy_p(t)}{dt} = b_y[x_p(t), y_p(t), z_p(t)]\frac{dz_p(t)}{dt} + \eta_{\perp y}(t)\,, \tag{74}$$

$$\frac{dz_p(t)}{dt} = \eta_{\parallel}(t)\,. \tag{75}$$

The V-Langevin equation makes use of the guiding center approach for small Larmor radii of the particles. Without collisions ($\eta_\perp = \eta_\parallel = 0$) the particle positions $\vec{r}_p(t)$ follow the magnetic field lines. Collisions cause diffusive motion along the zeroth order magnetic field (parallel direction) as well as deviations from the perturbed field lines in perpendicular direction.

The V-Langevin equation is a simplification of the A-Langevin equation (Acceleration Langevin equation)

$$\frac{d\vec{v}}{dt} = \frac{Ze}{mc}\vec{v} \times \vec{B} - \nu\vec{v} + \vec{a}\,, \tag{76}$$

which has to be used when finite Larmor radii become important.

Making use of the Taylor-Green-Kubo formula [88, 89, 90] in 3D

$$D = \frac{1}{3}\int_0^\infty d\tau \langle \vec{v}[\vec{x}(\tau),\tau]\cdot\vec{v}[0,0]\rangle\,, \tag{77}$$

we obtain for the magnetic diffusion coefficient in x-direction the estimate already used in the evaluation of the quasi-linear expression, namely

$$D_{m(agnetic)} = \int_0^\infty d\zeta \langle b_x[\vec{x}_\perp(\zeta),\zeta] b_x[\vec{x}_\perp(0),0]\rangle \sim b^2 L_{corr}\,. \tag{78}$$

But one should note that the Taylor-Green-Kubo formula contains the Lagrangian correlation function

$$L_{rs}[\zeta] = \langle b_r[\vec{x}_\perp(\zeta),\zeta] b_s[\vec{x}_\perp(0),0]\rangle \tag{79}$$

and not the (simpler) Eulerian correlation function. Reasonable assumptions for the latter are available, e.g.

$$B_{mn} = \langle b_m[\vec{x}+\vec{r}] b_n[\vec{x}]\rangle \approx b^2 \left[\left(1-\frac{r_\perp^2}{\lambda_\perp^2}\right)\delta_{mn} + \frac{r_m r_n}{\lambda_\perp^2}\right] \exp\left\{-\frac{z^2}{2\lambda_\parallel^2} - \frac{r_\perp^2}{2\lambda_\perp^2}\right\}\,. \tag{80}$$

The relation between the Lagrange correlation function and the Euler correlation function is a difficult task which, in principle, requires the knowledge of the exact dynamics. All discussions of the following pages are centered around that problem. Taking arguments from fluid dynamics, very often the Corrsin approximation [43] is applied in the form

$$L_{rs}(\zeta) \approx \int d^2 r_\perp B_{rs}(\vec{r}_\perp,\zeta)\langle\delta(\vec{r}_\perp - \vec{x}_\perp(\zeta))\rangle\,. \tag{81}$$

Here, the exact formula with an average over the whole right-hand side is approximated by performing the average only over the delta function. Then, the averaged propagator appears which can be further evaluated by making use of a cumulant expansion [22]. Within the Corrsin approximation, Balescu and coworkers were able to derive an integral equation for the Lagrangian form of the correlation function. Its solution, for example for magnetic field line wandering, yields

$$D_m \approx \begin{cases} b^2\lambda_\parallel & \text{for} \quad \lambda_\perp \to \infty \\ b\lambda_\perp & \text{for} \quad \lambda_\parallel \to \infty \end{cases} \tag{82}$$

which have been presented in the previous section on the basis of heuristic arguments..

Rechester and Rosenbluth have estimated the mean square displacement due to stochastic motion when no perpendicular displacements from the magnetic field lines occur, but the motion is diffusive along the magnetic field lines. Then one gets for the mean square displacement

$$\sigma^2 \sim t^\gamma \tag{83}$$

with a subdiffusive exponent $\gamma = \frac{1}{2}$. The line of argument is as follows

$$\sigma^2 = 2D_{magnetic}z, \quad z^2 \approx \chi_\parallel t, \quad \sigma^2 \approx 2D_{magnetic}\sqrt{2\chi_\parallel t} \quad \longrightarrow \quad \frac{\sigma^2}{t} \sim \frac{1}{\sqrt{t}} \to 0\,. \tag{84}$$

In general, however, deviations from the magnetic field lines do occur.

Using the quasi-linear magnetic diffusion coefficient, we can use random walk estimates for the particle diffusion coefficient. In the collisionless case we find

$$D_{\perp particle} \sim \frac{\langle(\Delta x)^2\rangle}{\Delta t} \approx \frac{D_{magnetic}\lambda_\parallel}{\lambda_\parallel / v_{thermal}} = D_{magnetic}\, v_{thermal} \approx b^2\lambda_\parallel v_{thermal}\,. \tag{85}$$

In the collisional fluid limit we use

$$D_{\perp particle} \sim \frac{\langle(\Delta x)^2\rangle}{\Delta t} \approx D_{magnetic}\frac{L_{corr}}{\tau_c} \approx b^2\lambda_\parallel^2 v_{collision} \sim b^2\, D_{\parallel particle} \tag{86}$$

since then $\lambda_{\parallel} \approx L_{corr} \approx v_{thermal}/v_{collision}$.

Rechester and Rosenbluth [29] also derived a diffusion formula for the case when deviations from the magnetic field lines occur. They approximated a typical mean-square-displacement

$$\langle(\Delta x)^2\rangle_{L_K} \approx b^2\, \lambda_{\parallel}\, L_K \tag{87}$$

when estimating the magnetic field line displacement over a Kolmogorov length L_K. The decorrelation time is assumed as

$$t_l \approx \frac{L_K^2}{\chi_{\parallel}} \tag{88}$$

leading to the particle diffusion coefficient

$$D_{particle} \approx \frac{\langle(\Delta x)^2\rangle_{L_K}}{t_l} \approx \frac{b^2\, \lambda_{\parallel}\, \chi_{\parallel}}{L_K} . \tag{89}$$

A more detailed calculation is summarized in the Appendix.

In the strongly collisional regime, one may use for the mean square displacement and the decorrelation time

$$\langle(\Delta x)^2\rangle_l \approx b^2\, \lambda_{\parallel} l \quad \text{with} \quad l^2 \approx \chi_{\parallel} t_l\,, \quad \text{and} \quad t_l \approx \frac{\lambda_{\perp}^2}{\chi_{\perp}} . \tag{90}$$

Combining these estimates,

$$D_{particle} \approx \frac{\langle(\Delta x)^2\rangle_l}{t_l} \approx \frac{b^2 \lambda_{\parallel}}{\lambda_{\perp}} \sqrt{\chi_{\parallel}\chi_{\perp}} \sim b\, K\, D_{Bohm}\,, \tag{91}$$

the particle diffusion coefficient is proportional to the fluctuation strength b, the Kubo number K, and the Bohm diffusion coefficient D_{Bohm}, where

$$K \approx \frac{b\lambda_{\parallel}}{\lambda_{\perp}}\,, \qquad D_{Bohm} \approx \frac{1}{16} \sqrt{\chi_{\parallel}\chi_{\perp}} \sim \frac{cT}{eB} . \tag{92}$$

We would like to emphasize that this diffusion coefficient should not be confused with the the result (72). For the latter case, percolation effects become important.

The just presented two diffusion coefficients belong to different collision regimes. Comparing the Rechester-Rosenbluth case

$$\langle(\Delta x)^2\rangle_{L_K} \approx D_m\, L_K = D_{particle}\, T_K\,, \quad T_K \approx \frac{L_K^2}{\chi_{\parallel}} \tag{93}$$

with the Kadomtsev-Pogutse expressions

$$\langle(\Delta x)^2\rangle_{l_{coll}} \approx D_m\, l_{coll} = D_{particle}\, t_l\,, \quad t_l \approx \frac{l_{coll}^2}{\chi_{\parallel}} \approx \frac{\lambda_{\perp}^2}{\chi_{\perp}} \to l_{coll} \approx \sqrt{\frac{\chi_{\parallel}}{\chi_{\perp}}}\, \lambda_{\perp}\,, \tag{94}$$

we can conclude

$$T_K \ll t_l : \nu < \Omega\, b K \quad \text{for Rechester-Rosenbluth} \tag{95}$$

and

$$T_K \gg t_l : \nu > \Omega\, b K \quad \text{for Kadomtsev-Pogutse} . \tag{96}$$

Percolation for K>1

The theories mentioned so far work very well (in their respective regimes) as long as no trapping occurs. Whether trapping becomes significant or not can be measured by the so called Kubo number. The Kubo number is the ratio between the distance traveled during an autocorrelation time and the correlation distance. If the latter is large, the system spirals around for some time within the highly correlated area, i. e. it is being trapped. For magnetic field lines we define the Kubo number

$$K = \frac{b\lambda_{\parallel}}{\lambda_{\perp}} \,. \tag{97}$$

From here we clearly recognize that the Kadomtsev-Pogutse formula $D_{magnetic} \sim b\lambda_{\perp} \sim K^0$ is problematic, since it has been derived for $\lambda_{\parallel} \longrightarrow \infty$ when $K \longrightarrow \infty$. The actual magnetic diffusion coefficient $D_{magnetic}$ should decrease with $K \sim \lambda_{\parallel}$ and not stay constant $\sim K^0$ as above. Before presenting some detailed calculation for that, let us present a simple argument [19].

We assume a model field which additively consists of a random part and a trapping component, such that the field lines follow

$$\frac{dx}{dz} = b_x^{trapping} + b_x^{random} \,, \qquad \frac{dy}{dz} = b_y^{trapping} + b_y^{random} \,, \tag{98}$$

where we model the trapping contribution by

$$\delta\vec{B}^{trapping} = \nabla \times a\hat{z} \,, \quad a \approx a_0 \exp\left\{-\frac{r^2}{2\sigma^2}\right\} \,. \tag{99}$$

To lowest order, the field line x-coordinate follows from

$$\frac{d^2x}{dz^2} \approx -\kappa^2 x \,, \quad \kappa = \frac{a_0}{B_0\sigma^2} \,. \tag{100}$$

Thus, the magnetic mean field displacement is approximately

$$\langle \Delta r^2 \rangle = \int\int dz' dz'' \left\langle b_r^{random}(z') b_r^{random}(z'') \right\rangle \sim 2z\, P_{xx}(\kappa) \tag{101}$$

with

$$P_{xx}(k) = \frac{1}{2\pi} \int B_{xx}(\Delta z) \exp\{-ik\Delta z\} d\Delta z \tag{102}$$

since $b_r^{random}(z) \sim b_x^{random}(z)\cos(\kappa a)$. Assuming a certain fluctuation spectrum, e.g. a Kolmogorov spectrum

$$P_{xx}(k) \approx \frac{C}{\left[1 + (k\lambda_{\parallel})^2\right]^{5/6}} \,, \tag{103}$$

we immediately find the radial diffusion coefficient

$$D_r \approx D_{quasilinear} \frac{P_{xx}(\kappa)}{P_{xx}(0)} \sim \begin{cases} D_{quasilinear} & \text{for} \quad \kappa\lambda_{\parallel} \ll 1 \,, \\ \kappa^{-5/3} \longrightarrow 0 & \text{for} \quad \kappa\lambda_{\parallel} \gg 1 \,. \end{cases} \tag{104}$$

The interpretation of κ in terms of the Kubo number is simple:

$$\kappa\lambda_{\parallel} = \frac{a_0\lambda_{\parallel}}{B_0\sigma^2} \sim \frac{a_0}{B_0\sigma} \frac{\lambda_{\parallel}}{\sigma} \sim b\frac{\lambda_{\parallel}}{\lambda_{\perp}} \sim K \tag{105}$$

From here we conclude that the diffusion coefficient decreases with $K \longrightarrow \infty$ as it should. However, the order of decrease depends on the spectrum; different forms can be used [91]. A rough sketch of the principal behavior is shown in Fig. 3

In a series of papers Vlad et al. [53, 55, 57, 62] developed a new procedure for the Lagrangian correlation functions, called the decorrelation trajectory method. It was designed for large Kubo numbers and can be regarded as a substitute for the Corrsin approximation. The method has been used within the guiding center approximation (when the so-called V-Langevin equation applies). Here, we shall adapt the DCT to the more general problem of the correlation functions of the A-Langevin equation. The perturbation field **b** is assumed to have only a two-dimensional structure. This condition is fulfilled for finite but small Larmor radii. The stochastic field **b** is generated by the scalar magnetic potential $\phi(\boldsymbol{x}, z)$,

$$\boldsymbol{b}(\boldsymbol{x}, z) = \nabla\phi(\boldsymbol{x}, z) \times \boldsymbol{e}_z \,, \tag{106}$$

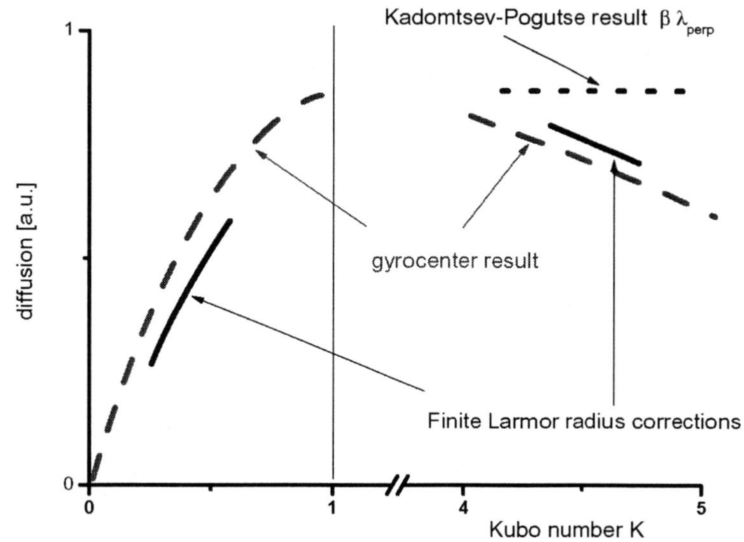

FIGURE 3. Sketch of the qualitative behavior of the K-dependence of diffusion and the influences of finite radii. For realistic parameters the deviations due to finite Larmor radii can be much larger than depicted here.

which we will call the flux function. The vector $\boldsymbol{x} = (x,y)$ refers to the perpendicular coordinates, whereas z can be regarded as the parallel component. Magnetic field lines follow from the flux function by the relation $d\boldsymbol{x}/dz = \boldsymbol{b}$. Within the DCT formulation we have in zeroth order of the Larmor radius expansion

$$L_{\mathrm{DCT}}^{(0)} = \int_{-\infty}^{\infty} P(\eta^0)\, P(\boldsymbol{b}^0)\, P(\phi^0)\, \eta^0 b_y^0 \langle\langle \eta_z(t) b_y[\boldsymbol{x}(t), z(t), t]\rangle_S\rangle_\perp \, d\eta_z^0 d\boldsymbol{b}^0 d\phi^0, \tag{107}$$

and in first order of the Larmor radius expansion

$$L_{\mathrm{DCT}}^{(1)} = \langle \frac{\rho_L^2}{v_t^2} \int_{-\infty}^{\infty} P(\eta_x^0)\, P(\eta_y^0)\, P(\eta_z^0)\, P(\boldsymbol{b}'^0)\, P(\phi^0)\, \eta^0 b_y'^0 \langle \eta_z(t) b_y'[\boldsymbol{x}(t), z(t), t]\rangle_S \, d\eta_x^0 d\eta_y^0 d\eta_z^0 d\boldsymbol{b}'^0 d\phi^0 \rangle_\perp. \tag{108}$$

Equations (107) and (108) are still exact since so far no approximation has been applied. Next, we assume Gaussian probabilities $P(x)$. We do not know the trajectory $\boldsymbol{r} = (\boldsymbol{x}, z)$. But the subensemble decomposition leads to a significant advantage: For each subensemble the values of ϕ^0 , η^0, and $\boldsymbol{b}^0$ are fixed. Within the DCT subensemble we can determine each trajectory for a mean field $\langle \boldsymbol{b}(\boldsymbol{x}, z)\rangle_S$ by solving the A-Langevin equation. The crucial simplification of that technique is that all contributions from the magnetic field are now non-stochastic values. The fictitious trajectory, along which a particle would travel if it is introduced into the subensemble magnetic mean field, is called the decorrelation trajectory.

The major approximation of the DCT [53] allowing the transition from the Eulerian to the Lagrangian expression, is to evaluate the formulas (107) or (108) by substituting for the unknown trajectory $\boldsymbol{x}(t)$ the decorrelation trajectory. This trajectory is introduced into the expression for the averaged field $\langle \boldsymbol{b}(\boldsymbol{x} = \boldsymbol{X}, z)\rangle_S$. With the trajectory $X(t)$ given by the A-Langevin equation, Eqs. (107) and (108) determine the Lagrangian velocity correlator. The first task is then to find the average $\langle \boldsymbol{b}(\boldsymbol{x}, z)\rangle_S$ in each subensemble defined by $\boldsymbol{b}^0$ and ϕ^0. Additionally, we need expressions for the averages occurring in the DCT equations.

The decorrelation trajectory itself is given by an A-Langevin equation

$$\ddot{\boldsymbol{X}}(t) = \dot{\boldsymbol{U}}(t) = \frac{Ze}{mc}\boldsymbol{U}(t) \times \{B_0 [b_0 \boldsymbol{e}_z + \langle \boldsymbol{b}(\boldsymbol{x}, z)\rangle_S]\} - \nu \boldsymbol{U}(t) + \boldsymbol{a}(t), \tag{109}$$

containing the non-stochastic subensemble average $\langle \boldsymbol{b}(\boldsymbol{x},z)\rangle_S$. The collisions $\boldsymbol{a}$ are still stochastic entries. Each subensemble produces an averaged **b**-field. One can calculate this average by using the conditional probability to find (within the subensemble defined by ϕ^0 and $\boldsymbol{b}^0$) a field $\boldsymbol{b}$ at the position $\boldsymbol{x}$,

$$\langle \boldsymbol{b}(\boldsymbol{x},z)\rangle_S = \int_{-\infty}^{\infty} d\boldsymbol{b}\boldsymbol{b}\, P(\boldsymbol{b},\boldsymbol{x}|\phi^0\boldsymbol{b}^0), \tag{110}$$

where one uses

$$P(\boldsymbol{b},\boldsymbol{x}|\phi^0\boldsymbol{b}^0) = \frac{\langle \delta(\boldsymbol{b}-\boldsymbol{b}(\boldsymbol{x},z,t))\delta\left(\phi^0-\phi(\mathbf{0})\right)\delta\left(\boldsymbol{b}^0-\boldsymbol{b}(\mathbf{0})\right)\rangle_S}{\langle \delta\left(\phi^0-\phi(\mathbf{0})\right)\delta\left(\boldsymbol{b}^0-\boldsymbol{b}(\mathbf{0})\right)\rangle_S}. \tag{111}$$

With the conditional probability we immediately find

$$\langle \boldsymbol{b}(x,y)\rangle_S = \begin{pmatrix} b_x^0\mathscr{E}_{xx} + b_y^0\mathscr{E}_{xy} + \phi^0\mathscr{E}_{\phi x} \\ b_x^0\mathscr{E}_{yx} + b_y^0\mathscr{E}_{yy} + \phi^0\mathscr{E}_{\phi y} \end{pmatrix},$$

where $\mathscr{E}$ designates the Euler correlations.

Summarizing: We can derive a subensemble decomposition of the Lagrangian correlation function for both, the magnetic field and for the velocity of a particle, depending on the averaged **b**-field in each subensemble. The average, $\langle \boldsymbol{b}(x,y)\rangle_S$, is given as a function of **x** and z. A similar result is obtained for the averaged derivative in the subensemble. Evaluating the DCT, quantitative results specifying the qualitative picture Fig. 3 can be obtained.

SUMMARY AND OUTLOOK

Performance projections to ITER rely on the H-mode global energy confinement time scaling, since models of local transport are not yet considered to be sufficiently accurate to replace scaling-based extrapolations [92]. There is confidence, based on wide-ranging experience in existing fusion devices, that the fusion performance of ITER will meet the reference target [93]. However, several key operational aspects remain the subject of focused research and development activities which aim to ensure the reliable operation of the device at or beyond its design capability. To develop plasma scenarios in which high fusion power production is combined with high confinement and plasma pressure, control of heat and particle fluxes, the fusion community will need to confront a range of challenges involving basic plasma physics understanding [93]. Stochastic transport definitely belongs to that category.

Stochastic transport is an interesting area with many applications in plasma physics. The general problems investigated in plasma stochasticity are also relevant in other fields, e.g. fluid dynamics and astrophysics.

Appendix: Details of the Rechster-Rosenbluth diffusion formula

In this appendix we briefly outline a more general deviation of the Rechester-Rosenbluth diffusion formula. In the presence of collisions and stochastic magnetic fields, three length scales are involved, which we may order in the form $\lambda_{mfp} \ll L_K < L_{diffusion}$. Since the collisional mean free path is the shortest length, one fist considers the collisional broadening up to the Kolmogorov length $L_K \approx \sqrt{\frac{2}{\pi}\frac{\lambda_\perp^2}{4b^2\lambda_\parallel}}$, assuming also diffusive motion caused by collisions along the magnetic field lines:

$$\left\langle (\Delta_x)^2 \right\rangle_{L_K} \approx \chi_\perp T_{L_K} = \chi_\perp \frac{L_K^2}{\chi_\parallel} \quad \text{for} \quad \nu T_{L_K} > 1\,. \tag{112}$$

Next field line divergence sets in with the Kolmogorov length as the e-folding length. The third scale $L_{diffusion}$ follows from the condition

$$\lambda_\perp^2 = \left\langle (\Delta x)^2 \right\rangle_{L_K} \exp\left\{\frac{2L_{diffusion}}{L_K}\right\} \rightarrow L_{diffusion} \approx L_K \ln\left(\frac{\lambda_\perp}{\sqrt{\langle(\Delta x)^2\rangle_{L_K}}}\right). \tag{113}$$

The particle diffusion coefficient can be obtained via

$$\left\langle (\Delta x)^2 \right\rangle_{L_{diffusion}} \approx D_m \, L_{diffusion} \, , \quad D_{particle} \approx \frac{\left\langle (\Delta x)^2 \right\rangle_{L_{diffusion}}}{\frac{L^2_{diffusion}}{\chi_{\parallel}}} \, . \tag{114}$$

Using these expressions, we get the Rechester-Rosenbluth formula in its original form [29]

$$D^{RR}_{particle} \approx \frac{D_m \chi_{\parallel}}{L_K \ln\left(\frac{\lambda_{\perp}}{L_K}\sqrt{\frac{\chi_{\parallel}}{\chi_{\perp}}}\right)} \sim \frac{D_m \chi_{\parallel}}{L_k} \sim \chi_{\parallel} \, b^2 \, K^2 \, , \tag{115}$$

where the approximations on the right are those which we have used previously.

ACKNOWLEDGMENTS

The present review is based on many discussions with Marcus Neuer and Andreas Wingen. The collaboration with IPP at FZ Jülich, especially with Sadrilla Abdullaev, Karl-Heinz Finken, and Bernd Unterberg, is gratefully acknowledged.

REFERENCES

1. J. Wesson, *Tokamaks*, Clarendon Press, Oxford, 2004.
2. R. J. Goldston, and P. H. Rutherford, *Introduction to Plasma Physics*, Institute of Physics, Philadelphia, 1995.
3. T. E. Evans, R. A. Moyer, K. H. Burrell, M. E. Fenstermacher, I. Joseph, A. W. Leonard, T. H. Osborne, G. D. Porter, M. J. Schaffer, P. Snyder, P. Thomas, J. Watkins, and W. West, *Nature Physics* **2**, 419 (2006).
4. R. D. Hazeltine, and J. D. Meiss, *Plasma confinement*, Addison-Wesley, Redwood City, 1992.
5. J. W. Connor, *Plasma Phys. Control. Fusion* **37**, 119 (1995).
6. R. Balescu, *Transport processes in plasmas: 2. Neoclassical transport theory*, North-Holland, Amsterdam, 1988.
7. A. A. Galeev, and R. Z. Sagdeev, *in: Handbook of Plasma Physics, Basic Plasma Physics I, A.A. Galeev and R.N. Sudan, eds., p. 679*, North-Holland, Amsterdam, 1984.
8. R. Balescu, *Aspects of anomalous transport in plasmas*, Institute of Physics Publishing, Bristol, 2005.
9. J. R. Jokipii, *Astrophys. J.* **183**, 1029 (1973).
10. F. Casse, M. Lemoine, and G. Pelletier, *Phys. Rev. D* **65**, 023002 (2001).
11. G. Qin, W. H. Matthaeus, and J. W. Bieber, *Astrophys. J* **578**, L117 (2002).
12. G. Qin, W. H. Matthaeus, and J. W. Bieber, *Geophys. Res. Lett.* **29**, 7 (2002).
13. W. H. Matthaeus, G. Qin, J. W. Bieber, and G. P. Zank, *Astrophys. J* **590**, L000 (2003).
14. D. Ruffolo, W. H. Matthaeus, and P. Chuychai, *Astrophys. J.* **614**, 420–434 (2004).
15. S. A. Khrapak, and G. E. Morfill, *Phys. Plasmas* **9**, 619 (2002).
16. A. Shalchi, J. W. Bieber, W. H. Matthaeus, and G. Qin, *Astrophys. J.* **616**, 617 (2004).
17. A. Shalchi, J. W. Bieber, W. H. Matthaeus, and R. Schlickeiser, *Astrophys. J.* **642**, 230 (2006).
18. J. W. Bieber, W. H. Matthaeus, A. Shalchi, and G. Qin, *Geophys. Res. Lett.* **31**, 101029 (2004).
19. P. Chuychai, D. Ruffolo, W. H. Matthaeus, and G. Rowlands, *Astrophys. J.* **633**, L49–L52 (2005).
20. R. Schlickeiser, *Cosmic ray astrophysics*, Springer-Verlag Berlin, Berlin, 2002.
21. R. Balescu, *Statistical Dynamics, Matter out of Equilibrium*, Imperial College Press, London, 2000.
22. W. D. McComb, *The Physics of Fluid Turbulence*, Oxford UP, New York, 1990.
23. T. E. Evans, R. A. Moyer, P. R. Thomas, J. G. Watkins, T. H. Osborne, J. A. Boedo, E. J. Doyle, M. E. Fenstermacher, K. H. Finken, R. J. Groebner, M. Grothand, J. H. Harris, R. J. L. Haye, C. J. Lasnier, S. Masuzaki, N. Ohyabu, D. G. Pretty, T. L. Rhodes, H. Reimerdes, D. L. Rudakov, M. J. Schaffer, G. Wang, and L. Zeng, *Phys. Rev. Lett.* **92**, 235003 (2004).
24. M. W. Jakubowski, O. Schmitz, S. S. Abdullaev, S. Brezinsek, K. H. Finken, A. Kraemer-Flecken, M. Lehnen, U. Samm, K. H. Spatschek, B. Unterberg, R. C. Wolf, and the TEXTOR team, *Phys. Rev. Lett.* **96**, 035004 (2006).
25. K. H. Finken, S. S. Abdullaev, M. F. M. de Bock, M. von Hellermann, M. Jakubowski, R. Jaspers, H. R. Koslowski, A. Kraemer-Flecken, M. Lehnen, Y. Liang, A. Nicolai, R. C. Wolf, O. Zimmermann, M. de Baar, G. Bertschinger, W. Biel, S. Brezinsek, C. Busch, A. J. H. Donnacutee, H. G. Esser, E. Farshi, H. Gerhauser, B. Giesen, D. Harting, J. A. Hoekzema, G. M. D. Hogeweij, P. W. Huettemann, S. Jachmich, K. Jakubowska, D. Kalupin, F. Kelly, Y. Kikuchi, A. Kirschner, R. Koch, M. Korten, A. Kreter, J. Krom, U. Kruezi, A. Lazaros, A. Litnovsky, X. Loozen, N. J. L. Cardozo, A. Lyssoivan, O. Marchuk, G. Matsunaga, P. Mertens, A. Messiaen, O. Neubauer, N. Noda, V. Philipps, A. Pospieszczyk, D. Reiser, D. Reiter, A. L. Rogister, M. Sakamoto, A. Savtchkov, U. Samm, O. Schmitz, R. P. Schorn, B. Schweer, F. C. Schueller, G. Sergienko, K. H. Spatschek, G. Telesca, M. Tokar, R. Uhlemann, B. Unterberg, G. V. Oost, T. V. Rompuy, G. V. Wassenhove, E. Westerhof, R. Weynants, S. Wiesen, and Y. H. Xu, *Phys. Rev. Lett* **94**, 015003 (2005).

26. K. H. Finken, S. S. Abdullaev, M. W. Jakubowski, M. F. M. de Bock, S. Bozhenkov, C. Busch, M. von Hellermann, R. Jaspers, Y. Kikuchi, A. Kraemer-Flecken, M. Lehnen, D. Schege, O. Schmitz, K. H. Spatschek, B. Unterberg, A. Wingen, R. C. Wolf, O. Zimmermann, and the TEXTOR Team, *Phy. Rev. Lett.* **98**, 065001 (2007).
27. J. Jokipii, *Astrophys. J.* **146**, 480 (1966).
28. J. R. Jokipii, and E. N. Parker, *The Astrophysical Journal* **155**, 799 (1969).
29. A. B. Rechester, and M. N. Rosenbluth, *Phys. Rev. Lett.* **40**, 38 (1978).
30. B. B. Kadomtsev, and O. P. Pogutse, *in: Plasma Physics and Controlled Nuclear Fusion Research. Proc. 7th. Int. Conf.* **1**, 549 (1978).
31. J. A. Krommes, C. Oberman, and R. G. Kleva, *J. Plasma Phys.* **30**, 11 (1983).
32. J. R. Myra, P. J. Catto, H. E. Mynick, and R. E. Duvall, *Phys. Fluids B* **5**, 1160 (1993).
33. G. Laval, *Phys. Fluids B* **5**, 711 (1993).
34. K. H. Spatschek, M. Eberhard, and H. Friedel, *Physicalia Mag* **20**, 85 (1998).
35. M. de Rover, A. M. R. Schilham, A. Montvai, and N. J. Lopes Cardozo, *Phys. Plasmas* **6**, 2443 (1999).
36. R. Kubo, *J. Math. Phys.* **4**, 174 (1963).
37. N. VanKampen, *Phys. Fluids* **19**, 11 (1996).
38. Y. Elskens, and D. Escande, *Microscopic dynamics of plasmas and chaos*, Institute of Physics Publishing, Bristol, 2003.
39. N. G. van Kampen, *Phys. Reports* **24**, 171 (1976).
40. R. Balescu, H. D. Wang, and J. H. Misguich, *Phys. Plasmas* **1**, 3827 (1994).
41. E. Vanden-Eijnden, and R. Balescu, *Phys. Plasmas* **3**, 874 (1996).
42. R. Balescu, *Phys. Rev. E* **51**, 4807 (1994).
43. S. Corrsin, *in: Atmospheric Diffusion and Air Pollution, edited by F.N. Frenkiel and P.A. Sheppard, p. 161*, Academic Press, New York, 1990.
44. E. Vanden-Eijnden, and R. Balescu, *J. Plasma Phys.* **54**, 185 (1995).
45. H. D. Wang, E. Vanden-Eijnden, F. Spineanu, J. H. Misguich, and R. Balescu, *Phys. Rev. E* **51**, 4844 (1995).
46. M. Vlad, F. Spineanu, J. H. Misguich, J.-D. Reuss, R. Balescu, K. Itoh, and S.-I. Itoh, *Plasma Phys. Control. Fusion* **46**, 1051 (2004).
47. M. Vlad, F. Spineanu, and J. H. Misguich, *Phys. Rev. E* **53**, 5302 (1996).
48. M. Vlad, F. Spineanu, J. H. Misguich, and R. Balescu, *Phys. Rev. E* **54**, 791 (1996).
49. M. B. Isichenko, *Plasma Phys. Control. Fusion* **33**, 795 (1991).
50. M. B. Isichenko, *Plasma Phys. Control. Fusion* **33**, 809 (1991).
51. M. B. Isichenko, *Rev. Mod. Phys.* **64**, 961 (1992).
52. J.-D. Reuss, and J. H. Misguich, *Phys. Rev. E* **54**, 1857 (1996).
53. M. Vlad, F. Spineanu, J. H. Misguich, and R. Balescu, *Phys. Rev. E* **58**, 7359 (1998).
54. M. Vlad, F. Spineanu, J. H. Misguich, and R. Balescu, *Phys. Rev. E* **61**, 3023 (2004).
55. M. Vlad, F. Spineanu, J. H. Misguich, and R. Balescu, *Phys. Rev. E* **63**, 066304–1 (2001).
56. M. Vlad, F. Spineanu, J. H. Misguich, and R. Balescu, *Nucl. Fusion* **42**, 157 (2002).
57. M. Vlad, F. Spineanu, J. H. Misguich, and R. Balescu, *Phys. Rev. E* **67**, 026406 (2003).
58. R. Balescu, M. Vlad, F. Spineanu, and J. Misguich, *Int. J. Quantum Chem.* **98**, 125–130 (2004).
59. M. Vlad, and F. Spineanu, *Phys. Rev. E* **70**, 056304 (2004).
60. O. G. Bakunin, *Rep. Prog. Phys.* **67**, 1 (2004).
61. O. G. Bakunin, *Chaos, Solitons and Fractals* **23**, 1703 (2005).
62. M. Vlad, and F. Spineanu, *Plasma Phys. Control. Fusion* **47**, 281–294 (2005).
63. M. Neuer, and K. H. Spatschek, *Phys. Rev. E* **73**, 026404 (2006).
64. M. Neuer, and K. H. Spatschek, *Phys. Rev. E* **74**, 036401 (2006).
65. K. H. Spatschek, *Theoretische Plasmaphysik*, Teubner, Heidelberg, 1990.
66. A. N. Kolmogorov, *Doklady Akademii Nauk SSSR* **30**, 301 (1941).
67. T. H. Dupree, *Phys. Fluids* **9**, 1773 (1966).
68. T. H. Dupree, *Phys. Fluids* **10**, 1049 (1967).
69. J. Weinstock, *Physics of Fluids* **12**, 1045 (1969).
70. J. Weinstock, *Physics of Fluids* **13**, 2308 (1970).
71. A. Hasegawa, *Advances in Physics* **34**, 1–42 (1985).
72. A. H. Boozer, *Rev. Mod. Phys.* **76**, 1071 (2004).
73. R. B. White, *The theory of toroidally confined plasmas*, Imperial College, London, 2001.
74. S. S. Abdullaev, K. H. Finken, and K. H. Spatschek, *Phys. Plasmas* **6**, 153–174 (1999).
75. S. S. Abdullaev, *J. Phys. A: Math. & Gen.* **32**, 2745–2766 (1999).
76. S. S. Abdullaev, *J. Phys. A: Math. & Gen.* **35**, 2811–2832 (2002).
77. S. S. Abdullaev, *Nuclear Fusion* **44**, 12–27 (2004).
78. A. Wingen, S. Abdullaev, K. H. Finken, M. Jakubowski, and K. H. Spatschek, *Nucl. Fusion* **46**, 941–952 (2006).
79. R. C. Davidson, *Methods in Nonlinear Plasma Theory*, Academic, New York, 1972.
80. M. Jakubowski, S. Abdullaev, K. Finken, and the TEXTOR team, *Nucl. Fusion* **44** (2004).
81. M. Jakubowski, S. Abdullaev, K. Finken, M. Lehnen, and the TEXTOR team, *J. Nucl. Mater.* **337-339**, 176 (2005).
82. E. da Silva, I. Caldas, R. Viana, and M. Sanjuán, *Phys. Plasmas* **9**, 4917 (2002).
83. S. S. Abdullaev, T. Eich, and K. H. Finken, *Phys. Plasmas* **8**, 2739 (2001).

84. E. Nusse, and J. A. Yorke, *Dynamics: Numerical explorations*, Springer, New York, 1998.
85. A. Wingen, K. H. Spatschek, and S. Abdullaev, *Contrib. Plasma Phys.* **45**, 500 (2005).
86. T. E. Evans, R. K. Roeder, J. A. Carter, and B. I. Rapoport, *Contrib. Plasma Phys.* **44**, 235 (2004).
87. T. E. Evans, R. K. Roeder, J. A. Carter, B. I. Rapoport, M. E. Fenstermacher, and C. J. Lasnier, *J. Phys.: Conf. Ser.* **7**, 174 (2005).
88. G. I. Taylor, *Proc. London Math. Soc.* **20**, 196 (1922).
89. M. S. Green, *J. Chem. Phys.* **19**, 1036 (1951).
90. R. Kubo, *J. Phys. Soc. Jpn.* **12**, 570 (1957).
91. P. Goldeich, and S. Sridhar, *The Astrophysical Journal* **438**, 763 (1995).
92. D. J. Campbell, *Phys. Plasmas* **8**, 2041 (2001).
93. D. J. Campbell, *ITER School 2007, these proceedings* (2007).

Kinetic simulations of turbulent fusion plasmas

Y. Idomura*, T.-H. Watanabe† and H. Sugama†

*Japan Atomic Energy Agency, Higashi-Ueno 6-9-3, Taitou, Tokyo 110-0015, Japan
†National Institute for Fusion Science / The Graduate University for Advanced Studies, Toki, Gifu 509-5292, Japan

Abstract. In this paper, advances in kinetic simulations of turbulent fusion plasmas are reviewed by focusing on physical and numerical models used in five dimensional (5D) gyrokinetic simulations. Theoretical frameworks of physical models from the 6D Vlasov equation to the 5D gyrokinetic equation are presented. Properties of numerical approaches, namely, particle and mesh approaches are described, and roles of numerical dissipation in collisionless turbulence simulations are discussed.

Keywords: Gyrokinetic theory, Eulerian simulation, Particle-In-Cell simulation

1. INTRODUCTION

Turbulence is a common phenomenon observed in most fluids, and is characterised by its chaotic behaviour and turbulent mixing leading to much larger transport than the molecular diffusion. Such turbulent transport is often observed also in a fusion plasma. Although a theoretical model for the transport due to Coulomb collisions in a stable, quiescent toroidal plasma was successfully established with taking into account particle orbits in a toroidal magnetic configuration [1, 2], the transport levels observed in the experiment normally exceeds predictions by this collisional transport model, and a dominating cause is considered as plasma turbulence driven by micro-instabilities, which are characterised by microscopic scale lengths typically given by the Larmor radius. Since the confinement performance is an important factor in determining the size and cost of a fusion reactor, understanding of plasma turbulence is one of critical issues in fusion science [3, 4]. However, because of its basic properties such as large degree of freedom, strong nonlinearity, and sensitivity to the initial condition, a theoretical treatment of turbulence is an extremely difficult issue, and even neutral fluid turbulence is still an formidable physical problem. In addition, plasma turbulence is even more complex phenomenon because of the following features. Firstly, multiple fluids (an electron fluid and other ion fluids) with considerably different mobility are coupled through electromagnetic fields and weak collisional interactions. Secondly, strong confinement fields provide highly anisotropic turbulent structures. Thirdly, inhomogeneities in density, temperature, and magnetic fields excite multiple micro-instabilities over wide spectral ranges. In particular, a toroidal magnetic configuration strongly affects the linear properties of micro-instabilities as well as nonlinearly evolved turbulent structures. Finally, because of weak collisionality in a high temperature core plasma, which is often called as collisionless, a particle distribution function is deviated from a local thermodynamic equilibrium. In addition, kinetic effects such as wave-particle resonant interactions and particle orbit effects play essential roles in collisionless plasma turbulence. Because of this collisionless character, conventional fluid models such as a magnetohydrodynamics (MHD) model becomes insufficient, and kinetic descriptions of a plasma in phase space are needed. As in a research of neutral fluid turbulence, theoretical treatments of plasma turbulence have been actively studied based on dimensional analysis techniques and renormalisation approaches using statistical theories, and their progress were discussed in the literature [5, 6]. However, because of the complexity of the phenomena, they have not matured to cover a broad parameter range in the experiment. On the other hand, a direct numerical simulation (DNS) is becoming important not only as a complementary approach to obtain physical understanding but also as an essential tool for predicting turbulent spectrum and transport in the experiment [7]. In contrast to neutral fluid turbulence described in three dimensional (3D) configuration space, a primitive kinetic model of plasma turbulence is given by using a particle distribution function in 6D phase space. Because of this huge requirement on computational resources, DNS of experimentally relevant plasma turbulence was prohibitive before developments of advanced simulation models and modern supercomputers. In particular, the gyrokinetic model simplified the problem from 6D to 5D and reduced a computational cost by making micro-turbulence simulations free from irrelevant high-frequency phenomena associated with particles' gyro-motion. In addition to the physical model, numerical methods have also been improved to simulate plasma turbulence efficiently using advanced particle and mesh approaches with

CP1013, *Turbulent Transport in Fusion Plasmas, First ITER International Summer School*
edited by S. Benkadda

field aligned coordinates. The rapid progress of computational power extremely enhanced capabilities of gyrokinetic simulations, and DNS of tokamak micro-turbulence has been established as an essential tool for studying tokamak turbulent transport. Recent gyrokinetic simulations have provided qualitative understanding of rich physics behind tokamak turbulent transport as well as quantitative estimations for key engineering issues such as the plasma size scaling of turbulent transport.

In this lecture, we devote much space to overviews of physical and numerical models, aiming at providing kind of recipes to develop gyrokinetic simulations for students who are going to work on the tokamak micro-turbulence based on this approach. The physical findings from recent gyrokinetic simulations are presented in the other lectures in this textbook, and comprehensive review of gyrokinetic simulations is also given in the literature [4, 7, 8, 9, 10, 11]. The manuscript is organised as follows. Section 2 presents theoretical frameworks of physical models from the 6D Vlasov equation to the 5D gyrokinetic equation. Section 3 discusses properties of numerical approaches used in recent gyrokinetic simulations. Section 4 describes issues of the numerical dissipation in collisionless gyrokinetic simulations. Finally, issues on the verification and validation are discussed in Sec. 5.

2. GYROKINETIC MODEL

2.1. KINETIC MODEL

In principle, a plasma can be completely described by the Newton-Maxwell system or the Klimontovich-Maxwell system. However, a fusion plasma consists typically of $\sim 10^{20}\mathrm{m}^{-3}$ of ions and electrons, and it is unrealistic to trace all of them even with present-day and foreseeable future computers. Therefore, instead of solving all the particle motion, a statistical approach is introduced to describe a plasma by a particle distribution function. In a high temperature fusion plasma with $\sim$ 10keV, the kinetic energy is much larger than the average potential energy between particles, which means that particles are weakly coupled with each other. In such a weakly coupled plasma, multiple particle correlations more than three particle interaction are neglected, and two particle interaction is reduced to a collision operator $C(f_s)$ for a single particle distribution function $f_s(\mathbf{q},\mathbf{p})$ in six dimensional (6D) phase space $\mathbf{Z}_{CC}=(\mathbf{q},\mathbf{p})$ [12], where $\mathbf{q}$ and $\mathbf{p}$ are the position and momentum of a particle, respectively. As a result, an evolution equation of f_s is given by the Boltzmann equation,

$$\frac{Df_s}{Dt} \equiv \frac{\partial f_s}{\partial t} + \{f_s, H_s\} = C(f_s), \tag{1}$$

where $\{\cdot,\cdot\}$ is the Poisson bracket in the canonical coordinates $(\mathbf{q},\mathbf{p})$,

$$\{F,G\} = \frac{\partial F}{\partial q_i}\frac{\partial G}{\partial p_i} - \frac{\partial F}{\partial p_i}\frac{\partial G}{\partial q_i}, \tag{2}$$

and $H_s(\mathbf{q},\mathbf{p})$ is the Hamiltonian of collisionless single particle motion,

$$H_s(\mathbf{q},\mathbf{p}) = \frac{1}{2m_s}\left|\mathbf{p} - \frac{e_s}{c}\mathbf{A}\right|^2 + e_s\phi. \tag{3}$$

Here, e_s and m_s are the charge and mass of the particle species s, c is the velocity of light, ϕ is the electrostatic potential, and $\mathbf{A}$ is the vector potential for the magnetic field $\mathbf{B}=\nabla\times\mathbf{A}$. In considering micro-turbulence, a collisionless model is often used, because the collision frequency, ν_{ss}, is much lower than characteristic frequencies of turbulent fluctuations such as drift waves ω_s^* and kinetic Alfvèn waves v_A/a (see Fig.1). In the collisionless limit with $C(f_s)=0$, Eq. (1) yields the Vlasov equation or the collisionless Boltzmann equation,

$$\frac{Df_s}{Dt} \equiv \frac{\partial f_s}{\partial t} + \{f_s, H_s\} = 0. \tag{4}$$

By taking the velocity moments of f_s, the particle density n_s and the current density $\mathbf{j}_s$ in the configuration space $\mathbf{q}$ are obtained as

$$n_s = \int f_s d^3p, \tag{5}$$

$$\mathbf{j}_s = e_s\int \mathbf{v} f_s d^3p, \tag{6}$$

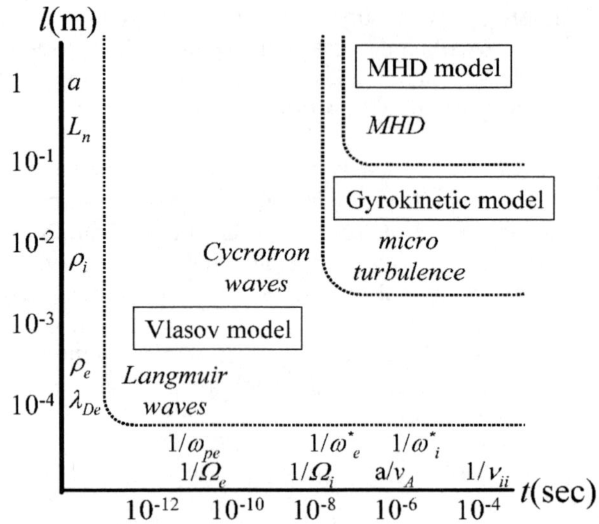

FIGURE 1. Enormous ranges of spatio-temporal scales in a fusion plasma and applicability of Vlasov, gyrokinetic, and MHD models. Here, ω_{ps} is the plasma oscillation frequency, Ω_s is the cyclotron frequency, ω_s^* is the diamagnetic rotation frequency, v_A is the Alfvèn velocity, v_{ii} is the ion-ion collision frequency, λ_{Ds} is the Debye length, ρ_s is the Larmor radius, L_n is the scale length of the equilibrium density profile, a is the plasma size, and s denotes the particle species.

where $\mathbf{v} = [\mathbf{p} - (e_s/c)\mathbf{A}]/m_s$. The electromagnetic fields, $\mathbf{E}$ and $\mathbf{B}$, are determined by substituting n_s and $\mathbf{j}_s$ to the Maxwell equations,

$$\nabla \times \mathbf{E} = -\frac{1}{c}\frac{\partial \mathbf{B}}{\partial t}, \tag{7}$$

$$\nabla \times \mathbf{B} = \frac{4\pi}{c}\sum_s \mathbf{j}_s + \frac{1}{c}\frac{\partial \mathbf{E}}{\partial t}, \tag{8}$$

$$\nabla \cdot \mathbf{E} = 4\pi \sum_s e_s n_s, \tag{9}$$

$$\nabla \cdot \mathbf{B} = 0. \tag{10}$$

The Vlasov-Maxwell system, Eqs.(4)-(10), gives the most fundamental description of a high temperature collisionless plasma.

2.2. GYROKINETIC MODEL

Although the Vlasov-Maxwell system is a reduced kinetic description compared to the Newton-Maxwell system, it involves enormous ranges of spatio-temporal scales as shown in Fig.1. Therefore, it is still prohibitive to simulate low frequency phenomena such as micro-turbulence and MHD waves via the Vlasov-Maxwell system. To avoid a direct treatment of multiple hierarchy spatio-temporal scales, a gyrokinetic model has been developed by eliminating high-frequency phenomena with $\omega > \Omega_s$ with keeping essential kinetic effects, where $\Omega_s = (e_s B_0)/(m_s c)$ is the cyclotron frequency. While the earlier gyrokinetic equation was formulated by gyro-averaging the Vlasov equation based on the recursive method [13], the modern gyrokinetic theory [14, 15, 16, 17, 18, 19, 20, 21] reviewed here has been developed based on the Hamiltonian or Lagrangian formalism with the Lie perturbation theory [22, 23]. The latter approach enables rigorous treatments of the symmetry and the conservation properties inherent to the first principles, which are essential to describe underlying physics and useful for a nonlinear simulation.

From experimental observations, tokamak micro-turbulence is considered to obey the gyrokinetic ordering in a smallness parameter ε_g,

$$\frac{\omega}{\Omega_s} \sim \frac{k_\parallel}{k_\perp} \sim \frac{e_s\phi}{T_s} \sim \frac{B_1}{B_0} \sim \frac{\rho_s}{L} \sim \mathcal{O}(\varepsilon_g), \tag{11}$$

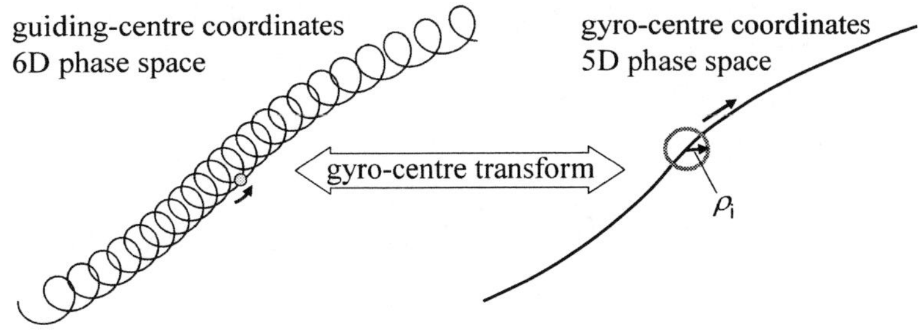

FIGURE 2. By applying the gyro-centre transform, fast gyro-motion is eliminated, and the problem is reduced from 6D to 5D with keeping kinetic effects such as the finite Larmor radius effect.

where ω is a characteristic frequency of micro-turbulence, $\mathbf{B}_0 = \nabla \times \mathbf{A}_0$ is the equilibrium field, $\mathbf{B}_1 = \nabla \times \mathbf{A}_1$ is the perturbed field, $\mathbf{b} = \mathbf{B}_0/B_0$ is the unit vector in the direction of $\mathbf{B}_0$, $k_\parallel = \mathbf{k}\cdot\mathbf{b}$ and $k_\perp = |\mathbf{k}\times\mathbf{b}|$ are parallel and perpendicular components of the wave vector $\mathbf{k}$, T_s is the temperature, $\rho_s = v_\perp/\Omega_s$ is the Larmor radius, L is a characteristic scale length of equilibrium profiles such as the density, $L_n = |\nabla \ln n_0|^{-1}$, the temperature $L_{ts} = |\nabla \ln T_{0s}|^{-1}$, and the magnetic field, $L_B = |\nabla \ln B_0|^{-1}$. As shown in Fig.2, the single particle motion under strong ambient fields consists of fast periodic gyro-motion and slow guiding-centre motion. Low frequency perturbations satisfying this ordering mainly affect the latter motion, and the magnetic moment, $\mu = m_s v_\perp^2/2B_0$, becomes an approximate adiabatic invariant. In analysing such a particle orbit, it is convenient to use the guiding-centre coordinates $\mathbf{Z}_{GC} = (t;\mathbf{R},u,\mu,\alpha)$, where $\mathbf{R}$ is the guiding-centre position, α is the gyro-phase angle, $u = v_\parallel + (e_s/cm_s)A_\parallel$ is the generalised parallel velocity, $A_\parallel = \mathbf{b}\cdot\mathbf{A}_1$, and $v_\parallel = \mathbf{v}\cdot\mathbf{b}$ and $v_\perp = |\mathbf{v}\times\mathbf{b}|$ are parallel and perpendicular components of the velocity, respectively. For a fusion plasma with finite but small β (typically $\beta \sim 1\%$), the perturbed field is given as $\mathbf{B}_1 = \nabla \times A_\parallel \mathbf{b}$, and the parallel component of $\mathbf{B}_1$ is often neglected compared to $\mathbf{B}_0$, where $\beta = (n_0 T_i + n_0 T_e)/(B_0^2/8\pi)$ is the ratio of the plasma kinetic pressure to the magnetic pressure. A transform of the Hamiltonian from $\mathbf{Z}_{CC}$ to $\mathbf{Z}_{GC}$ is systematically given by the guiding-centre transform [23], and we have

$$H_s = \frac{1}{2}m_s u^2 + \mu B_0 + e_s \Psi, \tag{12}$$

where $\Psi = \phi - uA_\parallel/c$ is the generalised potential.

We then separate fast periodic gyro-motion and slow guiding-centre motion by applying the secular perturbation theory to the perturbed Hamiltonian (12), and find new coordinates, the gyro-centre coordinates $\mathbf{Z}_{GY} = (t;\bar{\mathbf{R}},\bar{u},\bar{\mu},\bar{\alpha})$, where an approximate invariant μ becomes an exact invariant $\bar{\mu}$ and its conjugate variable $\bar{\alpha}$ is seen as an ignorable coordinate. The gyro-centre transform valid up to $\mathscr{O}(\varepsilon_g)$ is given as

$$\mathbf{Z}_{GY} = \mathbf{Z}_{GC} + \{\tilde{S},\mathbf{Z}_{GC}\} + \mathscr{O}(\varepsilon_g^2), \tag{13}$$

where $\{\cdot,\cdot\}$ is the Poisson bracket in the gyro-centre coordinates [14, 19],

$$\{F,G\} \equiv \frac{\Omega_s}{B}\left(\frac{\partial F}{\partial \alpha}\frac{\partial G}{\partial \mu} - \frac{\partial F}{\partial \mu}\frac{\partial G}{\partial \alpha}\right) + \frac{\mathbf{B}^*}{m_s B_\parallel^*}\cdot\left(\nabla F\frac{\partial G}{\partial u} - \frac{\partial F}{\partial u}\nabla G\right) - \frac{c}{e_s B_\parallel^*}\mathbf{b}\cdot\nabla F \times \nabla G, \tag{14}$$

and $B_\parallel^* = \mathbf{b}\cdot\mathbf{B}^*$ is a parallel component of $\mathbf{B}^* = \mathbf{B}_0 + (B_0\bar{u}/\Omega_s)\nabla\times\mathbf{b}$. The generating function $\tilde{S}$ is solved as

$$\tilde{S}(\mathbf{Z}_{GC},t) = \frac{e_s}{\Omega_s}\int^{\alpha}[\Psi - \langle\Psi\rangle_\alpha]\,d\alpha', \tag{15}$$

where the gyro-average is defined as $\langle\cdot\rangle_\alpha \equiv \oint \cdot\, d\alpha/2\pi$. This transform is formulated based on the Lie perturbation theory, so that the transform keeps an area preserving property of the Hamiltonian system up to arbitrary order in ε_g and the Poisson bracket becomes form invariant. Interested readers should refer to the aforementioned literature for further details. After the transform, α-dependent non-secular perturbations are absorbed by the generating function $\tilde{S}$, and the Hamiltonian $\bar{H}_s$ becomes independent of $\bar{\alpha}$,

$$\bar{H}_s = \frac{1}{2}m_s\bar{u}^2 + \bar{\mu}B_0 + e_s\langle\Psi\rangle_{\bar{\alpha}}. \tag{16}$$

We then have a reduced kinetic equation or a gyrokinetic equation, which describes an evolution of a guiding-centre distribution function $\bar{f}_s$ in 5D phase space,

$$
\begin{aligned}
\frac{D\bar{f}_s}{Dt} &\equiv \frac{\partial \bar{f}_s}{\partial t} + \{\bar{f}_s, \bar{H}_s\} \\
&= \frac{\partial \bar{f}_s}{\partial t} + \{\bar{\mathbf{R}}, \bar{H}\} \cdot \frac{\partial \bar{f}_s}{\partial \bar{\mathbf{R}}} + \{\bar{u}, \bar{H}\} \frac{\partial \bar{f}_s}{\partial \bar{u}} \\
&= 0, \qquad (17)
\end{aligned}
$$

where the nonlinear characteristics, $\dot{\mathbf{Z}}_{GY} = \{\mathbf{Z}_{GY}, \bar{H}\}$, are given as

$$\frac{d\bar{\mathbf{R}}}{dt} = \bar{u}\mathbf{b} + \frac{c}{e_s B_{\parallel}^*}\mathbf{b} \times \left(e_s \nabla \langle \Psi \rangle_{\alpha} + m_s \bar{u}^2 \mathbf{b} \cdot \nabla \mathbf{b} + \bar{\mu} \nabla B_0\right) \qquad (18)$$

$$\frac{d\bar{u}}{dt} = -\frac{\mathbf{B}^*}{m_s B_{\parallel}^*} \cdot \left(e_s \nabla \langle \Psi \rangle_{\bar{\alpha}} + \bar{\mu} \nabla B_0\right) \qquad (19)$$

$$\frac{d\bar{\mu}}{dt} = 0 \qquad (20)$$

The guiding-centre orbit (18) is determined by parallel thermal motion and perpendicular drift motion associated with the electromagnetic perturbation and the curvature and gradient of the magnetic field. In Eq. (19), the generalised parallel velocity suffers from acceleration forces associated with the electromagnetic perturbation and the gradient of the magnetic field. The latter force works as the magnetic mirror in the presence of two constants of motion in the unperturbed orbit, the energy $\varepsilon = m_s \bar{u}^2/2 + \bar{\mu} B_0$ and the magnetic moment $\bar{\mu}$. In Fig.3, the unperturbed orbits show complicated trajectories depending on the pitch angle $\xi = \sin^{-1}(v_{\perp}/v)$ at the outboard mid-plane, and are classified mainly into passing and trapped particles. While passing particles are characterised by the parallel motion and the perpendicular magnetic drift, trapped particles show the bounce motion in the weak field side, and their magnetic drift appears as the toroidal precession motion. On the poloidal plane, they show closed orbits, which corresponds to constant surfaces of the canonical toroidal angular momentum, $p_{\varphi} = -(e_s/c)\psi + m_s v_{\parallel}(\mathbf{B}_0 \cdot \nabla \varphi / |\nabla \varphi|^2)/B_0$, which is another constant of motion in the axisymmetric tokamak configuration. Therefore, the equilibrium solution of the gyrokinetic equation (17) is given as a function of three constants of motion, $f_{0s}(p_{\varphi}, \varepsilon, \bar{\mu})$. The parallel and perpendicular motion in the guiding-centre orbit is closely related to micro-instabilities, which are excited by density gradient, ion temperature gradient (ITG), and electron temperature gradient (ETG) through various wave-particle resonant interactions. Micro-instabilities are classified into slab modes, toroidal modes, and trapped particle modes, which are excited by the parallel motion of passing particles, the magnetic drift of passing particles, and the toroidal precession of trapped particles, respectively. Detailed classification of micro-instabilities can be found in the literature [24, 25, 26, 27]. On the other hand, these wave-particle interactions also works as a damping mechanism for smaller scale perturbations. Especially, high-$k_{\parallel}$ components suffer from the Landau damping due to the parallel motion of passing particles, which restricts turbulent structures to so-called flute-like structures with $k_{\parallel} \sim 0$.

Finally, the equation system is closed by imposing the self-consistency. f_s and $\bar{f}_s$ are related by the following pull-back transform valid up to $\mathscr{O}(\varepsilon_g)$,

$$f_s = \bar{f}_s + \{\tilde{S}, \bar{f}_s\} + \mathscr{O}(\varepsilon_g^2). \qquad (21)$$

By substituting f_s to Eqs. (8) and (9), the gyrokinetic Poisson-Ampère law is obtained as

$$-\nabla^2 \phi = \sum_s \left[4\pi e_s \int \delta \bar{f}_s \delta([\bar{\mathbf{R}} + \bar{\rho}_s] - \mathbf{x}) \mathscr{J}_s d^6 \bar{Z} - \frac{1}{\lambda_{Ds}^2}(\phi - \langle \bar{\phi} \rangle_{\alpha})\right], \qquad (22)$$

$$-\nabla_{\perp}^2 A_{\parallel} = \sum_s \left[\frac{4\pi e_s}{c} \int \bar{u} \delta \bar{f}_s \delta([\bar{\mathbf{R}} + \bar{\rho}_s] - \mathbf{x}) \mathscr{J}_s d^6 \bar{Z} - \frac{\omega_{ps}^2}{c^2} \langle \bar{A}_{\parallel} \rangle_{\alpha}\right], \qquad (23)$$

where λ_{Ds} is the Debye length, ω_{ps} is the plasma oscillation frequency, $\mathscr{J}_s = m^2 B_{\parallel}^*$ is the Jacobian of the gyro-centre coordinates, and $\langle \bar{\cdot} \rangle_{\alpha} = \int \langle \cdot \rangle_{\alpha} f_{0s} \delta([\bar{\mathbf{R}} + \bar{\rho}_s] - \mathbf{x}) \mathscr{J}_s d^6 \bar{Z} / n_0$. The displacement current in Eq.(8) is neglected for low frequency phenomena considered here. The second term in the r.h.s. of Eqs. (22) and (23), which comes from $\{\tilde{S}, \bar{f}_s\}$,

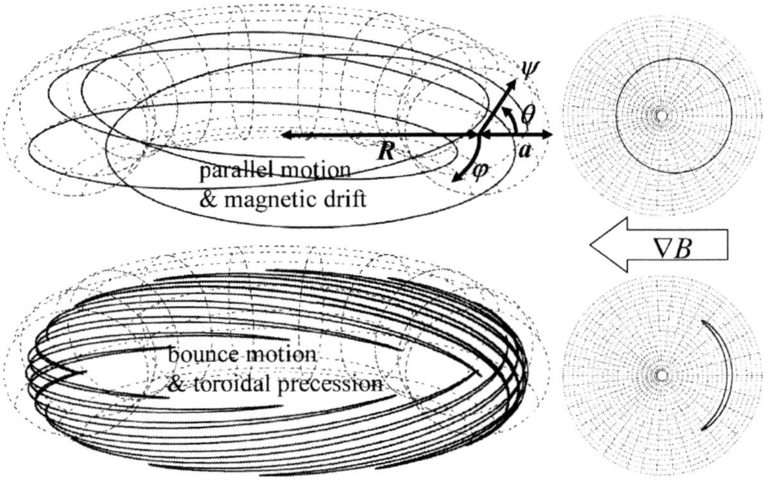

FIGURE 3. Typical unperturbed guiding-centre orbits of passing and trapped particles, and their projection on the poloidal cross section. The coordinates (ψ, θ, φ) show the radial direction, the poloidal angle, and the toroidal angle.

show gyrokinetic polarisation and magnetisation due to the finite Larmor radius effect. In deriving Eqs. (22) and (23), nonlinear polarisation and magnetisation effects, which are higher order in ε_g, are neglected. The gyrokinetic Vlasov-Maxwell system, Eqs. (17), (22), and (23), is a standard kinetic model to describe the micro-turbulence.

The most important feature of the gyrokinetic Vlasov-Maxwell system is that high frequency phenomena such as the Langmuir wave, ω_{ps}, and the cyclotron wave, Ω_s, are eliminated with keeping essential kinetic effects for low frequency micro-turbulence. From a point of view of the simulation, the limitations on the time step and the grid spacing are relaxed from $\Delta t\omega_{ps} < 1$ and $\Delta r < \lambda_{Ds}$ to $\Delta t\omega_s^* < 1$ and $\Delta r < \rho_s$, and the computational cost is significantly reduced. Another important feature is that the equation system keeps the conservation properties derived from the first principles or the original Vlasov-Maxwell system. Since the transform is designed to preserve the Hamiltonian nature of the system, the nonlinear characteristics, Eqs. (18), (19), and (20), given by Hamilton's equation satisfy the phase space conservation,

$$\nabla\cdot\left(\mathscr{J}_s\frac{d\bar{\mathbf{R}}}{dt}\right)+\frac{\partial}{\partial\bar{u}}\left(\mathscr{J}_s\frac{d\bar{u}}{dt}\right)=0. \tag{24}$$

By using the phase space conservation (24), the gyrokinetic equation (17) yields its conservative form,

$$\frac{\partial\mathscr{J}_s\bar{f}_s}{\partial t}+\nabla\cdot\left(\mathscr{J}_s\frac{d\bar{\mathbf{R}}}{dt}\bar{f}_s\right)+\frac{\partial}{\partial\bar{u}}\left(\mathscr{J}_s\frac{d\bar{u}}{dt}\bar{f}_s\right)=0. \tag{25}$$

In Eqs. (24) and (25), the gyrokinetic equation can be seen as the continuity equation of $\bar{f}_s$ transported by incompressible Hamiltonian flows in phase space. On the other hand, the gyrokinetic equation (19) is simply given by the Liouville equation, which conserves arbitrary functions of $\bar{f}_s$, $C(f)$, along nonlinear characteristics,

$$\frac{DC(\bar{f}_s)}{Dt}\equiv\frac{\partial C(\bar{f}_s)}{\partial t}+\{C(\bar{f}_s),\bar{H}_s\}=0. \tag{26}$$

Therefore, in addition to the particle number or the $L1$ norm, $N=\int f\mathscr{J}_s d^6Z$, the system has an infinite number of conserved quantities such as the $L2$ norm, $M=\int f^2\mathscr{J}_s d^6Z$, and the kinetic entropy, $S=-\int f\ln f\mathscr{J}_s d^6Z$. Another important conservation property is the total energy conservation derived from Eqs. (17), (22), and (23),

$$\frac{d}{dt}\left(E_k+E_\phi+E_{A_\parallel}\right)=0, \tag{27}$$

$$E_k=\sum_s\left[\int\left(\frac{1}{2}m_s\bar{u}^2+\bar{\mu}B_0\right)\bar{f}_s\mathscr{J}_s d^6\bar{Z}\right], \tag{28}$$

$$E_\phi = \frac{1}{8\pi}\int |\nabla\phi|^2 d^3x + \frac{1}{8\pi}\sum_s\sum_{\mathbf{k}} \frac{1}{\lambda_{Ds}^2}\left[1-\Gamma_0(k_\perp^2\rho_s^2)\right]|\phi_{\mathbf{k}}|^2, \quad (29)$$

$$E_{A_\parallel} = \frac{1}{8\pi}\int |\nabla_\perp A_\parallel|^2 d^3x + \frac{1}{8\pi}\sum_s\sum_{\mathbf{k}} \frac{\omega_{ps}^2}{c^2}\Gamma_0(k_\perp^2\rho_s^2)|A_{\parallel\mathbf{k}}|^2, \quad (30)$$

where $\phi = \sum_{\mathbf{k}}\phi_{\mathbf{k}}\exp(i\mathbf{k}\cdot\mathbf{x})$, $A_\parallel = \sum_{\mathbf{k}} A_{\parallel\mathbf{k}}\exp(i\mathbf{k}\cdot\mathbf{x})$, $\Gamma_0(b) = I_0(b)e^{-b}$, and I_0 is the zeroth order modified Bessel function. The conservation properties are important especially in studying nonlinear problems, where an analytic solution can hardly be found. The total particle and energy conservations are normally checked as a stringent test of the quality of the nonlinear simulation. In addition, the conservation properties are useful in developing robust and accurate nonlinear simulations, which often blow up without such physical constraints.

2.3. KINETIC FLUID MODEL

A fluid approach is the most commonly used approximation in describing turbulent fluids. A set of fluid equations is derived by taking the corresponding velocity moments of the Boltzmann equation. Since the electromagnetic perturbations are determined by the particle density and the current density, which are the zeroth and first order moments, a fluid model provides simple and useful means to analyse fluid phenomena, provided that a plasma is collisional and is close to a local thermodynamic equilibrium. The two component (ion and electron) fluid model, which is valid for a collisional plasma, was derived in Ref. [28]. Under the quasi-neutrality condition, $n_i = n_e$, and the limit of $\omega/\Omega_s \sim k_\perp\rho_s \sim \rho_s/L_n \sim (c/\omega_{ps})/L_n \ll 1$, this model is reduced to a well-known single fluid MHD model, which is commonly used also in space plasma physics. However, in a high temperature collisionless fusion plasma, the collision frequency is much smaller than that of phenomena of interest (see Fig.1), and a local thermodynamic equilibrium can not be assumed. In addition, the mean free path of particles $\sim$ 10km (particles move along the magnetic fields and rotate many times in the torus before suffering from the collisional effect) is much larger than the plasma size $\sim$ 1m, and the orbit effects due to the finite Larmor radius (see Fig.2) and trapped particles (see Fig.3) become important in analysing the micro-turbulence.

In general, the n-th order moment equation involves the $n+1$-th order moment term which comes from $\mathbf{v}\cdot\nabla f_s$ in the kinetic equation. A problem to close a set of fluid equations by approximating the $n+1$-th order moment term using lower order moments is called as a closure problem. In the MHD model, the adiabatic law is often used, or the heat flux in the energy equation is modelled by the thermal conductivity term. However, such models may break down in the collisionless limit, because complicated velocity space structures in a particle distribution function are not fully characterised by lower order moments. Therefore, conventional fluid models are not enough to describe the micro-turbulence.

In order to resolve this problem, a kinetic fluid model such as the gyro-Landau fluid model has been developed. The gyro-Landau fluid model consists of moment equations of the gyrokinetic equation with dissipative [29] or non-dissipative [30, 31, 32] kinetic closure models which takes into account the linear wave-particle resonant interactions or the linear Landau damping. The model has been extended including the magnetic drift motion [33], the finite Larmor radius effect [34], bounce-averaged trapped particles [35], and the closure model with optimisation techniques to reproduce the linear wave dispersion of micro-instabilities [35]. The advantage of the kinetic fluid model is its lower computational cost than the gyrokinetic model. With the aid of this advantage, transport models were constructed based on database of kinetic fluid simulations over a broad parameter range [36, 37]. However, a code benchmark project revealed that as for ion scale micro-turbulence driven by ITG (ITG turbulence), the gyro-Landau fluid model gave $2\sim3$ times higher transport levels than the gyrokinetic model [8]. This was partly explained by an insufficient modelling concerning to the trapped particle response to nonlinearly generated $E\times B$ zonal flows in the gyro-Landau fluid model [38], and efforts to fix this problem are ongoing [39, 40]. Although the kinetic fluid model is useful to study qualitative physics, further improvements of the model are needed for quantitative estimation of the turbulent transport.

3. NUMERICAL MODEL

3.1. PARTICLE APPROACHES

The Lagrangian approach based on a Particle-In-Cell (PIC) method [41, 42, 43] has been a major approach in simulating collective phenomena in a collisionless plasma. In the PIC method, a plasma is expressed by relatively small number of super-particles with $e_{SP} = \mathscr{M} e_s, m_{SP} = \mathscr{M} m_s$, and $n_{SP} = n_s/\mathscr{M}$, and their collective motion and the associated electromagnetic fields are treated in a straightforward manner using the Newton-Maxwell system for super-particles, which involves not only smoothly varying perturbations but also multiple particle interactions or collisions. In this sense, the system may be recognised as a collisional plasma described by the Boltzmann-Maxwell system, (1),(5)-(10). In the Boltzmann equation, the collective motion (the l.h.s. of Eq.(1)) is not changed by super-particles, while the collision term, which is characterised by the collision frequency $\nu_{ss} \sim \omega_{ps}/(n_0\lambda_{Ds}^3)$, is artificially enhanced by $\mathscr{M}$ times. It is noted that the Vlasov equation or the collisionless limit is obtained in the opposite limit with $\mathscr{M} \to 0$. In order to reduce the enhanced numerical collision, the PIC method uses a finite size particle model, which works as a low-pass Fourier filter to weaken sub-grid scale thermal fluctuations or a particle noise. This model may be valid for describing collective phenomena produced under an influence of the Debye shielding. Although the PIC method has been successfully applied to various kinetic phenomena in a collisionless plasma, it is essentially a model describing the Vlasov-Maxwell system and its application is limited for high frequency phenomena in a microscopic system. In a gyrokinetic particle model [44], the PIC method was extended for the gyrokinetic Vlasov-Maxwell system by replacing the Newton-Maxwell system with the nonlinear characteristics of the gyrokinetic equation and the gyrokinetic Poisson-Ampère law. In this model, limitations on the time step and the grid spacing are significantly relaxed, and low frequency phenomena can be simulated without a particle noise associated with gyro-motion. In spite of this low noise character, a remaining particle noise and its slow convergence property $\propto 1/\sqrt{n_{SP}}$ were still problematic in simulating low amplitude micro-turbulence (typically $\delta n_s/n_0 \sim e_s\phi/T_s < 0.01$) in a macroscopic system such as a full torus configuration. To overcome this difficulty, the δf method [45, 46] was proposed. In the δf method, a perturbed particle (or guiding-centre) distribution f_s is separated into an equilibrium distribution function f_{0s} and a fluctuating perturbation δf_s, and only δf_s is expressed by particles with the particle weight w_j, where w_j is defined as $w_j \equiv \mathscr{M} \int_{\mathscr{V}_j} \delta f_s d^6 Z$ and $\mathscr{V}_j$ is the phase space volume around the j-th particle. An evolution of δf_s is then solved along the nonlinear characteristics with self-consistent perturbed fields determined by Eqs. (22) and (23). The weighted particle is recognised as a Monte-Carlo sampling of δf_s over phase space, and is often called as a marker particle or a tracer particle. In the δf method, the particle weight is reduced roughly by $\delta f_s/f_{0s} \sim \delta n_s/n_0 \sim 0.01$, and the simulation system becomes closer to the collisionless limit than the conventional PIC method. An evolution equation of δf_s is obtained by substituting $f_s = f_{0s} + \delta f_s$ to the kinetic equation,

$$\frac{D\delta f_s}{Dt} = -\left\{f_{0s}, H_s\right\}. \tag{31}$$

Another simple way to evaluate δf_s is to use a constancy of f_s along the nonlinear characteristics,

$$\delta f_s(\mathbf{Z}(t)) = f_s(\mathbf{Z}(t_0)) - f_{0s}(\mathbf{Z}(t)), \tag{32}$$

where t_0 is an initial time. It was shown that both methods give almost the same results [47]. Since the δf method is a sort of Monte-Carlo integration, its statistical property can be improved by optimising a distribution of sampling points or marker particles [46, 48]. In Fig.6, comparisons of slab ITG turbulence simulations using the conventional (full-f) PIC method and the δf methods with Maxwellian and optimised marker particle distributions are shown. Compared with a full-f PIC simulation, δf PIC simulations converge with drastically smaller particle numbers. In addition, the convergence of a δf PIC simulation is accelerated by optimising a marker particle distribution so that it fits to δf_s at the end of simulation. Although toroidal particle simulations with the conventional PIC method were developed in earlier works [57, 58, 59], the δf method has become a standard method in most of recent gyrokinetic toroidal particle simulations [9, 60, 61, 62, 63, 64, 65].

In Eqs. (31) and (32), the δf method is constructed based on the conservation property of the Liouville equation, which is satisfied in an isolated system without any source and collision terms. Thus, the δf PIC simulation is essentially a transient decaying turbulence simulation, where the isolated system is relaxed from a linearly unstable initial condition to a nonlinearly saturated state through nonlinear evolution of micro-instabilities. Such decaying turbulence simulations of collisionless and sourceless system are still useful for studying transient nonlinear processes

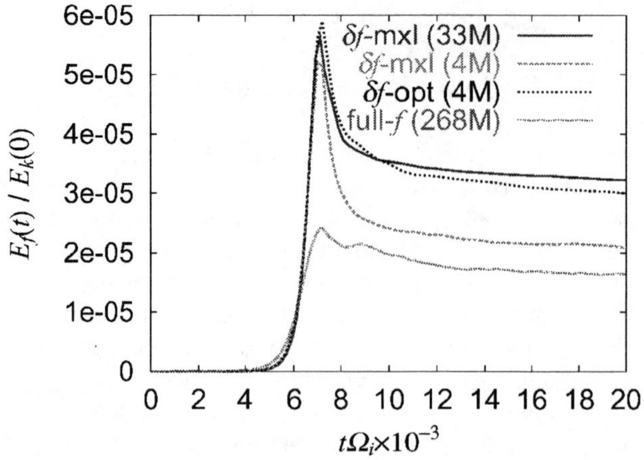

FIGURE 4. The time histories of the field energy observed in slab ITG turbulence simulations using (a) δf method with $\sim$ 33 million Maxwellian marker particles, (b) δf method with $\sim$ 4 million Maxwellian marker particles, (c) δf method with $\sim$ 4 million optimised marker particles, and (d) full-f method with $\sim$ 268 million particles.

coming from the Hamiltonian dynamics. However, as is discussed in Sec.4, from the viewpoint of the entropy balance relation [49, 50, 51], they do not reach at statistically steady states until the heat transport is almost quenched by stabilising effects such as zonal flows and quasi-linear flattening. Therefore, δf PIC simulations are not enough for estimating realistic turbulent transport coefficients for quasi-steady equilibrium profiles. To overcome this difficulty, extensions of the δf method including non-conservative effects such as the collision operator [52, 53, 54] and the heat source [55, 56] are in progress. So far, most of δf PIC simulations have been restricted to an electrostatic simulation with adiabatic electrons. Electromagnetic problems require more delicate treatments because of a particle noise coming from fast electron motion. Improvements of the δf method towards an electromagnetic model are also in progress [65, 66, 67, 68, 69, 70].

3.2. MESH APPROACHES

An alternative approach is a mesh (or Vlasov) approach, in which the particle distribution function defined on discrete grids in phase space is evolved using computational fluid dynamics (CFD) schemes. Mesh approaches are classified mainly into semi-Lagrangian and Eulerian approaches, and historically, a Vlasov simulation was pioneered using the former approach. The splitting method [71], which is one of semi-Lagrangian approaches, was the standard method used in earlier works on the 1D Vlasov-Poisson system. Let us consider the 1D electrostatic Vlasov equation in the canonical coordinates (q,p),

$$\frac{\partial f_s}{\partial t} + \frac{p}{m_s}\frac{\partial f_s}{\partial q} - e_s\frac{\partial \phi}{\partial q}\frac{\partial f_s}{\partial p} = 0, \tag{33}$$

which has the analytic solution given by $f_s(q,p,t) = f_s(q-(p/m_s)\Delta t, p+e_s\partial_q\phi\Delta t, t-\Delta t)$. In the splitting method, the Vlasov equation is evolved at each time step by tracing back this solution along the nonlinear characteristics. Here, one important property of Eq. (33) is that its nonlinear characteristics are Hamilton's equation given by a separable Hamiltonian, $H(q,p) = p^2/(2m_s) + e_s\phi(q) = T(p) + V(q)$. Since $\dot{q} = \partial_p T(p)$ and $\dot{p} = -\partial_q V(q)$ do not depend on q and p, and are uniform motions in the q and p directions, respectively, Eq. (33) may be split as follows,

$$f_s^*(q,p) = f_s^n(q-(p/m_s)\Delta t/2, p), \tag{34}$$
$$f_s^{**}(q,p) = f_s^*(q, p+e_s\partial_q\phi^*\Delta t), \tag{35}$$
$$f_s^{n+1}(q,p) = f_s^{**}(q-(p/m_s)\Delta t/2, p), \tag{36}$$

where the upper subscript shows n-th and intermediate time steps. In the scheme, each stage is given by a solution of a 1D advection equation given by an uniform motion, $\partial_t f + v\partial_x f = 0$. Since a mapping of the solution given by an uniform motion becomes a canonical transform, the splitting method is symplectic integration, which keeps the phase space conservation. The above splitting rule, which is similar to the leap-frog scheme used in the PIC method, has second order accuracy. Since the derivative terms are not explicitly evaluated as in a finite difference method, relatively large time step width, which is not limited by the numerical instability but by the numerical accuracy, is permitted. As an extension of this method for the gyrokinetic or drift-kinetic ($\rho_s \to 0$ limit) equation which is given by a non-separable Hamiltonian, the semi-Lagrangian method was developed [72], and it has been applied to gyrokinetic simulations [73, 74].

On the other hand, in the Eulerian approach, Eq. (33) is solved as a multi-dimensional partial differential equation, which is discretised as

$$\left[\frac{\partial f_s}{\partial t}\right]_{i,j} = -\frac{p_j}{m_s}\left[\frac{\partial f_s}{\partial q}\right]_{i,j} - e_s\left[\frac{\partial \phi}{\partial q}\right]_i \left[\frac{\partial f_s}{\partial p}\right]_{i,j}, \tag{37}$$

where i and j denote the indices of grids in the q and p directions, respectively. In Eq. (37), the derivative terms are estimated using finite difference and spectral methods. Although a spectral method is desirable to satisfy both numerical stability and accuracy, its application is limited for the perpendicular coordinates of a configuration space in flux-tube gyrokinetic simulations, and most of Eulerian gyrokinetic simulations are based on finite difference schemes. In solving the gyrokinetic equation, which is a nonlinear hyperbolic equation, using finite difference schemes, the most critical issue is to avoid numerical instabilities or spurious numerical oscillations produced by aliasing errors. The phase mixing is an universal feature of turbulent flows, and it continuously produce fine scale structures or filaments. The aliasing errors are inevitable in resolving such fine scale structures with finite discrete grids. To avoid numerical instabilities, two types of solutions have been proposed. One is an upwind type finite difference scheme and the other is a non-dissipative finite difference scheme. The former scheme smears out numerical oscillations by numerical dissipation which is inherent to upwind type operators. However, the numerical dissipation often eliminates a physically meaningful solution itself and violates the fist principles. To avoid strong numerical dissipation, less dissipative higher order accurate schemes such as higher order upwind schemes and total variation diminishing (TVD) type schemes are used in recent gyrokinetic simulations [75, 76, 77, 78, 79]. On the other hand, a non-dissipative finite difference scheme [80, 81] is designed so that it exactly conserves two invariants f and f^2, which bound the amplitude of numerical oscillations, and numerical instabilities are suppressed without numerical dissipation in a conservative gyrokinetic simulation [82]. Since the scheme keeps the relevant conservation properties, it is appropriate especially for studying long time behaviours of the micro-turbulence. It is noted that although conservative Vlasov simulations show excellent conservation properties even with poor grid resolution, the physical validity of their solutions should be carefully checked through physical verification tests as in the other numerical approaches.

A Vlasov simulation has the following advantages.

1. This approach is free of a noise problem inherent to a particle simulation.
2. The numerical resolution in phase space is kept constant or controlled in a long time simulation, while a distribution of sampling points or marker particles in the δf PIC method changes in time following nonlinear characteristics.
3. Straightforward treatments of non-conservative effects such as source and collision terms are possible.

Although the Eulerian approach has the above advantages which are essential for a long time simulation of experimentally relevant quasi-steady turbulent plasmas, this approach requires huge computational cost. Compared to a PIC simulation with $100 \sim 1000$ particles per a cell, a Vlasov simulation requires typically $\sim 10^4$ grids to resolve 2D velocity space. Because of this huge requirement on computational resources, a realistic micro-turbulence simulation via the Eulerian approach was difficult before developments of modern massively parallel supercomputers.

In Fig.5, a comparison of slab ITG turbulence simulations using a δf PIC simulation and a conservative Vlasov simulation with a non-dissipative finite difference scheme is shown [82]. In the comparison, two results show quantitative agreements during the initial nonlinear stage where both simulations conserve the total energy. In principle, particle and mesh approaches give the same solution provided that the conservation properties are satisfied. However, in the deep nonlinear stage, a δf PIC simulation shows spurious numerical heating due to enhanced collisions with growing particle weights, while a conservative Vlasov simulation keeps the total energy conservation for a long time. Therefore, conservative Vlasov simulations may be desirable for a long time turbulence simulation, while δf PIC simulations are useful for studying relatively short time scale transient phenomena with lower computational costs.

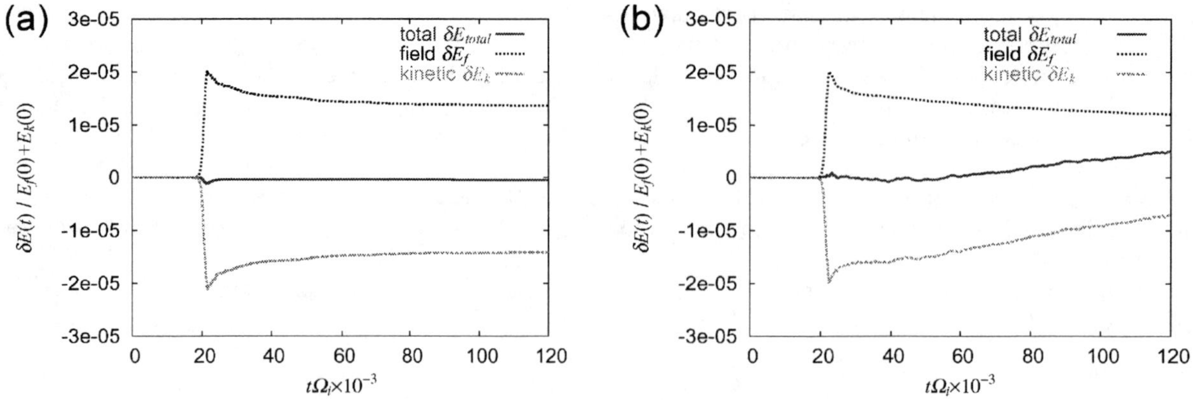

FIGURE 5. The time histories of variations of the field energy, the kinetic energy, and the total energy observed in slab ITG turbulence simulations using (a) conservative Vlasov simulation and (b) δf PIC simulation [82].

3.3. GLOBAL AND FLUX TUBE APPROACHES

A general description of the tokamak configuration consisting of nested magnetic surfaces labelled by the poloidal flux function ψ is written in a form [83],

$$\mathbf{B}_0 = \nabla\psi \times \nabla\left(q(\psi)\theta - \varphi\right), \tag{38}$$

where φ is the toroidal angle and θ is the straight-field line poloidal angle chosen so that the field line pitch, $q(\psi) = \mathbf{B}_0 \cdot \nabla\varphi / \mathbf{B}_0 \cdot \nabla\theta$, becomes constant on the magnetic surface. The flux coordinates are useful to represent flute like perturbations with $k_\parallel \simeq 0$ or $\mathbf{B}_0 \cdot \nabla h \simeq 0$. Since the tokamak configuration is periodic both in θ and φ, a perturbation $h(\psi, \theta, \varphi)$, can be expanded in a Fourier series,

$$h(\psi,\theta,\varphi) = \sum_{m,n} h_{mn}(\psi)\exp(im\theta - in\varphi), \tag{39}$$

where m and n are the poloidal and toroidal mode numbers. A flute like perturbation is then written as

$$\mathbf{B}_0 \cdot \nabla h = \mathbf{B}_0 \cdot \nabla\theta \sum_{m,n} i(m - nq(\psi))h_{mn}(\psi)\exp(im\theta - in\varphi) \simeq 0. \tag{40}$$

Since high $k_\parallel$ components far from the resonance condition, $m \simeq nq(\psi)$, suffer from strong Landau damping, a quasi 2D representation of flute like perturbations is possible,

$$h(\psi,\theta,\varphi) = \sum_{n} h_n(\psi,\theta)\exp(in[q(\psi)\theta - \varphi]), \tag{41}$$

where $h_n(\psi, \theta)$ is slowly varying in θ, reflecting variations of h in the direction of the magnetic field. Most of recent gyrokinetic simulations are efficiently implemented using this quasi 2D character. In addition, the magnetic moment $\bar{\mu}$ appears only parametrically in the gyrokinetic equation (17). Accordingly, in a discretised form, the gyrokinetic Vlasov-Maxwell system may be seen as a set of quasi 2D (ψ and φ) + 1D ($v_\parallel$) partial differential equations with different $\bar{\mu}$, which are coupled through the gyrokinetic Poisson-Ampère law.

The simplest and most realistic approach is a global model [9, 57, 58, 59, 60, 61, 63, 64, 74, 77, 79], in which the gyrokinetic Vlasov-Maxwell system is straitforwardly solved in a full torus geometry. Global gyrokinetic simulations are classified mainly into two approaches depending on treatments of equilibrium profiles. In a so-called δf simulation, the particle distribution function f_s is separated into f_{0s} and δf_s, and only δf_s is solved by forcing $f_{0s} + \delta f_s$ to be fixed (δf_s to be zero) on average by artificial sources. This approximation may be valid for estimating steady turbulent transport for given profiles. However, in studying open issues in tokamak experiments such as transient transport properties in the modulation experiment, the profile stiffness, and the formation of transport barriers [4], one may need to solve the turbulent transport and profile formations consistently for given heat sources. In addition, δf simulations can not be applied to the edge turbulence where the perturbation amplitude reaches at $\delta f_s/f_{0s} \sim \mathcal{O}(1)$. From these backgrounds, demand for a full-f simulation with relevant sources and collisions is increasing. Such a full-f simulation self-consistently treats relaxation and evolution processes of global equilibrium profiles due to the particle, momentum, and heat sources, and the turbulent transport. In the absence of these sources, a full-f simulation is reduced to an isolated system, which satisfies the conservation properties such as the conservations of the particle number, the momentum, the energy, and the entropy. A global approach involves multi-scale physics ranging from the small scale turbulent fluctuations to the large scale profile variations. This multi-scale character is essential to study nonlocal features of the micro-turbulence such as global profile effects, meso-scale $E \times B$ flows, and avalanche like transport events, which play important roles, e.g., in the plasma size scaling of turbulent transport. Global simulations have been significantly advanced with increasing computational resources, and most of global codes have a capability of simulating the electrostatic ion turbulence with adiabatic electrons in a full torus configuration. However, global simulations of multi-scale turbulence and electromagnetic turbulence with kinetic electrons are still prohibitive, and an alternative approach, a flux tube model, is useful especially in understanding physics in such frontier issues.

A flux tube gyrokinetic simulation [8, 65, 75, 76, 78] is a δf simulation in a local domain in the vicinity of a single field line. This model assumes a multi-scale expansion with respect to a smallness parameter $\rho^* = \rho_s/a$ [44] or complete scale separation between turbulent fluctuations and equilibrium profiles, retaining the lowest order terms in ρ^*. Because of this local approximation, both an equilibrium quantity A and its radial derivative $\partial A/\partial \psi$ are assumed to be constant in the calculation domain. In addition, the model is described in the field aligned coordinates with a periodic radial boundary condition with phase shift reflecting the magnetic shear or the radial variation of the field line pitch $q(\psi)$ [84]. These assumptions are physically sound, provided that the radial correlation length of turbulent fluctuations is much shorter than the equilibrium scale length or the size of calculation domain. The flux tube model is a non-conservative open system developed to study local quasi-steady turbulent transport with fixed background profiles, and the system satisfies so-called entropy balance relation [49, 50, 51] instead of the conservation properties in the global model. The advantages of this approach is as follows.

1. The computational cost is significantly reduced compared to the global model. Because of this advantage, flux tube simulations are widely used in experimental data analyses.
2. The model is useful especially in studying electron turbulence, electromagnetic turbulence, and multi-scale coupling of electron and ion turbulence, which require treatments of multiple spatio-temporal scales coming from kinetic electron dynamics.
3. Because of aforementioned simplifications, the model is well established, and results from different codes are converged well. In contrast, the global model still has some ambiguities concerning to equilibrium profiles, source models, and boundary conditions, which make cross benchmarking among codes very difficult.

The drawbacks of the flux tube model is obviously a lack of global profile effects. Especially, the plasma size dependence of the energy confinement time observed in the experiment can not be addressed using this model, because the model assumes an infinite plasma size from the beginning. In addition, recent global simulations disclosed formations of meso-scale structures in between the plasma size a and the Larmor radius ρ_s, which are sometimes beyond the limit of validity of the model. It is important to carefully check the applicability limit of each simulation model.

4. NUMERICAL DISSIPATION IN COLLISIONLESS MICRO-TURBULENCE SIMULATION

A collisionless micro-turbulence simulation is recognised as DNS of the Hamiltonian (incompressible) flow in 5D phase space given by the gyrokinetic equation. In a sense, this may correspond to simulating ideal fluids described by the Euler equation. Although issues related to the entropy balance relation were successfully addressed with non-dissipative Vlasov simulations based on an implicit symplectic scheme [85], in most of practical issues on the quasi-steady turbulent transport, numerical or physical dissipation is needed to avoid aliasing errors and numerical instabilities arising from filamentation or formation of fine scale structures in a long time simulation. In a particle simulation, similar course graining effects are naturally introduced by remaining numerical collisions among super-particles. Therefore, it is important to discuss effects of numerical dissipation on a collisionless micro-turbulence simulation. A plausible picture of a micro-turbulence simulation may be summarised as follows.

1. Micro-instabilities are excited by inhomogeneities in a plasma through various wave-particle interactions as discussed earlier.
2. Nonlinear mode coupling mainly due to the $E \times B$ nonlinearity transforms fluctuation field energy to linearly stable small scales typically given as $k_\perp \rho_s > 1$, where fluctuations are absorbed by the wave-particle interaction or the Landau damping.
3. The damped field energy is transferred to fine scales in the velocity space through the phase mixing, which comes from the ballistic motion or the parallel motion term in the kinetic equation. If one write the kinetic equation simply as $\partial_t f + v\partial_x f = 0$, an analytic solution of the ballistic mode, $f(x,v,t) = \sum_k f_{0k}(v)\exp(ik[x-vt])$, is obtained for an initial condition, $f(x,v,t=0) = \sum_k f_{0k}(v)\exp(ikx)$. Here, an effective wave number in the velocity space, kt, increases in time, and fine scale structures are continuously produced.
4. Such fine scale structures are eliminated by numerical or physical dissipation.

It is noted that form a comparison between the kinetic damping effect and the nonlinear effect, an effective Reynolds number of fusion plasmas is estimated to be $Re_{\rm eff} = 10 \sim 100$ [10], and the grid width is normally chosen as $\sim \rho_s$. However, a clear separation between an inertial spectral range and a dissipative spectral range as in neutral fluid turbulence has not been observed in tokamak micro-turbulence simulations, in which both damping and forcing are determined by the wave-particle interaction and are distributed in a broad spectral range. Although a number of convergence studies support that the grid resolution used in the present gyrokinetic simulations seems to be enough at least for estimating turbulent transport, further research is needed to understand structures of turbulent spectra.

In Refs. [86, 87], a role of the dissipation in determining quasi-steady turbulent transport was examined for the electrostatic ITG turbulence in a flux tube model without toroidal effects. In this limit, the gyrokinetic Vlasov-Poisson system yields

$$\partial_t \delta f_i + v_\parallel \nabla_\parallel \delta f_i + \mathbf{v}_{E\times B}\cdot\nabla\delta f_i$$
$$= \left[-\mathbf{v}_{E\times B}\cdot\nabla\ln n_0\left\{1+\eta_i\left(\frac{m_i v_\parallel^2/2+\mu B}{T_i}-\frac{3}{2}\right)\right\} - \frac{e}{T_i}v_\parallel\nabla_\parallel\langle\phi\rangle_\alpha\right] f_M + C(\delta f_i), \tag{42}$$

$$\frac{1}{\lambda_{Di}^2}\left[\phi - \langle\bar{\phi}\rangle_\alpha\right] + \frac{1}{\lambda_{De}^2}\left[\phi - \langle\phi\rangle_{k_\parallel=0}\right] = 4\pi e\int \delta f_i \delta([\mathbf{R}+\rho_i]-\mathbf{x})\mathcal{J}_i d^6 Z, \tag{43}$$

where a notation of the gyro-centre coordinates $\bar{}$ is ignored for simplicity, the equilibrium distribution function is given by the Maxwellian distribution, $f_M = n_0(2\pi m_i T_i)^{-3/2}\exp[-(m_i v_\parallel^2/2+\mu B_0)/T_i]$, $\mathbf{v}_{E\times B} = (c/B_0)\mathbf{b}\times\nabla\langle\phi\rangle_\alpha$ is the $E\times B$ drift velocity, $\eta_i = |\nabla\ln T_i|/|\nabla\ln n_0|$, and $C(\delta f_i)$ is a collision term. In Eq. (42), the higher order parallel nonlinearity, $\nabla_\parallel\langle\phi\rangle_\alpha\partial_{v_\parallel}\delta f_i$, is neglected compared to the $E\times B$ nonlinearity, $\mathbf{v}_{E\times B}\cdot\nabla\delta f_i$. In Eq. (43), the quasi-neutrality condition is imposed by neglecting the l.h.s of Eq. (22) compared to the ion polarisation term. An electron response is assumed to be adiabatic or the Boltzmann distribution, $n_e = n_0 + n_0 e\phi/T_e$, because the phase velocity of ion drift waves is much slower than the electron thermal velocity, $\omega_i^*/k_\parallel \sim v_{ti} \ll v_{te}$. It is noted that the adiabatic electron response, which comes from the parallel dynamics, vanishes for $k_\parallel = 0$ or magnetic surface averaged components, $\langle\phi\rangle_{k_\parallel=0}$. The entropy balance relation among the entropy variable δS, the fluctuation field energy W, the ion heat flux Q_i, and the collisional dissipation D is then derived by multiplying $\delta f_i/f_M$ to Eq. (42) and integrating it over phase space,

$$\frac{d}{dt}(\delta S + W) = Q_i + D_i, \tag{44}$$

$$\delta S = \frac{1}{2}\int\frac{\delta f_i^2}{f_M}\mathcal{J}_i d^6 Z, \tag{45}$$

$$W = \frac{1}{2}\sum_{\mathbf{k}}[1-\Gamma_0(k_\perp^2\rho_{ti}^2)]\left|\frac{e\phi_{\mathbf{k}}}{T_i}\right|^2 n_0 + \frac{1}{2}\frac{T_i}{T_e}\sum_{k_\parallel\neq 0}\left|\frac{e\phi_{\mathbf{k}}}{T_i}\right|^2 n_0, \tag{46}$$

$$Q_i = -\frac{1}{2T_i}\left[\int \mathbf{v}_{E\times B} n_0 \delta T_i d\mathbf{R}\right]\cdot\nabla\ln T_i, \tag{47}$$

$$D_i = \int C(\delta f_i)\frac{\delta f_i}{f_M}\mathcal{J}_i d^6 Z \tag{48}$$

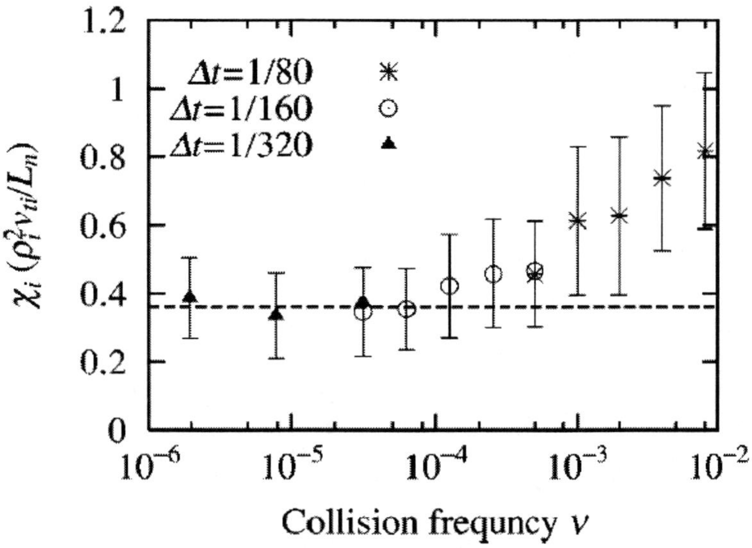

FIGURE 6. The turbulent heat transport coefficients $\chi_i/(\rho_{ti}^2 v_{ti}/L_n)$ observed in slab ITG turbulence simulations ($E \times B$ zonal flows are turned off) with different collision frequencies $\nu/(v_{ti}/L_n)$ [87].

where $\delta T_i = \int m v_{\parallel}^2 \delta f_i \mathscr{J}_i dv_{\parallel} d\mu / n_0$. In the collisionless limit with $D_i \to 0$, Eq. (44) suggests two scenarios. One is a statistically steady state with no heat flux, $Q_i = 0$, and $\dot{\delta S} = \dot{W} = 0$. The other is a quasi-steady turbulent state with $\dot{\delta S} = Q_i = const.$ and $\dot{W} = 0$. In the latter scenario, continuous growth of fine scale structures in the velocity space or higher order velocity moments contributes to monotonic increase of δS, while W and Q_i characterised by the lower order velocity moments stay in steady values. In Ref. [86], both scenarios were identified in ITG turbulence simulations with and without $E \times B$ zonal flows, which are nonlinearly generated from turbulent fluctuations. Although the former scenario is trivially satisfied in a linearly stable plasma, the simulation showed that it can be realized even in a linearly unstable plasma. In the simulation with $E \times B$ zonal flows, the heat flux Q_i is quenched in a nonlinearly saturated steady state, provided that $E \times B$ zonal flows grow strong enough to suppress the ITG turbulence. On the other hand, the simulation without $E \times B$ zonal flows showed the latter scenario with a quasi-steady heat flux, in which δS grows monotonically until fine scale structures reach at the limit of velocity space resolution. This suggests that a long time micro-turbulence simulation is prohibited without numerical or physical dissipation.

In Ref. [87], a relation between the dissipation D_i and the turbulent transport Q_i was studied by adding the collisional dissipation to the latter scenario. In this case, δS also keeps a quasi-steady value because of elimination of fine scale structures via the collisional dissipation, and the turbulent transport is balanced with the collisional dissipation, $Q_i + D_i = 0$. This relation looks as the turbulent transport is determined by the collision frequency. However, it was also found that in a weak collisional regime, the turbulent transport approaches to the collisionless limit asymptotically (see Fig.6). To understand this result, properties of the turbulent spectrum in the $v_{\parallel}$ space was investigated in detail, and the following properties were clarified.

1. The turbulent spectrum has a subrange of the $v_{\parallel}$ spectrum which is free from the entropy production and the collisional dissipation.
2. The turbulent cascade in the subrange is similar to that of the passive scaler convection beyond the Kolmogorov scale in turbulent fluids with a large Prandtl number [88].
3. If dissipation scales in the $v_{\parallel}$ space is small enough, the dissipation process does not affect lower order velocity moments, which dictate W and Q_i.

These findings support the hypothesis that forcing determines dissipation [50]. Therefore, a collisionless micro-turbulence simulation is possible with finite resolution available on the present-day supercomputers, provided that numerical dissipation in a Vlasov simulation or numerical collisions in a particle simulation are finite but small enough. A careful convergence study against grid size or a number of marker particles is essential to verify this point.

5. SUMMARY

With increasing computational resources and improved simulation models, kinetic simulations of collisionless turbulent plasmas have been established as an essential tool for studying tokamak turbulent transport. Recent gyrokinetic simulations succeeded in discovering rich physics behind tokamak turbulent transport, and their comparisons with selected experiments have been started. To answer critical engineering issues in the tokamak fusion research, demand for establishing quantitative prediction capabilities of the gyrokinetic simulation is increasing. For this purpose, the verification and validation of the simulation are becoming extremely important. In developing gyrokinetic simulations, firstly, one needs to recognise limitations of numerical schemes by estimating the numerical stability and accuracy, and then, proceeds to the verification processes. The most fundamental verification is to check the conservation properties and the entropy balance relation. In the next step, the physical verification processes in linear and nonlinear benchmark tests are examined. As for linear benchmark tests, there are standard test problems such as the linear ITG benchmark in Cyclone parameters [8] and the zonal flow damping test [38]. As for nonlinear benchmark tests, most of flux tube simulations check the heat transport level reported in the nonlinear ITG benchmark [8]. On the other hand, global simulations do not have such a standard case because of the aforementioned ambiguities in their simulation models, and cross code comparisons are still ongoing. Once the above verification processes are completed, one can proceed to the validation of the physical model through comparisons with the experiment, which would give feedback about missing physics. Although the steady ion heat transport caused by electrostatic fluctuations in conventional tokamak configurations has been intensively addressed so far, further validation (and code development) processes are needed to explain the outstanding issues in the tokamak experiment [4]. Such issues may require challenges for gyrokinetic simulations in extending capabilities to include multiple spatio-temporal scales and multiple physics. Towards reactor relevant micro-turbulence simulations, further advances in computer technology and simulation models are desirable.

ACKNOWLEDGMENTS

We would like to thank Dr. S. Benkadda for his support. One of the authors (Y.I.) is supported by the Japanese Ministry of Education, Culture, Sports, Science, and Technology (No. 18760647).

REFERENCES

1. F. L. Hinton and R. D. Hazeltine, Rev. Mod. Phys. **48**, 239 (1976).
2. S. P. Hirshman and D. J. Sigmar, Nucl. Fusion **21**, 1079 (1981).
3. ITER Physics Expert Groups on Confinement and Transport and Confinement Modelling and Database, Nucl. Fusion **39**, 2175 (1999).
4. ITPA Confinement Database and Modelling Topical Group and ITPA Pedestal and Edge Topical Group, Nucl. Fusion **47**, S18 (2007).
5. A. Yoshizawa *et al.*, Plasma Phys. Control. Fusion **43**, R1 (2001).
6. J. A. Krommes, Phys. Rep. **360**, 1 (2002).
7. W. M. Tang and V. S. Chan, Plasma Phys. Control. Fusion **47**, R1 (2005).
8. A. M. Dimits *et al.*, Phys. Plasmas **7**, 969 (2000).
9. L. Villard *et al.*, Plasma Phys. Control. Fusion **46**, B51 (2004).
10. P. H. Diamond *et al.*, Plasma Phys. Control. Fusion **47**, R35 (2005).
11. Y. Idomura, T.-H. Watanabe, and H. Sugama, Comptes Rendus Physique **7**, 650 (2006).
12. for example, S. Ichimaru, *Statistical Plasma Physics, Volume* I : *Basic Principles* (Addison-Wesley, Redwood City, California, 1992).
13. E. A. Frieman and L. Chen, Phys. Fluids **25**, 502 (1982).
14. A. J. Brizard and T. S. Hahm, Rev. Mod. Phys. **79**, 421 (2007).
15. D. H. E. Dubin *et al.*, Phys. Fluids **26**, 3524 (1983).
16. T. S. Hahm, W. W. Lee, and A. Brizard, Phys. Fluids **31**, 1940 (1988).
17. T. S. Hahm, Phys. Fluids **31**, 2670 (1988).
18. A. J. Brizard, J. Plasma Phys. **41**, 541 (1989).
19. A. J. Brizard, Phys. Plasmas **2**, 459 (1995).
20. H. Sugama, Phys. Plasmas **7**, 466 (2000).
21. S. Wang, Phys. Rev. E **64**, 056404 (2001).
22. J. R. Cary, Phys. Rep. **79**, 129 (1981); J. R. Cary and R. G. Littlejohn, Ann. Phys. **151**, 1 (1983).

23. R. G. Littlejohn, J. Math. Phys. **20**, 2445 (1979); R. G. Littlejohn, Phys. Fluids **24**, 1730 (1981); R. G. Littlejohn, J. Plasma Phys. **29**, 111 (1983).
24. B. B. Kadomitsev and O. P. Pogutse, *Reviews of Plasma Physics, Turbulence in Toroidal Systems* (Consultants bureau, New York, 1970), Vol. 5, Chapter 2.
25. A. B. Mikhailovskii, *Theory of Plasma Instabilities, Volume* 2 : *Instabilities of an Inhomogeneous Plasma* (Consultants bureau, New York, 1974).
26. W. M. Tang, Nucl. Fusion **18**, 1089 (1978).
27. W. Horton, Rev. Mod. Phys. **71**, 735 (1999).
28. S. I. Braginskii, *Reviews of Plasma Physics*, edited by M. A. Leontovich (Consultants Bureau, New York, 1965), Vol. I, p.205.
29. G. W. Hammett and F. W. Perkins, Phys. Rev. Lett. **64**, 3019 (1990).
30. N. Mattor and S. E. Parker, Phys. Rev. Lett. **79**, 3419 (1997).
31. H. Sugama *et al.*, Phys. Plasmas **8**, 2617 (2001).
32. W. Horton *et al*, in this book.
33. R. E. Waltz *et al.*, Phys. Fluids B **4**, 3138 (1992).
34. W. Dorland and G. W. Hammett, Phys. Fluids B **5**, 812 (1993).
35. M. A. Beer and G. W. Hammett, Phys. Plasmas **3**, 4018 (1996); M. A. Beer and G. W. Hammett, Phys. Plasmas **3**, 4046 (1996).
36. M. Kotschenreuther *et al.*, Phys. Plasmas **2**, 2381 (1995).
37. R. E. Waltz *et al.*, Phys. Plasmas **4**, 4046 (1997).
38. M. N. Rosenbluth and F. L. Hinton, Phys. Rev. Lett. **80**, 724 (1998).
39. M. Beer and G. Hammet, in *Proceedings of Varenna-Lausanne International Workshop* (Societa Italiana di Fisica, Editrice Compositori, Bologna, 1998).
40. H. Sugama and T.-H. Watanabe, Phys. Plasmas **14**, 022502 (2007).
41. C. K. Birdsall and A. B. Langdon, *Plasma Physics via Computer Simulation* (Adam Hilger, Bristol and New York, 1991).
42. R. W. Hockney and J. W. Eastwood, *Computer Simulation Using Particles* (Adam Hilger, Bristol and New York, 1988).
43. T. Tajima, *Computational Plasma Physics : With Applications to Fusion and Astrophysics* (Addison-Wesley, Redwood City, California, 1989).
44. W. W. Lee, Phys. Fluids **72**, 243 (1987); W. W. Lee, J. Comput. Phys. **72**, 243 (1987).
45. S. E. Parker and W. W. Lee, Phys. Fluids B **5**, 77 (1993).
46. A. Y. Aydemir, Phys. Plasmas **1**, 822 (1994).
47. S. J. Allfrey and R. Hatzky, Comput. Phys. Commun. **154**, 98 (2003).
48. R. Hatzky *et al.*, Phys. Plasmas **9**, 898 (2002).
49. W. W. Lee and W. M. Tang, Phys. Fluids **31**, 612 (1988).
50. J. A. Krommes and G. Hu, Phys. Plasmas **1**, 3211 (1994).
51. H. Sugama *et al.*, Phys. Plasmas **3**, 2379 (1996).
52. S. Brunner, E. Valeo, and J. K. Krommes, Phys. Plasmas **6**, 4504 (1999).
53. W. X. Wang *et al.*, Plasma Phys. Control. Fusion **41**, 1091 (1999).
54. S. Satake *et al.*, Nucl. Fusion **45**, 1362 (2005).
55. G. Hu and J. A. Krommes, Phys. Plasmas **1**, 863 (1994).
56. Z. Lin *et al.*, Phys. Rev. Lett. **88**, 195004 (2002); Z. Lin and T. S. Hahm, Plasmas **11**, 1099 (2004).
57. M. J. LeBrun *et al.*, Phys. Fluids B **5**, 752 (1993),
58. Y. Kishimoto *et al.*, in *Proceedings of the 15th International Conference on Plasma Physics and Controlled Nuclear Fusion research* (IAEA, Vienna, 1994).
59. G. Furnish *et al.*, Phys. Plasmas **6**, 1227 (1999).
60. S. E. Parker, W. W. Lee, and R. A. Santoro, Phys. Rev. Lett. **71**, 2042 (1993).
61. R. D. Sydora, V. K. Decyk, and J. M. Dawson, Plasma Phys. Control. Fusion **38**, A281 (1996).
62. A. M. Dimits *et al.*, Phys. Rev. Lett. **77**, 71 (1996).
63. Z. Lin *et al.*, Science **281**, 1835 (1998).
64. Y. Idomura, S. Tokuda, and Y. Kishimoto, Nucl. Fusion **43**, 234 (2003).
65. Y. Chen and S. E. Parker, J. Comput. Phys. **189**, 463 (2003).
66. H. Naitou *et al.*, Phys. Plasmas **2**, 4257 (1995).
67. W. W. Lee *et al.*, Phys. Plasmas **8**, 4435 (2001).
68. Z. Lin and L. Chen, Phys. Plasmas **8**, 1447 (2001).
69. A. Mishchenko, R. Hatzky, and A. Könies, Phys. Plasmas **11**, 5480 (2004).
70. Y. Nishimura, Z. Lin, and W. X. Wang, Phys. Plasmas **14**, 042503 (2007).
71. C. Z. Cheng and G. K. Knorr, J. Comput. Phys. **22**, 330 (1976).
72. E. Sonnendrücker *et al.*, J. Comput. Phys. **149**, 201 (1999).
73. M. Brunetti *et al.*, Comput. Phys. Commun. **163**, 1 (2004).
74. V. Grandgirard *et al.*, J. Comput. Phys **217**, 395 (2006).; V. Grandgirard *et al.*, Commun. Nonlinear Sci. Num. Simul. **13**, 81 (2008).
75. W. Dorland *et al.*, in *Proceedings of the 18th IAEA Fusion Energy Conference* (IAEA, Vienna, 2000).
76. F. Jenko, Comput. Phys. Commun. **125**, 196 (2000).
77. J. Candy and R. Waltz, J. Comupt. Phys. **186**, 545 (2003).

78. T. H. Watanabe and H. Sugama, Nucl. Fusion **46**, 24 (2006).
79. X. Q. Xu *et al.*, Nucl. Fusion **47**, 809 (2007).
80. A. Arakawa, J. Comput. Phys. **1**, 119 (1966).
81. Y. Morinishi *et al.*, J. Comput. Phys. **143**, 90 (1998).
82. Y. Idomura, M. Ida, S. Tokuda, and L. Villard, J. Comput. Phys. **226**, 244 (2007).
83. R. D. Hazeltine and J. D. Meiss, *Plasma Confinement* (Addison-Wesleyam, Redwood City, 1992).
84. M. A. Beer, S. C. Cowley, and G. W. Hammett, Phys. Plasmas **2**, 2687 (1995).
85. T.-H. Watanabe, H. Sugama, and T. Sato, J. Phys. Soc. Japan **70**, 3565 (2001).
86. T.-H. Watanabe and H. Sugama, Phys. Plasmas **9**, 3659 (2002).
87. T.-H. Watanabe and H. Sugama, Phys. Plasmas **11**, 1476 (2004).
88. G. K. Batchelor, Fluid Mech. **5**, 113 (1959).

Turbulence scaling laws and transport models

X. Garbet

Association Euratom-CEA, CEA/DSM/DRFC, CEA-Cadarache 13108 Saint Paul Lez Durance (France)

Abstract. This paper presents some elementary results concerning the dimensionless scaling laws of turbulent transport in fusion plasmas and the way they provide constraints on transport models.

Keywords: turbulence, plasmas, magnetic confinement, tokamaks
PACS: 52.55.-s,52.55.Fa,52.35.Ra

INTRODUCTION

The predictive accuracy of most of the 1D transport models is of the order of 20 percent [1]. One goal of the theory of turbulent transport is to improve this accuracy. The present lecture aims at showing that dimensionless scaling laws provides useful constraints on transport models. Indeed it appears that 3 main dimensionless numbers governs turbulent transport in magnetized plasmas, and therefore their confinement. Moreover one dimensionless parameter, the normalized gyroradius, plays a prominent role. This peculiarity greatly enhances the accuracy when extrapolating the present results to next step devices. Hence this methodology greatly simplifies a conventional multi-parameter approach. Nevertheless it bears some drawbacks that will be detailed in this lecture. A second advantage of scaling laws is that they allow to bridge the advances in turbulence simulations and the various measurements. This is particularly obvious for fluctuation measurements, which bring an efficient way to test transport models based on turbulence models. In spite of several progress, this approach is still far from being complete. Some open problems will be mentioned in this overview.

The paper is organized as follows. The first section describes the basics of a similarity principle for magnetized plasmas. The second section addresses the main dependencies and summarizes the principal theoretical and experimental findings. Finally the last section describes how theory can be tested by fluctuation measurements. A conclusion is then drawn.

SIMILARITY PRINCIPLE

Global and local scaling laws

Global scaling laws

The similarity principle was stated first by Kadomtsev [2] in the context of magnetized plasmas physics. In this seminal paper, the procedure was quite simple. It consisted in counting the dimensionless numbers for a pure electron/ion plasma. The result is summarized on Fig.1, which is extracted from Kadomtsev's paper. In fact, 3 types of dimensionless numbers appear. A first category, the most important one, are numbers which play a prominent role in scaling laws of the energy confinement time. These numbers are the normalized gyroradius $\rho_* = \rho_s/a$ (ρ_s is the ion Larmor radius, a is the minor radius), the collisionality $\nu_* = \nu_{ei}qR/\varepsilon_a^{3/2}v_{Te}$ (ν_{ei} is the electron-ion collision frequency, $\varepsilon_a = a/R$ the inverse aspect ratio, v_{Te} is the thermal electron velocity) and plasma beta $\beta = 2\mu_0 p/B^2$ (p is the total pressure and B the magnetic field). The second list is related to factors which are less sensitive, because they are fixed in a large class of plasmas, namely the aspect ratio R/a, the safety factor q, and the ratio of the electron and ion temperatures T_e/T_i. Obviously the aspect ratio does not vary in a given device, as it is a geometrical factor. Other factors of this type should be added, such as the elongation or the triangularity. The safety factor is supposed to be given in a typical Iter plasma (it ranges between 1 on the axis and 3 at 95 percent of the minor radius). Nevertheless it will be seen that it plays an important role. Finally the ratio T_e/T_i is constrained by the collisional equipartition process, which at high density implies that this ratio stays close to 1. The third category in table I covers numbers

CP1013, *Turbulent Transport in Fusion Plasmas, First ITER International Summer School*
edited by S. Benkadda

I. $\nu^* = qR/\lambda_{mfp}$ $\rho^* = \rho_c/a$ $\beta = 2\mu_0 p/B^2$

II. $A = R/a$ $\tau = T_e/T_i$ q

III. $\mu = m_e/m_i$ $N = n_e\lambda_d^3$

FIGURE 1. Dimensionless numbers in a magnetized plasmas (from [2])

which are irrelevant for a given plasma composition, namely the mass ratio and the number of electrons in the Debye length. In fact, the effective mass number does play a role in the confinement (this is the so-called isotopic effect). The Debye length λ_d is irrelevant for turbulence because the spatial scales of fluctuations are much larger than λ_d. In other words the Poisson equation reduces to the electro-neutrality condition, and it can be shown in this case that the Debye length does not appear explicitly in the equations. The main consequence of this analysis is that at fixed geometry, for a given plasma composition, the normalized confinement time $\omega_c\tau_E$ is some function of ρ_*,β and ν_* ($\omega_c = eB/m$ is the cyclotron frequency). This is not a trivial result, as it states that two very different machines with the same values of ρ_*,β and ν_* (but also same geometry, profile shape and boundary conditions) will be characterized by the same normalized confinement time. It turns out that this rule has been verified experimentally with some success. Some useful relations relate these dimensionless numbers to the plasma main parameters, i.e. the density n, temperature T, field B and minor radius a, namely

$$\rho_* \equiv \frac{T^{1/2}}{Ba}$$
$$\beta \equiv \frac{nT}{B^2}$$
$$\nu_* \equiv \frac{na}{T^2} \qquad (1)$$

where the symbol $\equiv$ means "scales as". These relations show that two very different plasmas may exhibit the same dimensionless numbers. Nevertheless, it turns out that the main parameter that varies when moving from the present devices to Iter (and further on to the reactor) is the normalized gyroradius ρ_*. The nominal values of β and ν_* are currently achieved, but rarely at the same time. This is illustrated by Fig.2, which shows the parameters reached on JET in the parameter space (ρ_*,ν_*) for 2 ranges of the normalized β_N (β_N is defined as $\beta_N = \frac{\beta}{I_p/aB}$, where I_p is the plasma current in MA and β is in percent). A remarkable fact is that the scaling law that has been used for designing Iter (the so called ITER98 IPB(y,2) scaling for the elmy H-mode), does respect the similarity principle stated above. More precisely it reads $\omega_c\tau_E \equiv \rho_*^{-2.7}\beta^{-0.9}\nu_*^{-0.0}$. In fact more recent experiments [4, 5] show that the scaling is closer to $\omega_c\tau_E \equiv \rho_*^{-3.0}\beta^{-0.0}\nu_*^{-0.35}$. This is not probably the last word in that matter as the determination of these exponents from experimental results turn out to be a very delicate task. Note that there is no reason for the scaling $\omega_c\tau_E$ to be a monomial function of the parameters ρ_*,β and ν_*. In fact several recent findings indicate that this is not the case at least for the collisionality. We will come back to the physical meaning of these exponents.

Local scaling laws

This line of research has been pursued by Connor and Taylor [3]. In this work, transformations that leave the equations invariant are identified and used to produce a set of *local dimensionless scaling laws*. The equations that describe a plasma turbulence are the Fokker-Planck (or Landau) equation coupled to the Maxwell equations. This analysis generalizes the previous Kadomtsev finding to local relations. For a given shape of profiles, the main result is that the diffusion coefficients scale as the Bohm coefficient T/eB up to an arbitrary function of ρ_*,β and ν_*. It is stressed here that this version of the similarity principle is much more powerful than the one for the global confinement time. Indeed the conditions for comparing normalized confinement times in two different devices are quite demanding, since the profile must be the same, and the boundary conditions as well. In particular it turns out that slight modifications of the geometry may lead to significant differences in the confinement time. Also edge plasmas involve atomic physics, which bring new dimensionless numbers into the game. Unfortunately, profile stiffness and turbulence spreading make the core turbulence quite sensitive to the conditions in the edge. Hence the confinement

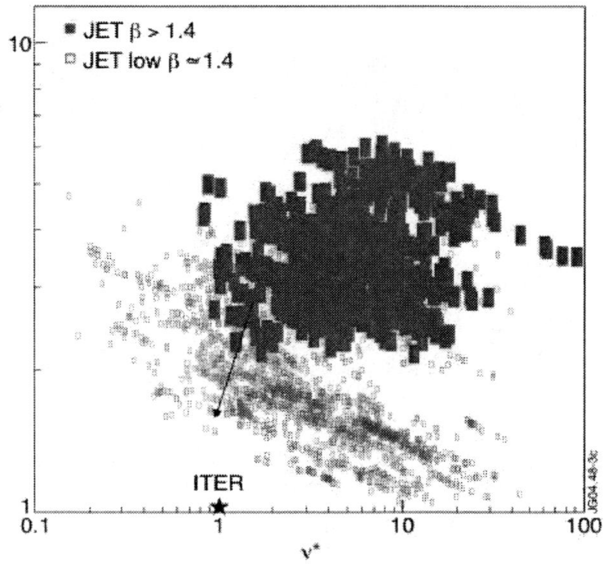

FIGURE 2. Performances reached on JET in the space ρ_* and ν_* within 2 intervals of normalized β_N

time is usually a rather rough quantity to assess. Its main advantage is its availability on all devices. Local quantities such as diffusion coefficients or level of fluctuations are much better as they are less sensitive to the plasma conditions in other places.

Physical meaning of scaling laws

Gyroradius scaling

The scaling with gyroradius is certainly the most studied subject in this area of research. The reason is that the dependence of scaling laws on this parameter is critical when extrapolating the present results to Iter, and a future reactor. An essential result is that this dependence is directly related to turbulence properties. Here fluctuations will be characterized by correlation length and time. Although this is a rather rough level of description, it allows to derive some simple relations between turbulence and confinement. The correlations length and time are defined as the efolding length ℓ_c and time τ_c of the correlation function

$$C_{nn}(\delta\mathbf{x},\delta t)=\frac{1}{V_{av}T_{av}}\int dt d^3\mathbf{x}\delta n(\mathbf{x},t)\delta n(\mathbf{x}+\delta\mathbf{x},t+\delta t) \quad (2)$$

where the integration volume V_{av} and time T_{av} are much larger than the correlation lengths and times, but smaller than the profile evolution time. Some simple physics can be done to get a clue. Most of linear growth rates scale as the minor radius divided by the sound speed c_s. Hence a reasonable guess is to assume a turbulence correlation time τ_c that behaves as a/c_s. Moreover the polarization drift introduces a characteristic scale for fluctuations which is nothing else than the gyroradius. Hence it is expected that the turbulence correlation length ℓ_c should scale as a gyroradius ρ_s. This is a huge difference with turbulence in neutral fluids where the only two scales are the system size and the dissipative scale. A simple random walk estimate then provides a diffusion coefficient $\chi\sim\ell_c^2/\tau_c$, which scales as $\chi=\chi_B\rho_*$, where $\chi_B=T/eB$ is the Bohm diffusivity. This scaling is called *gyroBohm scaling law*. It implies that $\omega_c\tau_E\equiv\rho_*^{-3.0}$, which agrees quite well with the scaling law obtained in the H-mode. In fact the situation is somewhat more complex when looking into the local transport analysis.

Local transport analysis finds a gyroBohm scaling law in most situations for electrons (i.e. in L and H-mode plasmas). The situation is more complex for ions. Indeed the ion diffusivity is found to behave as $\chi\equiv\chi_B\left[\rho_*\right]^{\alpha_\rho}$, where α_ρ ranges between -0.5 and 1, depending on plasma conditions [6]. GyroBohm scaling, i.e. $\alpha_\rho=1$, is found in H-mode plasmas for both ions and electrons, which agrees with the scaling found for the energy confinement time.

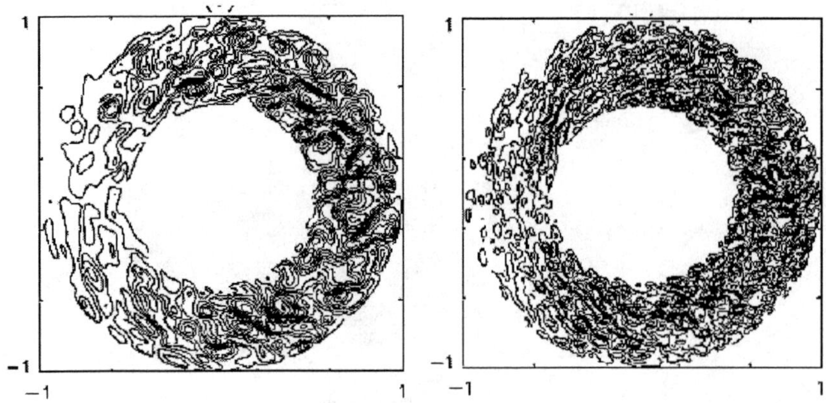

FIGURE 3. Contour lines of the electric potential in a poloidal plane for $\rho_* = 1/50$ (left) and $\rho_* = 1/100$ (right) (ITG turbulence, from [10]).

Concerning numerical simulations, most simulations are found to be in the gyroBohm regime [7, 8, 9, 10, 11] provided that the temperature gradient is well above the stability threshold and for low enough values of ρ_*. An example is shown on Fig.3, where it appears that decreasing ρ_* by a factor 2, the typical size of vortices is reduced by a similar factor. Close to the stability threshold and/or for large values of ρ_*, some breaking of gyroBohm scaling is usually observed (see for instance Fig.4). If most theoreticians agree on the conditions for getting a gyroBohm scaling law (small enough gyroradius and well developed turbulence), there is no consensus to explain the departure from gyroBohm scaling. The main hypothesis that have been pushed forward are the following:

- Global modes [12, 13]. This explanation relies on the width of linear eigenmodes, which scales as $(a\rho_s)^{1/2}$. Assuming a correlation ℓ_c that scales as this linear width, a random walk estimate leads to a Bohm scaling (if the correlation time τ_c is still assumed to scale as a/c_s). It appears that this explanation does not hold usually, except when the temperature gradient is very close to the threshold. The reason is that turbulence breaks the large scale linear structures to produce small scale vortices that scale as ρ_s. Close to the threshold however, i.e. in a regime where the system is weakly chaotic and keeps a memory of the linear structures, one finds that the correlation length do not scale any more as ρ_s.
- $E \times B$ shear flow [9, 14]. It is well known that the $E \times B$ shear flow reduces turbulent transport. A way to account for this effect is to introduce an effective growth rate γ_{eff} equals to $\gamma_{lin} - \gamma_E$ [7], where γ_{lin} is the linear growth rate that scales as c_s/a and $\gamma_E \simeq \partial_r E_r/B$ is the $E \times B$ flow shear rate. Assuming a thermal radial electric field $E_r \equiv T/eBa$, one finds that $\gamma_E \equiv \rho_* c_s/a$. Assuming again that $\tau_c \equiv 1/\gamma_{eff}$ and $\ell_c \equiv \rho_s$, it turns out that $\chi \equiv \chi_B \rho_* (1 - \alpha\rho_*)$, where the parameter α depends on plasma parameters, in particular on the distance to the threshold. This is an interesting example where the scaling law is not monomial. It can cover all situations, from gyroBohm to worst than Bohm. A gyroBohm scaling scaling is expected for low values of ρ_*, in agreement with simulations. Note that this formula is also connected to the formation of transport barriers since it predicts that for a large enough value of ρ_*, turbulent transport is quenched. This criterion is actually used on JET for characterizing internal transport barriers [15].
- Self-Organized Criticality [16]. When the plasma is stiff, the profiles tend to relax strongly towards the stability threshold, the relaxation takes place via avalanches, in a way similar to a sand pile. Avalanches are large scale events which might lead to a Bohm-like behavior.
- Turbulence spreading [17]. Turbulence spreading indicates the tendency for turbulence to spread from unstable to stable regions. An important consequence is that turbulence transport exhibits some degree of non locality (this is also called action at distance). This has obviously consequences on the level of fluctuations and hence on the scaling since the properties of turbulence at one place will depend on the parameters at other places. Quite interesting simple conservation arguments leads to a local scaling $\chi \equiv \chi_B \rho_* (1 - \alpha\rho_*)$, which is the same as for the $E \times B$ low shear effect. Hence these two effects can hardly be discriminated from the scaling of the diffusivity.

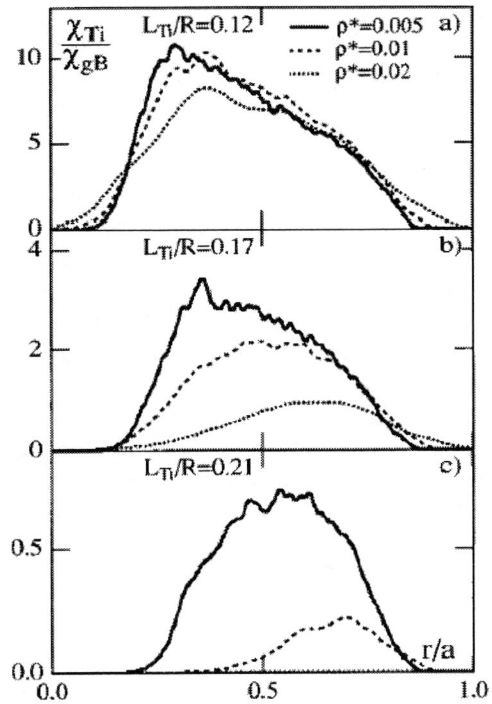

FIGURE 4. Calculated ion heat diffusivities normalized to the gyroBohm value for different values of ρ_* and temperature gradient length. The scaling departs from gyroBohm for large values of ρ_* and close to the threshold. From [9]).

β Scaling

The parameter β controls both the compression of magnetic surfaces (the Shafranov shift, which is stabilizing), and the excitation of electromagnetic instabilities [18]. Hence no dependence on β is expected if the turbulence is electrostatic and the Shafranov shift stabilization is negligible (typically well below the ideal MHD stability limit). As mentioned before, recent experiments [4, 5] indicate that the dependence on β is small on JET and DIII-D. On the other hand, a strong dependence is found on JT-60 [19] and Asdex-U [20]. Hence the situation is rather confusing. On the theoretical and modelling side, the situation is not much better. A robust result comes from linear stability [21]. It indicates that a low values of β, Ion Temperature Gradient (ITG) and Trapped Electron Modes (TEM) are stabilized by electromagnetic effects. In other words, the growth rate of ITG/TEM decreases with β. However this is a rather weak effect. Moreover, above a critical value β_c, Kinetic Ballooning Modes (KBM) are excited. This onset corresponds to a strong increase of the linear growth rate. On the basis of the linear stability analysis, one would expect a slight improvement of the confinement at low β, and a strong degradation at large $\beta > \beta_c$. This is actually what is observed in some numerical simulations [22, 23]. However these simulations do not agree on the critical value β_c above which the degradation occurs. In the simulation [22] β_c is of the order of half the ideal MHD value, while in [23] β_c is close to the MHD value. Obviously, this might be due to the different parameters chosen in these simulations. This might also explain the variety of results found experimentally. One key parameter that has been pushed forward to explain these differences is the geometry (in particular elongation and triangularity).

Collisionality scaling

The dependence on collisionality is also a controversial result. As mentioned before, the conventional ITER98-H scaling law gives a weak dependence on collisionality. More recent analysis indicates that the normalized confinement time does depend on collisionality. For instance, it was found on JET that at fixed ρ_* and β, $\omega_c\tau_E \equiv \nu_*^{-0.3}$ at low collisionality and $\omega_c\tau_E \equiv \nu_*^{-0.8}$ at high collisionality [29].

From the theoretical point of view, two opposite effects exist. The first one is the stabilization of trapped electron modes by collisions [24, 25]. This comes from the collisional detrapping of trapped electrons. This process occurs when the detrapping collision frequency ν_{ei}/ε becomes larger than $(k_\perp \rho_s)c_s/R$, which measures the electron precession frequency times a typical toroidal wave number. An effective collision frequency was defined in [30] as $\nu_{eff} = (5/2)^{1/2}\nu_{ei}R/(T_e/m_i)^{1/2}$, in the context of particle transport. It is stressed here that the parameter ν_{eff} differs quantitatively from the conventional parameter ν_*, which was historically defined in the context of neoclassical transport. However from the point of view of dimensionless analysis, there are the same. When $\nu_{eff} > 1$, TEM modes are stabilized by collisions. This is not an abrupt process as γ_{lin} behaves as $1/\nu_{eff}$. Also it turns out that the transition from ITG to TEM turbulence that occurs when increasing ν_{eff} is accompanied by a change of the sign of the phase velocity in the frame where the local radial electric field vanishes (different from the laboratory frame). The second physical effect goes in the opposite direction, and is associated to the collisional friction damping of zonal flows [26, 27, 28]. It is reminded here that zonal flows are fluctuations of the poloidal velocity, which reduce turbulent transport. Hence, when the collision frequency increases, zonal flows are damped, thus leading to a increase of the level of turbulent transport and a degradation of the confinement. Since it is found that the confinement time does decrease with collisonality, it looks like zonal flow damping is the dominant process. However this is contradictory with other observations which indicate that TEM collisional stabilization is important. One explanation is that zonal flow damping affects primarily ITG turbulence, and therefore the ion diffusivity, while TEM controls the electron diffusivity. Since the electron loss channel is subdominant in the total losses, the effect of collisions on TEM might be invisible when looking at the global confinement time in usual conditions. The fact that several scaling are found in the literature suggests that several mechanisms are involved.

FLUCTUATION MEASUREMENTS

Dimensionless scaling laws for fluctuations

It is shown here that fluctuation measurements provide a powerful way of testing dimensionless scaling law. An overview on this question can be found in [31]. This can be illustrated with the case of gyroBohm scaling, which is paradigmatic. Since gyroBohm scaling relies on correlation length and time ℓ_c and τ_c that scale respectively as ρ_s and a/c_s, one expect the correlation functions Eq.(2) to depend on $\delta x/\rho_c$ and $a\delta t/c_s$. Moreover the mixing-length estimate indicates that the level of density fluctuations should scale as $\delta n/n \equiv \rho_*$. To demonstrate this result, it is reminded that the mixing-length estimate can be derived by balancing the drive $\mathbf{v_E} \cdot \nabla n_{eq}$ with the non linear term $\mathbf{v_E} \cdot \nabla \delta n$ in the density evolution equation (n_{eq} is the equilibrium density). This implies that the density perturbation at scale ℓ behaves as $\delta n_\ell/n \simeq \ell/L_n$, where L_n is the density gradient length. This result can also be recovered with simple geometrical arguments, when assuming that all profiles are given, and gradient lengths scale as the minor radius a. In the case of gyroBohm scaling, all turbulent scales are related to ρ_s. Hence a consequence is that the level of density fluctuations should be of the form $\delta n/n \equiv \rho_*$ in the gyroBohm case. These considerations have consequences on the Fourier spectrum of fluctuations. In the general case, the spectrum of density fluctuations should be of the form

$$\frac{|\delta n_k|^2}{n^2} \equiv \ell_c^2 S(k\ell_c) F(\rho_*, \beta, \nu_*) \tag{3}$$

Here the spectrum is averaged over time. Also k is the perpendicular wave number. Indeed fluctuations in a strongly magnetized plasma are highly anisotropic. The parallel wave number $k_\parallel$ scales as a macroscopic length qR (q is the safety factor and R the major radius), and is much smaller than the perpendicular wave number $k_\perp$, which ranges between a fraction of cm^{-1} and several tens of cm^{-1}. Therefore fluctuations are quasi-two dimensional and we will identify k and $k_\perp$ in the following. Also we will assume that this quasi-2D turbulence is isotropic, which is almost true at high wave numbers, but not for the largest scales. The quantity $|\delta n_k|^2$ is called spectral density (noted usually S), and is nothing else than the Fourier transform of the spatial correlation function. For a quasi 2D turbulence, this relation reads

$$\frac{\langle \delta n^2 \rangle}{n^2} = \frac{1}{n^2} \frac{1}{V_{av} T_{av}} \int dt d^3\mathbf{x} \delta n^2(\mathbf{x}, t) \equiv \frac{1}{n^2} \int k dk |\delta n_k|^2 \equiv F(\rho_* \beta, \nu_*) \tag{4}$$

It is reminded here that the integration volume V_{av} and time T_{av} are much larger than the correlation lengths and times, but smaller than the profile evolution time. Thus δn^2 still contains some slow spatial and time variation. Assuming a monomial dependence, this scaling reads

$$\frac{\langle \delta n^2 \rangle}{n^2} \equiv [\rho_*]^{\alpha_\rho} [\beta]^{\alpha_\beta} [\nu_*]^{\alpha_\nu} \quad (5)$$

Again it is stressed that a monomial fit is a rather strong approximation, which might fail in some cases. In the case of gyroBohm scaling, one has $\alpha_\rho = 2$. Moreover the exponents α_β and α_ν can be related to the scaling coefficients for the diffusivity and energy confinement time. Indeed, the random walk diffusion coefficient reads as $\langle v_E^2 \rangle \tau_c$, where the ExB velocity v_E behaves as $k_{pol}\delta\phi/B$ (k_{pol} is some typical poloidal wave number and $\delta\phi$ is the perturbed electric potential). For low levels of fluctuations, it is expected that $e\delta\phi/T$ scales as $\delta n/n$ (note that this does not imply a Boltzmann response - it is rather a consequence of the gyrokinetic Poisson equation). Gyrobohm scaling law means that τ_c scales as a/c_s and that k_{pol} scales as $1/\rho_s$. It is then easy to find that Eq.(3) implies

$$\chi \equiv \chi_B \rho_* [\beta]^{2\alpha_\beta} [\nu_*]^{2\alpha_\nu} \quad (6)$$

and since $\omega_c \tau_E \equiv a^2/\chi$, one gets

$$\omega_c \tau_E \equiv [\rho_*]^{-3} [\beta]^{-2\alpha_\beta} [\nu_*]^{-2\alpha_\nu} \quad (7)$$

These expressions can be generalized to non gyroBohm scaling law by setting the explicit scaling for $k_{pol}\rho_s$ and $c_s\tau_c/a$.

Spatiotemporal properties of turbulence - Link with the ρ_*, collisionality and β scaling

Several authors have investigated the question of the wave number and frequency spectra of turbulence [32, 33]. Beam Emission Spestroscopy measurements of density fluctuations on the DIII-D tokamak [32] show that the spatial correlation functions depend on $\delta x/\rho_s$ only, where δx is the distance between two measurement points (Fig.5). In these experiments, only ρ_* was changed, while ν_* and β were held constant. In other words, the correlation length ℓ_c scales as the gyroradius, in agreement with the gyroBohm scaling prediction. In the same work, it is shown that the time correlation function depends on $a\delta t/c_s$ only, again in agreement with the gyroBohm expectation. In the same pulses, it was verified that the heat diffusivity is consistent with the fluctuations measurements, i.e. $\chi = \chi_B \rho_*$. Finally it was found that

$$\delta n/n \simeq \rho_*^{1.4\pm0.4} \quad (8)$$

which is (marginally) consistent with the mixing-length prediction. Similar results were obtained on Tore Supra, using a laser coherent scattering diagnostic [33]. The advantage of this diagnostic is its ability to provide a direct measurement of the wave number spectrum. It is found that the power spectrum decays as k^{-3} for low wave numbers, and typically as k^{-6} for large wave numbers. The change of slopes occurs typically for $k_c\rho_s \simeq 1$. The scaling $k_c \equiv \rho_s^{-1}$ was tested by changing the magnetic field or the ion mass number (Fig.6). Hence it appears that the spectrum is indeed some unique function of $k\rho_s$. A remarkable result is that the spectrum over the whole range of measured wave numbers can also be fitted by an exponential decay, typically $exp\{-4k\rho_s\}$. These results have been recently confirmed by measurements done on Tore Supra with a Doppler reflectometer [31]. Also it was found that the turbulence correlation time scale as a/c_s, as in DIII-D.

The scaling of fluctuations with ν_* and β is poorly documented. It was shown on Tore Supra that the level of fluctuations depends weakly on β, in agreement with the global confinement time [40]. Hence $\alpha_\beta = 0$ for these low β plasmas. This result is consistent with what was found on DIII-D and JET for the global confinement time. Concerning the collisionality dependence, it was shown on ASDEX-Upgrade that the phase velocity does depend on ν_{eff} [41]. In particular it changes sign when moving from ITG to TEM dominated plasmas.

Turbulence scaling properties

The theory of wave number spectrum in fluid turbulence has been intensively studied since the pioneering works of Kolmogorov, Obukov Landau and others. The interested readers may read existing overviews on this subject (for instance [34, 35, 36]). We summarize here the main results in view of existing measurements in magnetized plasmas. It is reminded that in 3D fluid turbulence, the longitudinal velocity perturbation at scale ℓ is given by the Richardson scaling

$$< \delta \mathbf{v}^2(\ell) > \equiv \varepsilon^{2/3} \ell^{2/3} \quad (9)$$

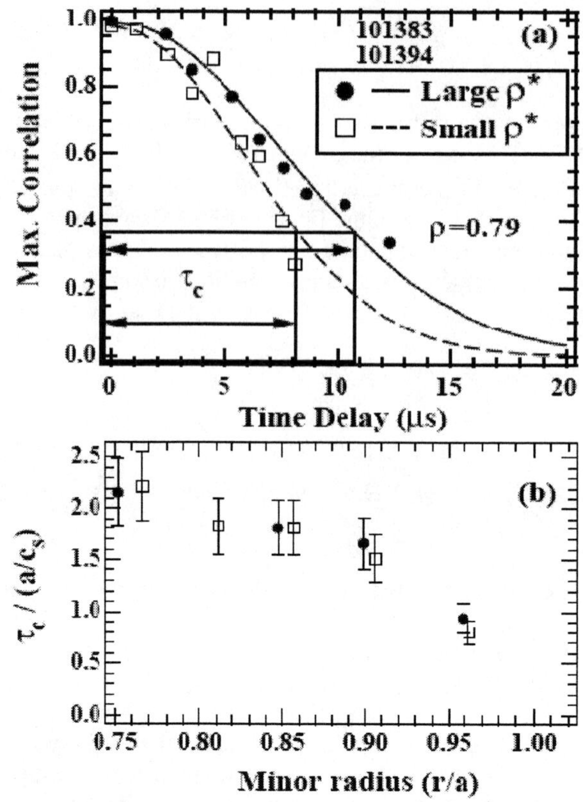

FIGURE 5. Scaling of correlation lengths and time for density fluctuations measured on DIII-D by BES (from [32]).

where the longitudinal velocity increment is defined as

$$\delta v(\ell) = (\mathbf{v}(\mathbf{r}+\ell) - \mathbf{v}(\mathbf{r})) \cdot \mathbf{n} \tag{10}$$

Here $\mathbf{n}$ is the unit vector oriented along the direction of ℓ, ε is the energy injected per time unit at large scales, and the bracket indicates an average over space and time. This scaling is valid in the inertial range, i.e. between the scale of energy injection and the dissipative scale. Note also that the eddy turnover time τ_ℓ is of the order of $\ell/\delta v(\ell)$ and scales as $\varepsilon^{-1/3}\ell^{2/3}$. The corresponding spectrum is the celebrated Kolmogorov scaling $E(k) \equiv \varepsilon^{2/3}k^{-5/3}$ (the spectral energy is defined as $1/2\langle\delta\mathbf{v}^2\rangle \equiv \int dkE(k)$). It is associated to a process of direct cascade, where energy is transferred nonlinearly from large to small scales, until it is dissipated by viscous effects. For a 2D turbulence, it is found that a dual cascade takes place. An inverse cascade of energy occurs for wave numbers smaller than the value k_{inj} at which energy is injected. In this process vortices coalesce to form larger scale structures, until the largest scale available is reached (unless some dissipative process occurs at large scale, like transport, collisional damping of zonal flows or Landau damping in magnetized plasmas). The scaling is of the Kolmogorov type (in particular the spectrum exhibits a slope $-5/3$, $\delta v(\ell) \equiv \varepsilon^{1/3}\ell^{1/3}$, and $\tau_\ell \equiv \varepsilon^{-1/3}\ell^{2/3}$). This inverse cascade is the consequence of the existence of two quadratic inviscid invariants (i.e. quantities which are conserved in absence of dissipation), namely the energy and the enstrophy (the enstrophy is defined as $\int d^2\mathbf{x}\,|\nabla\times\mathbf{v}|^2$). For wave numbers larger than k_{inj}, a direct cascade of enstrophy takes place. The corresponding wave number spectrum (Kraichnan spectrum) behaves as $E(k) \equiv \varpi^{2/3}k^{-3}$, where ϖ is the enstrophy injected per time unit at k_{inj} . In the region of direct cascade of enstrophy, the velocity increment scales as $\varpi^{1/3}\ell$ and the eddy-turnover time as $\varpi^{-1/3}$ (i.e.does not depend on ℓ). All these results apply to a homogeneous isotropic turbulence. However turbulence is not homogeneous in a tokamak, because of the transverse gradients. The archetype for a 2D electrostatic turbulence is the Hasegawa-Mima equation [37]

$$(\partial_t + \mathbf{v}_E \cdot \nabla) \cdot [\phi - \rho_s^2 \nabla^2 \phi] + \mathbf{V}_n^* \cdot \nabla\phi = -\mu\nabla^4\phi \tag{11}$$

where $V_n^* = T/eBL_n$ is the diamagnetic velocity and is oriented in the poloidal direction, and μ is some dissipation at small scales. Note that this equation also admits two inviscid invariants, which are the generalized energy

$$E = \frac{1}{2V_{av}} \int d^3\mathbf{x} \left(\phi^2 + |\nabla\phi|^2\right) \quad (12)$$

and enstrophy

$$\Omega = \frac{1}{2V_{av}} \int d^3\mathbf{x} \left(|\nabla\phi|^2 + \left|\nabla^2\phi\right|^2\right) \quad (13)$$

The shape of the spectrum has been investigated by Fyfe and Montogommery when ignoring the diamagnetic drift [38]. Their finding is in agreement with the prediction for dual cascade: the slope of the potential spectrum $< |\phi_{\mathbf{k}}|^2 >$ is $-14/3$ in the inverse energy cascade region and -6 in the direct enstrophy cascade range when $k\rho_s > 1$. Since the spectral energy behaves as $k^3 |\phi_{\mathbf{k}}|^2$, one recovers the Kolmogorov and Kraichnan spectra. These scaling for the electric potential spectrum can be shown to be valid also in the large wave number domain $k\rho_s < 1$ [39] (note however that the spectral energies are not the same in both regimes). Nevertheless the region $k\rho_s < 1$ corresponds to a situation where drift waves rather than vortices dominate the dynamics. This is also a situation where a high anisotropy between radial and poloidal directions develop, in particular via the formation of zonal flows. Thus an anisotropic spectrum is expected to develop.

The question is to link these classical results to observations in fusion devices. This link faces several difficulties. First the notion of inertial range is questionable in fusion plasmas, since instabilities exist at all relevant scales. Therefore no inertial range exists strictly speaking. Nevertheless, the maximum growth rate for ITG modes is reached for $k\rho_s \approx 0.3$. Thus, to some extent, one may assume that a significant part of the energy is injected at this scale. Second the link with fluid turbulence is not obvious. It appears however that for a quasi 2D turbulence, the electric potential plays the same role as the stream function for incompressible 2D fluid turbulence (the reason is the the $E \times B$ drift velocity reads as $\mathbf{v}_E = \mathbf{B} \times \nabla\phi/B^2 = -\nabla \times (\phi\mathbf{B}/B^2)$ if $\mathbf{B}$ is constant). Finally density rather than velocity fluctuations are usually measured in plasma turbulence. So a link must be established between density and potential fluctuations. One way is to consider the density as a passive scalar, i.e.

$$\partial_t n + \mathbf{v}_E \cdot \nabla n = D\nabla^2 n \quad (14)$$

where D is some molecular diffusion coefficient which acts at small scale. Note that $< n^2 >= \int dt d^3\mathbf{x} n^2/VT$ is an inviscid invariant. It is also stressed that there is no other quadratic invariant, so that the cascade of a passive scalar is always direct, even in 2D turbulence. Introducing the scalar dissipative rate $\varepsilon_n = -1/2\partial_t < n^2 >$, and reproducing the Kolmogorov argument, $< \delta n^2(\ell) >$ should scale as $\varepsilon_n \tau_\ell$ [34]. In the region of direct enstrophy cascade, one would then expect a spectral density of density fluctuations that behaves as $E_n(k) \equiv \varepsilon_n \varpi^{1/3} k^{-1}$ [34] (the spectral density is defined as $\langle \delta n^2 \rangle \equiv \int dk E_n(k)$). Since $E_n(k) \equiv kS(k)$, $S(k)$ should decay as k^{-2}. In the inverse cascade region, a similar arguments leads to $E_n(k) \equiv \varepsilon_n \varpi^{1/3} k^{-5/3}$ (same as energy Kolmogorov spectrum) and $S(k) \equiv k^{-8/3}$, which is not far from the observation, given the error bars [33]. The approximation of a passive scalar is acceptable to some extent for trapped electrons in a pure ITG turbulence. However it is not correct for passing electrons, whose response is essentially adiabatic for a low wave number ITG/TEM turbulence. Using a Boltzmann scaling, i.e. $n_k/n \equiv e\phi_k/T$, it is found that $S(k) \equiv E(k)/k^3$, i.e. $S(k) \equiv k^{-14/3}$ in the Kolmogorov range (inverse energy cascade) and $S(k) \equiv k^{-6}$ in the Kraichnan zone (direct enstrophy cascade). These spectra are clearly to steep when compared to measurements in the range $k\rho_s < 1$, but are in the correct range for $k\rho_s > 1$. As mentioned before, the region $k\rho_s < 1$ corresponds to scales where drift waves are dominant, turbulence is anisotropic, and zonal flows are generated. The shape of the spectrum is not well documented in this region.

Hence it looks like that the scaling that is the closer to the measurement for low wave numbers is the one for a passive scalar in the inverse energy cascade spectral region, which is somewhat surprising. In fact this problem is still open and it is rather clear at this stage that numerical simulations should be used to clarify this point. It must noted at this point that high order structure functions (i.e. moments $\langle \delta v(\ell)^p \rangle$) have been intensively used in fluid turbulence to characterize intermittency [35]. This technique has been rarely used by experimentalists in fusion plasma turbulence due to a lack of data. Some results exist on the theoretical side for passive scalars [42].

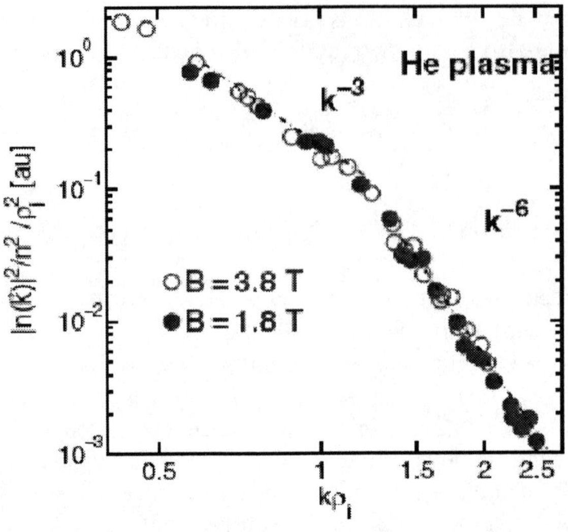

FIGURE 6. Spectrum of density fluctuations measured on Tore Supra by coherent laser scattering for two different values of the magnetic fields (from [33]).

CONSEQUENCES FOR TRANSPORT MODELS

Rationale for building transport models

Dimensionless scaling laws can be used to constrain transport models based on quasi-linear and mixing length assumptions. Examples of transport models based on this recipe are the Weiland [43], and GLF23 [44] models. In its simplest version, a mixing-length model just sets $\chi = \gamma_{lin}\ell_c^2$. Hence the scaling is gyroBohm if one chose a correlation length that scales as the gyroradius. Also the dependencies on ν_* and β will be essentially those given by the linear stability. Of course, actual models are more complex and more subtle variations can be accounted for. An alternative to transport models is based on the concept of "profile stiffness" [45]. This idea can be understood as follows. If the turbulent transport is very large when gradients cross the stability threshold, then the profiles should stay close to a marginal state, i.e. their gradients should stay close to the stability threshold. In practice, only part of the profile is close to marginal stability. This concept is nevertheless helpful to interpret experiments, when combined with linear stability analysis. An intermediate approach between predictive transport modeling and strong profile stiffness consists in using a semi-empirical critical gradient model (CGM). Such a model is characterized by 3 parameters only and will be used here to illustrate various concepts. Basically it consists in stating that the simplest diffusivity that corresponds to a gyroBohm collisionless electrostatic turbulence with a stability threshold of the form [46]

$$\chi = \chi_s \frac{T}{eB}\rho_* \left[-R\frac{dT}{Tdr} - \kappa_c\right] \tag{15}$$

where χ_s is called the stiffness factor and κ_c is the stability threshold. The identification of these parameters is made possible by analyzing experiments where the heating source is modulated [47]. Usually values of $\chi_s \approx 1$ and $\kappa_c \approx 6$ are found. Nevertheless more accurate models like the Weiland or GLF23 model should be used for a more detailed analysis.

Scaling with q, T_e/T_i, charge number and mass

Although a large part of the literature addresses the dependencies on ρ_*, ν_* and β, other parameters can play a role. In particular the safety factor q play an important role as it is well known that the confinement time is sensitive to the plasma current. In practice this appears as a dependence of the stiffness factor on the local value of q. Usually the best scaling in Eq.(15) is $\chi_s \sim q^{\alpha_q}$, where α_q ranges between 1 and 2 (a common choice is 3/2). The ratio of electron to ion temperatures and charge number Z_{eff} (defined as $Z_{eff} = \sum_{ions} n_i Z_i^2 / n_e$) also play a role. One reason is the dependence of the stability threshold for ITG modes on these two parameters. A simple formula is

$$\kappa_c = \kappa_{c0} \left(1 + C \frac{T_i}{Z_{eff} T_e} \right) \tag{16}$$

where κ_{c0} and C are numbers which depend on the density gradient length. The dependence is for example responsible for the so-called hot ion mode, which is a regime with improved confinement obtained when $T_i >> T_e$. Finally the question of mass dependence is briefly addressed. A gyroBohm scaling law predicts a heat diffusivity that behaves as $m_i^{1/2}$, and therefore a global confinement time that should scale as $m_i^{-1/2}$. On the other hand it is well known that a deuterium plasma is better confined than an hydrogen plasma. This appears as a favorable mass dependence in the Iter98 scaling law for the the global energy confinement time in H-mode. This finding is clearly in contradiction with a gyroBohm scaling law. There is no clear explanation for this behavior. It might come from the confinement of particles, which depend sensitively on the fueling, and therefore on the ionization length of neutrals (not the same for hydrogen and deuterium).

CONCLUSION

Dimensionless scaling laws appear as a powerful tool to constraint the models for confinement time and transport coefficients. A lot of progress has been made in the recent years in this field, thanks to dedicated experiments and measurements. In particular fluctuations measurements appear as an efficient tool to improve the accuracy in the determination of scaling coefficients. 3 main parameters appear to play a central role: the normalized gyroradius ρ_*, the ratio β of kinetic and magnetic pressures and the collisionality ν_*. There is now a consensus for predicting a gyroBohm scaling law for the dependence on ρ_*, at least if ρ_* is small enough, and for gradients well above the stability threshold. In this frame, all turbulence spatial lengths scale as a gyroradius, and all typical times scale as a/c_s. The diffusion coefficients scale as $\rho_* T/eB$ and $\omega_c \tau_E$ scales as ρ_*^{-3} (at fixed β and ν_*). Still there exists cases where the scaling does not follow the gyroBohm expectation (ions in the L-mode for instance). The reason for this departure is still unknown. The situation is less clear for the dependencies on β and ν_*. Experimental observations disagree on the dependence of the confinement time on β. From theory, it is expected that a weak dependence should be found at low values of β and a degradation above a critical value. However there is no clear prediction for the value of this critical β. As for collisions, the prediction is still to be done, since two antagonist effects compete, the stabilizing effect of collisions on TEM and the collisional damping of zonal flows. The situation will probably be clarified with development of codes able to simulate both ITG and TEM turbulence routinely. Also dedicated experiments should be able to clarify the situation. Finally, dimensionless scaling laws provide an efficient way of comparing numerical simulations with fluctuation measurements. In particular the scaling of wave number spectra and frequency spectra, or equivalently of the space and time correlation functions certainly deserve a more detailed analysis than what has been done up to now.

ACKNOWLEDGMENTS

Several contributions from the Task Force T at JET is acknowledged here. Recent discussions on the quasi-linear theory with Drs C. Bourdelle, D. Escande, P.H. Diamond have also been fruitful. The author also thanks the "Festival de Theory" for stimulating discussions on turbulence transport. Finally a large part of the chapter on fluctuations comes from interactions with Drs P. Hennequin, R. Sabot and F. Clairet.

REFERENCES

1. ITER Physics Expert Groups on Confinement and Transport and Confinement Modelling and Database, Nucl. Fusion **39**, 2175 (2000).
2. B.B. Kadomtsev, Sov. J. Plasma Phys. **1**, 295 (1975).
3. J.W. Connor, and J.B. Taylor, Nucl. Fusion **17**, 1047 (1977).
4. D.C. McDonald, et al. Plasma Phys. Cont. Fusion **46** A215 (2004).
5. C.C. Petty et al. Phys. Plasmas **11**, 2514 (2004).

6. C.C. Petty et al., Phys. Rev. Lett. **74**, 1763 (1995).
7. R.E. Waltz et al. Phys. Plasmas **1**, 2229 and 1995 Phys. Plasmas **2**, 2408 (1994).
8. P.N. Guzdar et al., Phys. Fluids B **5**, 3712 (1993).
9. X. Garbet, R.E. Waltz, Phys. Plasmas **3**, 1898 (1996).
10. M. Ottaviani and G. Manfredi Phys. Plasmas **6**, 3267 (1999).
11. Z. Lin et al., Phys. Rev. Lett. **88**, 195004 (2002).
12. J.W. Connor, J.B. Taylor, and H.R. Wilson Phys. Rev. Lett. **70**, 1803 (1993).
13. F. Romanelli and F. Zonca Phys. Fluids B **5**, 4081 (1993).
14. W. Dorland , et al. Theory of Fusion Plasmas (Proc. of the Joint Varenna-Lausanne Int. Workshop, Varenna 1996) (Editrice Compositori:Bologna), p.185.
15. G. Tresset et al., Nucl. Fusion 42, 520 (2002).
16. P.H. Diamond and T.S. Hahm, Phys. Plasmas **2**, 3640 (1995).
17. T.S. Hahm et al., Plasma Phys. Control. Fusion **46**, A323 (2004).
18. S.J. Camargo, B.D. Scott, and D. Biskamp , Phys. Plasmas **3**, 3912 (1996).
19. H. Urano et al., Nuclear Fusion, **46**, 781 (2006).
20. L. Vermare et al., Nuclear Fusion **47**, 490 (2007).
21. G. Rewoldt, W. M. Tang, and R. J. Hastie Phys. Fluids **30**, 807 (1987).
22. P. B. Snyder and G. W. Hammett, Phys. Plasmas **8**, 3199 (2001).
23. B.D. Scott, Plasma Phys. Control. Fusion **48**, B277 (2006).
24. B.B. Kadomtsev, O.P. Pogutse, Reviews of Plasma Physics, edited by M.A. Leontovitch (Consultant Bureau, New York, 1970) Vol. **5**, p. 249.
25. F. Ryter et al., Phys. Rev. Lett. **95**, 085001 (2005)
26. Z. Lin et al., Science **281**, 1835 (1998).
27. P H Diamond, S-I Itoh, K Itoh and T S Hahm Plasma Phys. Control. Fusion **47**, R35 (2005)
28. G. Falchetto and M. Ottaviani, Phys. Rev. Letters **92** 25002 (2004).
29. M.C. McDonald, private communication (2004).
30. C. Angioni et al., Phys. Plasmas **10**, 3225 (2003) and Phys. Rev. Lett. **90**, 205003 (2003).
31. P. Hennequin et al., Nucl. Fusion **46**, S771-S779 (2006).
32. G.R. McKee et al., Nucl. Fusion **41**,1235 (2001)
33. P. Hennequin et al., Plasma Phys. Control. Fusion **46**, B121 (2004).
34. M. Lesieur, "Turbulence in Fluids", Kluwer Academic Press (1987).
35. U. Frisch, "Turbulence", Cambridge University Press (1995).
36. S.B. Pope,"Turbulent Flows", Cambridge University Press (2000).
37. A. Hasegawa and K. Mima, Phys. Rev. Lett. **39**, 205 (1977).
38. D. Fyfe and D. Montgomery, Phys. Fluids **22**, 246 (1979).
39. M. Ottaviani and J. Krommes, Phys. Rev. Lett. **69**, 2923 (1992).
40. A. Sirinelli et al., submitted (2008).
41. G.D. Conway et al., Nucl. Fusion **46**, S799 (2006)
42. S. Futatani et al. , to appear in Phys. Rev. Letters (2008).
43. J. Weiland, "Collective Modes in Inhomogeneous Plasmas", IOP, 2000.
44. R.E. Waltz, G.M. Staebler, W. Dorland, et al., Phys. Plasmas **4**, 2482 (1997).
45. B. Coppi and N. Sharky, Nucl. Fusion **21**, 1363 (1981).
46. F. Imbeaux, F. Ryter and X. Garbet, Plasma Physics and Controlled Fusion **43**, 1503 (2001).
47. X. Garbet, P. Mantica, F. Ryter, G. Cordey, F. Imbeaux, C. Sozzi, A. Asp, V.Parail, R.Wolf, and the JET EFDA Contributors, Plasma Phys. Control. Fusion **46**, 1351 (2004) - Addendum Plasma Phys. Control. Fusion **47**, 957(2005).

Electron Temperature Gradient Mode Transport

W. Horton and J. -H. Kim
Institute for Fusion Studies, the University of Texas at Austin

G. T. Hoang
Association Euratom-CEA, CEA/DSM.DRFC

H. Park, S. M. Kaye, B. P. LeBlanc
Princeton Plasma Physics Laboratory, Princeton University

(Dated: December 5, 2007)

Anomalous electron thermal losses plays a central role in the history of the controlled fusion program being the first and most persistent form of anomalous transport across all toroidal magnetic confinement devices. In the past decade the fusion program has made analysis and simulations of electron transport a high priority with the result of a clearer understanding of the phenomenon, yet still incomplete. Electron thermal transport driven by the electron temperature gradient is examined in detail from theory, simulation and power balance studies in tokamaks with strong auxiliary heating.

PROPERTIES OF ELECTRON TRANSPORT

Electron temperature profiles respond differently than the density and ion temperature profiles to the influence of externally applied heating and particle sources. In the first toroidal experiments such as the model C-stellarator[1] and the first tokamak[2], it became clear that electron thermal transport could be characterized by an electron thermal diffusivity $\chi_e(\,\mathrm{m}^2/\mathrm{s})$ that was much higher than the value that given by collisional transport theory. Theses early devices showed a χ_e that roughly followed the diffusion rate given by Bohm [3], $D_B = \alpha_B(T_e/eB)$ with various values of the numerical coefficient α_B. In the model C-stellarator, for example, α_B was typically $1/50$ and in some lower temperature toroidal octopole experiment [4] records were reached with $\alpha = 1/100$ to $1/1000$. The scaling T_e/B of the diffusivity seemed secure.

In 1967 Artsimovich went to the Princeton Plasma Physics Laboratory to announce the break-through of reaching a record Lawson product of $n_e\tau_E$ at the Kurchatov Laboratory with the T-3 tokamak. Record confinement time of $\tau \simeq 20\,\mathrm{ms}$ were reached in the new T-3 tokamak at Kurchatov. Princeton quickly rebuilt their stellarator into an axisymmetric tokamak using the same ohmic heating transformer system and magnetic field power supplies. The new device, called the

CP1013, ***Turbulent Transport in Fusion Plasmas, First ITER International Summer School***
edited by S. Benkadda

Princeton Large Torus (PLT), reproduced the record energy confinement times $\tau_E \simeq 100\,\mathrm{ms}$ and went on to achieve record ion temperature after 2 MW of neutral beam heating NBI were added. The electron thermal transport still followed something close to the reduced ($\alpha_B \ll 1/16$) Bohm scaling with a numerical coefficient in the range of $\alpha_B \simeq 10^{-3}$.

Direct measurement of the source of the anomalous plasma transport due to the plasma fluctuations [5] requires very careful determination of the relative phase of density δn and potential $\delta\phi$ fluctuations, since to zeroth order they are in phase, implying no next transport. Measurements of the plasma decay time using a microwave interferometer, Langmuir probes and a total radial loss detector show that the plasma lifetime was roughly 10 times the Bohm diffusion time. The plasma lifetime was measured by the density decay rate after turning off the external drives. Thus the experiments were characterized by $\alpha_B \sim 1/160$ rather than the coefficient of $(1/16)(T_e/eB)$ in diffusion equation for the plasma density.

Modern tokamak experiments routinely have core heating power densities at or above of $P_e/V = 1\,\mathrm{MW/m^3}$ into the electrons. At a density of $5 \times 10^{19}\,\mathrm{m^{-3}}$ this corresponds to an energization rate of $P_e/n_e = 300\,\mathrm{KeV/s}$ giving $\delta T_e = 3\,\mathrm{KeV}$ in 10 ms of heating in the absence of losses. Typically the electron temperature will saturate from radial thermal transport to the edge plasma where the total heating power P_e is carried away by the limiters and/or the scrape-off-layer plasma. The standard model for the radial thermal loss is a hear flux $q_e = -n_e\chi_e(\mathrm{d}T_e/\mathrm{d}r)$ with typical modern tokamaks having $\chi_e \sim 5 - 50\,\mathrm{m^2/s}$.

For example, extensive electron transport scaling data is reported by Tore Supra experiments from scans in the externally applied heating power P_e from 0.75 MW to 7.5 MW as shown in Fig. 1. The corresponding electron temperature profiles are shown in Fig. 1a and the density n_e and the safety factor q profile in Fig. 1b. The steady state power balance condition then leads to the electron thermal flux q_e $[\mathrm{MW/m^2}]$ shown in Fig. 1c in the top (red) curves. The lower (black curves) in Fig. 1c show how much heat flux is accounted for by the radiation losses and the transfer of electron energy to lower temperature ions by electron-ion collisions. We see that more than 90% of the electron thermal loss in this set of 8 Tore Supra discharges is carried away by the anomalous or turbulent electron thermal diffusivity.

The Tore Supra data in Fig. 1 is from a classical circular cross section tokamak with $R/a = 2.4\,\mathrm{m}/0.8\,\mathrm{m}$ and a toroidally symmetric pump limiter. The maximum Ohmic heating occurs for the lowest temperature discharge and is approximately 0.75 MW. The data in Fig. 1 has RF heating up to 7.4 MW. More recent experiments have extended the data up to 15 MW heating with core T_e reaching 9 KeV.

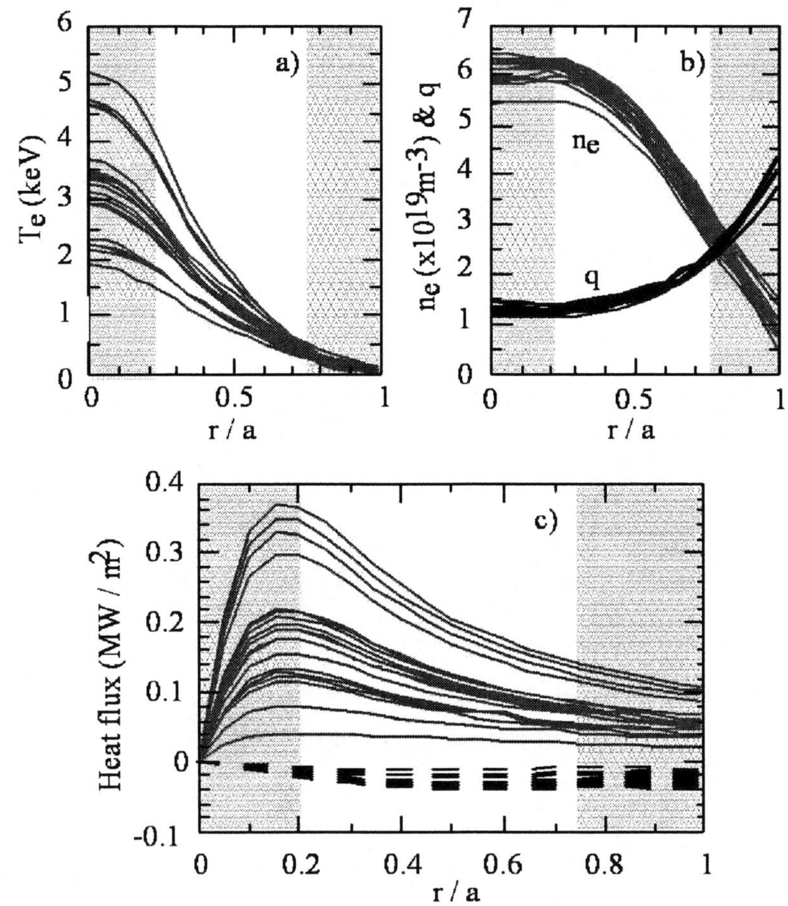

FIG. 1: Profiles of (a) The electron temperature T_e, (b) electron density n_e and safety factor q, and (c) heat fluxes (solid = FW coupled to electrons + Ohmic; dashed = radiation and electron-ion losses) in Tore Supra, $I_p = 0.6\,\mathrm{MA}$, $B = 2.2\,\mathrm{T}$ and $P_{\rm tot} = 1.5 - 7.5\,\mathrm{MW}$ ($P_{\rm OH} = 0.1 - 0.75\,\mathrm{MW}$ and $P_{\rm FW} = 0.75 - 7.4\,\mathrm{MW}$)[6]

The non-resonant fast wave in the ion cyclotron frequency range heats the thermal velocity part of the electron velocity distribution through electron Landau resonances with $\omega_{\rm RF} = k_{\|}^{\rm RF} v_{\|}$ Fast Wave heating. It is the RF heating mechanism favored by Stix [7] for reactors and is used extensively in both Tore Supra and the National Spherical Tokamak Experiment NSTX at PPPL. Compared with higher frequency Lower Hybrid heating and Electron Cyclotron heating, Fast Wave heating gives the least distortion from electron Maxwellian distribution and in this sense is closer to the Ohmic heating mechanism which is known to give the best confinement time properties of the electrons. The problem with Ohmic heating is that the electron heating mechanism becomes very weak at the electron temperatures required for fusion plasmas $T_e > 10\,\mathrm{KeV}$.

Fig. 2 shows a typical Fast Wave heating experiment from NSTX where the electron temperature

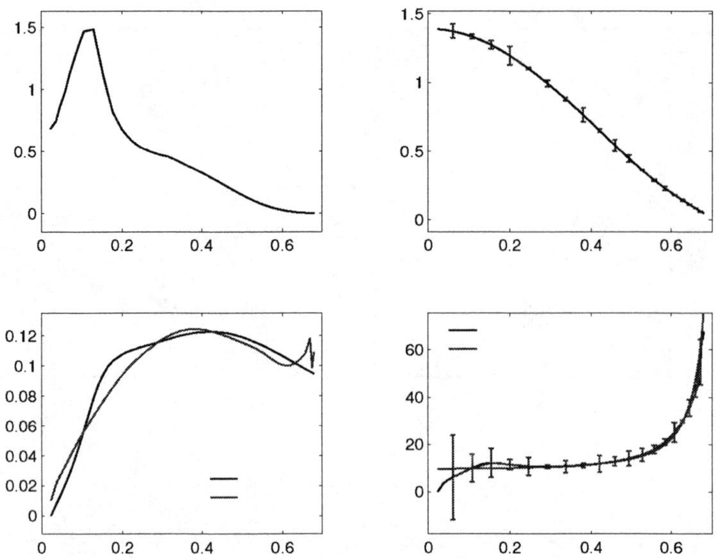

FIG. 2: NSTX profiles and power balance analysis for discharge 106194. From (a) is the electron absorbed RF power density, (b) the 20 cord 17ms Thomson scattering electron temperature profile with a 3% random error added to estimate the uncertainty in the electron power balance analysis, (c) the electron thermal fluxes, q_e from power balance (PB) and from the ETG model and (d) the respective thermal diffusivities with the error estimate for the power balance diffusivity.[8]

profile is measured with high resolution by a combination of high tech instruments. Fig. 2a shows the profile of the core deposited High Harmonic Fast Wave (HHFW) heating, Fig. 2b shows the corresponding electron temperature profile with estimates of the uncertainty from the red arrow bars, Fig. 2c the inferred electron thermal heat flux and compared with a theoretical model called the ETG model and discussed in Sec.. Finally, dividing the electron thermal flux by the electron temperature gradient yields in Fig. 2d the electron thermal diffusivity χ_e. The profile of a low χ_e in core and increasing rapidly toward the edge is a widely observed characteristic of toroidally confined plasmas and can be difficult to reconcile with the theoretical formulas for the anomalous diffusivity.

The resolution of the self consistency problem of the χ_e profiles with the theoretical formulas often requires using the theory formulas in the predictive-interpretative mode of an integrated transport model. These integrated transport models contain sufficient coupling between the multiple channels of power deposition, power flow and power loss to evaluate the performance of the theoretical formulas. Integrated transport codes for this purpose include the BALDUR code of the Lehigh University group, the TRANSP code at PPPL, the CRONOS code at Cadarache and the ASTRA code. An example of the evolution of the electron temperature profile with the ETG

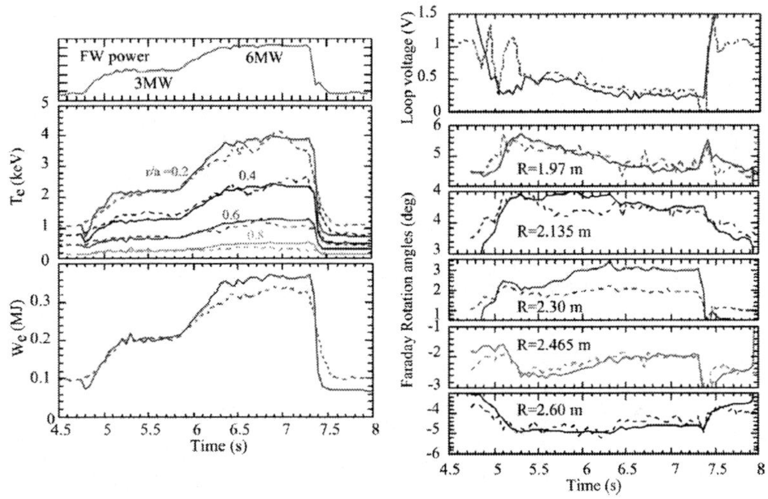

FIG. 3: Tore Supra FWEH discharge #18368 data (dashed curve) compared with the ETG model (solid curve). Top right panel shows the modeled loop voltage (solid) versus the measured voltage (dashed). Lower right panels give the poloidal magnetic field measured (dashed) and predicted (solid) from the Faraday Rotation diagnostics and simulation at four vertical cords.

model corresponding to the χ_e shown in Fig. 2d is shown in Fig. 3 using the CRONOS code in this predictive-interpretative mode in which data values for T_i and edge T_e parameters are used along with a comprehensive ray tracing code for the FW heating power deposition [9].

The electron thermal flux $q_e(r)$ is computed from power balance in the steady state by

$$\frac{1}{r}\frac{\partial}{\partial r}(rq_e) = P_e(r) - Q_{ei}(r) \tag{1}$$

where the electron thermal losses to the colder ions is given by

$$Q_{ei} = 3\frac{m_e}{m_i}\nu_{ei}n_e(T_e - T_i) \tag{2}$$

The electron thermal diffusivity from power balance is computed from

$$\chi_e = -\frac{q_e}{n_e\frac{\partial T_e}{\partial r}} \, . \tag{3}$$

The ETG has also been successfully used in comparing χ_e from modulated electron cyclotron heated (ECH) plasmas in the Frascati Tokamak named FTU. By modulating the envelop of the ECH power $P_{\rm ECH}(r,t)$ and measuring the corresponding variations in the temperature profile is one useful way to measure the thermal diffusivity $\chi_e(r)$. In Fig. 4 we show the results of this experiment reported by Jacchia et al. [10] for three different discharges. The results are explained adequately by the ETG model. Jacchia et al. [10] also compare the thermal diffusivity derived from steady state power balance with that obtained by applying heat pulses as shown in Fig. 5 where

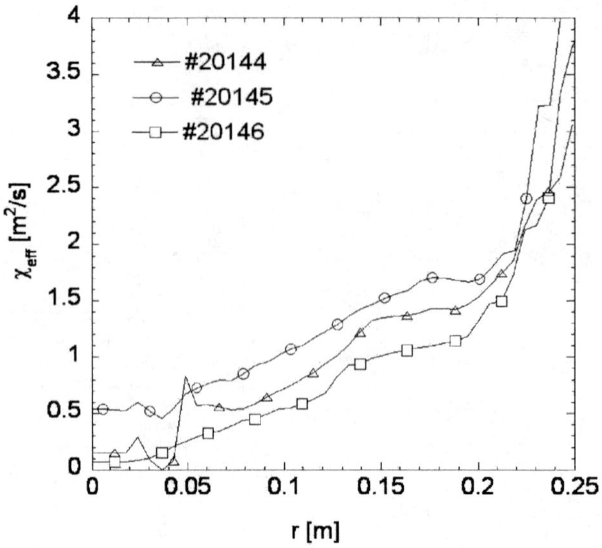

FIG. 4: FTU thermal transport coefficient χ_{eff} as given by the power balance analysis is shown vs radial position r.

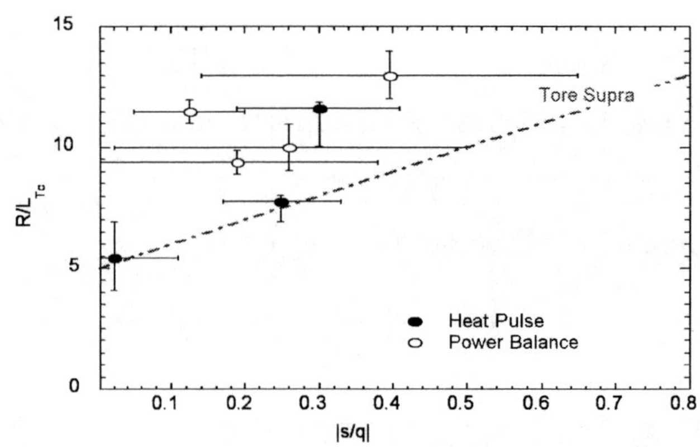

FIG. 5: Dimensionless parameter, R/L_{T_c} from FTU tokamak perturbative and power balance (open symbols) is shown vs the ratio s/q. The offset linear fit of Tore Supra [6] data ($R/L_{T_c} = 5 + 10s/q$) is also shown in the same figure for the sake of reference ($- \cdot -$). Error bars on data from perturbative analysis are obtained by propagating the uncertainty in the radial position of the step in χ_{HP} on the determination of $1/L_{Tc}$ and s/q.

the results are expressed in terms of the critical temperature gradient L_{Tc}/R versus the magnetic shear parameter s/q with the Tore Supra thermal balance analysis ($- \cdot -$). The ETG model also explains the key thermal transport properties of the electron cyclotron heated high T_e plasmas in the TCV tokamak [11].

Similar results are being observed by Asp et al. [11] for the very high power density electron cyclotron wave heating, reaching $6\,\text{MW/m}^3$ at the core of the TCV machine at Lausanne. Heat

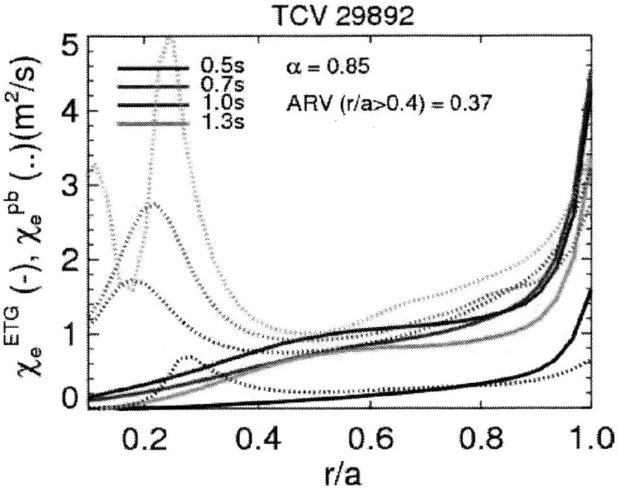

FIG. 6: The comparison between the electron gyro-Bohm heat diffusivity, χ_e^{ETG} based on the mixing length argument, (solid) with ETG time and spatial scales and the power balance heat diffusivity χ_e^{pb} (dotted) for each time slice $t = 0.5$(black), 0.7(red), 1.0(blue) and 1.3(green) in TCV 29892.

diffusivities χ_e^{ETG} based on the mixing length theory for ETG with length scale of electron gyro-radius ρ_e and time scale of the electron temperature gradient length divided by electron thermal velocity L_{T_e}/v_{T_e} is shown to be consistent with the power balance diffusivities χ_e^{pb} in the different phases of H-mode, above $r/a > 0.4$. Fig. 6 shows th model and power balance electron thermal diffusivities for the high P_e TCV discharge. The peak of heat diffusivities inside $r/a < 0.4$ shown in Fig. 6 are due to the MHD activity since $q_0 < 1$ and correlate with 3/2-neoclassical tearing mode islands.

We conclude that the ETG model is working for all four machines (TS, NSTX, TCV and FTU) and with different methods of determining the thermal diffusivity. Thus, the ETG model has been verified as an important theoretical model for explaining the anomalous electron thermal transport in toroidal systems.

Now, we turn to explaining the theoretical structure of the ETG model.

PHYSICAL MECHANISM OF THE ETG TURBULENT TRANSPORT

For drift wave turbulence from the space scales of the electron gyro-radius down to the ion gyro-radius, the effect of the local maximum and minimum of the electrostatic potentials is as shown in Fig. 7. At this scale length only the electrons are able to $\boldsymbol{E} \times \boldsymbol{B}$ drift and follow, with their guiding centers, the isopotential contour shown in Fig. 7 as circle. The ion gyro-radius is so large compared with the scale of the fluctuation that the electric field is averaged over the ion orbit to a

ETG Drift Waves

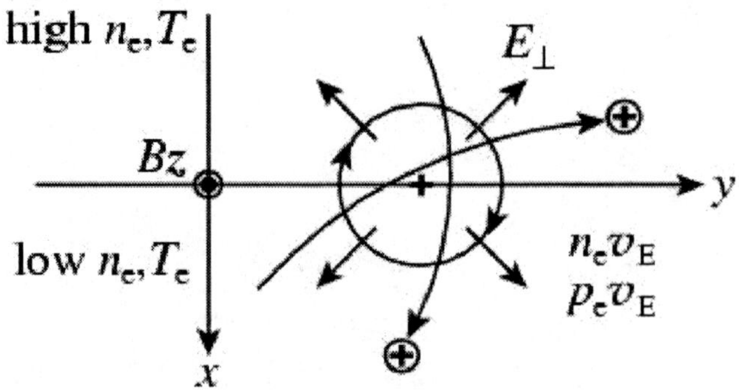

FIG. 7: Scheme of drit waves at the electron gyro-radius scale

small value. There only remains the local Boltzmann response of the ion density $\delta n_i = -n_i e\phi/T_i$ to first approximation in the wavelength $\lambda_\perp/\rho_i < 1$. Later on we can insert perturbatively the effect of small, resonant ion contributions from high energy ions that can resonate ($\omega_{\boldsymbol{k}}^{\mathrm{ETG}}$ with the high ETG wave frequency $\omega = k_y \rho_e v_e / L_{Te} \sim 0.3 v_e/L_{Te} \gg c_s/L_n$).

Owing to this high frequency-short scale length of the ETG modes the electrons that resonate with the waves are the thermal electrons rather than the low energy electrons as in the usual drift waves. The ETG mode does not depend on the electrons being magnetically trapped as in the trapped electron mode in Chap. 1. The mode is driven by the electron temperature gradient $\boldsymbol{\nabla} T_e$ and has both a slab version with an effective adiabatic gas constant of $\Gamma = 3$ and threshold $\eta_e = 2$ and a toroidal version with $\Gamma = 5/3$ and $\eta_e = 2/3$ for the case of weak magnetic shear and $k_y \rho_e < 1$.

These threshold formulas for the critical electron temperature-to-density gradient follow from the Nyquist analysis of the Vlasov-Poisson dispersion relation. The results are easily understood from the Carnot cycle analysis introduced in Chap. 1. We repeat the physical picture since it is important for understanding why the plasma collective modes in the ETG regime create long, extended radial structures with $\ell_c \gg \rho_e$ called "streamers". The increase of the released thermal-gradient energy by the turbulence with ℓ_c^2 follow from the maximum energy released by the Carnot engine operating between T_1 and T_2 separated by the correlation length ℓ_c shown in Fig. 3 of Chap. 1. The thermodynamic calculation requires the value of the effect gas constant Γ and the temperature-to-density gradient ratio $\eta_e = d\ln T_e/d\ln n_e$. The result gives the maximum energy that the spontaneous plasma convection can release in to the ETG mode and thus provides the useful upper bound on the ETG turbulence level. This released energy density W_{ETG} is a very small fraction the thermal energy density $3/2 n_e T_e$ of the electron component of the plasma. However, the power transfer $2\gamma W$ from the plasma to the waves of energy density W is high due to the growth rate of the instability being in the range of $\gamma \sim 0.3 v_e/(L_{Te} R)^{1/2} > 10^6\,\mathrm{s}^{-1}$.

GYROFLUID ETG MODEL

The basic electron temperature gradient turbulence may be described with three partial differential equations for the fluctuating electrostatic potential $\phi(\boldsymbol{x}, t)$, the fluctuating parallel vector potential $A_{\|}(\boldsymbol{x}, t)$ and the electron temperature $\delta T_e(\boldsymbol{x}, t)$ fluctuation. These are finite electron gyro-radius fluid equations that are valid in the wavenumber range $\rho_i^{-1} \leq k_\perp \leq \rho_e^{-1}$ where ρ_i is the ion gyro-radius and ρ_e is the electron gyro-radius. The fluid description parallel to $\boldsymbol{B}$ uses a closure for the chain form p_e to $\nabla_{\|} q_{e\|}$ to describe accurately the parallel electron Landau resonances $\tilde{\omega} = \omega - \omega_E - \omega_{De} = k_{\|} v_{\|}$ and the trapped electron response. The parallel electron heat flux $q_{\|}(\boldsymbol{k})$ is closed [12, 13] either in the adiabatic limit described by the adiabatic gas constant Γ or with Landau-fluid closure which is strongly dissipative with $\tilde{q}_{\|}(\boldsymbol{k}) = \chi_{e\|} |k_{\|}| v_e n_e \delta T_e(\boldsymbol{k})$, where $\chi_{e\|}$ is a suitable constant of order unity. Typically $\chi_{e\|}$ is chosen to match the associated growth rate $\gamma(\boldsymbol{k}_\perp, k_{\|}, \chi_{e\|})$ with the values derived from the kinetic dispersion relation that is free of the closure problem.

The ETG instability is robust once the critical gradient $(\eta_e > \Gamma - 1 + L_n/L_s(1 + T_e/T_i))$ is exceeded. Thus, the lack of a precise wave-particle resonances dissipation fluid closure does not cause a serious limit on the validity of the turbulence model. The advantage of the FLR-fluid description is that with the pseudo-spectral computer algorithm and large grid sizes (512×512, $0.025 < k_y \rho_e < 2.5$) run within a few hours of CPU time. Advanced gyrokinetic simulations [14, 15] confirm that large boxes with small $k_y \rho_e$ values are required to give converged values of $\chi_e t$.

The dominant feature of the nonlinear simulations is a strong inverse cascade to large perpendicular wavelengths that yield large enhancement (factors of 10-70) over the ETG weak turbulence mixing length formula $\gamma_k / k_\perp^2$ for the diffusivity. In the base-line core simulation for the NSTX 120967, we obtain a numerical coefficient of order 10-20 for the numerical factor in the gyro-Bohm formula. Fig. 8 shows the growth and saturation of the ETG turbulence spectrum for a mid-radius point in the NSTX discharge 120967. The figure shows clearly the exponential growth of modes at $k_y \rho_e \sim 0.3$ in the early stage and the nonlinear transfer to the low $k\rho_e \leq 0.1$ in the final nonlinear stage.

That the inverse cascade gives rise to the large enhancements of χ_e over the electron gyro-Bohm $\chi_e^{gB} = \rho_e v_e^2 / L_{T_e}$ value are reported in the many works (Nevins et al. [15], Li and Kishimoto

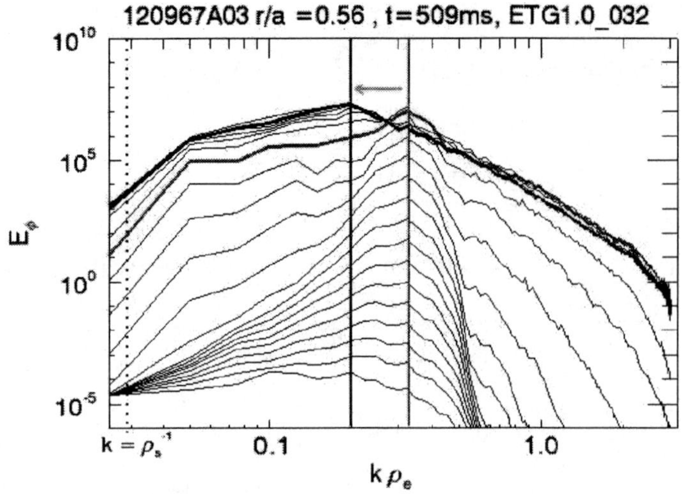

FIG. 8: Gyrofluid k-spectrum of perturbed electrostatic potential at $r/a = 0.56$ and $t = 509\,\mathrm{ms}$ for NSTX 120967 with parameters $\eta_e = -2.72$, $\epsilon_n = -0.858$ and $T_e = T_i$

[16, 17], Holland and Diamond [18]) and are due to extended radial coherent structures. The thermal flux $q_e(t)$ is typically intermittent as the large space scale coherent structures form and dissolve.

Both the gyrokinetic and the FLR-fluid simulation codes see structures with correlation lengths $l_x \le 100\rho_e$ and $l_y \le 30\rho_e$ that are intermittent in space and time as shown in Fig. 9 from the five-dimensional continuum GKV code [19]. To find a saturated nonlinear state requires a larger simulation box than for the corresponding ITG simulations. Watanabe used a box with $L_x = 292\rho_e$ and $L_y = 190\rho_e$ in Fig. 9. The transport in the saturated nonlinear state has $\langle\chi_e\rangle \simeq 20\chi_e^{gB}$ for $\eta_e = 3$, $s/q = 0.57$ and $T_e = T_i$.

The large radially extended coherent structures shown in Fig. 9 from the 5D GKV Watanabe code are similar to those seen in the FLR-Landau fluid simulation and produce the large enhancement of χ_e over the gyro-Bohm value.

Now we summarize the use of the electromagnetic fluid ETG turbulence simulation together with the theory to interpret the χ_e^{PB} in NSTX [20].

The electromagnetic ETG equation contains the electron scale c/ω_{pe} microtearing modes [21] through the nonlinear interaction of $\phi, A_\parallel$ and $\nabla^2 A_\parallel$. When there is an electron temperature fluctuation δT_e, the electron tearing mode transforms to the electron drift-tearing mode with the magnetic reconnection (MRC) component $\delta B_x = ik_y A_\parallel$. The parallel electron streaming along

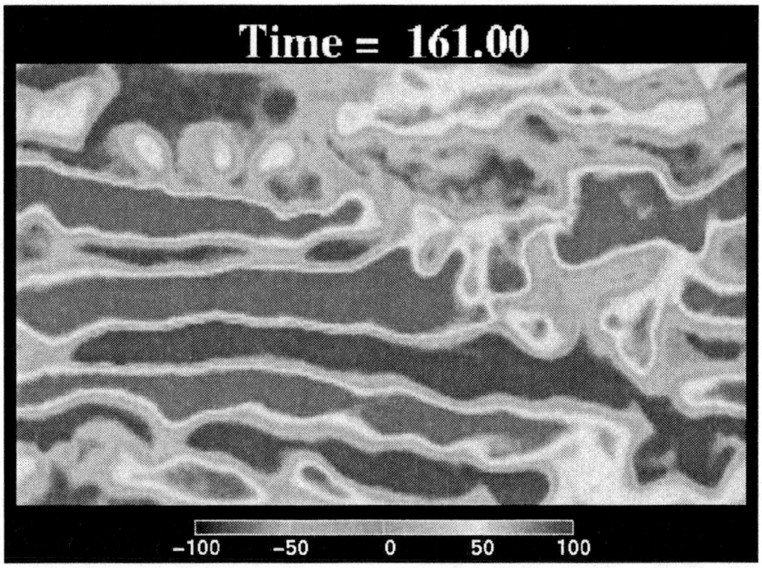

FIG. 9: Streamer formation in the ETG gyrokinetic-Vlasov GKV code simulation (T. Watanabe). The box size is $(L_x, L_y) = (292, 190)\rho_e$ and the magnetic shear $s =?$ and $q =?$.

$\delta\boldsymbol{B}$ gives an additional thermal flux. The main flux q_x^e remains from $\boldsymbol{E} \times \boldsymbol{B}$ convection given by $2\epsilon_n q_x^e$ where $q_x^e = n \langle \delta T_e \tilde{v}_x \rangle$ in the electron thermal flux. The nonlinear transfer between the MRC energy and the electron thermal energy occurs with the conservative coupling terms $-\delta J_{\|} \nabla_{\|} \delta p_e$ and $\delta p_e \nabla_{\|} \delta J_{\|}$ in the parallel acceleration equation and in the electron thermal equation.

Basic structure of the ETG model given in Horton et al. [8],Holland and Diamond [18] and Li and Kishimoto [16, 17] is

$$\hat{\boldsymbol{T}}\boldsymbol{\Phi} + \hat{\boldsymbol{L}}\boldsymbol{\Phi} = \hat{\boldsymbol{N}}(\boldsymbol{\Phi}, \boldsymbol{\Phi}) \tag{4}$$

where $\boldsymbol{\Phi} = (\phi, A_{\|}, T_e)$ is the field vector and $\hat{\boldsymbol{T}}$ are the first order time derivatives and $\hat{\boldsymbol{L}}$ and $\hat{\boldsymbol{N}}$ are linear and nonlinear spatial differential operators, respectively.

The operators in Eq. (4) are

$$\begin{aligned}
\hat{\boldsymbol{T}}_{11} &= (-1 + \nabla_\perp^2)\frac{\partial}{\partial t} \\
\hat{\boldsymbol{T}}_{22} &= \left(-\frac{\beta}{2} + \nabla_\perp^2\right)\frac{\partial}{\partial t} \\
\hat{\boldsymbol{T}}_{33} &= \frac{\partial}{\partial t} \\
\hat{\boldsymbol{T}}_{ij} &= 0 \quad \text{where} \quad i \neq j
\end{aligned}$$

$\hat{\boldsymbol{L}}_{11} = (1 - 2\epsilon_n + (1+\eta_e)\nabla_\perp^2)\frac{\partial}{\partial y}) - \mu\nabla^4$, $\hat{\boldsymbol{L}}_{12} = \nabla_\perp^2 \frac{\partial}{\partial z}$, $\hat{\boldsymbol{L}}_{13} = 2\epsilon_n \frac{\partial}{\partial y}$
$\hat{\boldsymbol{L}}_{21} = -\frac{\partial}{\partial z}$, $\hat{\boldsymbol{L}}_{22} = -\frac{\beta}{2}(1+\eta_e)\frac{\partial}{\partial y} + \frac{\eta}{\mu}\nabla_\perp^2$, $\hat{\boldsymbol{L}}_{23} = \frac{\partial}{\partial z}$
$\hat{\boldsymbol{L}}_{31} = (\eta_e - 4\epsilon_n(\Gamma - 1))\frac{\partial}{\partial y}$, $\hat{\boldsymbol{L}}_{32} = (\Gamma - 1)\nabla_\perp^2 \frac{\partial}{\partial z}$,
$\hat{\boldsymbol{L}}_{33} = 2\epsilon_n(\Gamma - 1)\frac{\partial}{\partial y} - \chi_\perp \nabla_\perp^2 - \chi_\parallel \frac{\partial^2}{\partial z^2}$

$$\begin{aligned}
\hat{\boldsymbol{N}}_1 &= \hat{\boldsymbol{N}}_{111} + \hat{\boldsymbol{N}}_{122} = -\left[\phi, \nabla_\perp^2 \phi\right] + \frac{\beta}{2}\left[A_\parallel, \nabla_\perp^2 A_\parallel\right] \\
\hat{\boldsymbol{N}}_2 &= \hat{\boldsymbol{N}}_{212} + \hat{\boldsymbol{N}}_{221} + \hat{\boldsymbol{N}}_{223} = -\left[\phi, \nabla_\perp^2 A_\parallel\right] - \frac{\beta}{2}\left[A_\parallel, \phi - T_e\right] \\
\hat{\boldsymbol{N}}_3 &= \hat{\boldsymbol{N}}_{313} + \hat{\boldsymbol{N}}_{322} = -\left[\phi, T_e\right] + (\Gamma - 1)\frac{\beta}{2}\left[A_\parallel, \nabla_\perp^2 A_\parallel\right]
\end{aligned}$$

where $[f,g] = \partial_x f \partial_y g - \partial_y f \partial_x g$ is the Poisson bracket arising from $\boldsymbol{v}_E \cdot \boldsymbol{\nabla}$ convection and $\delta \boldsymbol{B} \cdot \boldsymbol{\nabla}$ line bending. In deriving energy conservation the conservation properties for the double periodic boundary condition,

$$\begin{aligned}
< f\,[f,g] > &= \int f\,[f,g]\, d\boldsymbol{x} = 0 \\
< f[g,h] > &= < g[h,f] > = < h[f,g] > \\
< f[g,h] > &= - < g[f,h] > = - < h[g,f] >
\end{aligned}$$

are used. The double periodic boundary conditions eliminates the surface effects as in condensed matter physics.

The three energy component equations become

$$\begin{aligned}
\frac{\partial W_E}{\partial t} &= < \phi \hat{\boldsymbol{L}}_{11} \phi > + < \phi \hat{\boldsymbol{L}}_{12} A_\parallel > + < \phi \hat{\boldsymbol{L}}_{13} T_e > - < \phi \hat{\boldsymbol{N}}_{122} > \\
&= -\left\langle \mu |\boldsymbol{\nabla}_\perp \phi|^2 \right\rangle + \left\langle \phi \nabla_\perp^2 \frac{\partial A_\parallel}{\partial z} \right\rangle + 2\epsilon_n \left\langle \phi \frac{\partial T_e}{\partial y} \right\rangle - \frac{\beta}{2}\left\langle \phi \left[A_\parallel, \nabla_\perp^2 A_\parallel\right] \right\rangle \\
\frac{\partial W_B}{\partial t} &= - < \nabla_\perp^2 A_\parallel \hat{\boldsymbol{L}}_{21} \phi > - \left\langle \nabla_\perp^2 A_\parallel \hat{\boldsymbol{L}}_{22} A_\parallel \right\rangle - < \nabla_\perp^2 A_\parallel \hat{\boldsymbol{L}}_{23} T_e > + < \nabla_\perp^2 A_\parallel (\hat{\boldsymbol{N}}_{221} + \hat{\boldsymbol{N}}_{223}) > \\
&= -\left\langle \phi \nabla_\perp^2 \frac{\partial A_\parallel}{\partial z} \right\rangle - \frac{\eta}{\mu_0}\left\langle |\nabla_\perp^2 A_\parallel| \right\rangle + \left\langle T_e \nabla_\perp^2 \frac{\partial A_\parallel}{\partial z} \right\rangle + \frac{\beta}{2}\left\langle \phi \left[A_\parallel, \nabla^2 A_\parallel\right] \right\rangle - \frac{\beta}{2}\left\langle T_e \left[A_\parallel, \nabla^2 A_\parallel\right] \right\rangle \\
\frac{\partial W_{T_e}}{\partial t} &= \frac{1}{\Gamma - 1}\Big(- < T_e \hat{\boldsymbol{L}}_{31} \phi > - < T_e \hat{\boldsymbol{L}}_{32} A_\parallel > - < T_e \hat{\boldsymbol{L}}_{33} T_e > + < T_e \hat{\boldsymbol{N}}_{322} > \Big) \\
&= \left(\frac{\eta_e}{\Gamma - 1} - 4\epsilon_n \right) \left\langle \phi \frac{\partial T_e}{\partial y} \right\rangle - \left\langle T_e \nabla_\perp^2 \frac{\partial A_\parallel}{\partial z} \right\rangle - \left\langle \chi_\perp |\boldsymbol{\nabla}_\perp T_e|^2 + \chi_\parallel \left| \frac{\partial T_e}{\partial z} \right|^2 \right\rangle + \frac{\beta}{2}\left\langle T_e \left[A_\parallel, \nabla_\perp^2 A_\parallel\right] \right\rangle
\end{aligned}$$

where

$$\begin{aligned}
W_E &= \frac{1}{2}\left\langle \tau |\phi|^2 + |\boldsymbol{\nabla}_\perp \phi|^2 \right\rangle \\
W_B &= \frac{1}{2}\left\langle \frac{\beta}{2} |\boldsymbol{\nabla}_\perp A_\parallel|^2 + |\nabla_\perp^2 A_\parallel|^2 \right\rangle \\
W_{T_e} &= \frac{1}{2(\Gamma - 1)}\left\langle |T_e|^2 \right\rangle
\end{aligned}$$

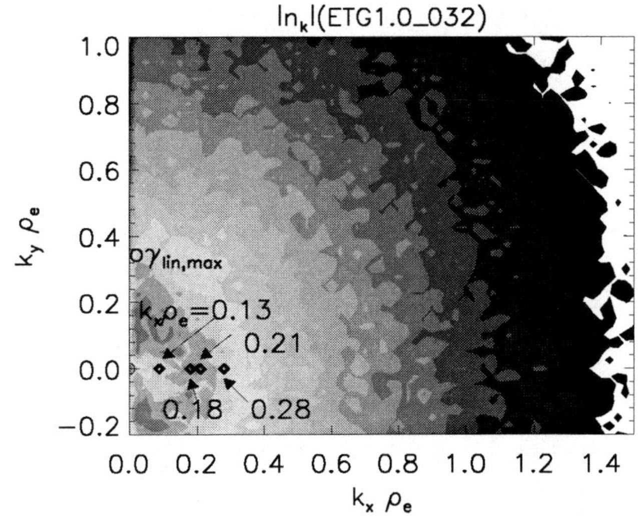

FIG. 10: Spectral distribution in saturated state of ETG simulation NSTX showing with diamonds the location of the measured $|\delta n_e(\boldsymbol{k})|$

The total energy evolves with

$$\frac{\partial}{\partial t}(W_E + W_B + W_{T_e}) = \left(\frac{\eta_e}{\Gamma - 1} - 2\epsilon_n\right)\left\langle \phi \frac{\partial T_e}{\partial y}\right\rangle$$
$$- \mu\left\langle|\nabla_\perp^2\phi|^2\right\rangle - \frac{\eta}{\mu_0}\left\langle|\nabla_\perp^2 A_\parallel|^2\right\rangle - \chi_\perp\left\langle|\boldsymbol{\nabla}_\perp T_e|^2\right\rangle - \chi_\parallel\left\langle\left|\frac{\partial T_e}{\partial z}\right|^2\right\rangle \quad (5)$$

In the saturated time-averaged steady state the time average of the RHS of Eq. (5) vanishes. Thus the sum of the four dissipation terms $(\mu, \eta/\mu_0, \chi_\perp, \chi_\parallel)$ balances the temperature gradient minus the critical gradient times the heat flux $q_e = \langle\phi\partial T_e/\partial y\rangle = -\langle E_y\delta T_e\rangle/B/$ which drives the turbulence in the first term on the RHS of Eq. (5) and in $\boldsymbol{k}$-space is

$$q_e = \sum_{\boldsymbol{k}} \frac{[\mathrm{Im}\, k_y\phi^*(\boldsymbol{k})\delta T_e(\boldsymbol{k})]}{B}. \quad (6)$$

VERIFICATION OF THE ETG MODEL WITH TOKAMAK DATA

A distinguishing characteristic of the ETG model is that the bulk of the electrons distribution function oscillates with the wave frequency so that there is a significant oscillating parallel electric current $\delta j_\parallel$. The resulting magnetic field δB_x while typically too small to add appreciably to the $\langle v_{\boldsymbol{E}\times\boldsymbol{B},x}\delta p_e\rangle$ thermal heat flux is a distinguishing feature of the mode. The internal magnetic fluctuations, however, were measured in Tore Supra with the cross-polarization scattering technique in which the incident extraordinary electromagnetic wave (X mode) is scattered by the nonlinear

Lorentz force from the δB_x fluctuation of the plasma into the outgoing ordinary electromagnetic mode (O mode). This diagnostic was developed in detail in the Tore Supra group by Zou et al. [22] and Colas et al. [23] from a proposal by Gressilon, and while not without some issues,the experiment confirms the presence of a spectrum of δB_x fluctuations at the level predicted by the ETG theory. The Colas et al. [23] experiments report that there is a correlation between the increasing intensity of these magnetic fluctuations with increasing levels of electron thermal flux q_e and the increasing electron temperature gradient $\boldsymbol{\nabla} T_e$. These are the key features of the ETG model, thus contributing the verification of the model.

Reflectometer measurements of the fluctuating electron density down to the ρ_e scale lengths have been reported by Kaye et al. [20]. The measurements and the correlation with turbulence theory modeling are shown in Fig. 10. While one sees from Fig. 10 that the part of the spectrum with largest amplitude (red) and the largest contribution to the thermal flux is not directly measured, theory gives a fixed relation between the maximum flux components and the measured $|n_{\boldsymbol{k}}|$ components of the spectrum. The new data showing significant reflected (backscattered) electromagnetic wave from the small scale electron density turbulence is found to be as predicted by the ETG model.

This area of research of the micron to millimeter scale range of plasma turbulence is an exciting new area recommended to incoming researchers in both experimental and simulations analysis of confined plasmas. The potential rewards to the search for fusion power from understanding better and controlling the turbulent electron transport is indeed great.

The record confinement shots toward achieving the highest fusion triple product and thus the highest value of the fusion power amplification factor Q_f are found in tokamaks where an internal electron thermal transport barrier has been created. While still a hot topic of research the electron internal transport barrier (ITB), as this reduction in q_e is called, is created by having a reversed safety factor $q(r)$ profile. Owing to their high-speed motion along the field lines, the electrons are the first to find the magnetic field structure has a region of vanishing magnetic shear. From number theory, the reversed profile makes a large gap in the density of the rational surfaces [24]. Such a gap in the density of rational surfaces is immediately reflected in the ETG mode structures and reduces the electron transport. Simple Toy Models of how this change in the spectral distribution of the turbulence and the immediate impact in reducing the electron transport are given in Balescu [25]. The simplicity of the models gives them a universality that may prove to be the best way to verify these models. In essence the standard model for the transition from well-confined orbits to chaotic electron orbits is changed when there is a region of reversed magnetic shear. A new family of more robust integral surfaces for the electron guiding center orbits is created as the magnetic shear goes through zero. Just before the reversal there can be regimes with extended radial zone of very low magnetic shear that are also known to produce better electron thermal confinement. These regimes known as Optimized Shear (OS) discharges were widely used in the JET tokamak to produce high Q fusion plasmas. Theoretically, these OS discharges have a wider spacing of the rational surfaces than the standard H mode discharges.

CONCLUSION

We show with examples from Tore Supra, NSTX, TCV and FTU how the electron thermal flux from power balance with several RF heating mechanisms is explained by the ETG [small scale electron temperature gradient driven turbulence model]. Earlier similar efforts to explain the anomalous electron fluxes with ion gyro-radius scale models have lead to serious problems with the qualitative inconsistencies between the data and the models. Thus, the ETG model has become the standard model for explaining electron transport in toroidal systems. In the future, the same model at higher plasma beta seems to be working for explaining the transport in the GAMMA-10 tandem mirror where the ITG modes are stabilized by the ion temperature anisotropy and the large value of T_i/T_e. The large T_i/T_e regimes in tokamaks have also indicated that the long wavelength ITG modes are stable yet the electron transport is the typical character explained by the ETG model.

Finally, we summarize that the ETG model produces large transport in the gyrokinetic simulations due to the strong inverse cascade of the quasi-2D nature of the turbulence. The very high density of the rational surfaces at this small millimeter scale of the turbulence and the inverse cascade allow the formation of radially extended convective electric field structures that give a large electron transport that explains much of the mystery of the five decade old problem of the high value and universal character of the electron transport that is called to special attention in Kadomtsev [26]. A full understanding of the problem remains for a future generation of plasma physicists and the ITER experiment's success in reaching the highest fusion Q_f values will certainly be strongly impacted by the value of the electron temperature that can be reached in its discharges. In the large ITER device ($R/a = 6\,\mathrm{m}/2\,\mathrm{m}$) the values of T_e and T_i will be closely coupled with T_e only slightly higher than T_i from the alpha particle fusion products being the dominant plasma heating power for $Q_f > 5$. The anomalous electron temperature transport, thus will be controlled and maintained to acceptable values as determined by integrated predictive code modelings using the ETG models for the base-line χ_e.

In such a large, complex plasma confinement machine as ITER the plasma system becomes not unlike a weather system in regards to their being many pathways and many feedback loops for the input powers and particle sources to find there wait to divertor chamber and walls of the vessel. Thus, there are exciting opportunities at many levels to contribute to the development of theory and modeling of ITER during the time of its construction and initial operation.

---

[1] K. M. Young, Phys. Fluids **10**, 213 (1967).

[2] L. A. Artsimovich, A. V. Gluchkov, and M. P. Petrov, JETP Lett **11**, 304 (1970).

[3] D. Bohm, E. H. S. Burhop, and H. S. W. Massey, in *The characteristics of electrical discharges in magnetic field*, edited by A. Guthrie and R. K. Wakerling (New York, McGraw-Hill, 1949).

[4] T. Ohkawa, M. Yoshikawa, R. E. Kribel, A. A. Schupp, and T. H. Jensen, Phys. Rev. Lett. **24**, 95 (1970).

[5] D. M. Meade and S. Yoshikawa, Phys. Fluids **10**, 2649 (1967).

[6] G. T. Hoang, C. Bourdelle, X. Garbet, G. Giruzzi, T. Aniel, W. H. M. Ottaviani, and P. Zhu, Phys. Rev. Lett. **87**, 125001 (2001).

[7] T. H. Stix, Nucl. Fusion **15**, 737 (1975).

[8] W. Horton, H. V. Wong, P. J. Morrison, A. Wurm, J. H. Kim, J. C. Perez, J. Pratt, G. T. Hoang, B. P. LeBlanc, and R. Ball, Nuclear Fusion **45** (2005).

[9] V. Basiuk, J. F. Artaud, F. Imbeaux, X. Litaudon, A. Becoulet, L.-G. Eriksson, G. T. Hoang, G. Huysmans, D. Mazon, D. Moreau, et al., Nucl. Fusion **43**, 822 (2003).

[10] A. Jacchia, F. D. Luca, S. Cirant, C. Sozzi, G. Bracco, A. Brushi, P. Buratti, S. Podda, and O. Tudisco, Nucl. Fusion **42**, 1116 (2002).

[11] E. Asp, J.-H. Kim, W. Horton, L. Porte, E. Fable, and O. Sauter, preprint (2007).

[12] G. W. Hammett and F. W. Perkins, Phys. Rev. Lett. **64**, 3019 (1990).

[13] H. Sugama, T. H. Watanabe, and W. Horton, Phys. Fluids **8**, 26172628 (2001).

[14] F. Jenko and W. Dorland, Phys. Rev. Lett. **89**, 225001 (2002).

[15] W. M. Nevins, J. Candy, S. Cowley, T. Dannert, A. Dimits, W. Dorland, C. Estrada-Milla, G. W. Hammett, F. Jenko, M. J. Pueschel, et al., Phys. Plasmas **13**, 122306 (2006).

[16] J. Li and Y. Kishimoto, Phys. Plasmas **9**, 1241 (2002), eTG, slab, gyrofluid.

[17] J. Li and Y. Kishimoto, Phys. Plasmas **11**, 1493 (2004).

[18] C. Holland and P. H. Diamond, Phys. Plasmas **9**, 3857 (2002).

[19] T. H. Watanabe and H. Sugama, Nucl. Fusion **46**, 24 (2006).

[20] S. Kaye, F. Levinton, D. Stutman, K. Tritz, H. Yuh, M. Bell, R. Bell, C. Domier, D. Gates, W. Horton, J. Kim, B.. LeBlanc H. Park, et al., Nucl. Fusion **47**, 499 (2007).

[21] J. W. Connor, C. M. Roach, R. J. Hastie, P. Helander, T. J. Martin, D. J. Applegate, N. Joiner, M. Reshko, S. Saarelma, W. D. Dorland, et al., IAEA Fusion Energy Conference,TH2/P2-2, Chengdu 16-21 October 2006.

[22] X. L. Zou, L. Colas, M. Paume, J. M. Chareau, L. Laurent, P. Devynck, and D. Gresillon, Phys. Rev. Lett. **75**, 1090 (1995).

[23] L. Colas, X. L. Zou, M. Paume, J. M. Chareau, L. Guiziou, G. T. Hoang, Y. Michelot, and D. Gresillon, Nucl. Fusion **38**, 903 (1998).

[24] A. D. Beklemishev and W. Horton, Phys. Plasmas **4**, 200 (1992).

[25] R. Balescu, *Aspects of anomalous transport in plasmas* (Institute of Physics Publishing, 2005), chap. 15, Series in Plasma Physics.

[26] B. B. Kadomtsev, *Tokamak plasma : a complex physical system* (Institute of Physics Pub, 1992).

Basics of turbulence computation for magnetically confined plasmas

Bruce D. Scott

Max-Planck-Institut für Plasmaphysik, Euratom Association, D-85748 Garching, Germany

Abstract. The ideas of low frequency dynamics driven by pressure gradients in magnetised plasmas are outlined. Fluid, gyrofluid, and gyrokinetic models are defined. Basic computational methods including coordinate representations are reviewed and a list of criteria for well posed cases is given. The consequences of nonlinearity vis-a-vis instabilities are shown. Other topics used as examples are scale separation, the self consistent equilibrium, and interaction with large scale flows.

Keywords: turbulence
PACS: 52.65.Tt, 52.35.Ra, 52.30.-q, 52.25.Fi

LOW FREQUENCY GRADIENT DRIVEN DYNAMICS

Measurement of thermodynamic quantities such as density and temperature of the charged species (typically, electrons and one main ion component) show them to exhibit fluctuations of order one percent, a bit below in the core and somewhat more in the edge region, and much more in the scrape-off layer bordering the LCFS: last closed magnetic flux surface [1, 2, 3, 4, 5]. For this purpose the "edge" is the last profile scale length within the closed magnetic surface region, bounded by the LCFS. The geometry of tokamaks [6], is a nested set of axisymmetric flux surfaces, with a toroidal current to support the poloidal magnetic field. The pitch parameter $q = B^\zeta/B^\theta$ giving the ratio of contravariant components of the magnetic field $\mathbf{B}$ typically rises from 1 to 2 in the core to 3 to 5 in the edge. The observed fluctuations are in the range of 10 to 200 kHz and 0.1 to 1 cm. Electron force balance tends to set the amplitude of electrostatic potential fluctuations $e\widetilde{\phi}/T_e$ commensurate with relative fluctuations $\widetilde{n}_e/n_e$ and $\widetilde{T}_e/T_e$ in the electron density and temperature (the latter in energy units). The perpendicular scale of motion then implies an ExB eddy velocity of $\mathbf{v}_E = (c/B^2)\mathbf{B}\times\nabla\widetilde{\phi}$. Power balance analysis indicates that fluctuations at this level are able to account for observed transport of particles and thermal energy, motivating the study of low frequency turbulence driven by the thermal gradient.

The basic velocities are the electron thermal and acoustic "sound" velocities V_e and c_s, given by $V_e^2 = T_e/m_e$ and $c_s^2 = T_e/M_i$, determined by the electron temperature and the electron and ion masses. The gyrofrequencies are $\Omega_e = eB/m_e c$ and $\Omega_i = eB/M_i c$, respectively. These define the electron gyroradius $\rho_e = V_e/\Omega_e$ and acoustic drift scale $\rho_s = c_s/\Omega_i$. The plasma is magnetised if these scales are much smaller than the profile scale lengths for any of n_e or T_e or the ion temperature T_i. The dynamics is "low frequency" if it is much slower than Ω_i, the slower gyrofrequency. These inequalities involving ρ_s or Ω_i are satisfied generally to about two orders of magnitude [7].

While ρ_e is usually regarded as arbitrarily small, the spectrum of perpendicular wave numbers does reach down to $k_\perp\rho_s > 1$, most sensitively in the vorticity and magnetic current fluctuations [8]. In moderate to large tokamaks $T_i \sim T_e$, so that the ion gyroradius ρ_i is comparable to ρ_s (the ratio is the square root of T_i/T_e). This implies that the turbulence spectrum reaches down to ρ_i, so that if ion temperature dynamics is treated the model must be able to address finite gyroradius (FLR) effects to arbitrary order [9, 10]. This motivates gyrofluid [11, 12] and gyrokinetic [13, 14, 15] computational models.

The reason while $k_\perp\rho_i \sim 1$ is encountered while dynamical frequencies always satisfy $\omega \ll \Omega_i$ is that the small parameter $\rho_s/L_\perp \ll 1$ enters all the perpendicular dynamics. Here, $L_\perp$ refers to the background scale length across the magnetic field, distinguished from the parallel scale length $L_\parallel$ set by the geometry (see below). The smallness of $e\widetilde{\phi}/T_e$ and the rest of the relative fluctuations follows. With $k_\perp\rho_s \sim 1$ but $e\widetilde{\phi}/T_e \ll 1$ we still have the ExB vorticity $\nabla\times\mathbf{v}_E$ small compared to Ω_i. So long as the plasma remains

CP1013, ***Turbulent Transport in Fusion Plasmas, First ITER International Summer School***
edited by S. Benkadda

magnetised these inequalities are well satisfied. Additionally, the commensurately small typical value of $L_\perp/L_\parallel$ motivates "flute mode ordering" in the fluctuations, with perpendicular and parallel wavenumbers satisfying $k_\parallel \ll k_\perp$ to between two and three orders of magnitude [16, 17, 18, 19].

In the above context a "moderate tokamak" is one with minor radius $a \sim 0.5\,\mathrm{m}$ and major radius aspect ratio R/a of about 3. Examples are ASDEX [20] and DIII-D [21]. The "large tokamak" environment starts with JET [22] at about $a = 1/meters$ and extends to ITER [23] about three times larger (both with aspect ratio close to 3). Moderate tokamaks have $L_\perp/\rho_s$ of at least 50 in the edge and large tokamaks have a/ρ_s of at least 300 in the core. The significance of these values will be discussed in the Sections on edge and core turbulence, below.

Dynamical Character

The parameters which define the character of the dynamics are a set of time or space scale ratios [9]. The low frequency results in the motion across the magnetic field splitting into the fast gyromotion of particles and the slow motion of the gyrocenters. The latter is referred to as drift motion, for which the fundamental example is the ExB drift at velocity $\mathbf{v}_E$. To this are added the magnetic drifts resulting from grad-B and curvature. Particle motion is generally thermal in all three dimensions with $v_\perp \sim v_\parallel \sim V_e$ for the electrons by example. The gyrocenter drifts are one order down in $\rho_s/L_\perp$ so that the ExB motion is in general subsonic.

The thermal reservoir has an energy density comparable to the electron pressure $p_e = n_e T_e$. This is much smaller than the magnetic field energy density. In drift dynamics the ratio is expressed in terms of the sound velocity and the Alfvén velocity v_A, yielding the parameter $\beta_e = 4\pi p_e/B^2$ which is the same as c_s^2/v_A^2. Tokamak plasmas have $\beta_e \ll 1$ with the consequence that there is not enough energy to dynamically compress the magnetic field, as the frequency inequality $\omega \ll k_\perp v_A$ is always satisfied. Perpendicular force balance $\nabla_\perp(\widetilde{p}_e + 4\pi B\widetilde{B})$ therefore remains strict in the fluctuations — ultimately this is the reason that the density and temperature can evolve independently. Parallel to the magnetic field this restriction does not apply and the flute mode inequality results in the general similarity $\omega \sim k_\parallel v_A$ so that the parallel motion of electrons is compressible and electromagnetic. The ultimate reason for $k_\parallel \ll k_\perp$ is related: if both $\omega \ll k_\parallel v_A$ and $\omega \ll k_\parallel V_e$ were satisfied then the electron dynamics would be wiped out by dissipation and if additionally $\omega \lesssim k_\parallel c_s$ then so would be the drift dynamics.

The sense of the coordinate geometry used to model the dynamics follows these properties. Two coordinates, $\{x, y\}$, are perpendicular and one, s, is parallel. Since $\mathbf{B}$ never vanishes in a tokamak it is possible to align these such that x is a flux surface label, y labels field lines within the flux surface, and s tracks the position along a field line [24]. This means that the contravariant components B^x and B^y are both zero while B^s remains finite. The parallel gradient is then given by $B\nabla_\parallel = B^s(\partial/\partial s)$. The ExB drift in the direction of ∇x (radially outward, across flux surfaces) is then proportional to $\partial\widetilde{\phi}/\partial y$. If fluctuations $\widetilde{p}_e$ and $\widetilde{\phi}$ are proportional then $\mathbf{v}_E \cdot \nabla\widetilde{p}_e$ vanishes. If on average there is no phase shift $\alpha_{p\phi}$ between $\widetilde{p}_e$ and $\widetilde{\phi}$ in the y-direction (the usual sense is $\nabla x \times \nabla y \cdot \nabla s > 0$) then the flux surface averaged radial flux $\langle\widetilde{p}_e v_E^x\rangle$ vanishes. Energetic drive is given by the product of $\langle\widetilde{p}_e v_E^x\rangle$ and $-\nabla p_e \sim (p_e/L_\perp)\nabla x$, so that if $\alpha_{p\phi} = 0$ there is neither net flux nor drive.

This motion is opposed by parallel dynamics [8]. The general Ohm's law is the same as the parallel force equation of the electrons. The electric field is split into static and dynamic pieces, $\widetilde{E}_\parallel = -(1/c)\partial\widetilde{A}_\parallel/\partial t - \nabla_\parallel\widetilde{\phi}$ giving the time and parallel space derivatives of the magnetic and electrostatic potentials, respectively. The Ohm's law contains the parallel electric and pressure forces, electron inertia, and resistive friction against the ions,

$$n_e e\left(\frac{1}{c}\frac{\partial\widetilde{A}_\parallel}{\partial t} + \nabla_\parallel\widetilde{\phi} + \eta_\parallel\widetilde{J}_\parallel\right) + \frac{m_e}{e}\frac{d\widetilde{J}_\parallel}{dt} = \nabla_\parallel\widetilde{p}_e \tag{1}$$

where $\frac{d}{dt}$ is the nonlinear ExB advective time derivative, $\eta_\parallel = 0.51 m_e\nu_e/n_e e^2$ is the parallel resistivity, ν_e is the electron collision frequency, and we assume the parallel current fluctuations $\widetilde{J}_\parallel = -n_e e\widetilde{v}_\parallel$ are carried mostly by the electrons with the parallel ion velocity $\widetilde{u}_\parallel$ remaining small. The terms involving $\nabla_\parallel$ represent the static forces. If resistivity, inertia, and induction are all small then the static forces have to balance. It is a property of closed magnetic surfaces with shear (given by $\partial q/\partial x$) that $\nabla_\parallel$ with localised fluctuations cannot vanish [25]. Therefore the static forces tend to cause $\widetilde{\phi}$ to line up with $\widetilde{p}_e$ (the physics is dissipative parallel Alfvén relaxation). The pressure force in the Ohm's law conserves energy against the parallel divergence $B\nabla_\parallel(\widetilde{v}_\parallel/B)$ in the electron continuity, which is why that is compressible. Charge conservation is maintained by ion polarisation (the small correction due to finite inertia upon the ExB motion, kept only through its divergence in the ion continuity).

The basic character of drift dynamics is then the dynamically incompressible, electrostatic motion across the magnetic field, advecting both profiles (the flux surface averaged component) and fluctuations (the departures), and compressible, electromagnetic motion along the magnetic field. The former is mostly ExB drift, and the latter is mostly competition with electron force balances with some additional ion dissipation through Landau damping or viscosity. The quantities in the Ohm's law may be normalised to yield [8]

$$\hat{\beta}\frac{\partial \widetilde{A}_\parallel}{\partial t}+\hat{\mu}\frac{d\widetilde{J}_\parallel}{dt}+C\widetilde{J}_\parallel=\nabla_\parallel(\widetilde{p}_e-\widetilde{\phi}) \tag{2}$$

Here, the time is normalised to $c_s/L_\perp$ while $\nabla_\parallel$ is scaled against $L_\parallel$, so that the normalised parameters contain the parallel/perp scale ratio as well as velocity or frequency ratios. These are given by

$$\hat{\beta}=\frac{4\pi p_e}{B^2}\left(\frac{L_\parallel}{2\pi L_\perp}\right)^2 \qquad \hat{\mu}=\frac{m_e}{M_i}\left(\frac{L_\parallel}{2\pi L_\perp}\right)^2 \qquad C=\hat{\mu}\frac{0.51\nu_e}{c_s/L_\perp} \tag{3}$$

where the factor of 2π is by convention (the ratio $k_\parallel L_\parallel=2\pi$ is taken). These parameters control the response of $\widetilde{\phi}$ to $\widetilde{p}_e$. The Alfvén dynamics tends to relax towards $n_e e\widetilde{\phi}=\widetilde{p}_e$, which is defined as the "adiabatic state" following the idea of quasistatic evolution of $\widetilde{\phi}$ following $\widetilde{p}_e$ through this equality. If $\hat{\beta}$ and $\hat{\mu}$ and C are all small then the electrons are close to adiabatic (neglecting trapping effects, which this introduction does not treat). In fact the distinction between the core and edge regions is precisely this: in the edge $L_\perp$ is so short that while $m_e/M_i \ll 1$ we have $\hat{\mu}>1$ since $L_\parallel/L_\perp \sim 10^3$ is typical. Here, note that for a model linear profile falling to zero at $x=x_0$ then $L_\perp=x_0-x$. Typical values of $\hat{\beta}$ and $\hat{\mu}$ and C in the edge are all larger than unity.

The competition between the drift motion and the parallel response is central to any scenario of this dynamics. A basic illustration is given in Fig. 1. In general the ExB motion is turbulent and fluctuations continually wash each other out. But unless large scale magnetohydrodynamic (MHD) instabilities are present the parallel response is always able to keep $\widetilde{\phi}$ and $\widetilde{p}_e$ correlated. The sense is of $\widetilde{\phi}$ to be tied to $\widetilde{p}_e$ rather than the other way around, because the response of $\widetilde{\phi}$ in the Alfvén dynamics is faster by a factor of $k_\perp^2\rho_s^2$ which is less than unity in most of the spectrum. The turbulence spectrum typically involves two decades $10^{-2}<k_\perp\rho_s<1$ of scales, with $\rho_i \sim \rho_s$ in general. The frequency range is wider because some of the flow responses are as slow as $2\pi c_s/L_\parallel$ which is closer to 10^{-3} less than the drift frequency $c_s/L_\perp$. The latter is the fastest frequency in the turbulence and is safely slower than Ω_i if the drift parameter $\rho_s/L_\perp$ indeed remains small.

GYROFLUID AND GYROKINETIC MODELS

The equations used to model this low frequency dynamics are derived using the above considerations, with $\rho_s \ll L_\perp$ the fundamental small parameter on which the others depend. The fundamental processes are advection by the ExB velocity of both the fluctuations and background gradients, and parallel dynamics mostly in the electrons but also in the ions. The parallel pressure gradient is commensurate to the other forces and hence to conserve energy the continuity equations include compressible parallel motion and, conserving charge, the perpendicular polarisation drift for the ions. Hence one thinks of continuity and parallel forces in all the species rather than a single fluid MHD model.

Since $\rho_i \sim \rho_s$ the values of $k_\perp\rho_i$ are not always small. ExB motion follows the gyroaveraged electrostatic potential, which is different for each species. While $\mathbf{v}_E=(c/B^2)\mathbf{B}\times\nabla\widetilde{\phi}$ the gyroreduced ExB velocity is $\mathbf{u}_E=(c/B^2)\mathbf{B}\times\nabla\widetilde{\phi}_G$. In kinetic models the particles are treated separately and the gyroaveraged potential is $\widetilde{\phi}_G=J_0(\widetilde{\phi})$, given by multiplication of each Fourier coefficient of $\widetilde{\phi}$ by $J_0(k_\perp v_\perp/\Omega_z)$ for species z. This follows from treating the particle as a charged ring of radius $v_\perp/\Omega_z$, the gyroradius. Gyrokinetic models are

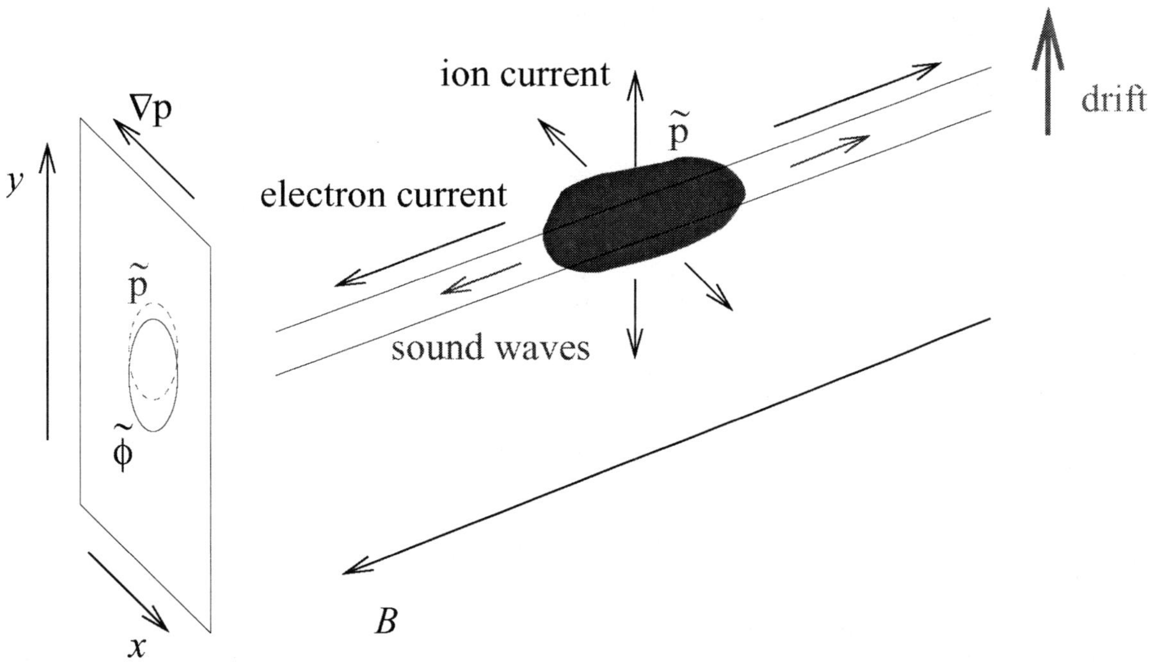

FIGURE 1. A sketch of drift Alfvén dynamics. An ExB vortex excites a pressure fluctuation (by advecting the profile or as part of the turbulence), or the other way around (by parallel forcing of an Alfvén response). Both have finite parallel gradient. The dissipative Alfvén response relaxes $\widetilde{\phi}$ towards $\widetilde{p}_e$ while advection in the gradient excites $\widetilde{p}_e$ phased ahead of $\widetilde{\phi}$. With $k_\parallel v_A$ comparable to or larger than $\mathbf{v}_E \cdot \nabla$ the structure is held together and drifts in the direction $\nabla p{\times}\mathbf{B}$.

based on this approximation [13, 14, 15], formalised in terms of Lie transforms [26, 27] and Hamiltonian field theory [28, 29]. Gyrofluid models are related to gyrokinetic ones in the same way fluid models are to the underlying kinetic theory: rather than following the gyrocenters separately a set of low order moments are followed [11, 12]. These always include the density, usually also the parallel velocity, sometimes also the parallel and perpendicular temperatures and each one's associated parallel heat flux (one, two, or six moments, with three dimensional models requiring at least two). The gyroreduced ExB velocity is also approximated with moments over the J_0 effects. The traditional one is to use a Padé approximant $\widetilde{\phi}_G = \Gamma_1 \widetilde{\phi}$ with $\Gamma_1 = [1-(\rho_z^2/2)\nabla_\perp^2]^{-1}$, where ρ_z is the species gyroradius. This form can be used even if the model geometry precludes the use of Fourier transforms.

Both gyrofluid and gyrokinetic models treat polarisation in a way different from MHD or fluid models. Instead of a polarisation velocity or current one has a polarisation density whose time derivative recovers the same charge conservation law as in the fluid models. This is due to the splitting between the contribution to the space density by the gyrocenter and a polarisation piece given by the part of the response which is different from the gyroaverage and depends on $\widetilde{\phi}$. Since the present purpose is numerics the equations are merely outlined. The simplest three dimensional gyrofluid model treats the gyrocenter density and parallel velocity of both the electrons and ions, with the isothermal simplification used for both species [9]. The continuity and parallel force equations for species z are given by

$$\frac{\partial \widetilde{N}_z}{\partial t} + \mathbf{u}_E \cdot \nabla[\widetilde{N}_z + N_0] + B\nabla_\parallel \frac{\widetilde{u}_{z\parallel}}{B} = \mathcal{K}\left(\widetilde{\phi}_G + \frac{T_z}{Ze}\widetilde{N}_z\right) \tag{4}$$

$$\frac{1}{c}\frac{\partial \widetilde{A}_\parallel}{\partial t} + \eta_\parallel \widetilde{J}_\parallel + \frac{M_z}{Ze}\left(\frac{\partial \widetilde{u}_{z\parallel}}{\partial t} + \mathbf{u}_E \cdot \nabla \widetilde{u}_{z\parallel}\right) + \nabla_\parallel\left(\widetilde{\phi}_G + \frac{T_z}{Ze}[\widetilde{N}_z + N_0]\right) = 2\frac{T_z}{Ze}\frac{M_z}{Ze}\mathcal{K}(\widetilde{u}_{z\parallel}) \tag{5}$$

where N denotes the logarithm of the density, $\nabla N_0 = (-1/L_\perp)\nabla x$ is the profile gradient, T_z and M_z and Ze are the species temperature, mass and charge, $\widetilde{J}_\parallel = \sum_{\mathrm{sp}} n_z Ze \widetilde{u}_{z\parallel}$ is the parallel current, and $\widetilde{\phi}_G$ is species dependent. The sums here and below are over species z. For ions and electrons $Z = \pm 1$ respectively. The differential operators are

$$\mathbf{u}_E \cdot \nabla f = [\widetilde{\phi}_G, f] \qquad \nabla_\parallel f = \frac{B^s}{B}\frac{\partial f}{\partial s} - \frac{1}{c}[\widetilde{A}_\parallel, f] \qquad \mathcal{K}(f) = -\frac{c}{B_0}[\log B^2, f] \tag{6}$$

where B_0 is a constant representative value of B, $\mathcal{K}$ represents magnetic curvature effects, with Poisson brackets and gyroaveraged potentials given by

$$[f,g] = \nabla f \cdot \frac{c\mathbf{B}}{B^2} \times \nabla g \qquad \widetilde{\phi}_G = \left(1 - \frac{\rho_z^2}{2}\nabla_\perp^2\right)^{-1} \widetilde{\phi} \tag{7}$$

The species gyroradii are given by $\rho_i^2 = (T_z M_z / T_e M_i)\rho_s^2$.

In addition are the polarisation and induction equations,

$$n_i \frac{e^2}{T_e}\rho_s^2 \nabla_\perp^2 \widetilde{\phi} + (1 - \rho_i^2 \nabla_\perp^2) \sum_{\text{sp}} (n_z Z e)\Gamma_1 \widetilde{N}_z = 0 \qquad \nabla_\perp^2 \widetilde{A}_\parallel + \frac{4\pi}{c}\sum_{\text{sp}} (n_z Z e)\widetilde{u}_{z\parallel} = 0 \tag{8}$$

in which $(n_z Z e)$ is a constant reference charge density for each species. The induction equation is just Ampere's law and $\widetilde{A}_\parallel$ is not gyroaveraged. The polarisation equation treats gyroaveraged gyrocenter densities and the ion polarisation given by the $\nabla_\perp^2$ operators and the finite $\widetilde{\phi}$. The actual space charge density is neglected. The time derivative of the polarisation equation recovers $\nabla \cdot \mathbf{J} = 0$ with all the contributions from $\nabla_\perp^2 \widetilde{\phi}$ and $\nabla_\perp^2 \widetilde{N}_z$ representing the divergence of the polarisation current. Note the appearance of ρ_s and T_e which is equivalent to ρ_i and T_i because $\rho_i^2/\rho_s^2 = T_i/T_e$.

The derivation of these equations is very involved. Gyrokinetic theory itself has been given a major review recently [30]. The original gyrofluid model [11, 12] was extended to include electromagnetic responses [31], corrected for energetic consistency [32], and then re-derived with a consistent path from the gyrokinetic one using the corresponding energy theorems [33]. The updated gyrofluid model was also shown to be consistent with the collision dominated fluid model in the appropriate limits [34].

Representation of the Geometry

Computational methods actually begin with the representation of the magnetic geometry, before the numerical scheme is considered. The first generation of models used one of two approaches. Two dimensional models used a heuristic treatment for the parallel dynamics, replacing it with a dissipative coupling model. In our terms that would be equivalent to taking

$$B\nabla_\parallel \frac{\widetilde{u}_{z\parallel}}{B} \rightarrow -D_z \left(\widetilde{N}_z + \frac{Ze\widetilde{\phi}_G}{T_z}\right) \tag{9}$$

in Eq. (4) with D_z constant, and then restricting the dynamics to the $\{x,y\}$ plane. If magnetic curvature and gyroaveraging were neglected, and T_i and D_i set to zero, this would reduce to something identical in content to the well-known Hasegawa-Wakatani model [18]. The grid in $\{x,y\}$ is equidistant and isotropic, with periodic boundary conditions. The resolution was made as high as possible, to try to cover the required two spectral decades even though in very early cases that was not possible.

Early three dimensional models [35, 36] were set up in a cylindrical geometry with $B = B_0$ but retaining toroidal effects through the model $\log B = (r/R_0)\cos\theta$ with R_0 a constant major radius. The coordinates were $\{r,\theta,\zeta\}$ representing the minor radius and poloidal and toroidal angles. No attempt was made to field-align the coordinates. Rather, a Fourier representation was used for the angles and a finite difference scheme used for the minor radius. In this representation $\nabla_\perp^2$ takes the usual cylindrical form and the remaining differential operators are

$$[f,g] = \frac{c}{rB_0}\left(\frac{\partial f}{\partial r}\frac{\partial g}{\partial \theta} - \frac{\partial g}{\partial r}\frac{\partial f}{\partial \theta}\right) \qquad \frac{B^s}{B}\frac{\partial}{\partial s} \rightarrow \frac{1}{qR_0}\left(q\frac{\partial}{\partial \theta} - \frac{\partial}{\partial \zeta}\right) \tag{10}$$

recalling $q = q(r)$. This form is still in wide use. Its major drawback is the need for high resolution in all three coordinates, since both B^θ and B^ζ are finite. Worse still, very large values of $k_\parallel$ are present on the grid, due to the requirement that rh_θ reach down towards ρ_s hence requiring well over 10^3 grid points in typical cases. Some help has been obtained by considering a triangle of components in Fourier poloidal/toroidal mode number space, but with even moderate shear in q the range of $k_\parallel L_\parallel$ values is still too large. Although these $k_\parallel$ values play no role in the dynamics, their presence on the grid requires filtering methods to remove the resulting large $k_\parallel v_A$ values. Hence the models using this representation are limited to very low resolution computations to avoid prohibitive time step constraints and/or CPU and memory resource demands.

Modern computations avoid the unnecessary resolution by aligning the coordinates to the magnetic field [24, 37, 25, 38]. This exploits the property that $k_\parallel \ll k_\perp$ in the dynamics, reducing the numerical grid values of $k_\parallel v_A$ or $k_\parallel V_e$ sufficiently that every degree of freedom in the computational domain can be temporally resolved without complication. Additionally the size of the computation is reduced since only a few tens of grid points are necessary in the s direction. The necessary transformation is

$$x = r/a \qquad y = q\theta - \zeta \qquad s = \theta \tag{11}$$

giving one example of how to make B^y vanish without approximation. This simplest version of field aligning has one remaining problem: the grid is no longer orthogonal in the perpendicular $\{x, y\}$ plane, since $g^{xy} = \nabla \cdot \nabla y$ is not only finite but dependent upon s. The resulting s dependence of nonlinear dissipative mixing by the turbulence has been shown to cause problems in representing the dynamics. So we go one step further and use a different transformation on every plane in s on which we need information, namely, the node positions $s = s_k$ in the grid. This is given by

$$x = r/a \qquad y_k = q(\theta - s_k) - \zeta \qquad s = \theta \tag{12}$$

which is a simple generalisation of Eqs. (11). These are called "straight metric" and "shifted metric" representations, respectively. In the latter, each member of the family of field aligned representations is equally, globally valid, but we exploit the property that g_k^{xy} is proportional to $s - s_k$ and is therefore zero where perpendicular drifts and $\nabla_\perp^2$ operators are evaluated. In the parallel derivatives, however, information on neighboring $\{x, y\}$ planes is needed, and these are on different coordinate systems. The resulting parallel derivative is found by

$$2h_s \frac{\partial f}{\partial s}(x, y_k, s_k) = f(x, y_{k+1} - \Delta^+, s_{k+1}) - f(x, y_{k-1} - \Delta^-, s_{k-1}) \tag{13}$$

where h_s is the equidistant grid spacing and $\Delta^\pm = \pm q\Delta s$ are the necessary shifts in the y-coordinate between the drift planes. Hence the name shifted metric for this treatment. It can be generalised without approximation to shaped tokamak geometry. In any of the above coordinate maps the choice of x is free. It can be r/a or $(r/a)^2$ or the volume enclosed by a flux surface or as in many models in use the toroidal flux enclosed by the surface.

All of the above models respect what is called global consistency, that is, the allowed values of $k_\parallel$ present on the grid are present in the actual torus [25]. The requirements for this follow from the fact that the θ and ζ domains are simultaneously periodic while $q(r)$ is sheared in Eqs. (10). This leads to the spectrum of $k_\parallel$ values given by $k_\parallel qR_0 = m - nq$ where $\{m, n\}$ are the poloidal/toroidal Fourier mode numbers. Since q is sheared, the values of $k_\parallel$ vanish only on "rational surfaces" where $q = m/n$ and then only for those particular degrees of freedom. Hence $\nabla_\parallel$ remains nonvanishing for any disturbance localised in both perpendicular dimensions (a flux surface averaged quantity obviously has $k_\parallel = 0$). Retaining these properties under approximation is the main point. In any localising truncation of the toroidal flux surfaces the correct discreteness of the $k_\parallel$ spectrum, and related to this the spacing of rational surfaces, must be kept. Without the correct properties of $\nabla_\parallel$ the character of the competition between drift and Alfvén dynamics is lost. The safe way to do such a globally consistent truncation is to decompose the grid in terms of $\{m, n\}$ values, keep only integral multiples of n (always reasonable in a tokamak, which is axisymmetric), express the boundary conditions in terms of the Fourier coefficients f_n for any scalar field f, and then rebuild the boundary conditions in terms of the (truncated) grid. The double periodicity is expressed as

$$f(r, \theta + 2\pi, \zeta) = f(r, \theta, \zeta + 2\pi) = f(r, \theta, \zeta) \tag{14}$$

on a $(2\pi)^2$ domain in $\{\theta,\zeta\}$. Using the coordinates in Eqs. (11), these become

$$f(x,y+2\pi,s)=f(x,y,s) \qquad f(x,y+2\pi q,s+2\pi)=f(x,y,s) \tag{15}$$

Axisymmetry in ζ yields a discrete spectrum for n with unit spacing. The same applies to the circuit in y while holding s fixed. The poloidal circuit changes both coordinates, however. Fourier transform of this condition yields

$$f_n(x,s+2\pi)=f_n(x,s)\exp[-i2\pi nq] \tag{16}$$

that is, the same amplitude with a radially dependent toroidal phase shift after one poloidal circuit. This is the condition that must be maintained, without approximation, however one chooses to represent the coordinates. Omitting the details, the straight metric representation is given by

$$x=r-a \qquad y=\frac{a}{q_a}(q\theta-\zeta) \qquad s=\theta \tag{17}$$

where a is a reference radius and $q_a=q(a)$, with boundary conditions

$$f(x,y+2\pi/K,s)=f(x,y,s) \qquad f(x,y+2\pi Sx,s+2\pi)=f(x,y,s) \tag{18}$$

where K is the fundamental wavenumber of the y-domain and $\hat{s}=d\log q/d\log r$ is the shear. This can also be generalised to shaped geometry with or without the shifted metric treatment.

Any of the above representations preserve global consistency. The remaining considerations have to do with resolution. Independent of physical scales, an requirement on the radial grid spacing follows indirectly from the distribution of rational surfaces [10]. Any degree of freedom experiences the magnetic shear if the boundary conditions force a degree of radial dependence that the grid can sense. This happens only if for all toroidal mode numbers the radial grid spacing is less than the rational surface spacing. For small scales the profile in $q(r)$ may be taken as linear with constant $\hat{s}$. Neighboring rational surfaces are at q values $(m\pm 1)/n$ and m/n, so that the spacing is $\Delta q=1/n$. This means

$$\Delta_{\mathrm{rs}}=\frac{1}{\hat{s}}\frac{r}{q}\Delta q=\frac{1}{\hat{s}k_y} \tag{19}$$

since $k_y=nq/r$. The requirement that the grid spacing be less than this for all wavenumbers means we have to use the maximum value of k_y, or π/h_y. Therefore we find that

$$h_x<\frac{1}{\pi\,\hat{s}}h_y \tag{20}$$

It is important to note that this requirement is independent of the representation as it is a property of the underlying physics. Unless sufficiently strong turbulence washes the dynamics out before parallel electron dynamics can propagate a distance $L_\parallel$, this means that typically the resolution in x must be four times that in y, assuming $\hat{s}\sim 1$. If not, then the physics of magnetic shear is not captured, since the parallel boundary conditions allow random values in the minimum values of $k_\parallel$. With the requirement met, turbulence propagating down magnetic field lines is smoothly sheared away and the parallel responses in the electrons retain their physical role.

The final consideration is the resolution itself [39, 8]. Assuming a proper representation of the geometry as outlined above, the computation need only further ensure it accesses the full spectral range at least close to $10^{-2}<k_\perp\rho_s<1$. The small scale end should actually reach past unity. The large scale end depends on the case. Edge turbulence is limited by the extent of the region over which edge parameters actually obtain. The domain on closed field lines should not be larger than $L_\perp$ (for experimental cases $L_\perp=L_{Te}$ should be set since T_e has the steepest gradient there). In some cases this can be as low as $20\rho_s$ although $60\rho_s$ is more typical. The size of the y-domain is set by physical extent or by the requirement that radial flows of any scale have room to reverse direction. In a tokamak with circular flux surfaces the physical domain size is $2\pi r/q$. The domain size in a computation should be either this or at least as large as $4L_\perp$. If all of these conditions are met then the typical grid sizes are 50 to 100 in x, at least 4 times that in y, and 16 or 32 along s depending on the case (shaped tokamaks require more). For core turbulence or if studying the linear/nonlinear transition then the required node count in x increases by four.

Numerical Schemes

Numerical schemes are quite different for different types of models. Here we concentrate on fluid models. For particle in cell methods [14, 15, 40, 41] or phase space continuum methods [42, 43, 44] for gyrokinetics the reader is directed elsewhere although it is important to note that the basic requirements on geometry and resolution are common.

The requirements of a numerical scheme for fluid computations are given by the properties of the differential operators in the equations. The typical mathematical structure is common to all models: a set of dynamical quantities is advanced in time, and then at the end of a timestep a set of field quantities are computed from them. In our example the first step is given by Eqs. (4,5) and the second by Eqs. (8). In the very simplest fluid model of all, incompressible Eulerian fluid dynamics in two dimensions, the first step advances the vorticity and the second finds the stream function (equivalently, one advances the velocity vector field and then imposes the incompressibility constraint by a method equivalent to projection). Hence the methods in use are all borrowed from fluid dynamics. We do have one additional property to watch for, however, in the three dimensional models: waves. Not only sound waves are present but also Alfvén waves. The particular difficulty of these in this setting is their anisotropic dispersion. Dispersion usually refers to propagation speed in a given direction dependent on the wavenumbers in that direction. In this case the propagation speed in the parallel direction depends on the values taken by $\nabla_\perp^2$, the perpendicular direction. This is an unusual sort of dephasing, for which not just any scheme will suffice. In particular, implicit schemes (wherein terms other than $\partial/\partial t$ are evaluated at the forward time step) do very badly for this type of dispersion.

An additional consideration is the presence of the Ohm's law: while $\widetilde{A}_\parallel$ is advanced the physical dissipation including turbulent mixing acts upon $\widetilde{J}_\parallel$. The scale dependence of this is given by Ampere's law which contains $\nabla_\perp^2$. Direct dissipation upon $\widetilde{A}_\parallel$ must be avoided; otherwise, the numerical resistivity becomes ever larger for ever smaller $k_\perp^2$ which is very unphysical.

In turbulence the physics may be very anisotropic at medium to large scale even in just the two perpendicular coordinates since gradient forcing, the zonal average, and magnetic curvature all act to select between them. On the other hand, at sufficiently small scales the ExB nonlinear advection (diffusive mixing) overcomes all the linear forcing phenomena. This imposes an isotropic tendency in x and y. When magnetic fluctuations are significant, the parallel gradient acts with wavelike character in all three dimensions. The numerical scheme must be able to capture these features, so it should treat derivatives in all the coordinates equally. Especially precluded are mixed Fourier/finite-difference representations or mixed timestep schemes in which linear terms are implicit and nonlinear terms are explicit (i.e., the nonlinearities are evaluated using information from the previous timestep and then the linear terms are all solved for at the next one). All of these will impose artificial character on the results.

Here we outline the best scheme found for these problems to date [45, 46]. It comes from fluid dynamics, but has high accuracy also for waves. The essential idea is to exploit the Poisson bracket property of all the perpendicular drift terms including the magnetic nonlinearities, to determine the spatial discretisation in $\{x,y\}$ or their equivalents [47]. Even the linear $\partial/\partial s$ terms can be done this way since $q^{-1}\partial f/\partial s = [\chi, f]$ for $\partial\chi/\partial x = q^{-1}$ and $\chi = \chi(x)$. The timestep is a multistep scheme using one evaluation of the right hand side per step, while expanding both left and right hand sides into a series of previous timesteps [48]. Among other things, this includes all mixed time/space derivatives in the equivalent Taylor expansion. Details and derivation of properties are deferred to the above two references. Herein we outline the basic ideas.

The timestep is given through the generic form $\partial f/\partial t = S$. The expansion is a variant on the Adams/Bashforth theme given by Karniadakis [48], with both sides expanded 3 timesteps deep in the history:

$$\sum_{j=1}^{3} \alpha_j \frac{f_0 - f_j}{\Delta t} = \sum_{j=1}^{3} \beta_j S_j \tag{21}$$

where Δt is an equidistant time step. The coefficients are determined by the requirement to capture the equivalent Taylor expansion,

$$\alpha_{1,2,3} = \quad 3 \quad -3/2 \quad 1/3 \qquad\qquad \beta_{1,2,3} = \quad 3 \quad -3 \quad 1 \tag{22}$$

The incorporation of an implicit dissipation piece L is straightforward,

$$\sum_{j=1}^{3}\alpha_j\frac{f_0-f_j}{\Delta t}+L(f_0)=\sum_{j=1}^{3}\beta_j S_j \tag{23}$$

where the factor of 6/11 comes from the inverse sum over the α_j. Although implicit techniques with wave dynamics should be avoided, there is no problem using this for the artificial diffusion one must do anyway to remove energy from the final factor of two in the spectrum beginning with wavenumbers in the vicinity of $k_y = (k_y)\text{max}/2$.

The spatial discretisation is that which is best for the Poisson brackets (often referred to as Jacobians in the literature). The three basic properties of antisymmetry of $[f,g]$ with respect to exchange of f and g, and energy and entropy conservation represented by multiplication by either f or g and integration over the domain to obtain zero up to boundary effects. The form for this was found by Arakawa by setting up a set of equivalent representations for $[f,g]$ based on contour integrals [47]. With an isotropically equidistant grid spacing of h these are evaluated at grid node 00 with + or − neighbours in xy plane, and are given by

$$J^{++}=\frac{1}{4h^2}[(f_{+0}-f_{-0})(g_{0+}-g_{0-})-(f_{0+}-f_{0-})(g_{+0}-g_{-0})] \tag{24}$$

$$J^{+\times}=\frac{1}{4h^2}[f_{+0}(g_{++}-g_{+-})-f_{-0}(g_{-+}-g_{--})-f_{0+}(g_{++}-g_{-+})+f_{0-}(g_{+-}-g_{--})] \tag{25}$$

$$J^{\times+}=\frac{1}{4h^2}[f_{++}(g_{0+}-g_{+0})-f_{--}(g_{-0}-g_{0-})-f_{-+}(g_{0+}-g_{-0})+f_{+-}(g_{+0}-g_{0-})] \tag{26}$$

$$J^{\times\times}=\frac{1}{8h^2}[(f_{++}-f_{--})(g_{-+}-g_{+-})-(f_{-+}-f_{+-})(g_{++}-g_{--})] \tag{27}$$

Linear combinations of these are considered, and then the three properties used as constraints, to find

$$[f,g]_{xy}=\frac{\partial f}{\partial x}\frac{\partial g}{\partial y}-\frac{\partial g}{\partial x}\frac{\partial f}{\partial y}=\frac{1}{3}\left(J^{++}+J^{+\times}+J^{\times+}\right) \tag{28}$$

With this combination one finds that the sum over nodes on a double periodic grid of either f or g times $[f,g]_{xy}$ vanishes and that exchange of f and g changes the sign, satisfying the three properties. This has the property of minimising the amount of grid-scale noise artificially created by the scheme. There is already a physical amount of this produced by the cascade dynamics from large to small scale (in thermal free energy, while flow energy cascades the other way). Use of this Arakawa scheme allows smaller values for the numerical grid-scale dissipation required to contain the cascade and ultimately lower values of artificial dissipation in the dynamically significant regions of the spectrum. For anisotropic grids the factor of h^2 is merely replaced by $h_x h_y$. For nonequidistant grids the formula should be re-derived. In practice the nonequidistant property is treated through an inhomogeneous metric while the grid in terms of transformed coordinates remains equidistant.

The combination of these two methods was found by Naulin [45] and thoroughly tested by Naulin and Nielsen [46]. Studies of local collisionless magnetic reconnection, which uses a set of equations equivalent to a subset of the model of Eqs. (4,5,8), found that the nonlinear transition could be reached with this scheme without any use of numerical dissipation (up to the stage where arbitrarily thin current sheets are nonlinearly generated) [49].

EDGE TURBULENCE AND THE MEANING OF NONLINEARITY

Edge and core turbulence in a tokamak are radically different in physical character. The reason for that is the relative strength of the "adiabatic response" (the parallel electron dynamics in response to what the ExB turbulence causes in the fluctuations). In the edge $\hat{\mu} > 1$ as defined in Eqs. (3), solely due to the scale ratio with large $L_\parallel = 2\pi q R_0$ and small $L_\perp$, so thermal electron motion is slower than the fastest part of the turbulence. Strong nonadiabaticity results: the difference $\widetilde{h}_e = \widetilde{p}_e - \widetilde{\phi}$ in normalised fluctuations is somewhat

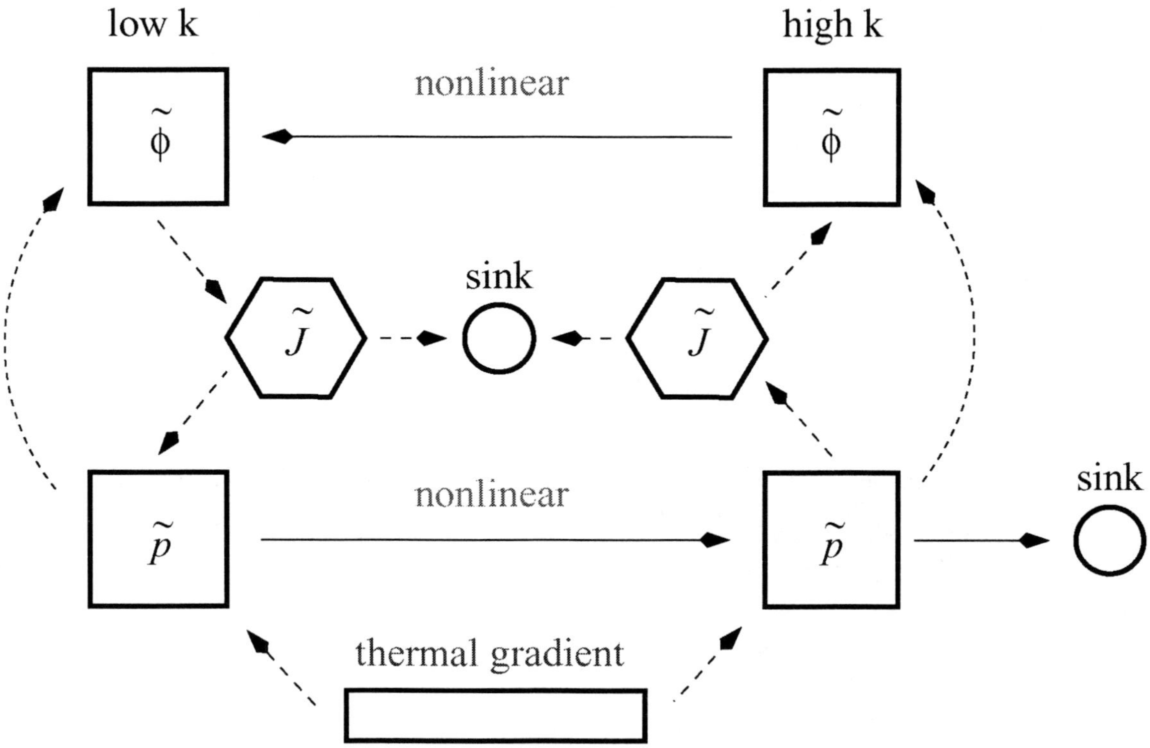

FIGURE 2. A sketch of the energetic transfer properties in drift Alfvén turbulence. The profile gradient drives the pressure fluctuations ($\widetilde{p}$) directly. The arrows labelled nonlinear give the cascade tendencies in $\widetilde{p}$ and in the potential fluctuations ($\widetilde{\phi}$), the latter representing ExB eddy energy. The arrows going through the parallel current fluctuations ($\widetilde{J}$) give the adiabatic response to the cascades. The two sinks are resistive dissipation (through $\widetilde{J}$) and turbulence mixing (cascade to subgrid scales through $\widetilde{p}$). The latter is the principal sink in three dimensional cases. These nonlinear processes are stronger than linear forcing (vorticity larger than drive rates throughout the spectrum).

smaller but still comparable to either $\widetilde{p}_e$ or $\widetilde{\phi}$. In the context of the gyrofluid model of Eqs. (4,5,8), $\widetilde{p}_e$ is $n_e T_e \widetilde{N}_e$ and $\widetilde{h}_e$ is $n_e(\widetilde{N}_e - e\widetilde{\phi}/T_e)$ in physical units. Note that $\widetilde{h}_e$ is conventionally normalised as a density. In the edge of moderate tokamaks $\hat{\beta} > 1$ so the adiabatic response is electromagnetic and $C > 1$ so it is also dissipative. In units of $c_s/L_\perp$ the Alfvén transit frequency v_A/qR_0 is given by $\hat{\beta}^{-1/2}$ and the thermal transit frequency V_e/qR_0 by $\hat{\mu}^{-1/2}$. The damping rate of shear Alfvén waves in the MHD limit is $(C/2\hat{\beta})k_\perp^2\rho_s^2$ that is, strongly scale dependent. All of these nonadiabatic/resistive features are stronger for $k_\perp^2\rho_s^2$ closer to unity, so the last decade in the spectrum up to unity is the most important one. At smaller scale the dissipation is stronger and at larger scale the gradient drive allowed by the adiabatic response is weaker.

In general the dynamical frequencies $c_s/L_\perp$ and V_e/qR_0 and v_A/qR_0 and ν_e are all comparable. By contrast the sound wave transit c_s/qR_0 and ion collision ν_i frequencies are slower by a factor of nearly 100. This puts ion dissipation mechanisms well outside the collisional fluid regime. The above two moment gyrofluid model may be used to learn the properties of edge turbulence but to do any experimental modelling the temperature dynamics of both species should be included and due to these time scales the temperature should not only be anisotropic but also the associated parallel heat fluxes should be kept as dynamical variables. Hence the set of moments available is either two per species for heuristic studies or six per species for serious modelling. The following considerations are however common to all edge turbulence models which satisfy the geometry and resolution issues mentioned above. The discussion will be kept to the two moment ("four field" describing four dependent variables) model of Eqs. (4,5,8).

The basic character of nonlinearity in the turbulence is transfer of free energy within the spectrum. The main effects are ExB advection of the vorticity, proportional to $k_\perp^2\rho_s^2\widetilde{\phi}$ or, equivalently in the gyrofluid model, the negative gyrocenter charge density given by $\Omega \equiv \widetilde{N}_e - \widetilde{N}_i$ (recall that the actual space charge density is set

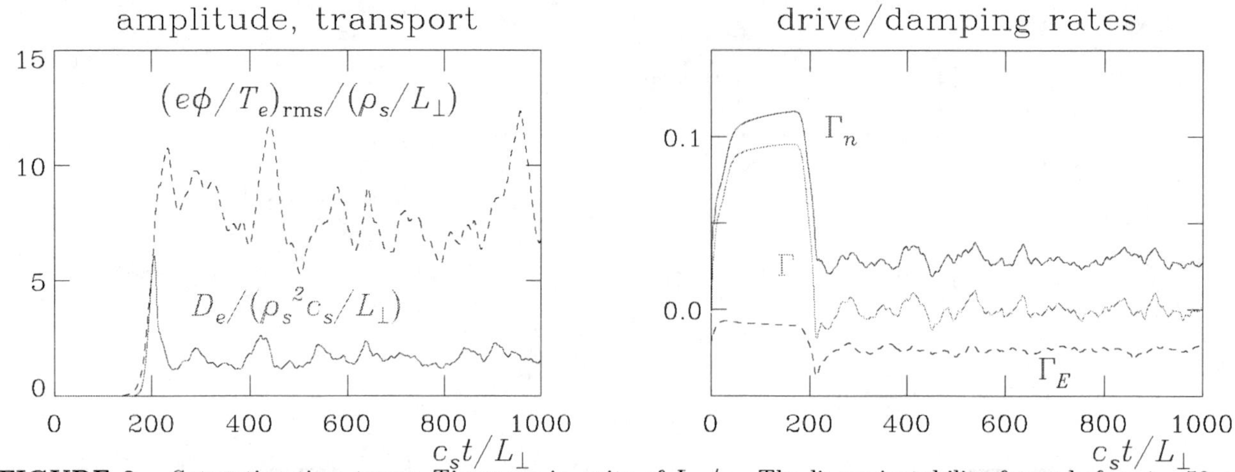

FIGURE 3. Saturation time traces. Times are in units of $L_\perp/c_s$. The linear instability forms before $t = 50$ and remains dominant up to the moment of saturation at which the total growth rate Γ sharply drops to zero, at about 200. Fully developed turbulence is after 300. The grid dissipation is labelled Γ_E and the gradient drive rate is Γ_n, defined in the text. In turbulence, Γ is much lower than the linear growth rate which is found to be about 0.1, while Γ_E rises to compensate, via cascade of $\widetilde{N}_e^2$ towards smaller scales. The RMS vorticity sharply rises to about 1.8 at saturation and then falls to a value of 0.74 in the turbulence, with a curve very similar in shape to the diffusivity D_e. The latter becomes stationary at a value of 1.7 in these gyro-Bohm units in the turbulence. The RMS amplitudes of $e\widetilde{\phi}/T_e$ and $\widetilde{N}_e$ are similar at all times.

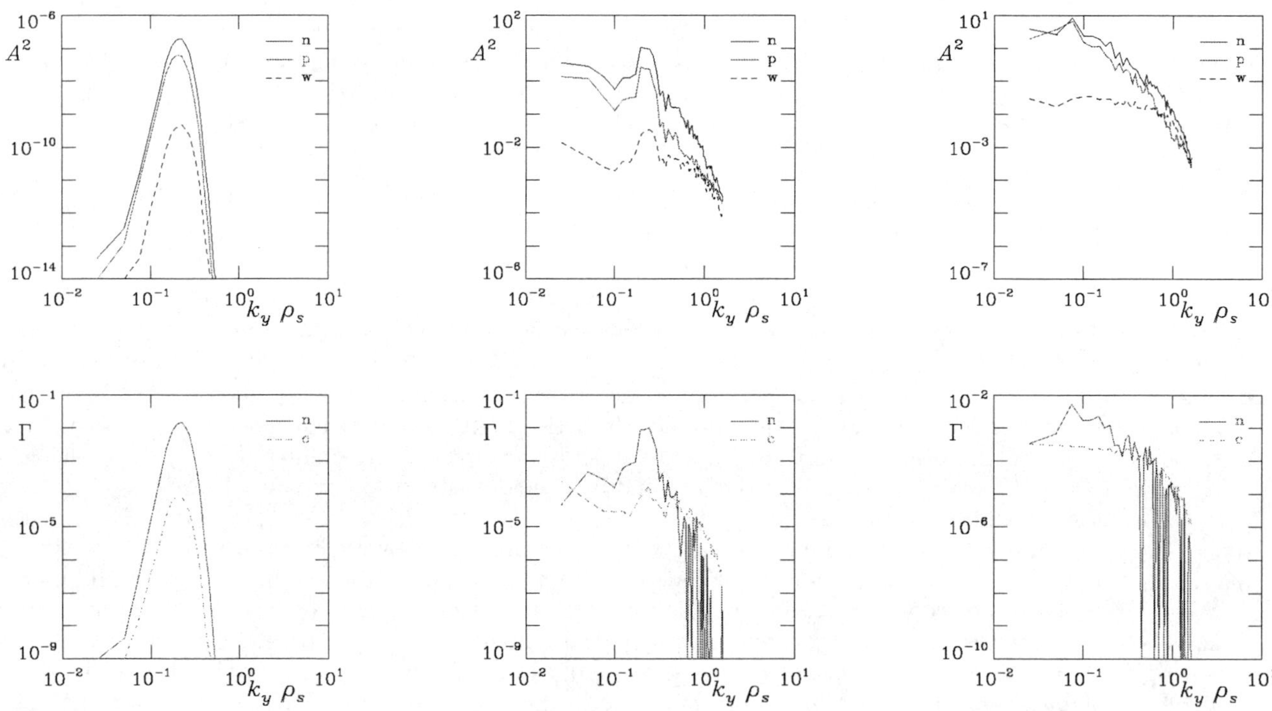

FIGURE 4. Spectra in linear growth (left), saturation (center), and fully developed turbulence (right) phases, at $c_s t/L_\perp = 100$ and 200 and 600, respectively. The top row gives the squared amplitudes of $e\widetilde{\phi}/T_e$ ('p') and $\widetilde{N}_e$ ('n') and Ω ('w') in units of $\rho_s^2/L_\perp^2$. The bottom row gives the gradient drive ('n') and resistive dissipation ('c') rates ('w') in units of $c_s/L_\perp$. At the moment of saturation all quantities still peak where the linear instability does, but later they have shifted to longer wavelength except Ω which is now mostly towards $k_y\rho_s = 1$. Note that the density of states is proportional to k_y.

to zero by the polarisation equation), and ExB advection of the pressure, here represented by $\widetilde{N}_e$ alone. The reason this combination, which in a gyrofluid model looks arbitrary, is chosen is that the vorticity is usually much smaller than either gyrocenter density and has different properties. In the equivalent fluid model we are dealing with $\nabla_\perp^2\widetilde{\phi}$ and $\widetilde{n}_e$. Most of the energy transfer is simply random exchange, tending to broaden the spectrum. On top of this however are the cascade tendencies, which are preferentially to larger scale for the vorticity nonlinearity (the 2D fluid property, "direct" cascade towards smaller scales for vorticity and "inverse" towards larger scales for eddy flow energy), and to smaller scale for the pressure/density (the passive scalar property, purely "direct" transfer). In strict passively advection by 2D turbulence, starting with equal spectra for $\widetilde{\phi}$ and $\widetilde{N}_e$, these nonlinearities cause the spectra to separate: $\widetilde{\phi}$ steepens and collects to low-$k_\perp$, while both $\widetilde{N}_e$ and Ω flatten out towards high-$k_\perp$.

The above two mechanisms describe the basic competition which is active in turbulence. The cascade tendencies cause the spectra to separate while the adiabatic response ties them together at both ends. Hence we find a basic transfer in the free energies represented by $k_\perp^2\widetilde{\phi}^2$ (the same as $-\widetilde{\phi}\Omega$ up to FLR corrections) and $\widetilde{N}_e^2$, towards low- and high-$k_\perp$, respectively, with the imbalances at both low- and high-$k_\perp$ compensated by the adiabatic response. This can be quantified and measured in a turbulence computation, as was done by Camargo [50]. A sketch of the processes is given in Fig. 2. The strength of these nonlinearly driven energetic transfer pathways typically scales as $k_\perp\rho_s$ times $c_s/L_\perp$, which is the same as the diamagnetic frequency. If growth rates of linear instabilities are smaller than this, then linear eigenmodes are sheared apart by the turbulent eddies faster than they can grow, and so they do not control the physics. This is quantified by comparing linear growth rates to the vorticity: if $\Omega > \gamma_L$ everywhere in the spectrum then the linear instabilities will be eliminated. Typical scaling for γ_L due to interchange forcing by magnetic curvature is the ideal interchange growth rate $\gamma_I = c_s/(RL_\perp)^{1/2}$ times a numerical constant in the range of 1/3, peaking at $k_\perp^2\rho_s^2 \sim C^{-1}$. For typically moderate $C < 10$ the vorticity scaling as $\Omega \sim k_\perp\rho_s(c_s/L_\perp)$ is larger than this. The reason is the very large values of $R/L_\perp$, typically at least 50, in the edge regions of moderate tokamaks.

Cursory illustration of the above properties can be obtained by starting the fluctuations at small amplitude, allowing the dominant linear instability to clearly form, and then observing changes that take place during saturation and later in the fully developed turbulence. In this way we obtain a proper scientific control case — comparison of linear instabilities and edge turbulence at equal parameter sets. This has been done in a lot of detail elsewhere [31, 51, 9, 52]. Here we follow Ref. [52], which contains the full details of the setup, while using the above two-moment gyrofluid model as in Ref. [9]. The parameters were $\hat{\beta} = 1$ and $\hat{\mu} = 5$ and $C = 7.65$ and $2L_\perp/R = 0.05$ and $M_i/m_e = 3670$ and $\hat{s} = 1$, on a domain of $(20\times 80)\pi\rho_s$ in $\{x, y\}$ and 2π in s with $128\times 128\times 16$ grid points. The resolution ratio h_y/h_x was four. The initial state was a random phase field in $\widetilde{N}_e = \widetilde{N}_i$ with amplitude $10^{-6}\rho_s/L_\perp$ in $\{x, y\}$ with parallel envelope $\exp(-s^2)$, initially aligned to the magnetic field. The other variables started at zero and then were determined by the model equations. The time step was $c_s\Delta t/L_\perp = 0.02$ and the run was carried for 5×10^4 steps to $c_st/L_\perp = 1000$. Time traces in the RMS amplitude of $e\widetilde{\phi}/T_e$ and the transport diffusivity $D_e = L_\perp\left\langle \widetilde{N}_e v_E^x\right\rangle$ are given in Fig. 3, scaled appropriately by factors of $\rho_s/L_\perp$. The total growth rate is given by $\Gamma = (1/2E)\partial E/\partial t$ where E is the total fluctuation free energy (most of E is in the thermal free energy whose density is $\sum_{\rm sp} n_zT_z\widetilde{N}_z^2/2$). The drive rate is given by $\Gamma_n = (1/2E)\sum_{\rm sp}(n_zT_z/L_\perp)\left\langle \widetilde{N}_z\widetilde{u}_E^x\right\rangle$, and the grid dissipation rate Γ_E is given by the action of numerical subgrid dissipation (cf. Ref. [32] for details). The linear growth phase begins when the dominant instability forms out of the initial bath and ends when the amplitude reaches of order $\rho_s/L_\perp$ at which $\nabla\widetilde{N}_e$ is comparable to ∇N_0. The amplitudes overshoot and then saturate. The amplitude of $\widetilde{N}_e$, not shown, is close to $e\widetilde{\phi}/T_e$ at all times. The RMS vorticity as a frequency in units of $c_s/L_\perp$ rises to an overshoot value of about 2.0 and then saturates at 0.74, which is much larger than any of Γ or Γ_n or Γ_E, showing the power of nonlinear mixing. This is the first indication that the turbulence is much more powerful than instabilities. Also, in saturation the nonlinear drive rate is significantly less than the original linear drive rates. This indicates much of the free energy is resident in degrees of freedom which are stable in a linear analysis. In the simple mixing level model of turbulence, by contrast, the vorticity is assumed to be comparable to the linear growth rate, which itself is assumed to be the nonlinear drive rate.

The spectra also show qualitative changes as the phase of turbulence is entered. These are given in Fig. 4. All quantities are Fourier decomposed in the y-direction and products are averaged over $\{x, s\}$, resulting in k_y-spectra. The squared amplitudes ("power spectra") in $\widetilde{N}_e$ and $e\widetilde{\phi}/T_e$ and Ω in units of $\rho_s/L_\perp$

are labelled ('n,p,w'), respectively, and the spectra of the gradient drive rate Γ_n and collisional damping rate $\Gamma_c = (1/2E)\eta_\parallel \widetilde{J}_\parallel^2$ are labelled ('n,c'), respectively. Spectra are given at $c_s t/L_\perp = 100$ and 200 and 600, respectively showing the linear phase, the moment of saturation where Γ reaches zero, and the fully developed turbulence phase where all time traces are statistically stationary. During the linear phase all quantities show a clear peak at the most unstable k_y-mode. At saturation this peak is still dominant. At late times the spectra become different, especially the vorticity which is flat out to $k_\perp \rho_s \sim 1$, where due to the density of states most of the activity is. The low-k_y collection of free energy and to a slightly lesser extent the resulting transport flux follows from the nonlinear transfer dynamics, not the most unstable linear mode.

Further properties of edge turbulence follow from these basic processes. Measurement of nonlinear transfer spectra finds the same results as in 2D computations, with $\widetilde{N}_z^2$ of both species, hence also the vorticity, showing a direct transfer out of longer wavelengths into shorter, while $(k_\perp \widetilde{\phi})^2$ opposite, showing an inverse transfer into longer wavelengths. The vorticity nonlinearity changes the balances among all the currents sufficiently that linearly stable degrees of freedom mutually destabilise in the presence of the profile gradient. This is ultimately why the fully developed turbulence loses memory of an initially present linear instability. Scaling of the turbulence finds a rising trend due to any of $\hat{\beta}$ or $\hat{\mu}$ or C which control the adiabatic response, mostly due to these nonlinear properties. The details of what is necessary to show these results are involved and therefore presented elsewhere (cursory review in Ref. [53], details in Refs. [31, 51, 9, 52]).

CORE TURBULENCE AND THE MEANING OF SCALE SEPARATION

Core turbulence in a moderate to large tokamak (with a/ρ_s at least about 150) has typical ratios of $R/L_\perp$ or order 10 or less so that $\hat{\mu} \ll 1$ and the adiabatic response is dominant. The temperatures in both species are of order 1 keV or larger, so that $C \ll 1$ also. The plasma beta is also small, with $\beta_e < 10^{-2}$ usually, so that $\hat{\beta} < 1$ also (departure from this leads to MHD instability). Hence the electrons are close to adiabatic. The physics is controlled either by trapped electrons or ions. For trapped electrons, the parallel motion is averaged out and so they have no adiabatic response. They can be treated as a separate species although simply setting the parallel dynamics to zero is problematic in a 3D model (arbitrary, unphysical parallel gradients result). Up to now all reasonable treatments of them are kinetic [43]. Turbulence controlled by the ion temperature dynamics can be treated with either kinetic or fluid models (ealier comparisons are given in Ref. [54]). For the ions the density gradient is not available as a drive, since the adiabatic response pins the phase shift between $\widetilde{N}_e$ and $\widetilde{\phi}$ to zero, and the difference $\widetilde{N}_e - \widetilde{N}_i$ is just Ω. Hence, any model of core turbulence using fluid methods must treat the ion temperature. Moreover, treating the temperature on the same level as the density leads to treating the associated parallel heat flux as a dynamical variable in the same way as the parallel velocity (ultimately, since $\omega \gg c_s/qR_0$ and ν_e is no larger than V_e/qR_0 and is often smaller). The complete gyrofluid model treats perpendicular and parallel temperatures separately (moments using $v_\parallel^2$ and $v_\perp^2$, respectively), each with its own associated heat flux, for a total of six moment equations per species with significant differences in the energetic character in each of three pairs of dependent variables [32].

Nevertheless, it is possible to understand the ideas of "ion temperature gradient" (ITG) core turbulence using a minimal model which can treat drive by ∇T_i through $\widetilde{T}_i$ fluctuations [55, 56]. Since the electrons are taken as adiabatic we need only electrostatic equations for $\widetilde{N}_i$, $\widetilde{u}_{i\parallel}$, $\widetilde{T}_i$, and (added here) also a parallel heat flux $\widetilde{q}_{i\parallel}$, and have them coupled to a polarisation equation. The four moment equations are

$$\frac{\partial \widetilde{N}_i}{\partial t} + \mathbf{u}_E \cdot \nabla[\widetilde{N}_i + N_0] + B\nabla_\parallel \frac{\widetilde{u}_{i\parallel}}{B} = \mathcal{K}\left(\widetilde{\phi}_G + \frac{T_i \widetilde{N}_i + \widetilde{T}_i}{e}\right) \tag{29}$$

$$\frac{M_i}{e}\left(\frac{\partial \widetilde{u}_{i\parallel}}{\partial t} + \mathbf{u}_E \cdot \nabla \widetilde{u}_{i\parallel}\right) + \nabla_\parallel \left(\widetilde{\phi}_G + \frac{T_i \widetilde{N}_i + \widetilde{T}_i}{e}\right) = 2\frac{T_i}{e}\frac{M_i}{e}\mathcal{K}\left(\widetilde{u}_{i\parallel} + \widetilde{q}_{i\parallel}\right) \tag{30}$$

$$\frac{\partial \widetilde{T}_i}{\partial t} + \mathbf{u}_E \cdot \nabla[\widetilde{T}_i + T_0] + B\nabla_\parallel \frac{\widetilde{u}_{i\parallel} + \widetilde{q}_{i\parallel}}{B} = \mathcal{K}\left(\widetilde{\phi}_G + \frac{T_i \widetilde{N}_i + 3\widetilde{T}_i}{e}\right) \tag{31}$$

$$\frac{M_i}{e}\left(\frac{\partial\widetilde{q}_{i\|}}{\partial t}+\mathbf{u}_E\cdot\nabla\widetilde{q}_{i\|}+\frac{V_i}{qR_0}\widetilde{q}_{i\|}\right)+\nabla_\|\left(\widetilde{\phi}_G+\frac{T_i\widetilde{N}_i+\widetilde{T}_i}{e}\right)=2\frac{T_i}{e}\frac{M_i}{e}\mathcal{K}\left(\widetilde{u}_{i\|}+3\widetilde{q}_{i\|}\right) \tag{32}$$

by direct analogy to Eqs: (4,5). The dependent variables are denoted by the tilde symbol. The factors of T_i multiplying $\widetilde{N}_i$ are normalisation constants. The dissipation term with V_i/qR_0 in Eq. (31) represents a simple model for Landau damping of parallel dynamics [57]. Some complications, particularly in FLR effects, have been left out, and the somewhat arbitrary factor of 3 in the latter two equations models the appearance of higher order moments in the hierarchy. Note also that in an electrostatic model $B\nabla_\|$ is just $B^s(\partial)/\partial s$, which does not act on the profile quantities N_0 and T_0. The gyroaveraged potential $\widetilde{\phi}_G$ and the other differential operators are defined as before. The polarisation equation is still

$$n_i\frac{e^2}{T_e}\rho_s^2\nabla_\perp^2\widetilde{\phi}+(1-\rho_i^2\nabla_\perp^2)\sum_{\mathrm{sp}}(n_zZe)\Gamma_1\widetilde{N}_z=0 \tag{33}$$

but now the model for the electron density is [11]

$$\widetilde{N}_e-\left\langle\widetilde{N}_e\right\rangle=\frac{T_e}{e}\left(\widetilde{\phi}-\left\langle\widetilde{\phi}\right\rangle\right) \tag{34}$$

where the angle brackets denote the flux surface average ("zonal," over y and s). This reflects the action of the adiabatic response, through $\nabla_\|$, only on the non-zonal component of $\widetilde{h}_e$.

One must be careful with $\left\langle\widetilde{N}_e\right\rangle$. This is an initial condition whose time derivative vanishes. However, if self consistent profiles in $\widetilde{N}_i$ and $\widetilde{T}_i$ are to be advanced in a global model (following vorticity as well as ion temperature), then the zonal averages of both $\widetilde{N}_e$ and $\widetilde{N}_i$ should be set equal in the initial state. Further, in a self consistent global model N_0 and T_0 are dispensed with and replaced by $\left\langle\widetilde{N}_i\right\rangle$ and $\left\langle\widetilde{T}_i\right\rangle$ in the initial state. On the other hand, if N_0 and T_0 are used as a separate background drive, then all of $\left\langle\widetilde{N}_e\right\rangle$, $\left\langle\widetilde{N}_i\right\rangle$, and $\left\langle\widetilde{T}_i\right\rangle$ should be initialised to zero.

The main point in presenting this example core turbulence model is more prosaic: simply to note that the same numerical methods can be used for both edge and core models because the terms in the equations and the order in which to treat them is the same. But the physics is rather different. This is why it is of advantage to develop numerical methods which do not assume the resulting character of the turbulence (dominance or absence of particular linear mode types, narrow ranges in the spectrum, etc).

In contrast to edge turbulence, the nonlinearities in core turbulence are rather weak. There is no turbulent density mixing (the lowest order fluid moment), only temperature and vorticity. Furthermore, the linear eigenmodes have a wave frequency spectrum $\Re[\omega(\mathbf{k})]$ which is strongly dispersive in k_y. Hence the linear instability is able to stamp its character on the turbulence even though the spectrum broadens, with the peak of all quantities (mainly, temperature and transport flux spectra) at roughly the same k_y. Fig. 6 shows the spectra in the linear and turbulence phases of a typical case using the six moment electromagnetic model (from Ref. [10]), with the peaks in the same place despite the general broadening in the turbulence.

Either this treatment or a fully electromagnetic one will admit core turbulence zonal flows [58, 59, 54, 60, 61, 62, 10, 56], described by the part $\left\langle\widetilde{\phi}\right\rangle$ which is allowed by polarisation (Eq. 33). The actual drive is through the vorticity nonlinearity, given by the vestige of $\mathbf{u}_E\cdot\nabla\widetilde{N}_i$ since $\mathbf{v}_E\cdot\nabla\widetilde{N}_e$ vanishes. The zonal component of this is related to the ExB Reynolds stress $\langle v_E^x v_E^y\rangle$ and the correlation of that with the zonal vorticity $\langle\Omega\rangle$ gives the energetic transfer rate from the ExB eddies (non-zonal $\widetilde{\phi}$) into the zonal flows. These zonal flows saturate due to a process called geodesic compression. In toroidal geometry a zonal ExB flow is not incompressible. It is faster on the outside of the torus due to the factor $1/B$. In the model this is represented by the curvature term $\mathcal{K}(\widetilde{\phi}_G)$ giving the ExB compression directly. The part of $\mathcal{K}$ which acts through $\partial/\partial x$ gives the compression of the zonal flow. The terms $\mathcal{K}(\widetilde{\phi}_G)$ and $\mathcal{K}(\widetilde{p}_i)$, with $\widetilde{p}_i=n_i\widetilde{T}_i+n_iT_i\widetilde{N}_i$, exchange energy conservatively between ExB zonal flow energy and thermal free energy in the part of $\widetilde{N}_i^2$ and $\widetilde{T}_i^2$ due to the component which is both non-zonal and axisymmetric (toroidal mode zero, poloidal mode nonzero, dominantly the $\sin\theta$ piece which lines up with $\mathcal{K}$ acting upon x). This latter component is called a

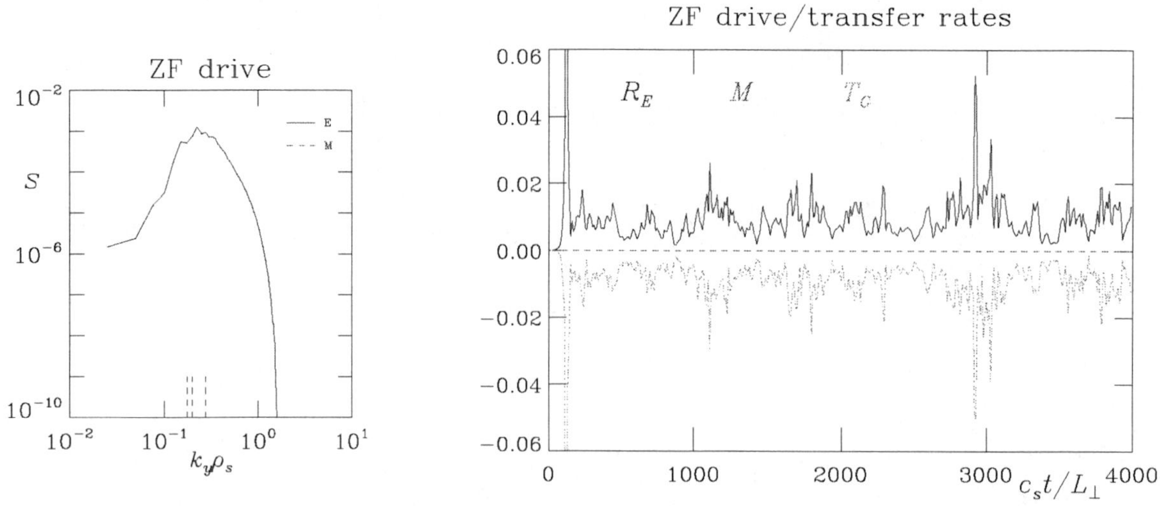

FIGURE 5. (left) Time-averaged spectrum of the zonal flow drive via Reynolds stress in adiabatic core ITG turbulence, positive at all wavelengths. (right) Time traces of the Reynolds stress drive (blue) and the geodesic compression sink (green), showing them in balance. The Maxwell stress (red) is zero because magnetic fluctuations are absent (from Ref. [10]).

sideband. The transfer of energy through this process is the same as the one involved in geodesic acoustic oscillations [63], which would obtain if these two terms were the only ones present. In the turbulence, however, mostly incoherent oscillations are present at all frequencies up to several times c_s/R and the process serves to take energy out of the zonal flows, saturating them. Again, this can be quantified and measured, as shown in Fig. 5 (from Ref. [10]). The zonal flow energy equation in the four moment model in Eqs. (29–33) is given by

$$\frac{\partial E_Z}{\partial t} = R_E + R_M - T_G \tag{35}$$

where the zonal flow energy, Reynolds stress drive, Maxwell stress drive, and geodesic curvature transfer are given by

$$E_Z = \frac{1}{2}\int dx \left[n_i e \left\langle \widetilde{\phi}_G \right\rangle \left\langle \widetilde{N}_i \right\rangle - n_e e \left\langle \widetilde{\phi} \right\rangle \left\langle \widetilde{N}_e \right\rangle \right] \tag{36}$$

$$R_E = \int dx\, n_i e \left\langle \frac{\partial \widetilde{\phi}_G}{\partial x} \right\rangle \left\langle \widetilde{N}_i \widetilde{u}_E^x \right\rangle \tag{37}$$

$$R_M = \int dx\, n_i e \left\langle \frac{\partial \widetilde{\phi}_G}{\partial x} \right\rangle \left\langle \widetilde{u}_{i\parallel} \widetilde{b}^x \right\rangle \tag{38}$$

$$T_G = \int dx \left\langle \frac{\partial \widetilde{\phi}_G}{\partial x} \right\rangle \omega_B \left\langle \left(\widetilde{\phi}_G + \frac{T_i \widetilde{N}_i + \widetilde{T}_i}{e} \right) \sin s \right\rangle \tag{39}$$

respectively, where $\mathcal{K}(x) = \omega_B \sin s$ with $\omega_B = 2L_\perp/R$ is the geodesic curvature (cf. Ref. [10]; in the six moment model the energy theorem is slightly different in detail). The Maxwell stress is zero because the magnetic fluctuations $\widetilde{b}^x$ vanish. The electron contributions to R_E and T_G vanish due to adiabaticity. The spectrum of R_E due to contributions from each k_y and the time traces of R_E and T_G are given in Fig. 5. This shows the robust Reynolds stress and the detailed balance with geodesic compression in the saturation of E_Z, which is statistically stationary. The geodesic compression holds the zonal flows to moderate amplitude, at which they moderate but do not suppress the turbulence.

The global scale rotation of the tokamak is not however determined by the turbulence. The influence of the turbulence upon the equilibrium scales as the square of the drift parameter $\rho_s/L_\perp$. But the geodesic compression processes (in all the equations, not just for zonal flow energy) scale with only one factor of $\rho_s/L_\perp$.

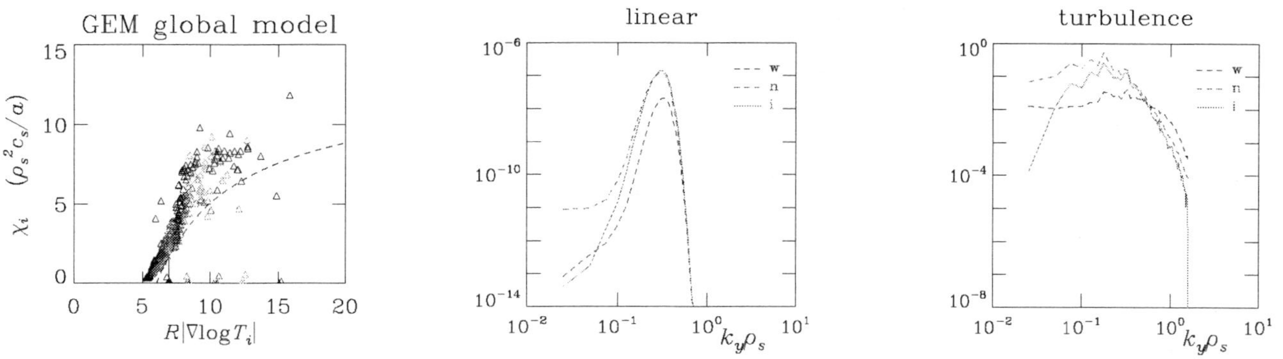

FIGURE 6. (left) Transport/gradient trajectories for various initial values of R/L_T, using the adiabatic electron global model. Although they have different histories, the flux spreads overlap when the gradient values are equal during different runs at different times. The dashed line gives the fit from Ref. [54]. (center and right) Spectra in linear growth and turbulence phases of the vorticity fluctuation amplitude (dashed) the density fluctuation amplitude (dashed, higher) and of the ion E-cross-B advective heat flux (solid) in the electromagnetic model (from Ref. [10]), showing that the main peaks remain at the strongest instability despite the general broadening and the shift of the vorticity to shorter wavelengths.

So with ever larger domains in units of ρ_s, the processes related to the turbulence and the maintenance of the equilibrium begin to separate. This is why the labels "moderate" and "large" tokamaks are significant. Computations of the entire core region (up to about $r/a = 0.9$) can be done at various values of ρ_s/a to study this. The scale of the radial layers of $\langle\widetilde{\phi}\rangle$ describing the zonal flows reaches up to about 20 or $30\rho_s$. If $L_\perp/\rho_s$ is only one to two times this, then the turbulence and profile scales overlap (the zonal flows are part of the turbulence, while the ExB rotation profile is part of the equilibrium).

This is a somewhat more restrictive condition than the traditional one derived from separating the turbulence and thermal transport time scales (which is what is meant by the term "gyro-Bohm convergence" [64, 65]). The RMS vorticity is of order $c_s/L_\perp$. The time required for turbulence to saturate is of order $100L_\perp/c_s$. The transport time scale τ_D is given by the square of the layer thickness divided by the diffusion coefficient. The latter scales as $\rho_s^2 c_s/L_\perp$, so that $\tau_D L_\perp/c_s \sim (L_\perp/\rho_s)^2$. The transport diffusion time would already be safely separate for $L_\perp/\rho_s \sim 30$. Due to that radial width of $\langle\widetilde{\phi}\rangle$ however the zonal flow saturation time is closer to $10^3 L_\perp/c_s$. This separates well from the transport time only for $L_\perp/\rho_s > 100$. For typical profiles $a/L_\perp \sim 2$, so that the moderate tokamak regime is reached only for a/ρ_s of 200 or larger. At this level the relaxation of the profile in a self consistent computation begins to be slow enough that sources can be left out of a computation. Hence the saturation times are all within a run time which is still short to the profile relaxation time. Beyond this, the large tokamak regime is reached when even the shape of the profile does not change during a computational run of 500 to $10^3 a/c_s$. In both cases the limit is about $L_\perp/\rho_s \approx 200$, so that if there is no shape to the profile $a/\rho_s \sim 200$ is sufficient, while a sinusoidal gradient profile (something like $a/L_\perp = 2\cos\pi[r/a - 1/2]$) will keep its shape over that run only for a/ρ_s larger than about 400. It is important to note that the actual numbers are dependent on the model and the parameters, and it is sensitive to the degree of shift of ExB eddy energy to larger scales. But these figures of 200 and 400 are a good guide in the case of the electromagnetic gyrofluid model.

The definition of transport scale separation is that the transport flux as a function of the actual gradients does not depend on the run history. That is, starting with several R/L_T (different from the nominal value of 2), the trajectories of ion heat flux $\widetilde{T}_i\widetilde{u}_E^x$ versus $R\left|\nabla\widetilde{T}_i\right|$, volume averaged over the domain $0.5 < r/a < 0.75$ have been found to actually overlap if a/ρ_s is at least about 200 [32]. In global cases with the six-moment gyrofluid model for ions with adiabatic electrons, the nominal gradient values were initialised at $R/L_T = 6.9$, 8.3, 10.4, 12.4, and 15.2 and allowed to relax. The resulting flux/gradient trajectories, shown in the left frame of Fig. 6, demonstrate that the flux as a function of the gradient is temporally local, independent of the previous history. Each run was done for about $1200a/c_s$. The diffusivity χ is defined by dividing the flux by the absolute gradient value and expressed in terms of $\rho_s^2 c_s/a$. The dashed line is a fit to the results from

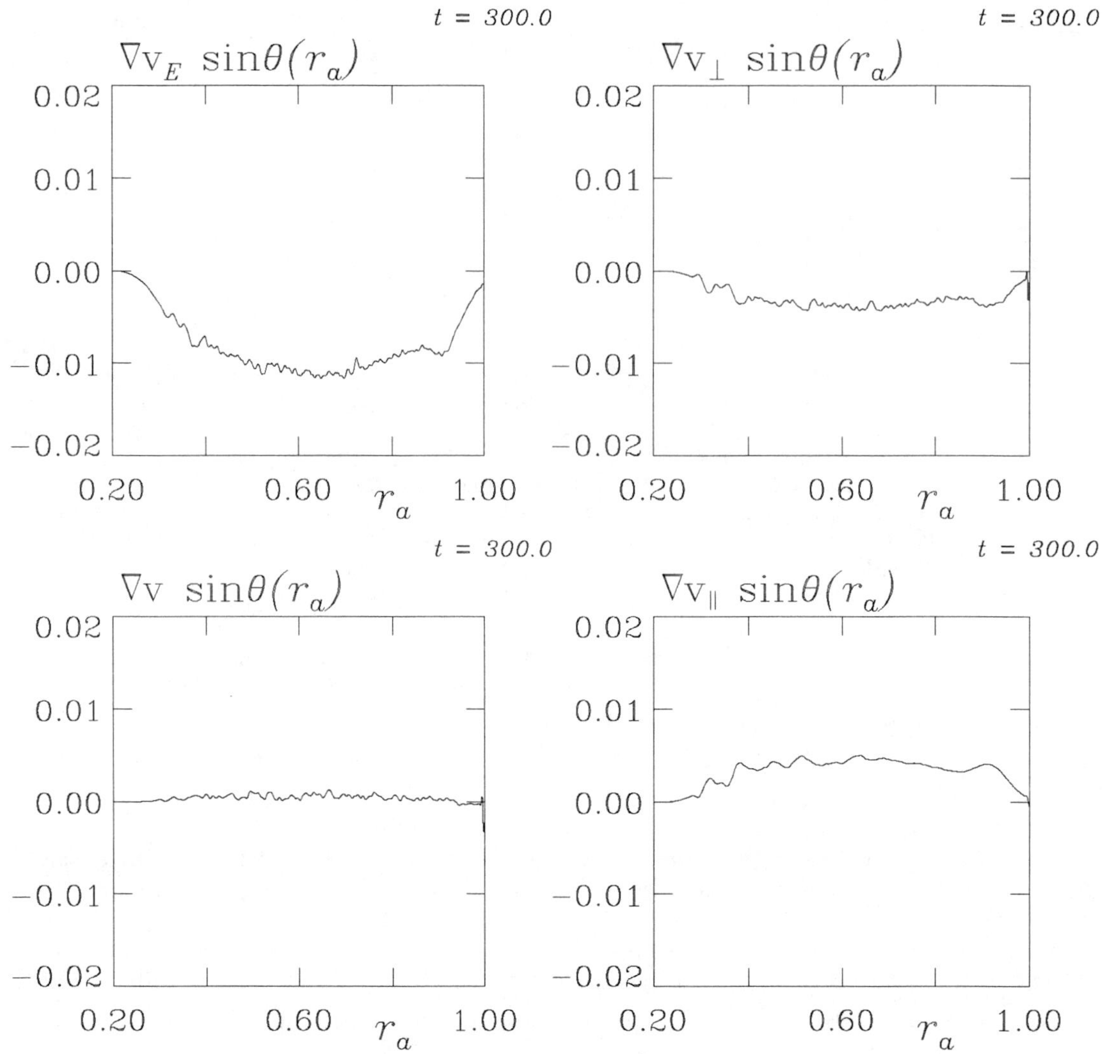

FIGURE 7. Ion flow sideband divergences, showing (clockwise from top left) the finite ExB, total drift, and parallel flow divergences and then the total flow divergence close to zero, using the electromagnetic global model. The coordinate is $r_a = r/a$ with $a = 200\rho_s$. The vertical scale is in units of c_s/a. The zonal flow layers of width and amplitude circa $10\rho_s$ and 10^{-3} are barely visible on even the total divergence profile. These are all measured as indicated in the text and the results represent scale separation between the turbulence and zonal flow combination and the background with finite flow profile.

a gyrokinetic model [54].

The above limits were found "numerically empirically" using runs with the electromagnetic six-moment gyrofluid model for both species with polarisation and induction equations. The values of a/ρ_s were in steps of a factor of two from 50 to 800. Roughly speaking, 200 is the moderate level, 400 is commensurate with JET, and 800 with ITER for typical parameters. The logarithmic gradients were set giving temperature and density scale lengths of $a/L_T = 2\cos\pi(r/a - 1/2)$ and $L_n = 3L_T$ for both species, and $n_e = n_i$ and $T_e = T_i$ were assumed. The values $\beta_e = 0.002$ and $m_e/M_i = 1/3670$ were taken. The rotation profile was found to be unaffected by the turbulence. This is quantified by measuring the $\sin s$ components of the terms in the ion continuity equation ($\left\langle \widetilde{N}_i \sin s \right\rangle$, specifically) while neglecting the nonlinearities. The divergences due to $\mathcal{K}$ and $\nabla_\parallel$, that is, the equilibrium pieces, are nonzero but cancel each other by a factor of at least ten compared to $\nabla \cdot \mathbf{v}_E$, for $a/\rho_s = 200$ or larger. This is shown in Fig. 7. For smaller values the temporal relaxation becomes apparent as a positive divergence layer with the same radial scale as the domain. On the level of the finite

$\nabla \cdot \mathbf{v}_E$ the zonal flow divergence layers are barely visible. By contrast, if $L_\perp/\rho_s < 50$ these zonal flow layers begin to dominate.

The salient feature of core turbulence is this scale separation. The processes setting up the equilibrium, including the ExB flow profile, are temporally separated from those involved in the turbulence, including the fluctuating, compressible zonal flows. The latter are too weak to influence the background profile and barely appear in the flow sideband profile diagnostic. The transport at given times is given by the profile gradient at those times, not from how the system is (or is not) driven. These results are possible only in computations in which this scale separation is actually present, in both dimensions perpendicular to the magnetic field. Even the qualitative "look and feel" of turbulence in tokamak core regions is only accessible with large computations done with high resolution.

OUTLOOK

The study of nonlinear phenomena in magnetised plasmas by means of numerical simulations is itself a young field, and the part of it which deals with turbulence is even younger. Well resolved three dimensional computations which treat the electron dynamics date back less than fifteen years. It is only natural that there are many subtle phenomena and even numerical pitfalls which often go unaddressed at the current time in the scientific literature. One hindrance has been the dependence on linear instability theory or MHD as the basis for constructing scenarios which contribute to the formulation of the computations. The understanding that beyond ensuring the proper coverage of the relevant scales down to the ion gyroradius the expected behaviour can undesirably influence the setup of the computations is very new and incomplete. A key concept is scale separation, which is the first thing to suffer when memory and CPU resource limitations cause one to relax the difficulty of the problem. Many concepts which have grown out of early work are in fact direct consequences of insufficient scale separation, particularly those which involve a direct role for the turbulence in establishing the axisymmetric equilibrium flow state. Beyond the criteria of geometry and resolution in setting up computations, the use of control cases to decide among physical mechanisms is too often lacking. Early computations typically had only one mode of instability which is then the same mechanism to maintain the turbulence. But when more than one are present, the hope that one can interpret results in terms of a favoured mechanism has to be supported by a control against that mechanism and any competing ones. Not only for turbulence drive and saturation, this is also the case with research into flow generation. Computations are only useful as a theoretical tool, rather than an interpretation aid, when the *a priori* selection of an expected outcome is fully avoided. Very often, the subtleties are counterintuitive, and the selection of favoured outcomes is subconscious. If the proper groundwork is done and all the hypotheses are fully tested, then these computations can continue to be expected to lead to new insight which will help the understanding of these very complicated phenomena in present and future devices in fusion energy research.

REFERENCES

1. W. L. Rowan, C. C. Klepper, C. P. Ritz, R. D. Bengston, K. W. Gentle, P. E. Phillips, T. L. Rhodes, B. Richards, and A. J. Wootton, Nucl. Fusion **27**, 1105 (1987).
2. A. J. Wootton, B. A. Carreras, H. Matsumoto, K. McGuire, W. A. Peebles, C. P. Ritz, P. W. Terry, and S. J. Zweben, Phys. Fluids B **2**, 2879 (1990).
3. M. Endler, H. Niedermeyer, L. Giannone, E. Holzhauer, A. Rudyj, G. Theimer, N. Tsois, and T. A. Team, Nucl. Fusion **35**, 1307 (1995).
4. G. R. McKee, C. C. Petty, R. E. Waltz, C. Fenzi, R. J. Fonck, J. E. Kinsey, T. C. Luce, K. H. Burrell, D. R. Baker, E. J. Doyle, et al., Nucl. Fusion **41**, 1235 (2001).
5. S. J. Zweben, J. A. Boedo, O. Grulke, C. Hidalgo, B. LaBombard, R. J. Maqueda, P. Scarin, and J. L. Terry, Plasma Phys. Contr. Fusion **49**, S1 (2007).
6. J. Wesson, *Tokamaks* (Oxford University Press, 1997).
7. R. D. Hazeltine and J. D. Meiss, *Plasma Confinement (Frontiers in Physics Vol 86)* (Addison-Wesley, 1992).
8. B. Scott, Plasma Phys. Contr. Fusion **39**, 1635 (1997).
9. B. Scott, Plasma Phys. Contr. Fusion **45**, A385 (2003).
10. B. Scott, Plasma Phys. Contr. Fusion **48**, B277 (2006).
11. W. Dorland and G. Hammett, Phys. Fluids B **5**, 812 (1993).

12. M. A. Beer and G. Hammett, Phys. Plasmas **3**, 4046 (1996).
13. W. W. Lee, Phys. Fluids **26**, 556 (1983).
14. A. M. Dimits and W. W. Lee, J. Comput. Phys. **107**, 309 (1993).
15. S. E. Parker and W. W. Lee, Phys. Fluids B **5**, 77 (1993).
16. F. L. Hinton and C. W. Horton, Phys. Fluids **14**, 116 (1971).
17. A. Hasegawa and K. Mima, Phys. Fluids **21**, 87 (1978).
18. M. Wakatani and A. Hasegawa, Phys. Fluids **27**, 611 (1984).
19. R. E. Waltz, Phys. Fluids **28**, 577 (1985).
20. O. G. for the ASDEX Upgrade Team, Nucl. Fusion **47**, S622 (2007).
21. M. R. W. for the DIII-D Team, Nucl. Fusion **47**, S543 (2007).
22. J. Pamela, J. Ongena, and J. E. Contributors, Nucl. Fusion **45**, S63 (2005).
23. K. Ikeda, Nucl. Fusion **47**, No 6, all (2007).
24. R. L. Dewar and A. H. Glasser, Phys. Fluids **26**, 3038 (1983).
25. B. Scott, Phys. Plasmas **5**, 2334 (1998).
26. T. S. Hahm, Phys. Fluids **31**, 2670 (1988).
27. T. S. Hahm, W. W. Lee, and A. Brizard, Phys. Fluids **31**, 1940 (1988).
28. H. Sugama, Phys. Plasmas **7**, 466 (2000).
29. A. Brizard, Phys. Plasmas **7**, 4816 (2000).
30. A. Brizard and T. S. Hahm, Rev. Mod. Phys. **79**, 421 (2007).
31. B. Scott, Phys. Plasmas **7**, 1845 (2000).
32. B. Scott, Phys. Plasmas **12**, 102307 (2005), `arXiv:physics/0501124`.
33. B. Scott, *Derivation via free energy conservation constraints of gyrofluid equations with finite-gyroradius electromagnetic nonlinearities*, submitted to Phys Plasmas, arXiv:0710.4899 (2007).
34. B. Scott, Phys. Plasmas **14**, 102318 (2007).
35. H. Sugama, M. Wakatani, and A. Hasegawa, Phys. Fluids **31**, 1601 (1988).
36. B. A. Carreras and P. H. Diamond, Phys. Fluids B **1**, 1011 (1989).
37. M. Beer, S. Cowley, and G. Hammett, Phys. Plasmas **2**, 2687 (1995).
38. B. Scott, Phys. Plasmas **8**, 447 (2001).
39. B. Scott, Phys. Fluids B **4**, 2468 (1992).
40. R. Hatzky, A. K. T. M. Tran, R. Kleiber, and S. J. Allfrey, Phys. Plasmas **9**, 898 (2002).
41. J. A. Heikkinen, S. Henriksson, S. Janhunen, T. P. Kiviniemi, and F. Ogando, Contrib. Plasma Phys. **46**, 490 (2006).
42. M. Kotschenreuther, G. Rewoldt, and W. M. Tang, Comput. Phys. Comm. **88**, 128 (1995).
43. T. Dannert and F. Jenko, Phys. Plasmas **12**, 07239 (2005).
44. J. Candy and R. E. Waltz, **13**, 032310 (2006).
45. V. Naulin, Phys. Plasmas **10**, 4016 (2003).
46. V. Naulin and A. Nielsen, SIAM J Sci Comput **25**, 104 (2003).
47. A. Arakawa, J. Comput. Phys. **1**, 119 (1966), repr. vol 135 (1997) 103.
48. G. E. Karniadakis, M. Israeli, and S. A. Orszag, J. Comput. Phys. **97**, 414 (1991).
49. B. Scott and F. Porcelli, Phys. Plasmas **11**, 5468 (2004).
50. S. Camargo, D. Biskamp, and B. Scott, **2**, 48 (1995).
51. B. Scott, New J. Phys. **4**, 52 (2002).
52. B. Scott, Phys. Plasmas **12**, 062314 (2005).
53. B. Scott, Plasma Phys. Contr. Fusion **49**, S25 (2007).
54. A. M. Dimits, G. Bateman, M. A. Beer, B. I. Cohen, W. Dorland, G. W. Hammett, C. Kim, J. E. Kinsey, M. Kotschenreuther, A. H. Kritz, et al., Phys. Plasmas **7**, 969 (2000).
55. M. Ottaviani and G. Manfredi, Phys. Plasmas **6**, 3267 (1999).
56. G. L. Falchetto, M. Ottaviani, X. Garbet, and A. Smolyakov, Phys. Plasmas **14**, 082304 (2007).
57. G. W. Hammett and F. W. Perkins, Phys. Rev. Lett. **64**, 3019 (1990).
58. R. E. Waltz, G. D. Kerbel, and J. Milovich, Phys. Plasmas **1**, 2229 (1994).
59. Z. Lin, T. S. Hahm, W. W. Lee, W. M. Tang, and R. B. White, Science **281**, 1835 (1998).
60. T. S. Hahm, K. H. Burrell, Z. Lin, R. Nazikian, and E. J. Synakowski, Plasma Phys. Contr. Fusion **42**, A205 (2000).
61. G. L. Falchetto and M. Ottaviani, Phys. Rev. Lett. **92**, 025002 (2004).
62. N. Miyato, Y. Kishimoto, and J.-Q. Li, Phys. Plasmas **11**, 5557 (2004).
63. N. Winsor, J. Johnson, and J. Dawson, Phys. Fluids **11**, 2448 (1968).
64. Z. Lin and T. S. Hahm, Phys. Plasmas **11**, 1099 (2004).
65. R. E. Waltz and J. Candy, Phys. Plasmas **12**, 072303 (2005).

LIST OF PARTICIPANTS

AGOSTINI Matteo Consorzio RFX Corso Stati Uniti 4 Padova Italy
AGULLO Olivier CNRS- Université de Provence, PIIM 13397 Marseille Cedex 20, France
ARNAS C CNRS- Université de Provence, PIIM 13397 Marseille Cedex 20, France
BANARJEE Santanu Institute Plasma Research Bhat, Gandhinagar 382 428 Gandhinagar India
BELTRAME Philippe Max-Planck Institute Dresden Germany
BEYER Peter CNRS- Université de Provence, PIIM 13397 Marseille Cedex 20, France
BESSE Nicolas LPMIA Univ Poincaré Nancy 1 54506 Vandoeuvre les Nancy Cedex, France
BISAI Nirmal kumar Institute Plasma Research Bhat, Gandhinagar 382 428 Gandhinagar, India
BOSE Mridul Jadavpur Univ Kolkata Depart physics Kolkata-700032, India
CALDAS Iberé Luiz Inst Physics Univ of Sao Paulo CP 053 15 970 Sao Paulo Brazil
CASATI Alessandro Association EURATOM-CEA DRFC Bât. 507 CEA-Cadarache 13108 Saint Paul-Lez-Durance Cedex - France
CHANDRA Debasis CNRS- Université de Provence, PIIM Case 321 13397 Marseille cedex 20
CLAIRE Nicolas Turbulence Plasma CNRS- Université de Provence, PIIM 13397 Marseille cedex 20
DEWHURST Joseph, Centre for Fusion, Space and Astrophysics,

Physics Department, University of Warwick, Coventry,CV4 7AL,United Kingdom.
DOVEIL Fabrice CNRS- Université de Provence, PIIM 13397 Marseille Cedex 20
DUNAI Dániel KFKI-Research Institute for Particle and Nuclear Physics P.O.B. 49, H-1525, Budapest, Hungary
ERIKSSON Annika Dept of Radio et Space Science Chalmers Univ of Technology SE 412 Göteborg, Sweden
FARGE Marie LMD-IPSL-CNRS, Ecole Normale Superieure, 24, rue Lhomond, 75231 Paris Cedex 5.France
FEDORCZAK Nicolas CEA Cadarache (DRFC) St Paul Lez Durance Cedex, France
FUHR Guillaume CNRS- Université de Provence, PIIM 13397 Marseille Cedex 20, France
FUTATANI Shinpei Kyoto University Japan
GUINEE A.Richard Department of Electronic Engineering, Cork Institute of Technology, Cork, Ireland
GUSTAFSON Kyle University of Maryland Dept Physics College Park Maryland, USA
GUIMARAES Zwinglio Inst Fisica da USP Rua do Matao Trav R, 187 cidada Univ CEP 05508-090 Sao Paulo SP, Brazil
GRANSTEDT Erik Princeton University PPPL MS 30 PO BOX 451 Princeton, NJ08543, USA
HIGHCOCK Edmund Imperial College London Pembroke College Cambridge, United Kingdom
HORACEK Jan Institute of Plasma Physics, Za Slovankou 3, 182 00 Prague 8, Czech Republic
IMAZAWA Ryota

University of Tokyo Tokyo Japan
INOMOTO Michiaki Center for Atomic and Molecular Technologies Graduate School of Engineering Osaka University, Japan
KOCAN Martin CEA CADARCHE, DRFC/SIPP/GIPP Association CEA-EURATOM Saint Paul Lez Durance 13108 France
KOMINIS Yannis National Technical University of Athens, Association EURATOM- Hellenic Republic, Greece
KONZETT Stefan Institute for theoretical physics Innsbruck / AG Plasmaphysik Innsbruck Austria
KUBIC Martin Czech Technical Univ in Prague Fac Of Nuclear Sciences Phys Engineering Det Physics Brehova 7, 115 19 Praha 1, Czech Republic
LECONTE Michael CNRS- Université de Provence, PIIM 13397 Marseille Cedex 20, France
LERCHE Ernesto LPP - ERM/KMS (Association EURATOM Belgian State) 32, Av. de la Renaissance, B1000, Brussels - Belgium
MACOR Alexandro CNRS- Université de Provence, PIIM 13397 Marseille Cedex 20, France
MAHMOOD Muhammad Ansar Chalmers Univ Tech Depart Signals Systems Chalmers Univ Tech SE 412 96 Gothenburg, Sweden
MARANDET Yannick CNRS- Université de Provence, PIIM 13397 Marseille Cedex 20, France
MORISSET Sabine CNRS- Université de Provence, PIIM 13397 Marseille Cedex 20, France
MORO Alessandro Istituto di Fisica del Plasma Consiglio Nazionale delle Ricerche EURATOM-ENEA-CNR Association Via Cozzi, 53- 20125 Milano (ITALY)
MURAGLIA Magali CNRS- Université de Provence, PIIM 13397 Marseille Cedex 20, France

NAKATA Motoki
National Institute for Fusion Science
322-6 Oroshi-Cho
Toki
GIFU 509 5292 Gifu, Japan

NICOL Ruth
Centre for Fusion, Space and Astrophysics
Department of Physics
The University of Warwick
Coventry, CV4 7AL, United Kingdom

NUYEN VAN YEN Romain
ENS Paris
LMD-IPSL-CNRS, Ecole Normale Superieure,
24, rue Lhomond, 75231 Paris Cedex 5.

NISHIMURA Seiya
Kyushu University, Fukuoka, Japan

PAMELA Stanislas
Imperial College, London, United Kingdom

PLUNK Gabriel..
University of California at Los Angeles
Los Angeles, CA USA

REBONT Cyril
CNRS- Université de Provence, PIIM
13397 Marseille Cedex 20, France

ROSATO Joel
CNRS- Université de Provence, PIIM
13397 Marseille Cedex 20, France

RIBIERO Celso
Instituto de Fisica da Universidade de Sao Paulo
Rua Cariris 406, Sao Paulo, SP,
Brazil CEP 05422-020
Sao Paulo Brazil

SANDBERG Ingmar
Laboratory of Plasma, Electron Beam & Non-linear Optics
School of Electrical and Computer Engineering
National Technical University of Athens
Iroon Polutehniou 9, Zografou
157 73 Athens, Greece

SATO Masahiko
National Institute for Fusion Science
322-6 Oroshi-Cho
Toki
GIFU 509 5292 Gifu Japan

SCARIN Paolo
Consorzio RFX
Corso Stati Uniti 4
Padova, Italy

SEVERO F.Jose Helder
Institution Plasma Physics Laboratory, Institute of Physics,
University of Sao Paulo

Rua do Matão, Travessa R,
187 Cidade Universitária
Sao Paulo Country Brazil

SHIMIZU Akihiro
National Institute for Fusion Science
322-6 Oroshi-Cho
Toki
GIFU 509 5292 Gifu Japan

SCHUBERT Martin
LPMIA, UMR CNRS No 7040
Université Henri Poincaré,
54506 Vandoeuvre-lès-Nancy Cedex, France

THOMAS Cédric
CNRS- Université de Provence, PIIM
13397 Marseille Cedex 20, France

TSUCHIYA Hayato
National Institute for Fusion Science
322-6 Oroshi-Cho
Toki
GIFU 509 5292 Gifu Japan

VARSHNEY Sanjeev Kumar
Institute for Plasma Research
Near Indira Bridge, Bhat
Gandhinagar India

VOSLION Thibaut
CNRS- Université de Provence, PIIM
13397 Marseille Cedex 20, France

WANG Eric
University of California at Los Angeles, USA

WICKS Robert
Centre for Fusion Space and Astrophysics
University of Warwick
Physics Department,
Coventry, United Kingdom

YATSENKO Nataliya
IT Department, Kharkov National University
53 Pushkinskaya Street
Kharkiv, Ukraine

ZAKHAROV Vasily
RRC Kurchatov Institute, Moscow Russia

ZIMMER Clotide
CEA
Bagnols sur Cèze, France

Author Index

W

Y